Central Regulation of the Endocrine System

Central Regulation of the Endocrine System

Edited by
Kjell Fuxe
Tomas Hökfelt
and
Rolf Luft

Karolinska Institute
Stockholm, Sweden

PLENUM PRESS · NEW YORK AND LONDON

Library of Congress Cataloging in Publication Data

Nobel Symposium, 42d Stockholm, 1978.
  Central regulation of the endocrine system.

  Includes index.
  1. Hypothalamo-hypophyseal system — Congresses. 2. Pituitary hormone releasing
factors — Congresses. 3. Neurotransmitters — Congresses. 4. Neurons — Congresses. I.
Fuxe, Kjell, 1938-   II. Hökfelt, Tomas. III. Luft, Rolf, 1914-   IV. Title.
[DNLM: 1. Neurons — Physiology — Congresses. 2. Neurosecretion — Congresses.
3. Peptides — Congresses. W3 NO369 42d 1978p/ WL102.5 N744 1978p]
QP188.H9N6 1978   599'.01 42   78-27000

ISBN-13: 978-1-4684-3398-2   e-ISBN-13: 978-1-4684-3396-8
DOI: 10.1007/978-1-4684-3396-8

Proceedings of Nobel Foundation Symposium 42 on Principles of the Central Regulation
of the Endocrine System held in Stockholm, Sweden, June 7–9, 1978

© 1979 Plenum Press, New York
Softcover reprint of the hardcover 1st edition 1979

A Division of Plenum Publishing Corporation
227 West 17th Street, New York, N.Y. 10011

PUBLISHED NOBEL SYMPOSIA

1 • Muscular Afferents and Motor Control—*Edited by Ragnar Granit*
2 • Prostaglandins—*Edited by Sune Bergström and Bengt Samuelsson*
3 • Gamma globulins—*Edited by Johan Killander*
4 • Current Problems of Lower Vertebrate Phylogeny—*Edited by Tor Ørvig*
5 • Fast Reactions and Primary Processes in Chemical Kinetics—*Edited by Stig Claesson*
6 • Problems of International Literary Understanding—*Edited by Karl Ragnar Gierow*
7 • International Protection of Human Rights—*Edited by Asbjörn Eide and August Schou*
8 • Elementary Particle Theory—*Edited by Nils Svartholm*
9 • Mass Motions in Solar Flares and Related Phenomena—*Edited by Yngve Öhman*
10 • Disorders of the Skull Base Region—*Edited by Carl-Axel Hamberger and Jan Wersäll*
11 • Symmetry and Function of Biological Systems at the Macromolecular Level—*Edited by Arne Engström and Bror Strandberg*
12 • Radiocarbon Variations and Absolute Chronology—*Edited by Ingrid U. Olsson*
13 • Pathogenesis of Diabetes Mellitus—*Edited by Erol Cerasi and Rolf Luft*
14 • The Place of Value in a World of Facts—*Edited by Arne Tiselius and Sam Nilsson*
15 • Control of Human Fertility—*Edited by Egon Diczfalusy and Ulf Borell*
16 • Frontiers in Gastrointestinal Hormone Research—*Edited by Sven Andersson*
17 • Small States in International Relations—*Edited by August Schou and Arne Olav Brundtland*.
20 • The Changing Chemistry of the Oceans—*Edited by David Dyrssen and Daniel Jagner*
21 • From Plasma to Planet—*Edited by Aina Elvius*
22 • ESR Applications to Polymer Research—*Edited by Per-Olof Kinell and Bengt Rånby*
23 • Chromosome Identification Technique and Applications in Biology and Medicine—*Edited by Torbjörn Caspersson and Lore Zech*
24 • Collective Properties of Physical Systems—*Edited by Bengt Lundqvist and Stig Lundqvist*
25 • Chemistry in Botanical Classification—*Edited by Gerd Bendz and Johan Santesson*
26 • Coordination in the Field of Science and Technology—*Edited by August Schou and Finn Sollie*
27 • Super-Heavy Elements—*Edited by Sven Gösta Nilsson and Nils Robert Nilsson*
28 • Somatomedins and Some Other Growth Factors—*Edited by Rolf Luft and Kerstin Hall*
29 • Man, Environment, and Resources—*Edited by Torgny Segerstedt and Sam Nilsson*
30 • Physics of the Hot Plasma in the Magnetosphere—*Edited by Bengt Hultqvist and Lennart Stenflo*
31 • The Impact of Space Science on Mankind—*Edited by Tim Greve, Finn Lied, and Erik Tandberg*
33 • Molecular and Biological Aspects of the Acute Allergic Reaction—*Edited by S. G. O. Johansson, Kjell Strandberg, and Börje Uvnäs*
34 • Structure of Biological Membranes — *Edited by Sixten Abrahamsson and Irmin Pascher*
36 • Plasma Physics: Nonlinear Theory and Experiments—*Edited by Hans Wilhelmsson*
40 • Biochemistry of Silicon and Related Problems—*Edited by Gerd Bendz and Ingvar Lindqvist*
42 • Central Regulation of the Endocrine System — *Edited by Kjell Fuxe, Tomas Hökfelt, and Rolf Luft*

Symposia 1-17 and 20-22 were published by Almqvist & Wiksell, Stockholm and John Wiley & Sons, New York; Symposia 23-25 by Nobel Foundation, Stockholm and Academic Press, New York; Symposium 26 by the Norwegian Nobel Institute, Universitetsforlaget, Oslo; Symposium 27 by Nobel Foundation, Stockholm and Almqvist & Wiksell International, Stockholm; Symposium 28 by Academic Press, New York; Symposium 29 by Nobel Foundation, Stockholm and Trycksaksservice AB, Stockholm; and Symposia 30, 31, 33, 34, 36, 40, and 42 by Plenum Press, New York.

Preface

According to the classical concept of Geoffrey Harris the pituitary
gland is controlled by the brain by means of blood-borne chemical
messengers produced by central neurons. The recent isolation and
structural characterization of several such messengers by Roger
Guillemin and Andrew Schally and their collaborators brought the
final proof for this hypothesis. This also meant that the extensive
knowledge collected in the field of neurobiology now became highly
relevant for the endocrinologists. For this reason it was felt
important to organize a symposium which brought together experts
in the fields of neurobiology and endocrinology. The idea was to
focus the attention on neuronal mechanisms, particularly those
related to chemical transmission, which may be of importance for
the central regulation of hormonal secretion patterns.

We would like to express our sincere gratitude to the Nobel Founda-
tion for supporting the organization of the Nobel Symposium 42 on
"Principles of the Central Regulation of the Endocrine System".
We would also like to express our thanks to all participants, to
Professor Carl-Gustaf Bernhard, Permanent Secretary of the Royal
Academy of Sciences, for making the facilities of the Academy
available to us, and to the Symposium secretaries Mrs. Gun Hultgren,
Mrs. Lena Persson and Mrs. Ulla-Britt Wedin.

It is a pleasure to acknowledge the generous financial support from
the Nobel Foundation and its Nobel Symposium Committee through
grants from the Tercentenary Foundation of the Bank of Sweden,
and from the Swedish Medical Research Council, KABI AB, Stockholm,
and ASTRA AB, Södertälje.

Kjell Fuxe
Tomas Hökfelt
Rolf Luft

Contents

Opening address – R. Luft 1

Session I. THE PEPTIDERGIC NEURON

        Chairmán: C.H. Sawyer

        Reporter: C. Kordon

Neuroendocrine regulation: The peptidergic neuron
    Introduction and historical background – C.H.
Sawyer 3

The biochemistry of peptidergic neurons

Hypothalamic hormones regulating pituitary (and
    other) functions: Their physiology and bio-
    chemistry as well as recent studies with
    their synthetic analogues – A.V. Schally,
    A. Arimura, D.H. Coy, A.J. Kastin, C.A.
    Meyers, T.W. Redding, K. Chihara, W.Y. Huang,
    R.C.C. Chang, E. Pedroza and J. Vilchez-
Martinez 9

Presynaptic mechanisms in peptidergic neurons

Neurotransmitters and neuropeptides: Distribution
    patterns and cellular localization as
    revealed by immuno-cytochemistry – T. Hökfelt,
    O. Johansson, Å. Ljungdahl, J. Lundberg, M.
    Schultzberg, K. Fuxe, M. Goldstein, H.
    Steinbusch, A. Verhofstad and R. Elde 31

Activation of release – and mechanism of release
    of neurohypophyseal hormones – N.A. Thorn,
    J.T. Russel, C. Torp-Pedersen and M. Treiman 49

Studies on the release and degradation of
    hypothalamic releasing hormones by the
    hypothalamus and other CNS areas in vitro –
    S.L. Jeffcoate, N. White, G.W. Bennett, J.A.
    Edwardson, E.C. Griffiths, R. Forbes and
    J.A. Kelly   61

*Postsynaptic mechanisms in peptidergic
    transmission*

            Chairman: P. Greengard

            Reporter: J.P. Changeux

The role of GTP in the coupling of hormone
    receptors and adenylate cyclase – M. Rodbell
    and W. Schlegel   71

Interactions between hypothalamic and peripheral
    hormones at the anterior pituitary level –
    F. Labrie, J. Drouin, L. Lagace, L. Ferland,
    M. Beaulieu, V. Raymond and J. Massicotte   85

Peptide and neurotransmitter receptors in the
    brain: Regulation by ions and guanyl
    nucleotides – S.H. Snyder   109

Neurophysiology of hypothalamic peptidergic
    neurons – L.P. Renaud, Q.J. Pittman and
    H.W. Blume   119

The opioid receptors and their ligands – L.
    Terenius   137

Session II.   TRANSMITTER AND NEUROPEPTIDE SYNAPTIC
               MECHANISMS

            Chairman: U.S. Von Euler

            Reporter: G. Sedvall

Regulation of the β-adrenergic receptor in the
    pineal gland and red cell membranes – J.
    Axelrod   149

Cyclic nucleotide and protein phosphorylation
    mechanisms in the central nervous system –
    P. Greengard   157

Contrasting principles of synaptic physiology:
    Peptidergic and non-peptidergic neurons -
    F.E. Bloom   173

Regulation of neuropeptide release in rat
    brain - L.L. Iversen and T.M. Jessell   189

Studies on interactions of epinephrine neuronal
    systems with other neuronal systems -
    M. Goldstein, A. Sauter, F. Hata, J.Y. Lew,
    Y. Baba, J. Engel, K. Fuxe and T. Hökfelt   209

Effect of peptides on brain monoamines and on
    gross behaviour - A. Carlsson, J.A. Garcia-
    Sevilla and T. Magnusson   223

Session III.   HORMONAL CONTROL OF PEPTIDERGIC
               NEURONS

               Chairman: S. Reichlin

               Reporter: P. Eneroth

Current approaches to steroid hormone-cell
    interactions - E.-E. Baulieu   239

Steroid hormone receptors in brain and
    pituitary - B.S. McEwen, P.G. Davis,
    L.C. Krey, I. Lieberburg, N. MacLusky
    and E. Roy   261

Feedback effects on central mechanisms
    controlling neuroendocrine functions -
    L. Martini, F. Celotti, H. Juneja,
    M. Motta and M. Zanisi   273

Pituitary neuropeptides and behavior -
    D. De Wied   297

Sexual differentiating actions of steroids
    on the hypothalamopituitary-liver axis -
    J.-Å. Gustafsson, P. Eneroth, B. Haglund,
    T. Hökfelt, A. Mode, P. Skett and
    Ö. Wrange   315

Session IV.   INTERACTIONS BETWEEN HORMONES AND
               NEUROTRANSMITTERS IN THE CONTROL
               OF PEPTIDERGIC NEURONS

               Chairman: J. Axelrod

               Reporter: C.H. Sawyer

Neurotransmitters in the control of anterior
    pituitary function - S.M. McCann, L.
    Krulich, S.R. Ojeda, A. Negro-Vilar and
    E. Vijayan   329

Neurotransmitter mechanisms in the control of
    the secretion of hormones from the anterior
    pituitary - K. Fuxe, K. Andersson, A.
    Löfström, T. Hökfelt, L. Ferland, L.F.
    Agnati, M. Pérez De La Mora, R. Schwarcz,
    P. Eneroth, J.-Å. Gustafsson and P. Skett   349

Session V.   CONTROLS OF PEPTIDERGIC NEURONS IN
               HUMANS

               Chairman: R. Luft

               Reporter: S. Reichlin

Control of peptidergic neurons in humans. An
    introduction - R. Luft   381

Studies of the role of dopamine in the control
    of prolactin and gonadotropin secretion
    in humans - S.S.C. Yen   387

Neurotransmitter control of growth hormone and
    prolactin secretion - E.E. Müller, D. Cocchi,
    V. Locatelli, E.A. Parati and P. Mantegazza   417

Clinical neuroendocrine relationships in normal
    and disordered prolactin secretion - G.M.
    Besser, T. Yeo, G. Delitala, A. Jones, W.A.
    Stubbs, J.A.H. Wass and M.O. Thorner   457

Session VI.   REPORTERS' OVERVIEWS

               Chairman: K. Fuxe

Role and regulation of neuropeptide neurons -
    C. Kordon   473

Some principles of neuronal regulation at the
    postsynaptic level - J.P. Changeux   489

Reporter's remarks - G. Sedvall   507

Overview of Session IV: Interactions between
    hormones and neurotransmitters in the
    control of peptidergic neurons - C.H.
    Sawyer   515

Discussion of clinical neuroendocrinology
    section - S. Reichlin   521

Participants   537

Index   539

OPENING ADDRESS

Rolf Luft

Department of Endocrinology
Karolinska Hospital
S-104 01 Stockholm, Sweden

The Nobel Symposia, of which this is number 42, were started
in 1965 with a symposium on "Muscular Afferents and Motor Control."
Over the years, it has become a tradition to organize Nobel
Symposia differently from the usual congresses so that more time is
devoted to exchange of ideas and concepts than to presentation of
data. Also, by tradition, there are certain prerequisites to be
met in planning a Nobel Symposium: the area of research to be
discussed should be of immediate interest; the "state of the arts"
in the particular area should have developed to a level that makes
it appropriate and necessary to have mutual exchange among the
scientists involved; the organizers should have special expertise
in the field. We believe that this Symposium meets these requirements.

The present Symposium is concerned with the "Principles for
the Central Regulation of the Endocrine System." This certainly
is a rapidly developing area of immediate interest, where we are
compelled by an avalanche of new knowledge. We may discuss how it
all started, but I agree with Dr. Seymour Reichlin, one of the
distinguished participants of this Symposium, that we should
consider the landmark to be the report in 1901 by Dr. Alfred
Fröhlich of Vienna, on what came to be known as Fröhlich's
syndrome. It predicted the nature of hypothalamic function and
gave a lead to the anatomists and physiologists. We may visualize
other landmarks: the discovery of the portal circulation between
the hypothalamus and pituitary gland by Popa & Fielding in 1931;
Ernst Scharrer's introduction in the 1930's of the concept of
neurosecretion, that nerve cells could have a secretory function,
and that the secretory neuron stands as the link between neural
activity and endocrine regulation; the contributions by Geoffrey

Harris in the 1940's and 1950's to our understanding of the significance of the portal blood supply as the information link from
the hypothalamus to the pituitary; the work of Sawyer and coworkers
in the late 1940's which pointed to the neurotransmitters as
controllers of anterior pituitary function; the identification of
the structure of oxytocin in 1950 and vasopressin in 1954 by du
Vigneaud and collegues; and finally, in 1969-73 the efforts of
Guillemin and Schally and their groups who first isolated and synthesized the hypothalamic hormones that regulate the secretion of
the anterior pituitary gland.

It has been the intention of the organizers of this Symposium
to gather people working in the frontline of this area of research:
those dealing with the flow of information via the biogenic amine
neural pathways arising in the lower brain stem and the hypothalamus to the specialized peptidergic pathways, and from there to
the hypophysis and its hormone release. The clinical relevance of
this basic research will be dealt with, especially the nature of
hypothalamic-pituitary control in man and the recognition of
hypothalamic-pituitary conditions with disturbances of the normal
pattern of information transmission from the CNS to the endocrine
glands.

The organizers of this Symposium - Kjell Fuxe, Tomas Hökfelt
and myself - welcome you all, and express the conviction that you
will find these three days in Stockholm stimulating and of great
value.

NEUROENDOCRINE REGULATION: THE PEPTIDERGIC NEURON

INTRODUCTION AND HISTORICAL BACKGROUND

Charles H. Sawyer

Department of Anatomy and Brain Research Institute
UCLA School of Medicine
Los Angeles, California 90024 U.S.A.

The award of the 1977 Nobel Prize in Physiology or Medicine to two neuroendocrinologists for their research on hypothalamic releasing and inhibitory hormones (22) signals an increased appreciation of brain-endocrine interactions and sets the stage for this 42nd Nobel Symposium on "Principles for central neuronal regulation of the endocrine system".

The historical background of this symposium may be dated to the first half of this century when the Austrian pharmacologist Otto Loewi and the British pharmacologist Henry Dale were demonstrating that chemical transmission of the nervous impulse was effective at nerve endings in the autonomic and peripheral nervous systems, achievements for which they shared the 1936 Nobel Prize (33). They found that acetylcholine and adrenaline were potent chemical mediators not only at sites of slow transmission such as the heart, smooth muscle and glands but also at ganglionic synapses and neuromuscular junctions in skeletal muscle, and Dale (8) coined the terms "cholinergic" and "adrenergic" for different types of synaptic and neuroeffector transmission in 1933. In the early 1930's the young Swedish investigator U.S. von Euler visited Dale's laboratory in England and with John Gaddum demonstrated the existence of what appeared to be a peptidergic neurotransmitter which they named "Substance P". Back in Sweden von Euler continued to study the nature of the humoral mediators and in the 1940's he demonstrated that the adrenergic transmitter at sympathetic postganglionic endings was largely noradrenaline (36), work for which he was to share the 1970 Nobel Prize with Axelrod and Katz.

Also in the 1940's Dale's associates Feldberg and Vogt (12) summarized the evidence that cholinergic transmission was playing

3

an important role at certain synapses in the central nervous system. Later Vogt (35) demonstrated large amounts of noradrenaline in the dog hypothalamus and in 1959 Carlsson (7) showed that dopamine, which had previously been considered only a precursor of noradrenaline and adrenaline, was concentrated in the basal ganglia. In the early 1960's the Swedish investigators Hillarp and Falck (11) developed an exquisitely sensitive fluorescence method for the histochemical localization of monoamines, and in the hands of Fuxe, Hökfelt, Bjorklund, Ungerstedt and others this technique has provided an abundance of valuable information on the central monoamines (see 14, 25).

The 1930's witnessed a wealth of original research on the anatomy and physiology of the hypothalamus and its relationship to the posterior pituitary gland or neurohypophysis, summarized in the 1940 Volume 20 of the Research Publications of the Association for Research in Nervous and Mental Disease. Previously it had been thought that nerve impulses from the supraoptic (SO) and paraventricular (PV) nuclei controlled the secretion of the neurohypophysial hormones vasopressin and oxytocin from gland cells in the posterior pituitary (13). At the ARNMD meeting Ernst and Berta Scharrer (30) summarized their histological evidence that hypothalamic nerve cells, especially the magnocellular components of the SO and PV nuclei were themselves secretory, and they suggested that hormones might be produced in such cells. The neurosecretion hypothesis received a lift when Bargmann and his students applied the Gomori chromalum hematoxylin stain (developed for pancreas study) to the brain and found the whole supraoptico- and paraventricular-neurohypophysial systems differentially stained. Bargmann and Scharrer published a joint paper (2) summarizing this new concept in the American Scientist in 1951 and Ingram (19) accepted the new interpretation in his Ciba Review. At that stage in midcentury the octapeptide structures of oxytocin and vasopressin were established by du Vigneaud (9), another achievement recognized by the award of a Nobel Prize (in Chemistry, 1955).

The 1930's and 1940's were also a time for speculation that secretion of the newly described anterior pituitary trophic hormones (32) might be controlled by humoral mediators from the hypothalamus. This was first proposed by Hinsey and Markee (17) in 1933 at a time when Popa and Fielding (23) were still convinced that their recently discovered portal system conducted blood upward from the pituitary to the hypothalamus. The hypothalamo-hypophysial direction of flow in the living toad was observed in the laboratory of another future Nobel Prize winner, Houssay (18), but the paper was lost in the literature for many years, and it was the histological studies of Wislocki and King (37) which convinced many endocrinologists of the downward flow and the potential of this vascular route. It was on the basis of their work that we undertook

experiments in Markee's laboratory on neurohumoral control of
adenohypophysial function, experiments which revealed an adrenergic
link (29) which turned out to be a central rather than a peripheral
mediator (see Sawyer, 27). Meanwhile in England Green and Harris
were establishing the neurovascular concept by demonstrating the
pituitary portal system in all classes of vertebrates, by observing
the downward direction of bloodflow in the living frog and later in
the rat, and by cutting the pituitary stalk and watching the portal
vessels regenerate and restore function. With Jacobsohn Harris
demonstrated that pituitary transplants regained complete function
only if placed under the median eminence where portal vessels could
reach them (see Harris, 15). This was confirmed by Nikitovitch-
Winer and Everett, who transplanted the pituitary under the kidney
capsule and then back under the median eminence (see 10, 28).

With the acceptance of the neurovascular concept biochemists
started their search for the identity of the humoral mediators or
"releasing factors" as they came to be called, and most prominent
among these researchers were our new Nobel Laureates Guillemin and
Schally. In the 1950's it appeared that the adrenocorticotropin
releasing factor (CRF) would be the first to be analyzed but its
instability has prevented analysis to this day (May, 1978). In
the early 1960's McCann (see McCann, 20) and Harris (see Harris,
16) independently demonstrated the existence of a specific gonadotropin releasing factor (LRF) and they as well as Schally and
Guillemin concentrated on its structural analysis. The simpler
tripeptide structure of thyrotropin releasing factor (TRF) was the
first to submit to full analysis and synthesis almost simultaneously and independently by Guillemin's and Schally's laboratories
in 1969 (4, 6), and its name was changed to TRH - "H" for "hormone". The structure of LRH or LHRH as a decapeptide was first
announced two years later in 1971 by Schally's laboratory (21).
Guillemin's laboratory registered a first in 1973 in announcing
the structure of somatostatin as a tetradecapeptide (5) which,
among other actions, inhibits the secretion of pituitary growth
hormone.

In recent years several other neural peptides including Substance P and the endorphins and enkephalins have yielded to structural analysis. The same is true of enzymes involved in synthesis
of the catecholamines. This information has permitted synthesis of
the peptides in pure form and the production of specific antibodies
to them for use in radioimmunoassay and immunohistochemistry. The
precise immunocytochemical localization of releasing hormones, peptides and enzymes in the brain has been described authoritatively
in recent books (14, 24, 25) and will be among the subjects discussed in this symposium.

The availability of the peptides in pure form also permits
their local application to neurons in electrophysiological

experiments. Single unit recordings from antidromically identified neurons and the effects of microiontophoretic application of
peptides and amines on their firing patterns have been studied by
Renaud in Canada, Moss in the United States, Dyer, Dyball and Cross
in Britain and Kawakami and Sakuma in Japan (see 26). Dr. Renaud
will report on this subject later in the morning.

Finally, mention should be made of recent findings relative to
the "hypophysiotrophic area" (HA) of the hypothalamus. Halász and
associates (see Szentágothai et al., 34) described the HA as the
region in the basal hypothalamus which would maintain the function
and morphology of small implants of pituitary tissue, and it was
thought that proximity to neurons and their fibers producing
releasing hormones explained the hypophysiotrophic influence.
Recent immunohistological studies of Sétáló et al. (31) have questioned this hypothesis and the current interpretation is that upward flow of deep portal vessels from the median eminence tufted
capillaries supplies the HA with releasing hormones. Torok described such vessels in the early 1960's (see 34) and recent careful
studies by Ambach, Palkovits and Szentágothai (1) have confirmed
their existence. Such vessels provide an anatomical route for the
ultra-short-loop feedback action of releasing hormones on hypothalamic function. Bergland and Page (3) have recently described other
deep venous channels in the monkey's infundibular stalk by which
anterior pituitary secretions can be conveyed to the neurohypophysis and thence to the brain.

These observations reflect some of the highlights of the historical background of this symposium. If we were to single out the
contributions of one scientist whose research laid a foundation for
the remarkable achievements of Schally and Guillemin and for this
symposium it would have to be the inspired work of Geoffrey Harris
(15, 16, 26). But for his untimely demise he would undoubtedly
have shared the Nobel Prize with Guillemin and Schally, and we are
sure that they and other neuroendocrinologists would join us in
paying this tribute to him.

REFERENCES

1. Ambach, G., M. Palkovits and J. Szentágothai. Acta morph. Acad.
   Sci. hung. 24:93-119, 1976.

2. Bargmann, W. and E. Scharrer. Am. Sci. 39:255-259, 1951.

3. Bergland, R.M. and R.B. Page. Endocrinology 102:1325-1328, 1978.

4. Bøler, J., F. Enzman, K. Folkers, C.Y. Bowers and A.V. Schally.
   Biochem. Biophys. Res. Commun. 37:705-710, 1969.

5. Brazeau, P., W. Vale, R. Burgus, N. Ling, M. Butcher, J. Rivier and R. Guillemin. Science 179:77-79, 1973.

6. Burgus, R., T.F. Dunn, D. Desiderio and R. Guillemin. C R Acad. Sci. (D) (Paris) 269:1870-1873, 1969.

7. Carlsson, A. Pharmacol. Rev. 11:490-493, 1959.

8. Dale, H.H. J. Physiol. Lond. 80:10-11, 1933.

9. du Vigneaud, V. Harvey Lectures 1954-55, pp. 1-26.

10. Everett, J.W. In: Pioneers in Neuroendocrinology, edited by J. Meites, B.T. Donovan and S.M. McCann, pp. 97-109. Plenum Press, New York, 1975.

11. Falck, B., N.A. Hillarp, G. Thieme and A. Torp. J. Histochem. Cytochem. 10:348-354, 1962.

12. Feldberg, W. and M. Vogt. J. Physiol. Lond. 107:372-381, 1948.

13. Fisher, C., W.R. Ingram and S.W. Ranson. Diabetes Insipidus and the Neurohumoral Control of Water Balance. Edwards Brothers, Ann Arbor, Michigan, 1938.

14. Ganong, W.F. and L. Martini (Editors). Frontiers in Neuro-endocrinology Vol. 5. Raven Press, New York, 1978.

15. Harris, G.W. Neural Control of the Pituitary Gland. Edward Arnold, London, 1955.

16. Harris, G.W. J. Endocr. 53:ii-xxiii, 1972.

17. Hinsey, J.C. and J.E. Markee. Proc. Soc. Exp. Biol. Med. 31: 270-271, 1933.

18. Houssay, B.A., A. Biasotti and R. Sammartino. Compt. Rend. Soc. Biol. Paris 120:725-727, 1935.

19. Ingram, W.R. Ciba Clin. Symp. 8:117-156, 1956.

20. McCann, S.M. New England J. Med. 296:797-802, 1977.

21. Matsuo, H., Y. Baba, R.M.G. Mair, A. Arimura and A.V. Schally. Biochem. Biophys. Res. Commun. 43:1334-1339, 1971.

22. Meites, J. Science 198:594-596, 1977.

23. Popa, G.T. and U. Fielding. J. Anat. Lond. 67:227-232, 1933.

24. Porter, J.C. (Editor). Hypothalamic Peptide Hormones and Pituitary Regulation. Plenum Press, New York, 1977.

25. Reichlin, S., R.J. Baldessarini and J.B. Martin (Editors). The Hypothalamus, Res. Publ. Assoc. Nerv. Ment. Dis. Vol. 56. Raven Press, New York, 1978.

26. Sawyer, C.H. Neuroendocrinology 17:97-124, 1975.

27. Sawyer, C.H. Acta Biol. Acad. Sci. Hung. 28:11-23, 1977.

28. Sawyer, C.H. Biology of Reproduction 18:325-328, 1978.

29. Sawyer, C.H., J.E. Markee and W.H. Hollinshead. Endocrinology 41:395-402, 1947.

30. Scharrer, E. and B. Scharrer. In: The Hypothalamus, Res. Publ. Assoc. Nerv. Ment. Dis. Vol. 20, pp. 170-174. Hafner Publishing Co., New York, 1940.

31. Sétáló, G., S. Vigh, N. Hagino and B. Flerkó. Acta morph. Acad. Sci. hung. 24:79-91, 1976.

32. Smith, P.E. Am. J. Anat. 45:205-273, 1930.

33. Sourkes, T.L. Nobel Prize Winners in Medicine and Physiology 1901-1965. Abelard-Schuman, New York, 1966.

34. Szentágothai, J., B. Flerkó, B. Mess and B. Halász. Hypothalamic Control of Anterior Pituitary. Akadémiai Kiadó, Budapest, 1968.

35. Vogt, M. J. Physiol. Lond. 123:451-481, 1954.

36. von Euler, U.S. In: The Neurosciences: Paths of Discovery, edited by F.G. Worden, J.P. Swazey and G. Adelman. MIT Press, Cambridge, Mass., pp. 180-187, 1975.

37. Wislocki, G.B. and L.S. King. Amer. J. Anat. 58:421-472, 1936.

HYPOTHALAMIC HORMONES REGULATING PITUITARY (AND OTHER) FUNCTIONS: THEIR PHYSIOLOGY AND BIOCHEMISTRY AS WELL AS RECENT STUDIES WITH THEIR SYNTHETIC ANALOGUES

A.V. Schally, A. Arimura, D.H. Coy, A.J. Kastin,
C.A. Meyers, T.W. Redding, K. Chihara, W.Y. Huang,
R.C.C. Chang, E. Pedroza and J. Vilchez-Martinez

Veterans Administration and Tulane University School of
Medicine, New Orleans, Louisiana, USA

INTRODUCTION

It is well established now that the hypothalamic releasing and inhibitory hormones produced in the neuronal cell bodies regulate the secretory activity of the anterior pituitary gland. The discovery of several of these hormones and their isolation, structural identification and synthesis furnished the evidence for the theory of neurohumoral control of the pituitary gland put forward by Harris (1) and others.

Some hypothalamic peptide hormones are also produced in the extra-hypothalamic brain areas (where they may serve as peptidergic synaptic neurotransmitters and where they can modulate neural and behavioral events (2) and in endocrine-like cells of non-neural tissues (3). Thus, somatostatin is present in discrete cells of the pancreas, gastric mucosa, duodenum, jejunum and may, through paracrine control, play an important role in the regulation of endocrine pancreas and gastrointestinal tract. In this short review we shall report some of the latest biochemical and physiological findings on hypothalamic hormones and their analogues, especially their effects on anterior pituitary function and at other levels of the organism.

The effects of these peptides on brain functions and behavior and their ultrastructural localization, mechanism of action and clinical significance have been reviewed by us previously (2,4,5, 6) and will be discussed by others in this volume.

CORTICOTROPIN RELEASING FACTOR (CRF)

Recently, the work on the purification of CRF has been re-sumed in several laboratories. The results of Cooper et al. (7) and Jones et al. (8) along with our recent work appear to confirm the hypothesis concerning the existence of several molecules with CRF activity, put forward by us more than 17 years ago (9). Our laboratory is in the final stages of isolating, sequencing and synthesizing several peptides with CRF activity.

In our most recent studies (10), hypothalamic extracts from nearly one-half million pig hypothalami were first separated into 14 fractions by preparative gel filtration on Sephadex G-25. Sig-nificant CRF activity was found near the void volume (fraction 2, R_f=0.77), in intermediate fractions (fraction 8, R_f=0.4) with mol wt about 1,000, and even in the fractions 11 and 12 which had low R_f (0.3) and also contained catecholamines.

High mol wt CRF-active fractions from Sephadex with R_f=0.77 were purified by column chromatography on CMC. CRF activity was eluted in fractions with a conductivity of 7,000-10,000 μMHOS. The CRF-active area, still contaminated with ACTH-like materials, was subjected to countercurrent distribution (CCD) in a system of 0.1% acetic acid : butanol : pyridine = 11:5:3. The CRF active fractions, K=1.1, released ACTH in vivo and in vitro in doses of 1 μg. These fractions were repurified with chromatography on SE-Sephadex to yield a CRF active peptide, detectable in doses of 0.1 μg/ml in vitro and devoid of any inherent ACTH activity. After rechromatography on SE-Sephadex, some CRF fractions were active in vitro in doses of 0.01 μg. This material was repurified by gel-filtration on Sephadex G-50 and partition chromatography. CRF activity was completely lost after 16 hr digestion with trypsin and partially destroyed by thermo-lysin. The results indicate that CRF activity in this fraction is due to a basic polypeptide and this CRF has been tentatively named large MW CRF.

Fraction 8 with R_f of 0.4 from Sephadex containing interme-diate MW CRF was also purified by chromatography on CMC (10). CRF activity was found in well-separated acidic fractions, neutral frac-tions and basic fractions. The basic fractions had the highest CRF activity, increasing ACTH release in vitro 10-100 fold. However, those fractions had the highest contamination with ACTH-like peptides and required further purification by CCD and SE-Sephadex. The highly purified material devoid of ACTH-like activity is now active in vitro in doses of 3 mg/ml and shows an excellent dose-response relationship. Attempts are being made to identify structurally the active material in basic fractions, which is most likely to repre-sent the physiological CRF.

Similar methods were used for the purification of small mol wt or retarded CRF from Sephadex (fraction 11, 12 with $R_f=0.3$). Catecholamines are present in this fraction and evidence was obtained that they can stimulate ACTH release in vitro. However, after countercurrent distribution in 0.1% acetic acid: 1-butanol : pyridine=11:5:3 (K=5-8), gel filtration on Sephadex G-15 and partition chromatography or ion-exchange chromatography on SE-Sephadex, more than 15 milligrams of a tetradecapeptide with CRF activity were obtained. Its amino acid sequence was determined as Phe-Leu-Gly-Phe-Pro-Thr-Thr-Lys-Thr-Tyr-Phe-Pro-His-Phe (10, 11). This tetradecapeptide was then synthesized, but subsequently it was found that this amino acid sequence is identical with the residues No. 33-46 of α-chain of porcine hemoglobin and hence this tetradecapeptide is unlikely to have an origin different from porcine hemoglobin and most probably is an artifact of extraction and isolation. Nevertheless, it is interesting that the synthetic and natural tetradecapeptide stimulates the release of ACTH in monolayer cultures of mouse pituitary cells in doses of 0.1-6.4 µg and from rat pituitary quarters in doses of 0.5-3 µg. However, it is inactive in vivo even in doses of 100 µg. Some preliminary evidence also indicates that some CRF fractions may release β-lipotropin and β-endorphin in addition to ACTH (Labrie et al., in preparation). It is difficult if not impossible at present to interpret the finding of multiple CRF activities, but it is possible that high mol wt fractions represent pro-CRF (a precursor of CRF), and the intermediate mol wt the physiological CRF (CRH). A part of the CRF activity of low mol wt fractions is due to catecholamines and part to the tetradecapeptide formed from porcine hemoglobine (11). It is hoped that further work will result in the isolation of these CRF's and in the determination of their physiological role.

THYROTROPIN RELEASING HORMONE (TRH)

Synthetic pyroGlu-His Pro-NH$_2$ has been shown to stimulate thyrotropin (TSH) release in all mammals studied, including mice, rats, nutria (myocaster coypus), sheep, goats, cows, humans and birds (12, 13). TRH will liberate prolactin in animals and humans (14-17), but it remains to be established whether this effect is physiological. TRH can also increase the synthesis of TSH and prolactin. Recently it has been reported that TRH will stimulate colonic activity in rabbits (18). This response, which may explain side effects such as nausea and gastric cramps seen occasionally after administration of TRH, appears to originate in the CNS and is mediated by cholinergic receptors.

The interactions of TRH with centrally acting drugs have been the subject of many investigations (2, 13). The antidepressant activity of TRH in several animal models, which cannot be attributed to its effect on the pituitary-thyroid axis, as well as other neuropharmacological studies (13) suggest that TRH may act on the brain and the spinal cord (2,19,20). This possible role of TRH as a neurotransmitter in the CNS is supported by its presence in significant concentration in the extra-hypothalamic brain areas of various classes of vertebrates examined, including rat, chicken, snake, frog, tadpole, salmon and lamprey (20). Studies on the generation of antibodies to TRH and the development of sensitive and specific radioimmunoassays (RIA) for this hormone (20-25), permitted also the detection of immunoreactive TRH in blood, cerebrospinal fluid and urine of rat and man (13,21,26).

ANALOGS OF TRH

Relatively few analogs of TRH have been synthesized, as compared with LH-RH and somatostatin, perhaps because early work (27-29) soon indicated that the structural requirements of this small peptide are quite stringent. Thus, most analogs (12,13,26-29) have little TRH activity. However, $[(N^{3im}\text{-Me-His})^2]$-TRH has 3-10 times greatest activity than the natural product (28) and $[\beta\text{-(pyrazolyl-1)-Ala}^2]$-TRH (30) is also more active than TRH in rats and mice. Recently it has also been shown that $[(N^{3im}\text{-Me-His})^2]$-TRH is a more potent TSH and prolactin releaser in humans than TRH (31). The activities of TRH analogs for stimulating thyrotropin release and prolactin production are quite similar (32) and a good correlation exists between binding affinity and these biological activities (13).

There do appear to be differences between the CNS and pituitary receptors for TRH as indicated by some dissociation between TRH and CNS activity, since some analogs such as homo-pyroGlu-His-Pro-NH2 and homo-pyroGlu-His-Thioproline-NH2 are equipotent with TRH in stimulating TSH, but have 4-15 times as much activity on the CNS (33).

PROLACTIN RELEASING FACTOR (PRF)

PRF activity appears to be present in the hypothalami of mammals and birds. A part of the PRF activity in the extracts of hypothalami of domestic animals is clearly due to TRH which, under specific conditions, can greatly stimulate prolactin secretion in vivo and in vitro in various mammals, including humans (12-17, 34). It has also been reported recently that administration of antisera to TRH will greatly lower prolactin levels in rats (35). This finding, if confirmed, might support the concept of a physiological role of TRH in the control of prolactin secretion.

However, there is also evidence that a PRH different from TRH exists. Prolactin and TSH release induced by TRH can be dissociated under various physiological and clinical conditions. The work of several groups suggests that partially purified hypothalamic fractions, apparently different from TRH, can still stimulate the release of prolactin from rat pituitaries in vivo and in vitro (12, 13,36). The search for PRF, in our own experience, has been made difficult by the presence in hypothalamic fractions of substances which increase the release of prolactin, such as TRH, and those which inhibit it, such as catecholamines and gamma-aminobutyric acid (37,38). These compounds can, under certain conditions, obscure or neutralize other PIF and PRF activities. However, we have established that several other fractions with different physicochemical properties from TRH, as shown by different partition coefficients on CCD, basicity on CMC and SE-Sephadex, and behavior on molecular sieving on Sephadex, can also stimulate the release of prolactin from rat pituitary fragments from monolayer cell cultures (A.V. Schally, A. Arimura, T.W. Redding, unpublished). Further purification and identification of these materials is in progress.

A variety of natural substances of CNS and extra-CNS origin can augment prolactin release but their effects can be explained by an action mediated via the CNS (34). Similarly, the opiate peptides β -endorphin and Met-enkephalin and some of their analogs will also release prolactin, if given into the cerebral ventricle, but their effects are exerted on the CNS centers and therefore, they do not represent the physiological PRF, the chemistry of which remains to be elucidated.

PROLACTIN RELEASE-INHIBITING FACTOR (PIF)

Although the presence of PIF activity in hypothalamic extracts was demonstrated many years ago (for reviews, see 12,13,26,34), the nature of PIF is still unclear. Many substances are present in the brain or hypothalamic extracts which can inhibit prolactin secretion in vitro and in vivo (12,13,26,34,39,40,41).

It has also been demonstrated that the catecholamines influence the release of prolactin (39). Our recent work showed that hypothalamic catecholamines inhibit prolactin release by a direct action on the pituitary. PIF activity present in acetic acid extracts of pig hypothalami was purified by gel filtration on Sephadex G-25, extraction with phenol, chromatography on CMC columns, countercurrent distribution, rechromatography on Sephadex-G-25 and partition chromatography (37). Some of the highly purified fractions which strongly inhibited the release of prolactin in vitro and in vivo were found to contain much noradrenaline and dopamine. The magnitude of inhibition of release of prolactin was related to noradrenaline content. Synthetic noradrenaline and dopamine, in doses of 10-100 ng also strongly inhibited the

release of prolactin <u>in vitro</u> (40). When dopamine and noradrenaline were dissolved in a fresh 5% glucose solution and infused into
the hypophysial vessel of the rat, prolactin secretion was significantly suppressed as compared to the glucose-infused group (40).
The suppressive effect of catecholamines was dose-dependent (40).
These results (40,39) showed that catecholamines, purified from
hypothalamic tissue or synthetic, inhibit the release of prolactin
by an action exerted directly on the pituitary gland. However,
whether any catecholamine represents a physiological prolactin release inhibiting hormone remains to be elucidated.

Another hypothalamic substance with PIF activity, the effect
of which, in contrast to catecholamines, cannot be blocked by
perphenazine is gamma-aminobutyric acid (GABA) (38). GABA was isolated by us from pig hypothalami by rechromatography on Sephadex
G-25, CCD in two different solvent systems, free-flow electrophoresis, and chromatography on triethylaminoethyl cellulose (38).
Natural and synthetic GABA inhibited prolactin release <u>in vitro</u>
from isolated rat pituitary halves in doses as low as 0.1 ug/ml.
The extent of inhibition was proportional to the dose, natural and
synthetic GABA possessing identical PIF activity. Similarly, synthetic GABA suppressed prolactin release in monolayer cultures of
rat pituitary cells and inhibited the TRH-stimulated prolactin secretion. GABA also had PIF activity <u>in vivo</u>, although large doses
were needed for an effect. The results indicate GABA can inhibit
prolactin release by a direct action on the pituitary gland, but
again it is not known whether this effect is physiological or
pharmacological, the doses needed to obtain an effect being rather
large (38). Our observation that GABA inhibits prolactin release
was confirmed by recent <u>in vitro</u> work (41), <u>in vivo</u> studies
(Libertun <u>et al</u>., in preparation) and clinical observations
(Guitelman <u>et al</u>., in preparation) and pharmacological investigations (Müller <u>et al</u>., in preparation).

Other compounds with PIF activity different from catecholamines
and GABA, but still unidentified are present in pig hypothalamic
extracts. One of them appears to be a polyamine and the other a
polypeptide (A.V. Schally, A. Arimura, T.W. Redding, unpublished).
Somatostatin has some inhibitory effect on prolactin secretion,
particularly <u>in vitro</u> in monolayer cultures of pituitary cells (13),
but its presence can be easily detected by RIA and it has been absent in some fractions with PIF activity. Further work is needed
to identify the physiological PIF.

MELANOCYTE STIMULATING HORMONE (MSH)
-RELEASE INHIBITING FACTOR (MIF)

There is agreement now that the release of MSH from the pars
intermedia of the pituitary gland is controlled by hypothalamic

stimulating factor (MRF) and inhibitory (MIF), the latter having a predominant role (2,12,13,26). However, considerable confusion still exists as to the identity of MIF and MRF. The first prospective MIF was isolated from bovine hypothalami and identified as H-Pro-Leu-Gly-NH2 in this laboratory (42,43). The group of Walter (44) originally observed H-Pro-Leu-Gly-NH2 could be formed by incubating oxytocin with an enzyme present in hypothalamic tissue and that this tripeptide inhibits MSH release in the rat. Our studies with Pro-Leu-Gly-NH2 in 1971 in animal models of parkinsonism and depression introduced the concept of extra endocrine actions (CNS effects) of hypothalamic peptides (2,45), since these effects could be obtained in hypophysectomized animals.

H-Pro-Leu-Gly-NH2 has been shown to be effective alone or in conjunction with L-DOPA in Parkinson's disease and clinical trials have been conducted (2).

Several analogs of MIF have been synthesized recently. These peptides include H-Pro-N-isobutyl-Gly-Gly-NH2 and two stereoisomers of both H-Pro-N-Me-Leu-Gly-NH2 (LL) and (LD) and Pro-N-Me-Leu-Ala-NH2 (LLD) and (LDD) (46). All analogs except LL potientated behavioral effects of L-DOPA in mice and all antagonized fluphenazine induced catalepsy in rats. In these tests, L-Pro-N-Methyl-D-Leu-Gly-NH2 was the most active after parenteral or oral administration and it also affected brain catecholamine turnover (46).

LUTEINIZING HORMONE (LH)-AND FOLLICLE STIMULATING HORMONE (FSH) -RELEASING HORMONE (LH-RH/FSH-RH)

The isolation (47,48), determination of structure (49,50), and synthesis (51) of porcine LH-RH/FSH-RH by our group opened up vast new areas to physiological, immunological, biochemical, behavioral, immunohistological, veterinary and clinical investigations into reproduction. In addition, the interest in possible medical applications of LH-RH analogs caused a real explosion in the field of their synthesis. These investigations are so extensive, that only the latest key developments will be described.

Actions on the Pituitary

Porcine LH-RH corresponding in structure to (pyro)Glu-His-Trp-Ser-Tyr-Gly-Leu-Arg-Pro-Gly-NH2 was first synthesized by us (49, 51) and then by many other groups (12,13). This made the new hormone readily available for a variety of studies. The structure of ovine LH-RH was later shown to be identical with the porcine hormone (52) and subsequent biochemical and immunological studies indicated that bovine, human and rat LH-RH probably have the same structure (12,53). Our original observations that natural porcine LH-RH and the synthetic decapeptide release both LH and FSH (12, 47,48) have been confirmed in and extended to a variety of animals, including rats, mice, rabbits, golden hamsters, mink, spotted skunk, impala, rock hyrax, sheep, cattle, pigs, horses, monkeys and humans (4,5,53). The fact that ovulation has been induced in most of these species with LH-RH demonstrates that this decapeptide can release enough FSH to induce follicular maturation. Increases in sex steroid levels in blood have also been reported after administration of LH-RH (15,18). LH-RH is also active in birds such as domestic fowl and pigeons, and in some species of fish such as brown trout and carp (53). This indicates also that species specifity does not occur with LH-RH.

The evidence that the decapeptide corresponding in structure to porcine LH-RH also has FSH-RH activity in vivo and in vitro is now indisputable. Recent results indicate that LH-RH decapeptide represents the bulk of FSH-RH activity in the hypothalamus (54). Various immunological results (see below) strongly support the importance of LH-RH as the physiological FSH-releasing hormone. Our present view is that the LH-RH decapeptide may be the only FSH releasing hormone. This view is strongly supported by recent studies of Cattanach et al.(55), who showed that LH-RH deficient mutant mice have a reduced pituitary and plasma content of FSH, in addition to LH, and immature primary and accessory sex organs.

Investigations of the routes of administration in rats and humans revealed that LH-RH is effective not only after intravenous, subcutaneous, intramuscular and intracarotid injection, but also after intravaginal, intrarectal, intranasal, cutaneous (on the skin in dimethyl sulfoxide), and even oral administration (53). However, the doese required for suitable effects by extravascular routes are 100-10,000 times larger than the parenteral.

Immunological Studies with LH-RH

A variety of immunological studies have been carried out with LH-RH. Antisera to LH-RH have been produced in rats, rabbits, guinea pigs, sheep and humans, and several radioimmunoassays were developed (12,53,56,57). Passive immunization of rats with LH-RH antiserum prevented the preovulatory surge of LH and FSH and blocked ovulation (53,57). In hamster there was an arrest of follicular maturation, reduction in serum estradiol levels and suppression of LH surge and of ovulation (58). Reduction in serum FSH in addition to LH in rats after passive or active immunization with LH-RH supports the physiological role of the peptide in the regulation of FSH secretion (53). LH-RH is also necessary for normal implantation of fertilized ova and maintenance of pregnancy, since administration of LH-RH antisera to rats delays the former or terminates the latter, depending upon the time of injection (59, 60).

The availability of antisera to LH-RH made possible various studies on localization of LH-RH by radioimmunoassays or immunocytochemical methods (6,26,53) which will be reviewed by others. The significant LH-RH content of extrahypothalamic brain areas, and other findings, suggest that in addition to being the regulator of the release of LH and FSH, LH-RH might act as a neurotransmitter (2,26,61). LH-RH has indeed been shown to excite sexual behavior in rats and to modulate the electrical activity of neurons in the CNS (61). This confirms the concept of extra-endocrine effects of hypothalamic peptides first postulated by us (2,45) and indicates the possibility that LH-RH may act as a neurotransmitter concerned with mating behavior.

LH-RH has been detected by bioassay and by radioimmunoassay in the hypophysial portal blood of rats and monkeys and in peripheral circulation of women at mid-cycle ovulatory LH surge (26, 53) and in post-menopausal women. LH-RH, like other hormones, appears to be released in pulsatile fashion.

Analogs of LH-RH

The synthesis of many hundreds of LH-RH analogs was made possible by the use of rapid solid-phase techniques. The biological activities of these peptides have provided much insight into the role played by individual amino acids in preserving overall conformation and binding affinity to pituitary receptor sites and in triggering gonadotropin release.

Since the mechanism of action of LH-RH involves interaction with pituitary plasma membrane receptors, an in vitro assay using highly purified pituitary cell membrane preparations was also developed (62) in this laboratory. We have determined that the number of binding sites for LH-RH is approximately 2.36 pM/mg of protein with a high affinity constant of 7.1×10^9 M-1. We have also observed that the agonist [D-Trp-6] -LH-RH, and the antagonist [D-Phe-2,D-Trp-3,D-Phe-6]-LH-RH compete with LH-RH for its pituitary plasma membrane receptors, displacing the [^{125}I] -LH-RH more strongly than its parent hormone. Therefore, both stimulatory and inhibitory analogs of LH-RH apparently exert their action through the same pituitary plasma membrane receptors as those of LH-RH. The more potent and long-acting effect of such superactive analogs as [D-Trp-6] -LH-RH could be due to their higher ability to bind to the pituitary LH-RH receptors. Similarly, the blockade of ovulation of several inhibitory analogs of LH-RH, such as [D-Phe-2,

D-Trp-3,D-Phe-6 -LH-RH, is explained by the same mechanism of binding to LH-RH pituitary receptors.

Agonists. Peptides with greater activities and, therefore, greater potential therapeutic usefulness than LH-RH have resulted from changes at two positions in the decapeptide. Replacement of glycineamide in position 10 by certain alkylamine groups such as CH_3CH_2NH- and CF_3CH_2NH- result in analogs which are 3-9 times more active than LH-RH (63,64). Replacement of glycine in position 6 D-alanine (65) increases gonadotropin-releasing activity 6-fold and this is perhaps due to the stabilization of the preferred binding conformation of LH-RH. The activities of D-amino acid-6- analogs increase with the size of the side-chains and the D-leucine-, D-phenylalanine-, and D-tryptophan- peptides increase in activity in that order to an in vivo limit in the region of 15 times the activity of LH-RH (66).

A logical extension of this work was the synthesis and exami-nation of peptides containing both types of modification and, as predicted, [D-Ala-6,des-Gly-NH$_2$^{10}] -LH-RH ethylamide, [D-Leu-6, des-Gly-NH$_2$^{10}] -LH-RH ethylamide, and [D-Ser(Bu^t)^6,des-Gly-NH$_2$^{10}]-LH-RH ethylamide were found to be roughly 20-100 times more active than LH-RH in immature male rats and for induction of ovulation (67,68). Furthermore, this type of analog exhibited prolonged activity in vivo. Experiments with [^{125}I]-labeled D-Ala-6- and D-Leu-6-ethylamide peptides indicate that they do not have in-creased plasma half-lives and that their greater potencies and protracted gonadotropin responses are due to increases in binding affinity and in the time the analogs remain bound to pituitary tissue (69).

Since one injection of the superactive analogs [D-Ala-6, des-Gly-NH$_2$^{10}] -LH-RH ethylamide, [D-Leu-6,des-Gly-NH$_2$^{10}] -LH-RH ethyl-amide, [D-Ser(Bu^t)^6,des-Gly-NH$_2$^{10}] -LH-RH ethylamide, and [D-Trp^6]-LH-RH induces protracted stimulation of release of LH and FSH in humans lasting as long as 24 hours, they maybe more convenient and practical to use than LH-RH, which has occasionally been given three time daily for therapeutic purposes (4,5,53,68). Moreover, these analogs are active not only after parenteral but also intra-nasal, intravaginal, intrarectal and oral administration if suit-able doses are given (4,5,53).

Paradoxical Antifertility Effects of Large Doses of LH-RH and its Agonists. Either LH-RH (10-1,000 µg) or the superactive ana-logs such as [D-Ala-6,des-Gly-NH$_2$^{10}] -LH-RH ethylamide (20-100 ng) have been found to cause premature ovulation and inhibit mating and pregnancy when administered acutely the day of diestrus or pro-estrus to cyclic rats (70,71). This temporary antifertility effect was ascribed to the induction of ovulation at the physiologically "wrong time" (70). Prolonged treatment of immature female rats with 1-10 ug/day superactive analogs [D-Ser(Bu^t)^6,des-Gly-NH$_2$^{10}]-LH-RH ethylamide, [D-Trp^6]-LH-RH inhibited ovarian growth and maturation, produced dose-related delays in the vaginal opening, absence of cycling and in mature rats caused cessation of cycling, atrophy of ovaries, reduction in uterine weight and in estrogen production (68,72). There is evidence that [^{125}I]-labeled [D-Leu^6, des-Gly-NH$_2$^{10}]-LH-RH ethylamide is specifically bound to ovarian receptors. It is therefore possible that these superactive analogs, and presumably LH-RH itself, could have a direct inhibitory influence on ovarian growth. Recently we have also determined that rats treated with 1 μg [D-Trp^6]-LH-RH daily for 10 days showed decreased number of uterine binding sites for estradiol. This is the first proof for a direct effect of [D-Trp^6]-LH-RH on the uterus, and a demonstration that this analogue can exert extra-pituitary as well as pituitary actions.

Administration of [D-Ser(Bu^t)^6,des-Gly-NH$_2$^{10}]-LH-RH ethyl-amide in doses of 50-200 μg/kg body weight to male rats and dogs reduced testosterone content and caused atrophy of the testes, seminal vesicles and prostrate gland (68). Recent studies (73) have shown that daily administration of as little as 8 ng of [D-Leu^6,des-Gly-NH$_2$^{10}]-LH-RH ethylamide three times a day for one week to male rats results in a 30% reduction of testicular LH/hCG and prolactin receptors with a maximal reduction of 80% at 40 ng. FSH receptor levels were not affected, but testosterone levels were reduced. Equally dramatic is the ability of large doses of LH RH (100-1,000 μg/day) and smaller but still pharmacological doses (1 6 μg/day) of the superactive analogs to block implantation and terminate gestation when given daily postcoitally to rats (74-76). This effect is also dose-dependent and appears to be directly related to the hypersecretion of LH, functional luteolysis and/or inhibition of progesterone secretion (73,76). In view of these paradoxical antifertility effects of large doses of LH-RH and long acting superactive analogs, caution must be exercised in devising clinical protocols.

Inhibitory. It is now clear that only certain modifications to positions 1, 2, and 3 appear to result in LH-RH antagonists and it is considered that this N-terminal portion of the decapeptide constitutes its active center with respect to gonadotropin release (4,5,12,53). The remaining C-terminal portion of the molecule is involved primarily in the binding process and in preserving overall confirmation. As in the case of the corresponding agonist peptides, the substitution of D-Ala, D-Leu, D-Phe, or D-Trp in position 6 of a particular inhibitor increases antagonist activity dramatically in that order (4,5). Another important discovery (77) in the antagonist field was that [D-Phe^2]-LH-RH was far more effective than

[des-His^2]-LH-RH in vivo and in vitro. At present, all of the
most active inhibitory peptides are based on this modification (4,
5). Analogs such as [D-Phe^2,D-Phe^6] - and [D-Phe^2,D-Trp^6]-LH-RH
are effective in single doses of about 6 mg/kg.

Tryptophan in [D-Phe^2,D-Phe^6]-LH-RH can be replaced by phenyl-
alanine, proline or D-Trp with a reduction in residual agonist
activity. [D-Phe^2-Phe^3-D-Phe^6]-LH-RH, [D-Phe^2,D-Trp^3,D-Phe^6]-LH-RH
(78) and D-Phe^2,Pro^3,D-Phe^6 -LH-RH completely blocked ovulation
at single doses of about 1 mg per rat. [D-Phe^2,D-Trp^3,D-Phe^6]-LH-RH
can suppress the liberation of LH and FSH in response to exogenous
LH-RH in humans and lower the elevated levels of gonadotropins in
patients with low or absent gonadal function (79). Replacement of
L-(pyro)-Glu by D-pyro(Glu) further increases the inhibitory activity of the antagonistic analog and should significantly lower the
doses required for clinical effects. Work is presently in progress
in this laboratory on certain branched chain and dimeric forms of
some of these inhibitory peptides which are up to four times more
active than the best ones reported here. The progress being made
in this area, may possibly lead to development of new birth control
methods.

GROWTH HORMONE-RELEASING FACTOR (GH-RF)

The secretion of growth hormone (GH) from the anterior pituitary is regulated by a dual system of hypothalamic control, one
inhibitory and one stimulatory (12,13,26).

The stimulatory effect on GH release of some hypothalamic
fractions (80) appears to be due to a GH-RF, first detected more
than 13 years ago (12,13,26), and which, under some conditions,
might predominate over the inhibitory action of somatostatin because of the short half-life of the latter. Despite the intense
effort by several groups, the nature of the physiological GH-RF is
still unknown. However, using antisera to somatostatin or columns
of Sepharose linked to anti-somatostatin-gamma-globulin to eliminate somatostatin and its assumed precursors, we have obtained new
evidence for the existence of GH-RF and we have purified several
fractions with GH-RF activity by gel filtration on Sephadex, CCD
and chromatography on CMC and on partition columns. These fractions
stimulated the release of growth hormone from monolayer cultures of
rat anterior pituitaries at a dose of 1 μg/ml and increased plasma
GH in mice pretreated with chloropromazine-morphine-Nembutal. Their
repurification, testing as well as structural approach combined with
synthetic are continuing.

SOMATOSTATIN

Somatostatin (growth hormone release-inhibiting hormone; GH-RIH) was isolated from ovine, (81) and subsequently porcine (82) hypothalami for its ability to inhibit the release of pituitary growth hormone (GH). The primary structure of this peptide is identical in both species (83,84): H-Ala-Gly-Cys-Lys-Asn-Phe-Phe-Trp-Lys-Thr-Phe-Thr-Ser-Cys-OH. Synthetic somatostatin was found to produce a remarkable array of actions on diverse hormones and other substances from many tissues. These discoveries gained physiological significance when we found somatostatin-like immuno-reactivity in high concentrations within those tissues affected by the hormone (85). Thus, in addition to the hypothalamus, somato-statin appears to be localized in pancreas, stomach, gut and brain.

Larger, highly basic forms of somatostatin found in pig hypo-thalami are biologically and immunologically active (82). We also recognized two types of immunoreactive somatostatin in extracts from rat pancreas, stomach and duodenum (85). These molecules may all represent precursors of somatostatin.

Somatostatin was synthesized by several groups (4,5,13,83,84), thus providing sufficient quantities of the pure peptide for exami-nation of its biological activities.

Pituitary

The inhibitory action of somatostatin on both basal and stimu-lated secretion of GH in a variety of in vitro and in vivo assays was demonstrated in several species, including humans (4,5,13,86, 87). Both the reduced (linear) and oxidized (cyclic) forms are active (13). A physiological role for somatostatin in the regula-tion of GH secretion is supported by our observations that passive immunization in rats with anti-somatostatin elevates basal GH levels and prevents the stress-induced decrease of GH (88).

Somatostatin also inhibits the TRH-induced secretion of TSH, but not prolactin, in vitro and in vivo, and it could play a phy-siological role in the regulation of TSH secretion (89). The re-lease of prolactin in vitro is also decreased by somatostatin (13).

Pancreas

Somatostatin inhibits both basal and stimulated secretion of insulin and glucagon by a direct action on the β and α pancreatic islet cells (13,90).

Somatostatin affects the exocrine pancreas as well, inhibiting pancreatic bicarbonate and protein secretion (91).

Stomach and Gut

The observation by our group (87,92) that the rise in blood sugar after oral administration of glucose was delayed by administration of somatostatin, prompted us (93) to examine the effect of somatostatin on the gastrointestinal tract and gastrin. Both basal gastrin and that released in response to food were inhibited by somatostatin. Somatostatin inhibited gastric acid and pepsin in response to pentagastrin, as well as the gastric acid response to food and hypoglycemia in cats with gastric fistulae, indicating a direct action by somatostatin on the parietal and peptic cells (94). Thus, a direct action of somatostatin on exocrine secretion was established.

Subsequently, somatostatin has been found to inhibit the release of secretin, pepsin, cholecystokinin/pancreozymin, gastric inhibitory polypeptide, vasoactive intestinal polypeptide, and motilin (91,94).

Central Nervous System

Plotnikoff et al. (95) were the first to systematically examine its actions on the behavior of mice to determine its CNS effects. They found that somatostatin potentiates, but does not antagonize the behavioral effects (hyperactivity) induced by L-DOPA. Other CNS effects have also been reported (2,13).

Analogs of Somatostatin

Somatostatin is of little therapeutic value since it has many actions and a short biological half-life (4,5,13,96). Analogs are being developed for their enhanced, selective, prolonged and antagonistic activities. The analogs also provide insight into structure function relationships of somatostatin in its various target tissues.

Rivier et al. (97) reported, and we later confirmed (4,5,98) that D-Trp[8]-somatostatin was 6-8 times as active as somatostatin in inhibiting GH release. We showed that this analog had a time course of inhibition identical to that of somatostatin in vivo, indicating that the increased potency was related to conformational effects rather than prolonged biological half-life. This analog also inhibits GH more selectively than pentagastrin-induced gastric

acid secretion (98). Multiple substitutions are sometimes additive; [N-Tyr,D-Trp[8]]- and [D-Ala[2],D-Trp[8]]-somatostatin are each
about 10 times more active than somatostatin in suppressing GH
release (4,5). Recently, we found that the D- and L- diastereo-
isomers of [5F-Trp[8]]-, [6F-Trp[8]]- and [5Br-Trp[8]]-somatostatin were
all more active than somatostatin in the in vitro GH assay (99).
[D-5F-Trp[8]]- and [D-5Br-Trp[8]]-somatostatin were 25 and 30 times
more active than somatostatin in that assay and may be long-acting.

The first analog with selective actions was des[Ala[1],Gly[2],
Asn[5]]-somatostatin which inhibited insulin, and to a lesser extent
GH, without affecting glucagon release (100,101). The same investigators later reported an analog which retained about 20% GH-re-
lease-inhibiting activity, but did not affect insulin or glucagon
levels.

Recently, we (102,103) and others (104) observed that [D-Cys[14]]-
somatostatin and [D-Trp[8],D-Cys[14]]-somatostatin selectively inhibited GH and glucagon more than insulin release. In our hands, the
latter analog had a ratio of 22.1 for the selective inhibition of
glucagon over insulin, and 100:1 for that of GH over insulin (102).

We have also observed that [Phe[4]]-somatostatin is 2 to 6 times
more active than somatostatin in suppressing GH release in vitro,
but only 3% and 16% as active as somatostatin in suppressing the
in vivo release of insulin and glucagon, respectively.

Attempts to prepare analogs with prolonged activity have been
less successful. Many D-amino acid substitutions have been made,
either singly or in combination, to suppress enzymatic degradation
but this has not yielded long-acting derivatives. Veber et al.
(105) prepared two nonreducible cyclic analogs of the somatostatin
ring portion by replacing the sulfur atoms with methylene groups.
These analogs retained 50% of somatostatin's GH-release inhibiting
activity, and had full potency to inhibit pentagastrin-induced
gastric secretion. A 30-minute protracted activity was observed.
Sarantakis et al. (106) has reported a nonreducible cyclic analog
with a shortened ring structure which significantly suppresses
plasma GH for four hours after injection, but high doses are required. In preliminary experiments, we have observed that somatostatin analogs with extra atoms in the 38-membered cyclic portion
of the molecule inhibit GH release in vivo for up to six hours,
following an initial decline in suppression at 1-2 hours (Meyers
et al., unpublished). Long-acting analogs with selective activities may be useful in the treatment of juvenile diabetes, diabetic
retinopathy, peptic ulcers, and other conditions.

ACKNOWLEDGEMENTS

Some basic studies originating in this laboratory and described in this review were supported in part by the Veterans Administration, and NIH Research Grants AM-07467, AM-18370, and NIH Contract NICHDD 72-2741, NIH Research Grant AM-09094 and NIH Contract HD-06555.

REFERENCES

1. Green, J.D. and G.W. Harris. J. Endocrinol. 5: 136-146, 1947.
2. Kastin, A.J., L.H. Miller, C.A. Sandman, A.V. Schally and N.P. Plotnikoff. In: Essays of Neurochemistry and Neuropharmacology, Vol. 1, edited by M.B.H. Youdim, W. Lovenberg, D.F. Sharman and J.R. Lagnado. London:Wiley, 1977 pp. 139-176.
3. Hökfelt, T., S. Efendic, C. Hellerstrom, C. Johansson, R. Luft and A. Arimura. Acta Endocrinol. 80 (Suppl. 200): 5-41, 1975.
4. Schally, A.V., D.H. Coy and C.A. Meyers. Ann. Rev. Biochem. 47: 89-128, 1978.
5. Schally, A.V., D.H. Coy, C.A. Meyers and A.J. Kastin. In: Hormonal Proteins and Peptides, Vol. 4, edited by C.H. Li. In Press. 1978.
6. Sétáló, G., B. Flerko, A. Arimura and A.V. Schally. Int. Rev. Cytol. 55: In Press. 1977.
7. Cooper, D.M.F., D. Synetos, R.B. Christie and D. Schulster. J. Endocrinol. 71: 171-172, 1976.
8. Jones, M.T., B. Gillham and E.W. Hillhouse. Fed. Proc. 36: 2104-2109, 1977.
9. Schally, A.V., R.N. Andersen, H.S. Lipscomb, J.M. Long and R. Guillemin. Nature 188: 1192-1193, 1960.
10. Schally, A.V., T.W. Redding, K. Chihara, W.Y. Huang, R.C.C. Chang, W.H. Carter, D.H. Coy and M. Saffran. The Endocrine Society, Submitted. Abstract, 1978.
11. Schally, A.V., W.Y. Huang, T.W. Redding, A. Arimura, D.H. Coy, K. Chihara, R.C.C. Chang, V. Raymond and F. Labrie. Submitted.
12. Schally, A.V., A. Arimura and A.J. Kastin. Science 179: 341-350, 1973.
13. Vale, W., C. Rivier and M. Brown. Ann. Rev. Physiol. 39: 473-527, 1977.
14. Jacobs, L., P. Snyder, J. Wilber, R. Utiger and W.J. Daughaday. J. Clin. Endocrinol. Metab. 33: 996-998, 1971.
15. Tashijan, A., N. Barowski and D. Jensen. Biochem. Biophys. Res. Commun. 43: 516-523, 1971.
16. Debeljuk, L., T.W. Redding, A. Arimura and A.V. Schally. Proc. Soc. Exp. Biol. Med. 142: 421-423, 1973.

17. Bowers, C.Y., H. Friesen, P. Hwang, H. Guyda and K. Folkers. Biochem. Biophys. Res. Commun. 45: 1033-1041, 1971.
18. Smith, J.R., T.R. laHann, R.M. Chestnut, M.A. Carino and A. Horita. Science 196: 660-661, 1977.
19. Martin, J.B., L.P. Renaud and P. Brazeau: Lancet 7931: 393-395, 1975.
20. Jackson, I.M.D. and S. Reichlin. Endocrinology 95: 854-862, 1974.
21. Hökfelt, T., K. Fuxe, O. Johansson, S. Jeffcoate and N. White. Neurosci. Lett. 1: 133-139, 1975.
22. Bassiri, R.M. and R.D. Utiger. Endocrinology 90: 722-727, 1972.
23. Oliver, C., J.P. Charvet, J.L. Codaccioni, J. Vague and J.C. Porter. Lancet 1: 873, 1974.
24. Koch, Y., T. Baram and M. Fridkin. FEBS Lett. 63: 295-298, 1976.
25. Mitsuma, T., Y. Hirooka and N. Nihei. Acta Endocrinol. 83: 225-235, 1976.
26. Reichlin, S., R. Saperstein, I.M.D. Jackson, A.E. Boyd, III and Y. Patel. Ann. Rev. Physiol. 38: 389-424, 1976.
27. Hoffman, K. and C.Y. Bowers. J. Med. Chem. 13: 1099-1101, 1970.
28. Rivier, J., W. Vale, M. Monahan, N. Ling and R. Burgus. J. Med. Chem. 15: 479-482, 1972.
29. Bowers, C.Y., A. Weil, J.K. Chang, H. Sievertsson, F. Enzmann and K. Folkers. Biochem. Biophys. Res. Commun. 40: 683-691, 1970.
30. Coy, D.H., Y. Hirotsu, T.W. Redding, E.J. Coy and A.V. Schally. J. Med. Chem. 18: 948-949, 1975.
31. Sowers, J.R., J.M. Hershman, H.E. Carlson, A.E. Pekary, A.W. Reed, M.G. Nair and C.M. Baugh. J. Clin. Endocrinol. Metab. 43: 856-860, 1976.
32. Sievertsson, H., S. Castensson, O. Lindgren and C.Y. Bowers. Acta Pharma. Suec. 11: 67-76, 1974.
33. Hirschmann, R.F. In: Proc. Int. Symp. Med. Chem. 5th, Paris, July 1976, edited by J. Matthieu, 1977, In Press.
34. Meites, J. and J.A. Clemens. In: Vitamins and Hormones, edited by R.S. Harris, P.L. Munson, E. Dizfalusy and J. Glover. New York:Academic, 1972, pp. 165-221.
35. Koch, Y., G. Goldhaber, I. Fireman, U. Zor, J. Shani and E. Tal. Endocrinology 100: 1476-1478, 1977.
36. Boyd, A.E., E. Spencer, I.M.D. Jackson and S. Reichlin. Endocrinology 99: 861-871, 1976.
37. Schally, A.V., A. Dupont, A. Arimura, J. Takahara, T.W. Redding, J. Clemens and C. Shaar. Acta Endocrinol. 82: 1-14, 1976.
38. Schally, A.V., T.W. Redding, A. Arimura, A. Dupont and G.L. Linthicum. Endocrinology 100: 681-691, 1977.
39. MacLeod, R.M. Endocrinology 85: 916-923, 1969.
40. Takahara, J., A. Arimura and A.V. Schally. Endocrinology 95: 462-465, 1974.

41. MacLeod, R.M. Proc. Soc. Exp. Biol. Med. In Press, 1978.
42. Schally, A.V. and A.J. Kastin. Endocrinology 79: 768-772, 1966.
43. Nair, R.M.G., A.J. Kastin and A.V. Schally. Biochem. Biophys.
    Res. Commun. 43: 1376-1381, 1971.
44. Celis, M.E., S. Taleisnik and R. Walter. Proc. Natl. Acad.
    Sci. USA 68: 1428-1433, 1971.
45. Plotnikoff, N.P., A.J. Kastin, M.S. Anderson and A.V. Schally.
    Life Sci. 10: 1279-1283, 1971.
46. Failli, A., K. Sestanj, H.U. Immer and M. Gotz. Arzneim.-
    Forsch. Drug Res. 27: 2286-2289, 1977.
47. Schally, A.V., A. Arimura, Y. Baba, R.M.G Nair, H. Matsuo,
    T.W. Redding, L. Debeljuk and W.F. White. Biochem. Biophys.
    Res. Commun. 43: 393-399, 1971.
48. Schally, A.V., R.M.G. Nair, T.W. Redding and A. Arimura.
    J. Biol. Chem. 246: 7230-7236, 1971.
49. Matsuo, H., R.M.G. Nair, A. Arimura and A.V. Schally. Biochem.
    Biophys. Res. Commun. 43: 1134-1139, 1971.
50. Baba, Y., H. Matsuo and A.V. Schally. Biochem. Biophys. Res.
    Commun. 44: 459-463, 1971.
51. Matsuo, H., A. Arimura, R.M.G. Nair and A.V. Schally. Biochem.
    Biophys. Res. Commun 45: 822-827, 1971.
52. Burgus, R., M. Butcher, M. Amoss, N. Ling, M. Monahan, J.
    Rivier, R. Fellows, R. Blackwell, W. Vale and R. Guillemin.
    Proc. Natl. Acad. Sci. USA 69: 278-282, 1972.
53. Schally, A.V., A.J. Kastin and D.H. Coy. Int. J. Fertil. 21:
    1-30, 1976.
54. Schally, A.V., A. Arimura, T.W. Redding, L. Debeljuk, W.
    Carter, A. Dupont and J.A. Vilchez-Martinez. Endocrinology
    98: 380-391, 1976.
55. Cattanach, B.M., C.A. Iddon, H.M. Charlton, S.A. Chiappa and
    G. Fink. Nature Vol. 269: 338-340, 1977.
56. Kerdelhue, B., M. Jutisz, D. Gillessen and R.O Studer. Biochem.
    Biophys. Acta. 297: 540-548, 1973.
57. Koch, Y., P. Chobsieng, U. Zor, M. Fridkin and H.R. Lindner.
    Biochem. Biophys. Res. Commun. 55: 623-629, 1973.
58. de la Cruz, A., A. Arimura, K.G. de la Cruz and A.V. Schally.
    Endocrinology 98: 490-497, 1976.
59. Arimura, A., N. Nishi and A.V. Schally. Proc. Soc. Exp. Biol.
    Med. 152: 71-75, 1976.
60. Nishi, N., A. Arimura, K.G. de la Cruz and A.V. Schally.
    Endocrinology 98: 1024-1030, 1976.
61. Moss, R.L. Fed. Proc. 36: 1978-1983, 1977.
62. Pedroza, E., J.A. Vilchez-Martinez, J. Fishback, A. Arimura
    and A.V. Schally. Biochem. Biophys. Res. Commun. 79: 234-238,
    1977.
63. Fujino, M., S. Kobayashi, M. Obayashi, T. Fukuda, S. Shinagawa,
    I. Yamakazi, R. Nakagami, W.F. White and R.H. Rippel. Biochem.
    Biophys. Res. Commun. 49: 698-705, 1972.

64. Coy, D.H., J.A. Vilchez-Martinez, E.J. Coy, N. Nishi, A. Arimura and A.V. Schally. Biochemistry 14: 1848-1851, 1975.
65. Monahan, M., M.S. Amoss, H.A. Anderson and W. Vale. Biochemitry 12: 4616-4620, 1973.
66. Coy, D.H., J.A. Vilchez-Martinez, E.J. Coy and A.V. Schally. J. Med. Chem. 19: 423-425, 1976.
67. Vilchez-Martinez, J.A., D.H. Coy, A. Arimura, E.J. Coy, Y. Hirotsu and A.V. Schally. Biochem. Biophys. Res. Commun. 59: 1226-1232, 1974.
68. Sandow, J., W. von Rechenberg, W. König, M. Han, G. Jerabek and H. Fraser. In: Proc. Eur. Colloq. Hypothal. Horm., 2nd. Tübingen, July 26-28, 1976. Hamburg:Verlag Chemie. In Press.
69. Reeves, J.J., G.K. Tarnavsky, S.R. Becker, D. Coy and A.V. Schally. Endocrinology 101: 540-547, 1977.
70. Banik, U.K. and M.L. Givner. Fertil. Steril. 27: ,078-1084, 1976.
71. Beattie, C.W. and A. Corbin. Biol. Reprod. 16: 333-339, 1977.
72. Johnson, E.S., R.L. Gendrich and W.F. White. Fertil. Steril. 27: 853-860, 1976.
73. Auclair, C., P.A. Kelly, F. Labrie, D.H. Coy and A.V. Schally. Biochem. Biophys. Res. Commun. 76: 855-862, 1977.
74. Humphrey, R.R., B.L. Windsor, J.R. Reel and R.A. Edgren. Biol. Reprod. 16: 614-621, 1977.
75. Corbin, A., C.W. Beattie, J. Yardley and T.J. Foell. Endocrine Res. Commun. 3: 359-376, 1976.
76. Arimura, A., E. Pedroza, J.A. Vilchez and A.V. Schally. Endocr. Res. Commun. In Press.
77. Rees, R.W.A., T.J. Foell, S. Chais and N. Grant. J. Med. Chem. 17: 1015-1019, 1974.
78. de la Cruz, A., D.H. Coy, J.A. Vilchez-Martinez, A. Arimura and A.V. Schally. Science 191: 195-197, 1976.
79. Gonzalez-Barcena, D., A.J. Kastin, D.H. Coy, K. Nikolics and A.V. Schally. Lancet 2: 997-998, 1977.
80. Schally, A.V., S. Sawano, A. Arimura, J. Barrett, I. Wakabayashi and C.Y. Bowers. Endocrinology 84: 1493-1506, 1969.
81. Brazeau, P., W. Vale, R. Burgus, N. Ling, M. Butcher, J. Rivier and R. Guillemin. Science 179: 77-79, 1973.
82. Schally, A.V., A. Dupont, A. Arimura, T.W. Redding, N. Nishi, G.L. Linthicum and Schlesinger, D.H. Biochemistry 15: 509-514, 1976.
83. Rivier, J., P. Brazeau, W. Vale, N. Ling, R. Burgus, C. Gilon, J. Yardley and R. Guillemin. C.R. Acad. Sci. Paris 276: 2737-2740, 1973.
84. Coy, D.H., E.J. Coy, A. Arimura and A.V. Schally. Biochem. Biophys. Res. Commun. 54: 1267-1273, 1973.
85. Arimura, A., H. Sato, A. Dupont, N. Nishi and A.V. Schally. Science 189: 1007-1009, 1975.

86.   Hall, R., G.M. Besser, A.V. Schally, D.H. Coy, D. Evered,
      D.J. Goldie, A.J. Kastin, A.S. McNeilly, C.H. Mortimer,
      C. Phenekos, W.M.G. Tunbridge and D. Weightman. Lancet 2:
      581-584, 1973.

87.   Besser, G.M., C.H. Mortimer, D. Carr, A.V. Schally, D.H. Coy,
      D. Evered, A.J. Kastin, W.M.G. Tunbridge, M.O. Thorner and
      R. Hall. Br. Med. J. 1: 352-355, 1974.

88.   Arimura, A., W.D. Smith and A.V. Schally. Endocrinology 98:
      540-543, 1976.

89.   Arimura, A. and A.V. Schally. Endocrinology 98: 1069-1072,
      1976.

90.   Alberti, K.G.M.M., S.E. Christensen, J. Iverson, K. Seyer-
      Hansen, N.J. Christensen, Aa. Prange-Hansen, K. Lundbaek
      and H. Orshov. Lancet 2: 1299-1301, 1973.

91.   Konturek, S.J., J. Tasler, W. Obtulowicz, D.H. Coy and A.V.
      Schally. J. Clin. Invest. 58: 1-6, 1976.

92.   Mortimer, C.H., D. Carr, T. Lind, S.R. Bloom, C.N. Mallinson,
      A.V. Schally, W.M.G. Tunbridge, L. Yeoman, D.H. Coy, A.J.
      Kastin, G.M. Besser and R. Hall. Lancet 1: 697-701, 1974.

93.   Bloom, S.R., C.H. Mortimer, M.O. Thorner, G.M. Besser, R.
      Hall, A. Gomez-Pan, V.M. Roy, R.C.G. Russell, D.H. Coy,
      A.J. Kastin and A.V. Schally. Lancet 2: 1106-1109, 1974.

94.   Besser, G.M. and C.H. Mortimer. In: Frontiers in Endocrinology,
      edited by L. Martini and W.F. Ganong. New York:Raven, 1976,
      pp. 227-254.

95.   Plotnikoff, N.P., A.J. Kastin and A.V. Schally. Pharmacol.
      Biochem. Behav. 2: 693-696, 1974.

96.   Hall, R. and A. Gomez-Pan. In: Advances in Clinical Chemistry,
      Vol. 18, New York:Academic Press, 1976, pp. 173-212.

97.   Rivier, J., M. Brown and W. Vale. Biochem. Biophys. Res.
      Commun. 65: 746-751, 1975.

98.   Coy, D.H., E.J. Coy, C. Meyers, J. Drouin, L. Ferland, A.
      Gomez-Pan and A.V. Schally. Endocrinology 98: 305A, 1976.

99.   Meyers, C.A., D.H. Coy, W.Y. Huang, A.V. Schally and T.W.
      Redding. Biochemistry: In Press, 1978.

100. Efendic, S., R. Luft and H. Sievertsson. FEBS Letters 58:
      302-305, 1975.

101. Sarantakis, D., W.A. McKinley, I. Juanakais, D. Clark and
      N.H. Grant. Clin. Endocrinol. 5s: 275-278, 1976.

102. Meyers, C., A. Arimura, A. Gordin, R. Fernandez-Durango, D.H.
      Coy, A.V. Schally, J. Drouin, L. Ferland, M. Beaulieu and
      F. Labrie. Biochem. Biophys. Res. Commun. 74: 630-636, 1977.

103. Gordin, A., C. Meyers, A. Arimura, D.H. Coy and A.V. Schally.
      Acta Endocrinol. 86: 833-841, 1977.

104. Brown, G.M., G.R. Van Loon, B.C.W. Hummel, L.J. Grota, A.
      Arimura and A.V. Schally. J. Clin. Endocr. Metab. 44: 784-
      790, 1977.

105. Veber, D.F., R.G. Strachan, S.J. Bergstand, F.W. Holly, C.T.
      Homnick, R. Hirschmann, M.A. Torchinana and R. Saperstein.
      J. Amer. Chem. Soc. 98: 2367-2369, 1976.
106. Sarantakis, D., J. Teichman, E.L. Lieng and R.L. Fenichel,
      Biochem. Biophys. Res. Commun. 73: 336-346, 1976.

NEUROTRANSMITTERS AND NEUROPEPTIDES: DISTRIBUTION PATTERNS AND CELLULAR LOCALIZATION AS REVEALED BY IMMUNOCYTOCHEMISTRY

T. Hökfelt, O. Johansson, Å. Ljungdahl, J. Lundberg, M. Schultzberg, K. Fuxe, Dept. of Histology, Karolinska Institute, Stockholm, Sweden,

M. Goldstein, Dept. of Psychiatry, New York University Medical Center, New York, N.Y., U.S.A.,

H. Steinbusch, A. Verhofstad, Dept. of Anatomy and Embryology, Katholieke Universiteit, Nijmegen, The Netherlands and

R. Elde, Dept. of Anatomy, University of Minnesota, Minneapolis, U.S.A.

INTRODUCTION

During the last years dramatic advances have been made with regard to the possibility to analyse the nervous system, not at least the hypothalamus, with histochemical methods. This is in part due to the fact that immunohistochemical techniques have been employed in the analysis and identification of specific neuron systems. The immunohistochemical approach is characterized by its unique general applicability, since principally any substance can be traced against which antibodies can be raised. Immunohisto-chemistry was originally introduced more than thirty years ago by Coons and collaborators (see 1). It was successfully applied to transmitter related problems by Geffen et al. (2) who identified catecholamine neurons with antibodies to dopamine-β-hydroxylase, the enzyme converting dopamine to noradrenaline. Now the phase has been reached where the full capacity of this technique can be exploited in studies on the nervous system. This is a consequence of the fact that only now have a larger number of enzymes and peptides been purified and structurally characterized in a way that they can be utilized as proper antigens for raising antisera. Thus, the dramatic advancements in protein and peptide biochemistry, accomplished e.g. by Schally (see this book), Guillemin and their

31

collaborators, have opened up entirely new possibilities for the
neurohistochemists. Furthermore, recent improvements in the immunocytochemical field including i.a. peroxidase labelled antibodies
(see 3,4) have not only extended the sensitivity of the technique
but also allowed ultrastructural analysis. The immunohistochemical
analysis of various neuron systems represent a continuation and
extension of the methods and work carried out to map cholinergic
neuron with cholinesterase techniques (5, see 6) and monoamine
neuron with formaldehyde induced fluorescence (7-9).

In the present short article no detailed description of the
different neuron systems can be given but some recent findings will
be summarized. The monoamine and peptide systems in the hypothalamus
and in other parts of the nervous system have recently been reviewed
elsewhere (see 10-12).

METHODOLOGY

The results summarized in this article have mainly been obtained
with the indirect immunofluorescence technique, essentially as
originally described by Coons and collaborators (see 1). The tissues
have been fixed by perfusion via the ascending aorta with ice-cold
formalin. This approach seems necessary in order to prevent diffusion
of cytoplasmic enzymes such as tyrosine hydroxylase and also of
smaller peptides. The possibility to localize also very small molecules with immunohistochemistry is discussed in this paper. As first
shown by the Dutch workers Steinbusch, Verhofstad and Joosten (13)
antibodies directed to serotonin can be used to identify neurons
containing this indoleamine. Also in this case formalin fixed tissue
has been used demonstrating that at least part of the serotonin
escapes diffusion out from the neuron during this procedure. Whether
or not also other small molecules such as the catecholamines can be
traced in this way is at present under investigation by Steinbusch
and colleagues.

Results obtained with immunohistochemical techniques should be
interpreted with some caution. It is well known that the antisera
raised against a certain peptide may cross-react with structurally
related substances. It seems therefore preferable to use terms such
as "substance P-like immunoreactivity", etc. The evidence so far,
however, indicates that, with some exceptions, each peptide neuron
system described here contains only one peptide and, thus, represents
a unique subsystem.

MONOAMINES

Catecholamines

The hypothalamic catecholamine neurons have been extensively
studied histochemically, initially with the Falck-Hillarp technique
(see 14-16). Several cell groups have been distinguished forming
essentially one dorsal (groups A11 and A13) and one ventral (groups
A12 and A14) system (classification according to Dahlström and Fuxe
(17). With the immunohistochemical approach tyrosine hydroxylase
could be identified in these systems but in addition also in small
cell groups and single cells in several other hypothalamic nuclei
(18,19). Of particularly interest was a large group of very small
neurons located over the mamillary bodies and extending into mesencephalon thus forming an almost continuous band to the so called
A10 dopamine cell group of the ventral tegmental area. Sladek
(20) have recently described dopamine containing subependymal
cells. So far there is no clear histochemical evidence for intrahypothalamic noradrenaline or adrenaline systems (21,22, see 10)
although some biochemical studies (23) indicate the possible existence of adrenaline cell bodies in the hypothalamus. Thus, the
vast majority of noradrenaline and adrenaline nerve terminals in
the hypothalamus have an extrahypothalamic origin probably arising
mainly in the lower brain stem.

5-Hydroxytryptamine

As described early by Fuxe (24,25) the hypothalamus in general
receives a rich 5-hydroxytryptamine (serotonin) innervation. The
detailed knowledge of these systems has, however, been lacking. This
has been due to the fact that the Falck-Hillarp formaldehyde fluorescence technique has been considerably less sensitive for serotonin than for catecholamines (26). Moreover, particularly in areas
with high densities of catecholamine nerve terminals it is difficult
to discover the rather weakly yellow fluorescent serotonin systems.
Such an area is the median eminence which contains a very dense
catecholamine, mainly dopamine innervation. Kordon and collaborators
(see 27) obtained early evidence for the existence of serotonergic
mechanisms in the control of the anterior pituitary hormone secretion. Also according to biochemical evidence serotonin exists in the
median eminence (see 28). These findings could be supplemented by
indirect histochemical approaches such as autoradiography (29),
ultrastructural analysis of the median eminence after treatment with
neurotoxins (30) and electron microscopic immunocytochemistry and
fluorescence histochemistry (31). Final evidence for the existence
of serotonin nerve terminals in the external layer of the median
eminence has now been obtained with the indirect immunofluorescence
technique and antiserum directed against serotonin (Steinbusch et al.,
in preparation). A moderately dense network of serotonin positive
fibers are present mainly in the lateral parts of the external

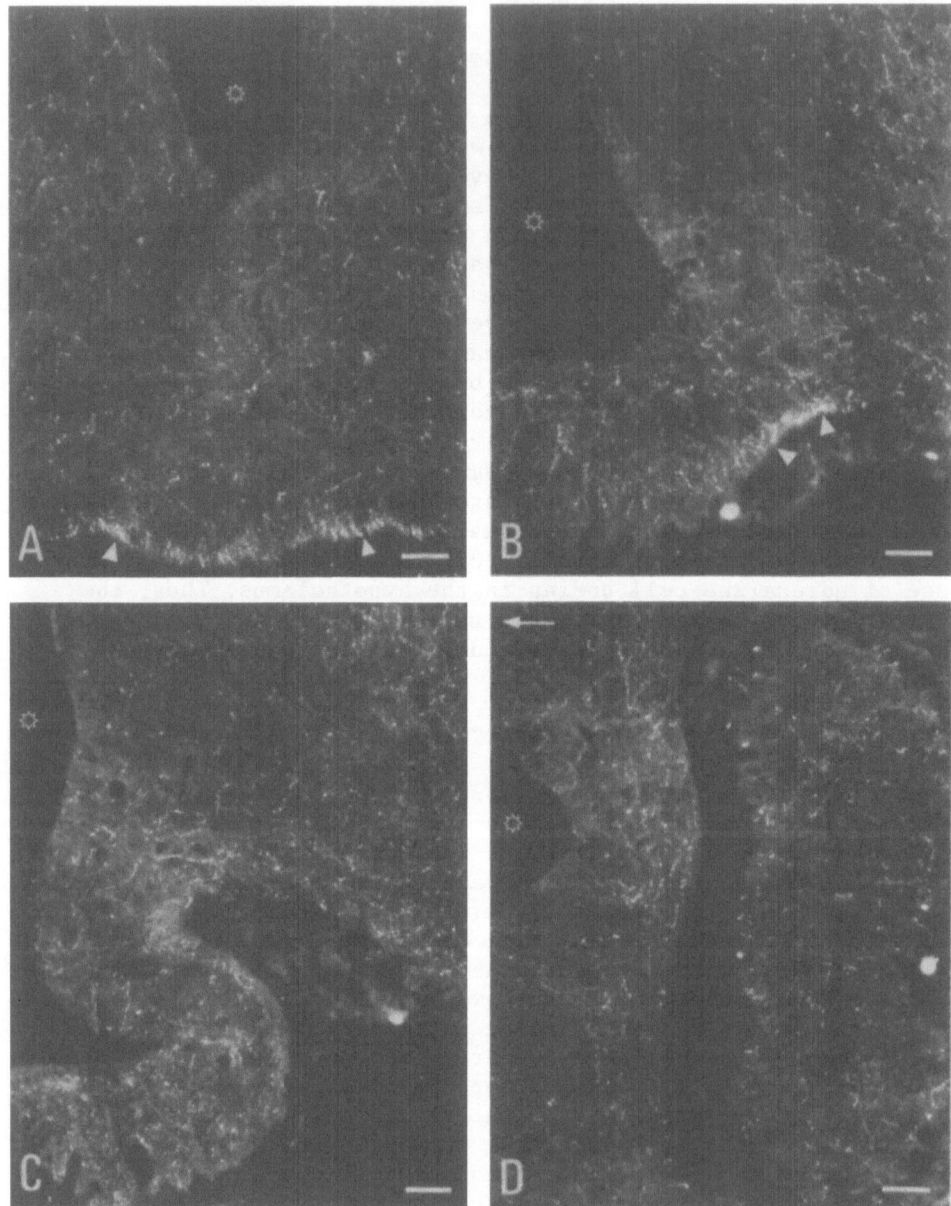

Fig. 1 A-D. Immunofluorescence micrographs of four different levels
of the rat median eminence (A rostral → D caudal) after incubation
with antiserum to 5-hydroxytryptamine. Immunofluorescent nerve terminals are found at all levels of the median eminence, both in the
internal and external (arrow heads) layers. Asterisks indicate third
ventricle. Arrow in D points dorsally. Bars indicate 50 μm.

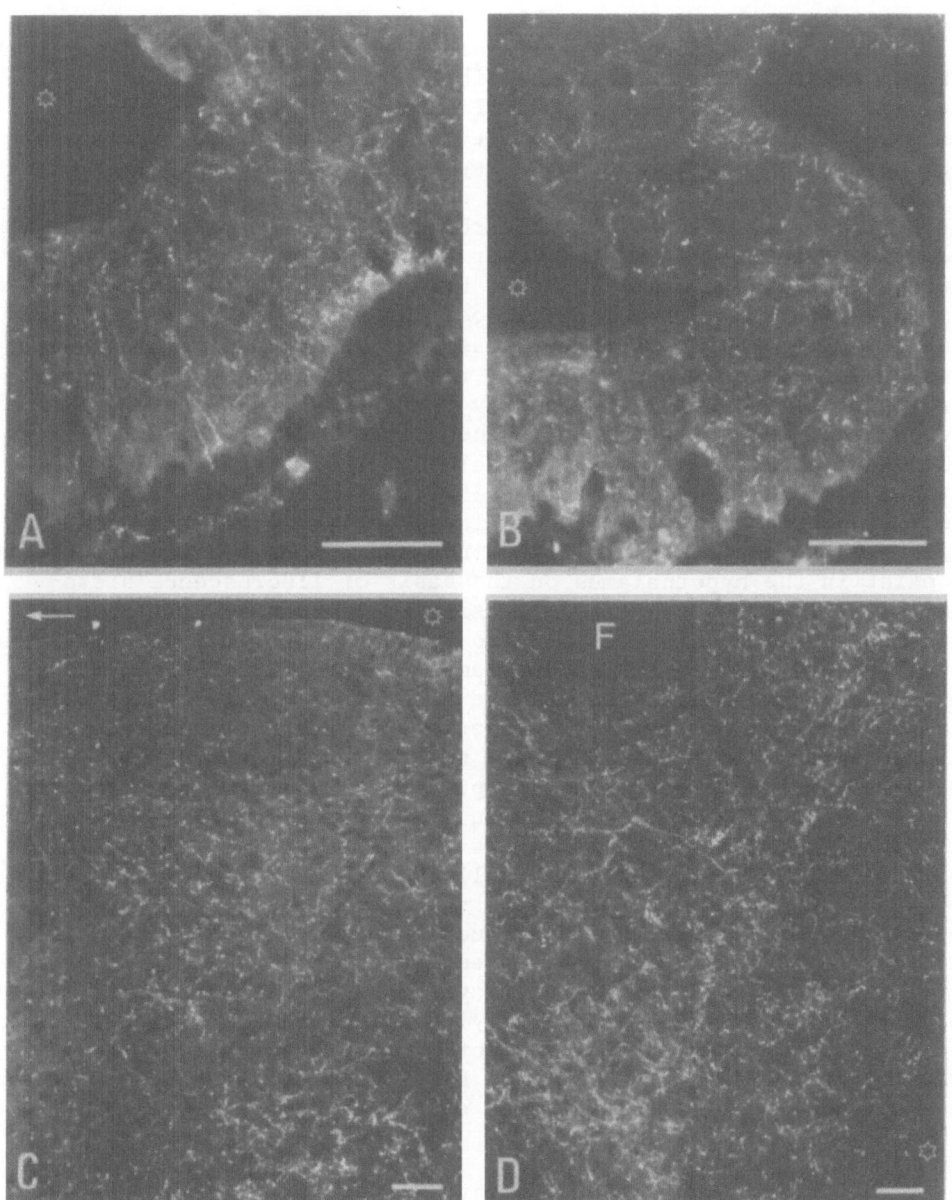

Fig. 2 A-D. Immunofluorescence micrographs of (A, B) the lateral part of the median eminence, (C) the arcuate nucleus and (D) the perifornical area-lateral hypothalamus after incubation with anti-serum to 5-hydroxytryptamine. Immunofluorescent nerve terminal net-works of varying densities are seen. Asterisks in A-C indicate third ventricle. Asterisk in D indicates ventral surface of the brain. F = fornix. Bars indicate 50 μm.

layer of the median eminence (Figs. 1A-D, 2A,B). In the most rostral
parts the entire external layer of the median eminence contains such
fibers. Serotonin immunoreactive fibers are also present in the
internal and subependymal layers (Fig. 1A-D; 2A,B). No serotonin
immunoreactivity could be identified with certainty in the tanycytes
(32). A further point of controversy has been the possible existence
of the intrahypothalamic serotonin containing cell bodies. According to the original data of Dahlström and Fuxe (17) serotonin cell
bodies were found only in the lower brain stem giving rise to i.a.
the hypothalamic terminal systems. Using antibodies to dopadecarboxylase a number of hypothalamic cell groups could be identified
(see 10). Since this antiserum cross-reacts with the decarboxylase
present in serotonin neurons (33), these cell groups were potential
serotonin cell groups, particularly since they lacked tyrosine
hydroxylase, the first enzyme in the catecholamine synthesis. These
cells have also been shown to be able to decarboxylate dopa (34).
Using autoradiography Chan-Palay (35) and Beaudet and Descarries
(36) also observed several hypothalamic cells accumulating and
storing exogonous serotonin. Serotonin neurons have also been claimed
to exist in the hypothalamus on the basis of fluorescence histochemical investigations (37,38). Using the serotonin antiserum
raised by Steinbusch et al. (13) so far no immunoreactive cell bodies
have been observed in the hypothalamus of rats (Steinbusch et al.,
in preparation). These studies have been carried out both on untreated rats as well as on rats pretreated with intraventricular
injections of colchicine. The last treatment is known to markedly
increase the levels not only of peptides and enzymes in cell bodies
but also of serotonin in neurons in other parts of the brain, e.g.
in the lower brain stem (39). It is therefore our opinion that the
possible existence of serotonin cell bodies in the hypothalamus has
to be further investigated. The intrahypothalamic cell systems
described with the various techniques above may contain another
indoleamine. Such neurons may possess an uptake mechanism also for
serotonin and a decarboxylating enzyme cross-reacting with the antiserum raised against dopa decarboxylase. The studies with serotonin
antiserum also revealed dense networks of serotonin fibers in many
other hypothalamic nuclei (Fig. 2C,D) as will be described in detail
in a forthcoming paper (Steinbusch et al., in preparation).

OTHER SMALL PUTATIVE TRANSMITTERS

Comparatively little is known about the exact distribution of
systems containing putative transmitters such as acetylcholine,
certain amino acids and histamine. The reason for this is the lack
of histochemical techniques suitable to locate these substances.
Cholinergic systems have been traced with acetylcholinesterase
staining but, as pointed out in many articles, the significance of
this technique may be discussed (see 6). Attempts have been made to
trace amino acid neurons with autoradiography but technical problems

have resulted in difficulties with regard to interpretation of the
results (see 40,41) For GABA neurons, however, a more fortunate
situation exists since Roberts and colleagues have succeeded in
purifying glutamate decarboxylase, the GABA synthesizing enzyme.
Using antiserum to this enzyme probable GABA neurons have been
traced in various parts of the nervous system (see 42). Dense networks of glutamate decarboxylase positive fibers can be observed
in most hypothalamic nuclei including the external layer of the
median eminence (see 10).

NEUROPEPTIDES

Central peptide neurons were early identified on the basis of
immunocytochemistry by Livett et al. (43; see also 44) using antiserum to neurophysin. The first small peptide successfully traced
with immunohistochemistry was luteinizing hormone releasing hormone
(LHRH) (45-47) and the work in this field has recently been reviewed
by Sternberger and Hoffman (48). Numerous research groups agree on
the distribution of nerve fibers in the external layer of the median
eminence, but controversy has arisen over the issue of the localization of LHRH cell bodies (see 48). Originally, such cell bodies could
only be observed in the preoptic area after special treatments (e.g.
colchicine) (45,46). More recently, Hoffman and collaborators (see
48) have been able to demonstrate LHRH positive neurons also in the
arcuate nucleus employing a certain antiserum (so called Sorrentino
antiserum F). Of the three hypothalamic hormones so far isolated and
structurally characterized - LHRH,thyrotropin releasing hormone
(TRH) and somatostatin - the former has the most limited distribution
in the brain. Both somatostatin and TRH appears to be localized at
all levels of the central nervous system including the spinal cord.
The highest concentrations also of these two peptides are, however,
found in the external layer of the median eminence. Particularly
somatostatin is distributed both in the medial and lateral parts
whereas TRH is preferentially located in the medial aspects. Thus,
these three hypothalamic hormones are clearly located in nerve endings close to the portal vessels, which is in good agreement with
their role as hypothalamic releasing (or inhibiting) hormones. In
view of the wide spread occurrence particularly of TRH and somatostatin it is, however, likely that these peptides must play a further
role(s) in the nervous system function, possibly as modulator or
neurotransmitter, as suggested e.g. by electrophysiological studies
(49,50). Such a role has also been proposed for numerous other small
peptides such as substance P and the enkephalins which are present
in many parts of the nervous system and appear to satisfy several
of the transmitter criteria (see 51).

The three hypothalamic hormones LHRH, TRH and somatostatin are
not the only substances present in the median eminence. In the foregoing dopamine, noradrenaline, adrenaline, serotonin and GABA con-

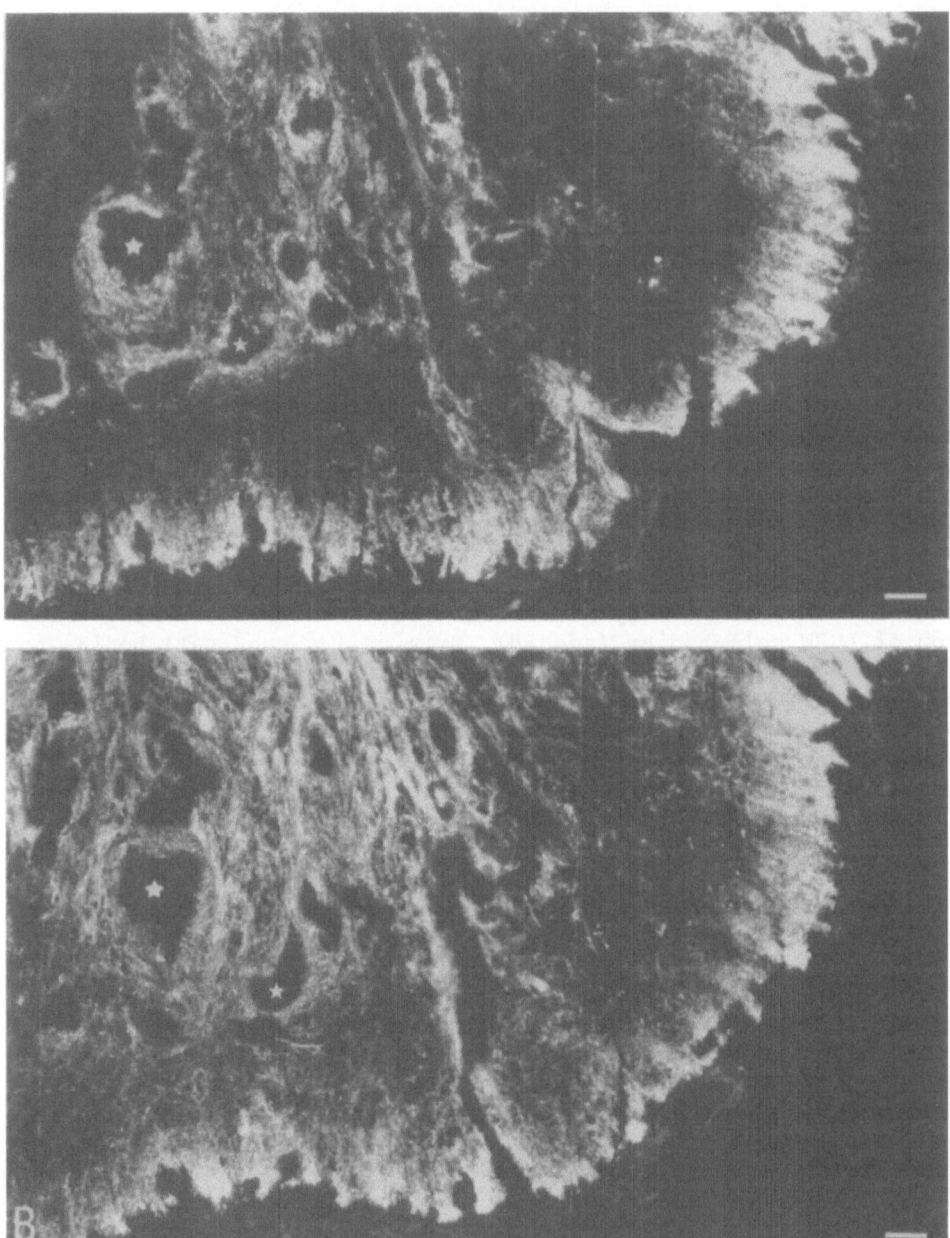

Fig. 3. Immunofluorescence micrographs of semiconsecutive sections of the eminentia medialis of monkey after incubation with antiserum to (A) substance P and (B) somatostatin. Dense fiber networks containing immunoreactive substance P and somatostatin are seen, often closely related to blood vessels (asterisks). Bars indicate 50 μm.

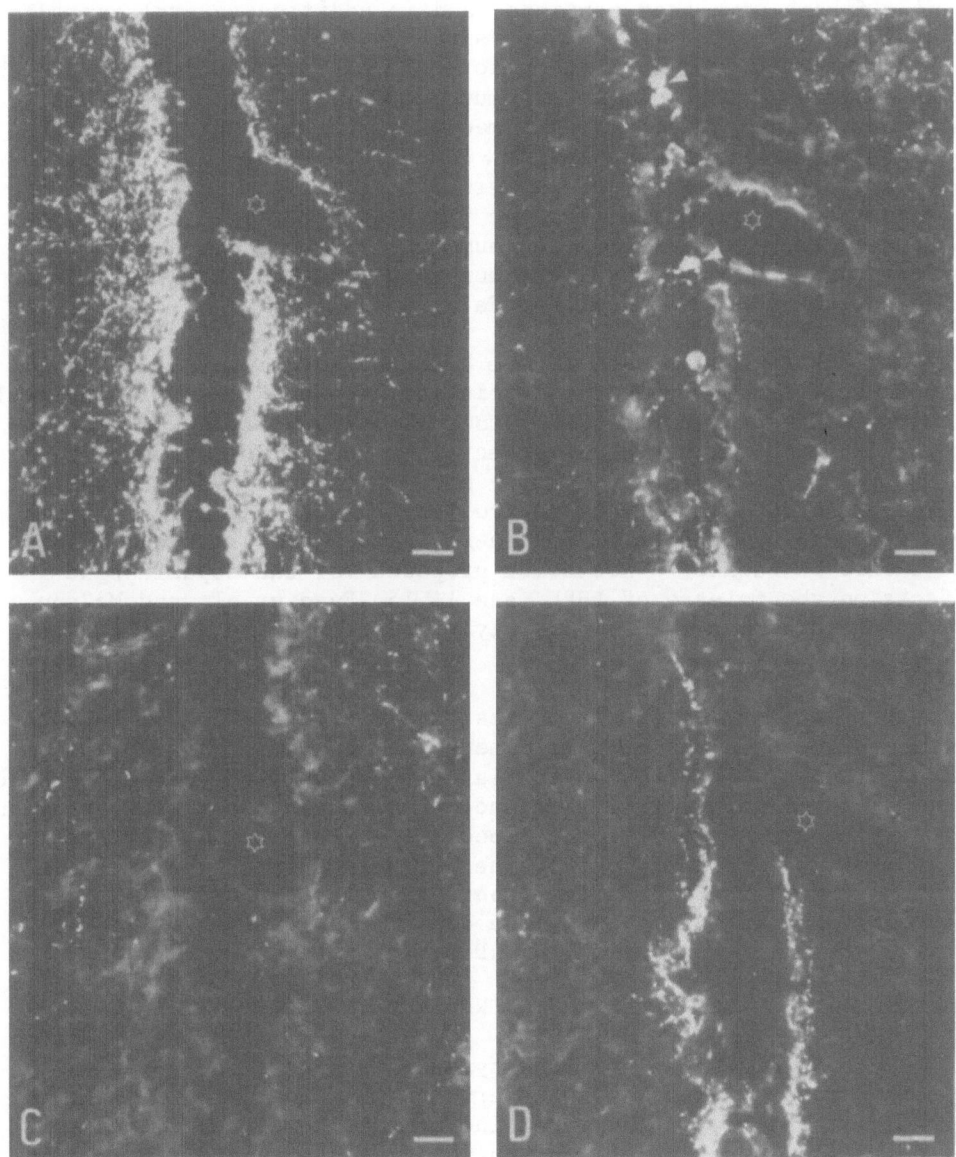

Fig. 4 A-D. Immunofluorescence micrographs of consecutive sections of the organum vasculosum laminae terminalis after incubation with antiserum to (A) LHRH, (B) tyrosine hydroxylase, (C) dopamine-β-hydroxylase and (D) somatostatin. LHRH-positive (and to a lesser extent somatostatin-positive) fibers are seen, partly close to the surface and to blood vessels (asterisks) but only few catecholamine fibers overlap with the peptide containing terminals. Arrow heads in B point to unspecifically stained structures. Bars indicate 50 μm.

taining fibers have been mentioned and in addition several peptides
have been observed. Thus, neurophysin-, oxytocin- and vasopressin-
(52-55), enkephalin- (10,19), angiotensin II- (56), gastrin- (probably cholecystokinin-) (10) like immunoreactivities have been observed
in the external layer of the rat median eminence. In primates - but
only to a small extent in the rat - substance P positive fibers are
found in large numbers in the external layer of the median eminence
(Fig. 3A) (57). As judged from the immunohistochemical studies the
substance P fibers are almost as numerous as the somatostatin
positive ones (Fig. 3B) but less numerous than e.g. the dopamine,
LHRH and TRH fibers (unpublished data).

      All substances present in the external layer of the median
eminence are from morphological point of view candidates for humoral
factors, which can be released into the portal vessels and can control anterior pituitary hormone secretion according to the concept
of Harris (58). Alternatively they may act at the level of the
median eminence, possibly by controlling the release of the "true"
releasing and/or inhibiting hormones. This can be achieved by an
axo-axonic influence, a mechanism proposed for the dopaminergic
effects on the release of LHRH (see 14,15, Fuxe et al. and McCann
et al., this symposium) or possibly via an action on glial elements
(tanycytes).

      The immunocytochemical studies on the distribution of LHRH and
dopamine fibers in the median eminence have shown a close overlap
between these two systems in the lateral parts of the median eminence
(10,59). The findings form a morphological basis for the interaction
between the two systems as described above. In contrast, the second
brain area rich in LHRH positive nerve endings, the organum vasculosum of the lamina terminalis, contains only few dopamine and noradrenaline fibers (Fig. 4A-C). The density of the somatostatin
positive fibers is shown in Fig. 4D for comparison.

COEXISTENCE OF AMINES AND PEPTIDES

      In the median eminence many types of peptides and putative
transmitters are found in dense nerve terminal networks. In the
lateral external layer the monoamine boutons have been calculated
to constitute approximately one third of all nerve endings (60). For
somatostatin positive nerve endings a similarly high figure (30%)
has been given by Pelletier et al. (61), which appears to be in good
agreement with the very high density of somatostatin positive fibers
seen with immunofluorescence. Thus, these two types of fibers may
occupy almost 2/3 of all nerve endings leaving only 1/3 for all
remaining types of fibers. It is therefore reasonable to ask whether
or not each substance, in fact, is stored in separate neuron systems.
In view of the small size of nerve terminals and their high density
in the median eminence, it is not possible to directly decide at the
light microscopic level whether or not two (or more) substances co-

exist in the same fibers. It appears, however, as if each substance
has its own characteristic pattern both with regard to distribution
and amounts. A more reliable approach is to compare the localization
of cell bodies after incubation with different antisera. Possible
sites where two substances may coexist are the paraventricular
nucleus where in addition to the oxytocin/vasopressin/neurophysin
containing cells many enkephalin and a few TRH immunoreactive and
dopamine cells are found. In the arcuate nucleus enkephalin (and
endorphin) positive cells have been observed in addition to LHRH
(see 48) and dopamine cells. In no case has a careful comparative
study been performed to determine a possible coexistence. In the
periventricular area of the anterior hypothalamus both dopamine and
somatostatin cells are found but they appear to constitute separate
neuron populations (10).

In several other parts of the central nervous system coexistence
of a "classical" small transmitter and a biologically active peptide
has, however, been demonstrated extending the APUD (Amine content
and/or Precursor Uptake and Decarboxylation) concept of Pearse (62)
from endocrine cells to neurons and indicating the possibility that
one neuron may synthesize and store more than one transmitter. In
several sympathetic ganglia a large proportion of the noradrenergic
ganglion cells contain somatostatin-like immunoreactivity (63). In
the lower medulla oblongata the majority of the 5-hydroxytryptamine
neurons contain substance P-like immunoreactivity (64,65). In our
laboratory this was demonstrated after incubation of consecutive
sections with substance P antiserum (Fig. 5A) and, as a marker for
serotonin neurons, antibodies either to the indoleamine itself (Fig.
5B) (13) or to dopadecarboxylase (66). The latter antiserum appears
to cross-react with the decarboxylating enzyme in serotonin neurons
(33) (and in addition probably with an enzyme in some other types
of neurons). Recently enkephalin-like immunoreactivity has been
demonstrated both in some peripheral noradrenergic principal ganglion
and small intensely fluorescent (SIF) cells (67) and in adrenal gland
cells (Fig. 5C,D) (68). Thus the number of examples of coexistence
of a peptide with putative transmitter or modulator role and a biogenic amine in neurons rapidly increases. Perhaps this situation is
sooner the rule than rare exceptions?

In most of the studies cited above coexistence of peptide and
amine could only be demonstrated in cell bodies. This, of course, is
unsatisfying since it tells us little about a possible transmitter
or modulator role of the substances. In fact, cell somata immunoreactivity may only demonstrate a peptide as part of a larger (precursor?) molecule, which never leaves the cell body nor plays a
functional role in the neuron. In one case, however, indirect
evidence suggest that both peptide and amine are concomitantly
present also in nerve terminals, i.e. the sites where a release may
occur. Thus, after pretreatment of rats with the neurotoxins 5,6- or

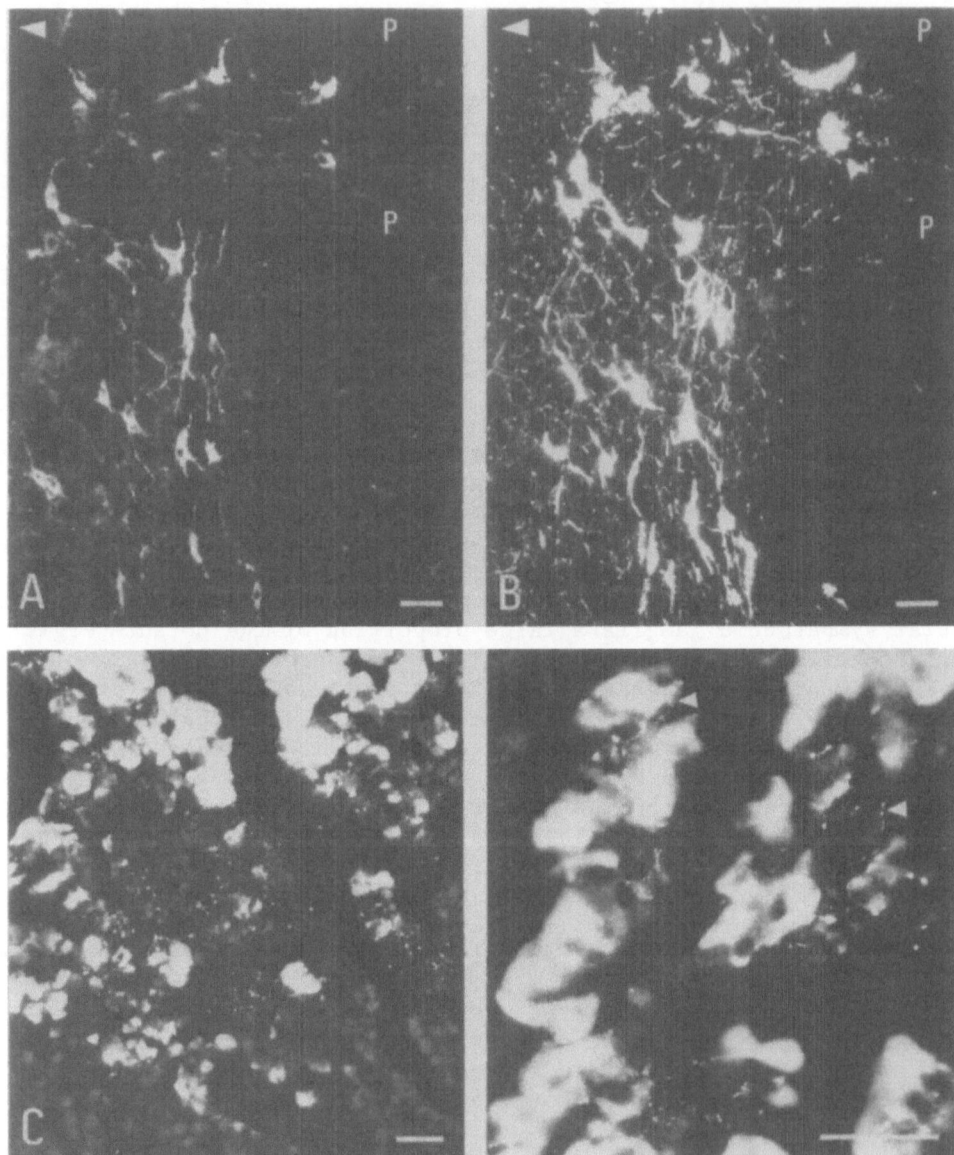

Fig. 5 A-D. Immunofluorescence micrographs of (A, B) the rat raphe
magnus nucleus (consecutive sections) and (C, D) the guinea pig
adrenal gland after incubation with antiserum to (A) substance P,
(B) 5-hydroxytryptamine and (C, D) methionine-enkephalin. (A, B) Substance P and 5-hydroxytryptamine are often present in the same cells.
Arrow heads point dorsally. P = Pyramidal tract. (C, D) Many medullary gland cells and some nerve terminals (arrow heads in D) are
enkephalin immunoreactive. Bars indicate 50 μm.

5,7-dihydroxytryptamine not only all serotonin nerve terminals in the spinal cord but also the majority of the substance P positive fibers in the ventral horn of the cord disappeared (65). The neuro-toxins 5,6- and 5,7-DHT are known to destroy serotonin neurons although unspecific actions also have been observed (69-72). Our findings suggest a descending "5-HT-SP" pathway from the medullary raphe nuclei to the ventral (and perhaps dorsal) horn of the spinal cord and that, in fact, 5-HT and SP may be released together from nerve endings. Further studies are, however, necessary to obtain full evidence for this view.

The possible coexistence of a peptide and an amine in the same nerve terminals raises the question of the subcellular storage site(s) for these substances. It is generally accepted that mono-amines mainly are stored in vesicles (granules). Early evidence by von Euler (73) indicated that also substance P may be present in subcellular storage particles. It is well known that at least two major types of vesicles are present in most nerve endings, small vesicles (diameter about 500 Å) and large vesicles (diameter about 1000 Å). With conventional fixation techniques (e.g. glutaraldehyde-osmium tetroxide) the large vesicles mostly contain an electron

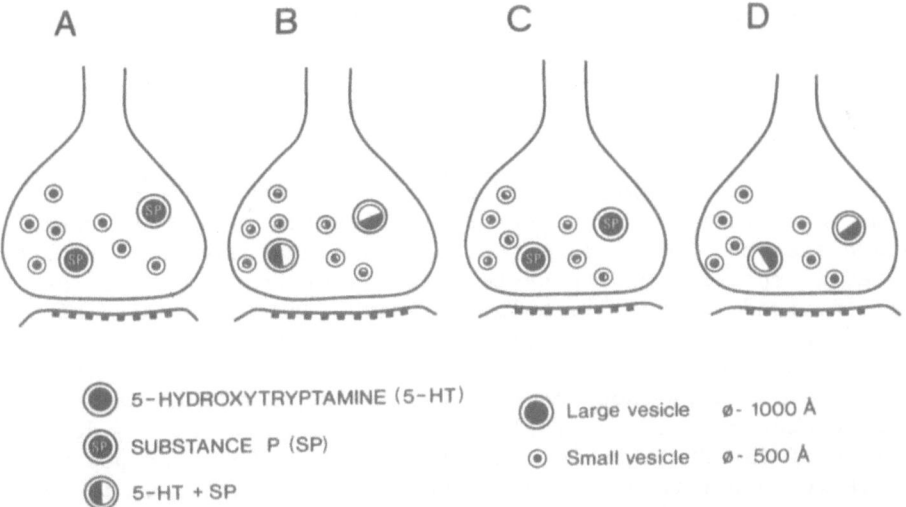

Fig. 6. Schematic illustration of various hypothetical alternatives for storage compartments for an amine (5-hydroxytryptamine) and a peptide (substance P) coexisting in the same nerve ending. See text for further details.

dense core whereas the small ones appear empty (low electron density). From theoretical point of view several alternatives for storage
of coexisting peptide and amine may be considered as partly indicated
in Fig. 6A-D, where serotonin and substance P have been chosen as
examples. Serotonin may be stored in small and substance P in large
vesicles (or vice versa) (alt. A). Both substances may be stored in
both types of vesicles (alt. B). Only substance P (alternatively
serotonin) may be stored in the large vesicles and the small ones
may contain both substances (or vice versa) (alt. C). Finally, the
large vesicles may contain both substances and the small ones only
serotonin (alternatively substance P) (alt. D). On the basis of
available data some suggestions with regard to the true storage
sites may be given. Electron microscopic cytochemical studies have
shown that serotonin (and catecholamines) is present both in small
and large vesicles (74,75). So far substance P (and other peptides)
has been demonstrated with certainty only in large vesicles (76-78),
although its presence in small vesicles (or extravesicularly) cannot be excluded. From these data the D-alternative may at present
seem to be the most likely one. Clearly, electron microscopic immunocytochemistry and subcellular studies are necessary to obtain the
correct answer to this problem. Furthermore, various types of
pharmacological treatments may give important clues. It is of interest that in preliminary experiments substance P stores did not
appear to be affected by a large dose of reserpin, which totally
depleted the serotonin stores. These immunohistochemical findings
must, however, be supplemented by quantitative radioimmunological
determination of substance P levels.

SUMMARY

     On the basis of immunohistochemistry the distribution and
cellular localization of some putative transmitters and neuropeptides are discussed. Dopamine, noradrenaline and adrenaline have been
localized in specific neuron systems with antisera raised to the
enzymes involved in the catecholamine synthesis. The tracing of
serotonin neurons with antibodies to the amine itself as introduced
by Steinbusch et al. (13) shows that also very small molecules can
be visualized with immunocytochemistry. As an example the distribution of serotonin nerve terminals in the median eminence is described.
Comparison of the distribution patterns of monoamines and peptides
have revealed coexistence of these two types of substances in the
same neuron in several places of the nervous system, e.g.of serotonin and substance P in the raphe nuclei of the lower brain stem and
of noradrenaline and somatostatin in some peripheral sympathetic
ganglia. These findings indicate that one neuron may synthesize,
store and release more than one substance involved in transmission
and/or modulation at chemical synapses.

ACKNOWLEDGEMENTS

This work was supported by grants from the Swedish Medical Research Council (04X-2887), Magnus Bergvalls Stiftelse, Knut and Alice Wallenbergs Stiftelse, by a grant from the U.S. Department of Health, Education and Welfare (1F01 HL 18994-01), through the National Heart and Lung Institute, by United States Public' Health Service Grants MH-02717 and MH-25504-3, and by National Science Foundation Grant CB-8465. The skillful technical assistance of Mrs. W. Hiort, Mrs A. Tovi-Nygårds, Miss G. Norell and Miss L. Persson is gratefully acknowledged.

The results reviewed in this article have been obtained with antisera generously supplied by several laboratories. Some results have previously been published in original articles and more completely in two recent review articles (10,11). The substance P antiserum was supplied by Dr. G. Nilsson, Department of Pharmacology, Karolinska Institute, Stockholm, Sweden, the enkephalin and endorphin antisera by Prof. L. Terenius, Department of Pharmacology, Uppsala University, Uppsala, Sweden and the LHRH antiserum by Dr. S.L. Jeffcoate, Department of Endocrinology, Chelsea Hospital for Women, London, UK.

REFERENCES

1. Coons, A.H. In General Cytochemical Methods, pp. 399-422, Ed. Danielli, J.F., Academic Press, New York, 1958.
2. Geffen, L.B., Livett, B.G. and Rush, R.A. J. Physiol. (Lond.) 204, 593-605, 1969.
3. Nakane, P. In Karolinska Symposia on Research Methods in Reproductive Endocrinology Vol. 3, pp. 190-204, Ed. Piczfalusy, E., Forum, Copenhagen, 1971.
4. Sternberger, L.A. Immunocytochemistry. Prentice Hall, Englewood Cliffs, N.J., 1974.
5. Koelle, G.B. and Friedenwald, J.S. Proc. Soc. Exptl. Biol. Med. 70, 617-622, 1949.
6. Silver, A. The Biology of Cholinesterases. North Holland Publ. Comp., Oxford, Amsterdam, 1974.
7. Falck, B. Acta physiol. scand. 56, Suppl. 197, 1-25, 1962.
8. Falck, B., Hillarp, N.-Å., Thieme, G. and Torp, A. J. Histochem. 10, 348-354, 1962.
9. Carlsson, A., Falck, B. and Hillarp, N.-Å. Acta physiol. scand. 56, Suppl. 196, 1-28, 1962.
10. Hökfelt, T., Elde, R., Fuxe, K., Johansson, O., Ljungdahl, Å., Goldstein, M., Luft, R., Efendić, S., Nilsson, G., Terenius, L., Ganten, D., Jeffcoate, S., Rehfeld, J., Said, S., Perez de la Mora, M., Possani, L., Tapia, R., Teran, L. and Palacios, R. In The Hypothalamus, pp. 69-135, Ed. Reichlin, S., Baldessarini, R.J. and Martin, J.B., Raven Press, New York, 1978.

11. Elde, R. and Hökfelt, T. In <u>Frontiers in Neuroendocrinology</u>
     <u>Vol. 5</u>, pp. 1-33, Ed. Ganong, W.F. and Martini, L., Raven Press,
     New York, 1978.
12. Müller, E.E., Nisticò, G. and Scapagnini, U. <u>Neurotransmitter</u>
     <u>and Anterior Pituitary Function</u>, Academic Press, New York, 1977.
13. Steinbusch, H., Verhofstad, A. and Joosten, H. <u>Neuroscience</u>,
     1978, in press.
14. Fuxe, K. and Hökfelt, T. In <u>Frontiers in Neuroendocrinology</u>
     <u>Vol. 1</u>, pp. 47-96, Ed. Ganong, W.F. and Martini, L., Oxford
     University Press, New York, 1969.
15. Hökfelt, T. and Fuxe, K. In <u>Brain-Endocrine Interaction. Median</u>
     <u>Eminence: Structure and Function</u>, pp. 181-223, Ed. Knigge, K.M.,
     Scott, D.E. and Weindl, A., Karger, Basel, 1972.
16. Björklund, A. and Nobin, A. <u>Brain Res.</u> <u>51</u>, 193-205, 1973.
17. Dahlström, A. and Fuxe, K. <u>Acta physiol. scand.</u> <u>62</u>, Suppl. 232,
     1-55, 1964.
18. Hökfelt, T., Johansson, O., Fuxe, K., Goldstein, M. and Park,
     D. <u>Med. Biol.</u> <u>54</u>, 427-453, 1976.
19. Hökfelt, T., Johansson, O., Elde, R., Ljungdahl, Å., Fuxe, K.
     and Goldstein, M. In <u>Cell Biology of Hypothalamic Neurosecretory</u>
     <u>Processes</u>, Ed. Vincent, J.D. and Kordon, C., 1978, in press.
20. Sladek, J.R. <u>Brain Res.</u> <u>142</u>, 165-173, 1978.
21. Swanson, L.W. and Hartman, B. <u>J. Comp. Neurol.</u> <u>163</u>, 467-505,
     1975.
22. Hökfelt, T., Fuxe, K., Goldstein, M. and Johansson, O. <u>Brain</u>
     <u>Res.</u> <u>66</u>, 235-251, 1974.
23. Brownstein, M.J., Palkovits, M., Tappaz, M.L., Saavedra, M.
     and Kizer, J.S. <u>Brain Research</u> <u>117</u>, 287-295, 1976.
24. Fuxe, K. <u>Z. Zellforsch.</u> <u>65</u>, 573-596, 1965.
25. Fuxe, K. <u>Acta physiol. scand.</u> <u>64</u>, Suppl. 247, 39-85, 1965.
26. Jonsson, G. <u>Progr. Histochem. Cytochem.</u> <u>2</u>, 299-334, 1971.
27. Kordon, C., Javoy, F., Vassent, G. and Glowinski, J. <u>Europ. J.</u>
     <u>Pharmacol.</u> <u>4</u>, 169-174, 1968.
28. Saavedra, J.M., Palkovits, M., Brownstein, M. and Axelrod, J.
     <u>Brain Res.</u> <u>77</u>, 157-165, 1974.
29. Calas, A., Alonso, G., Arnauld, E. and Vincent, J.D. <u>Nature</u>
     <u>250</u>, 241-243, 1974.
30. Baumgarten, H.G. and Lachenmayer, L. <u>Z. Zellforsch.</u> <u>147</u>, 285-
     292, 1974.
31. Ajika, K. and Ochi, J. <u>J. Anat.</u>, 1978, in press.
32. Sladek, J.R., Jr. and Sladek, C.D. <u>Cell Tiss. Res.</u> <u>186</u>, 465-
     474, 1978.
33. Hökfelt, T., Fuxe, K. and Goldstein, M. <u>Brain Res.</u> <u>62</u>, 461-469,
     1973.
34. Lidbrink, P., Jonsson, G. and Fuxe, K. <u>Brain Res.</u> <u>67</u>, 439-456,
     1974.
35. Chan-Palay, V. <u>J. Comp. Neurol.</u> <u>176</u>, 467-494, 1977.
36. Beaudet, A. and Descarries, L. <u>Neurosci. Abst.</u> <u>2</u>, 479, 1976.
37. Smith, A.R. and Ariens Kappers, J.A. <u>Brain Research</u> 86, 353-
     371, 1975.

38. Kent, D. and Sladek, J.R. Neurosci. Abst. 2, 493, 1976.
39. Hökfelt, T., Ljungdahl, Å., Steinbusch, H., Verhofstad, A., Nilsson, G., Brodin, E., Pernow, B. and Goldstein, M. Neuroscience 3, 517-538, 1978.
40. Iversen, L. and Schon, F. In New Concepts in Transmitter Regulation, pp. 153-193, Ed. Mandel, A. and Segal, D., Plenum Press, New York,1973.
41. Hökfelt, T. and Ljungdahl, A. In The Use of Axonal Transport for Studies of Neuronal Connectivity, pp. 249-305, Ed. Cowan, W.M. and Cuénod, M., Elsevier, Amsterdam, 1974.
42. Roberts, E. In Neuroscience Symposia Vol. 1, pp. 123-138, Ed. Ferrendelli, J.A., McEwen, B.S. and Snyder, S.H., Society for Neuroscience, Bethesda, Md., 1975.
43. Livett, B.G., Uttenthal, L.O. and Hope, D.B. Proc. R. Soc. Phil. Trans. B. 261, 371-378, 1971.
44. Zimmerman, E.A., Hsu, K.G., Robinson, A.G., Carmel, P.W., Frank, A.G. and Tannenbaum, M. Endocrinology 92, 931-940, 1973.
45. Leonardelli, J., Barry, J. and Dubois, M.P. C.R. Acad. Sci. (D) (Paris) 276, 2043-2046, 1973.
46. Barry, J., Dubois, M.P. and Poulain, P. Z. Zellforsch. 146, 351-366, 1973.
47. Calas, A., Kerdelhué, B., Assenmacher, I. and Jutisz, M. C.R. Acad. Sci. (D) (Paris) 277, 2765-2768, 1973.
48. Sternberger, L.A. and Hoffman, G.E. Neuroendocrinology 25, 111-128, 1978.
49. Dyer, R.G. and Dyball, R.E.J. Nature 252, 486-488, 1974.
50. Renaud, L.P., Martin, J.B. and Brazeau, P. Nature 255, 233-235, 1975.
51. Otsuka, M. and Takahashi, T. Ann. Rev. Pharmacol. Toxicol. 17, 425-439, 1977.
52. Watkins, W.B. Cell Tiss. Res. 155, 201-210, 1974.
53. Silverman, A.J. and Zimmerman, E.A. Cell Tiss. Res. 159, 291-301, 1975.
54. Dierickx, K., Vandesande, F. and DeMey, J. Cell Tiss. Res. 168, 141-151, 1976.
55. Dube, D., Leclerc, R. and Pelletier, G. Am. J. Anat. 147, 103-108, 1976.
56. Fuxe, K., Ganten, D., Hökfelt, T. and Bolme, P. Neurosci. Lett. 2, 229-234, 1976.
57. Hökfelt, T., Pernow, B., Nilsson, G. Wetterberg, L., Goldstein, M. and Jeffcoate, S.L. Proc. Natl. Acad. Sci. USA 75, 1013-1015, 1978.
58. Harris, G.H. Neural Control of the Pituitary Gland. Edward Arnold, London, 1955.
59. Agnati, L., Fuxe, K., Hökfelt, T., Goldstein, M., Jeffcoate, S.L. and Elde, P. J. Histochem. Cytochem. 25, 1222-1236, 1977.
60. Ajika, K. and Hökfelt, T. Brain Res. 57, 97-117, 1972.
61. Pelletier, G., Labrie, F. Arimura, A. and Schally, A.V. Am. J. Anat. 140, 445-450, 1974.
62. Pearse, A.G.E. Z. Krebsforsch. 84, 1-18, 1975.

63. Hökfelt, T., Elfvin, L.-G., Elde, R., Schultzberg, M., Gold-
     stein, M. and Luft, R. Proc. Natl. Acad. Sci. USA 74, 3587-
     3591, 1977.
64. Chan-Palay, V., Jonsson, G. and Palay, S.L. Proc. Natl. Acad.
     Sci. USA 75, 1582-1586, 1978.
65. Hökfelt, T., Ljungdahl, Å., Steinbusch, H., Verhofstad, A.,
     Nilsson, G., Pernow, B. and Goldstein, M. Neuroscience 3, 517-
     538, 1978.
66. Goldstein, M., Fuxe, K. and Hökfelt, T. Pharmacol. Rev. 24,
     293-309, 1972.
67. Schultzberg, M., Hökfelt, T., Terenius, L., Elfvin, L.-G.,
     Lundberg, J.M., Brandt, J., Elde, R. and Goldstein, M. Neuro-
     science, 1979, in press.
68. Schultzberg, M., Lundberg, J.M., Hökfelt, T., Terenius, L.,
     Brandt, J., Elde, R.P. and Goldstein, M. Neuroscience, 1978,
     in press.
69. Baumgarten, H.G., Björklund, A., Lachenmayer, L., Nobin, A.
     and Stenevi, U. Acta physiol. scand. Suppl. 373, 1971.
70. Baumgarten, H.G., Björklund, A., Lachenmayer, L. and Nobin, A.
     Acta physiol. scand. Suppl. 391, 1973.
71. Daly, J., Fuxe, K. and Jonsson, G. Brain Research. 49, 476-
     482, 1973.
72. Daly, J., Fuxe, K. and Jonsson, G. Res. Comm. Chem. Pathol.
     Pharmacol. 7, 175-187, 1974.
73. Euler, U.S. von and Gaddum, J.H. J. Physiol. (Lond.) 72, 74-
     87, 1931.
74. Hökfelt, T. Z. Zellforsch. 91, 1-74, 1968.
75. Tranzer, J.-P., Thoenen, H., Snipes, R.L. and Richards, J.G.
     Progr. Brain Res. 31, 33-46, 1969.
76. Chan-Palay, V. and Palay, S.L. Proc. Natl. Acad. Sci. USA 74,
     4050-4054, 1977.
77. Hökfelt, T., Johansson, O., Kellerth, J.-O., Ljungdahl, Å.,
     Nilsson, G., Nygårds, A. and Pernow, B. In Substance P Nobel
     Symposium 37, pp. 117-145, Ed. Euler, U.S. von and Pernow, B.,
     Raven Press, New York,1977.
78. Pickel, V.M., Reis, D.J. and Leeman, S.E. Brain Res. 122,
     534-540, 1977.

ACTIVATION OF RELEASE AND MECHANISM OF RELEASE OF NEUROHYPOPHYSEAL

HORMONES

Niels A. Thorn, James T. Russell, Christian
Torp-Pedersen and Marek Treiman

Institute of Medical Physiology C
University of Copenhagen
Copenhagen, Denmark

Mechanism of Activation of the "Classical" Oxytocinergic and
Vasopressinergic Neurons in the Hypothalamus

These neurons were some of the first to be shown to secrete
hormones. They have their perikaryal region in the nucleus supra-
opticus and paraventricularis and end in the neurohypophysis (some
neurons or branches go to the median eminence). They release their
hormones to the blood. It has been shown that they have recurrent
collaterals (Kandel, 1964; Dreifuss and Kelly, 1972; Negoro and
Holland, 1972). Thus a transmitter function must also be considered
for them. Recently neurons with the same hormone and/or binding
protein have been demonstrated in other parts of the central nervous
system, e.g. the spinal cord (Buijs et al., 1978; Sofroniew and
Weindl, 1978; Summy-Long et al., 1978), but it has not been proved
that hormone is released from these neurons. It may be mentioned
that several other functions for vasopressin than the antidiuretic
have been suggested, among them a role in memory and behaviour
(deWied and Gispen, 1977; deWied, this volume). The origin of vasopressin responsible for such effects is not known.

There seems to be fairly uniform agreement about the physiological mechanism of activation of the oxytocinergic cells (reflexes,
primarily from papilla mammae). Excitatory adrenergic α-receptors
appear to form a component of the natural reflex. Inhibitory β-adrenergic receptors outside the reflex can modulate the release
(Lincoln et al., 1977; Tribollet et al., 1978).

For the "classical" vasopressinergic neurons the problem of
activation is more complex. Whereas there seems to be an element of

reflex input from volume receptors in the large vessels and the
heart (see, e.g., Share, 1968; Gauer, 1978) the question of activation through osmolarity or sodium concentration receptors and
their relative significance, as well as their location, is still
subject to discussion.

According to Verney's "classical" hypothesis (Verney, 1947)
"osmoreceptors" localized in or close to the supraoptico-neurohypo-
physeal pathways are activated by an increase of the osmolarity of
plasma, which Verney usually produced by a bolus injection of 1.3 M
NaCl into a carotid artery, a procedure which can be said not to
mimic the physiological conditions very well.

Activations of the hypothalamus-neurohypophyseal system incubated in vitro have been made with large increases of osmolarity in
the medium for a short time (Eggena and Thorn, 1970; Eggena and
Polson, 1974) (also an unphysiological approach) or small increases
for a period of 24 h (Sladek and Knigge, 1977).

Mainly on the basis of experiments in goats, among other things
demonstrating lack of antidiuretic effect of sucrose-induced elevation of cerebrospinal fluid osmolarity, a hypothesis was formulated
(see Andersson, 1977) that periventricular sodium receptors are an
essential element in the activation of antidiuretic hormone release.

In such experiments it should be remembered that isolated nerve
endings kept in solutions dominated by sucrose are very little
reactive (Sperk and Baldessarini, 1977) and that sucrose is an inhibitor of many enzyme systems. McKinley et al. (1978) have expanded
Andersson's thesis in a very recent series of experiments on ewes.
They suggest that both osmoreceptors and sodium receptors regulate
ADH secretion. The basis was i.a. that intraventricular 0.35 M NaCl
in artificial cerebrospinal fluid (containing the normal electrolytes)
caused a greater antidiuretic effect than intraventricular 0.7 M
sucrose in the same solution. Experiments in which small and slow
increases in the sodium chloride dependent osmolarity of intracarotid
blood were produced were made in a limited number of unanesthetized
dogs by Verney. Such experiments would seem to be the most relevant
basis for Verney's hypothesis about osmoreceptors in the hypothalamus.
However, Bie (see survey, 1978), using anaesthetized dogs and bilateral infusion in a very careful series of experiments, did not find
any antidiuretic effect of hyperosmolar solutions administered this
way.

Experiments on whether vasopressinergic neurons themselves are
sensors of plasma osmolarity or sodium concentration have been done
mainly using bolus injections of hyperosmolar NaCl into a carotid
artery or into localized areas of hypothalamus. On the basis of such
experiments (which obviously have their limitations) it has been
suggested (Bridges and Thorn, 1970; Peck and Blass, 1975; Hayward,

1977) that the neurosecretory neurons are not themselves the sensors. Such a concept seems to be supported by recent findings (Halter et al., 1977) in a patient in whom the volume control of ADH secretion was dissociated from the osmosensor control (there was very little response to hyperosmolarity of the plasma, whereas a reflex release of ADH due to hypovolemia was intact).

A number of other stimuli (e.g. pain) can participate in the input on the vasopressinergic neurons. In pathological situations a large number of factors can increase or reduce the input on the vasopressinergic neurons (Thorn, 1970; Miller and Moses, 1977).

The nature of the synapses involved from the presumed osmo- and/or sodium receptors to the vasopressinergic neurons as well as the nature of transmitters acting on the vasopressinergic neurons on the whole is not settled. On the basis of a set of pharmacological blocking experiments of diuresis reactions to intracarotid hyper-osmolar sodium chloride (in a bolus injection) in rats, Bridges and Thorn (1970) suggested a model as shown Fig. 1 in a modified form.

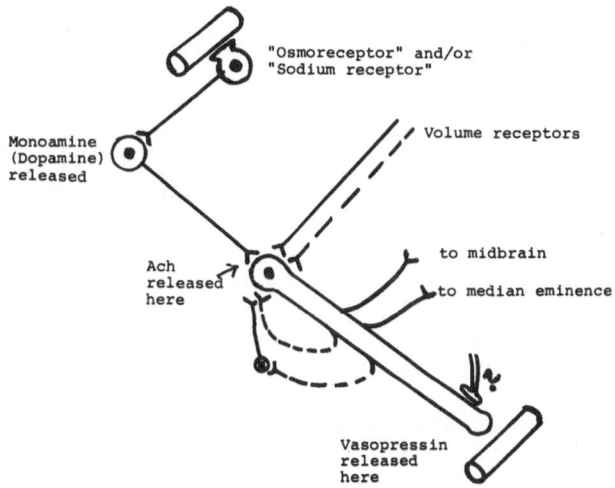

Fig. 1 Input on vasopressinergic neuron

The transmission in the inhibitory neurons synapse on the SON cell from volume receptors might be the β-adrenergic. Bridges et al. (1976) found that dopamine induced hormone release in a rat hypothalamus in vitro preparation, which, however, had the nerve endings in the neurohypophysis severed. Bridges et al. summed up the evidence that acetylcholine is a transmitter affecting the vasopressinergic neurons. In the discussion of their paper a further elaboration is found on possible other transmitters in the system.

Weitzman et al. (1977) demonstrated that β-endorphin stimulated secretion of arginine vasopressin in rabbits in vivo. Mathison and
Lederis (1977) demonstrated an inhibitory effect of cAMP on vasopressin release in vitro.

Mechanism of Release of Vasopressin

The mechanism of release from other neurosecretory neurons seems
to resemble the release from the neurohypophyseal ones (Thorn et al.,
1978; Terry and Martin, 1978; Iversen, this volume).

Release Probably Occurs by Exocytosis

The main evidence for this is that exocytosis can be seen, e.g.
in freeze cleaving pictures of isolated neurohypophyses or isolated
nerve endings after strong stimulation (Gratzl et al., 1977). Release
to the surrounding medium of vasopressin + binding protein as well
as lack of release of cytoplasmic markers (Edwards et al., 1973)
further support the hypothesis. It may be added that exocytosis is
generally believed to be the mechanism of secretion in other nerve
endings (Llinas and Heuser, 1977).

Significance of Extracellular Ca^{2+} for Release

On stimulation Ca^{2+} in very small amounts is transported over
the plasma membrane. This Ca^{2+} transport is essential for release.
It is especially studies with Ca^{2+} channel blocking substances which
support this hypothesis (for a detailed discussion see Russell and
Thorn, 1974; Thorn et al., 1978)

Calcium Regulating Systems in the Neurohypophysis

Russell and Thorn (1975) have described several Ca^{2+} regulating
processes in the neurohypophysis, among them a mitochondrial uptake.

An ATP dependent uptake to a microsomal fraction was also
described by Russell and Thorn. Experiments by Kendrick et al. (1977)
support the suggestion that a system exists in nerve ending endoplasmic reticulum that actively accumulates Ca^{2+}. The affinity of
this system for Ca^{2+} might be relevant in short time regulations.
The presence of a Ca^{2+} stimulated ATPases such as isolated and characterized in whole brain by Sæ rmark and Vilhardt (submitted) should be
considered. Ca^{2+} transporting ATPases might be localized to the
endoplasmic reticulum and/or the plasma membrane.

Processes Following Ca^{2+} Messenger Signal

The question of what happens beyond the stimulation-produced
increase in intracellular calcium concentration seems to be one of the
most challenging problems in the secretion mechanism today. Attempts

have been made to elucidate some of these problems by making an
"artificial exocytosis" model. This, so far, has not been succesful,
but a fusion of secretory vesicles has been demonstrated by exposing
isolated secretory granules to physiological changes of the surrounding Ca^{2+} concentration.

Fusion of Neurosecretory Granules

     The intracellular concentration of free calcium in the nerve
terminals is probably 50-150 nM. It thus follows that studies on
effects of increases of calcium concentration should be done starting
with very low concentrations (using Ca^{2+}-buffers).

     In our laboratory in collaboration with Dahl and Gratzl and
extending their studies on other systems(Gratzl et al., 1977), isolated secretory vesicles were incubated in a simple medium (Na-
cacodylate, 10mM, pH 7.0; sucrose, 250 mM; and EGTA, 1mM) and the
effects of increasing the calcium concentration in the medium from
less than 10^{-7} to 10^{-3} M on fusion of granules were studied.

     Fusion between vesicles was demonstrated by the fact that the
freeze cleavage plane was continuous from one to the other vesicle
both in membrane P and E faces. With increasing calcium concentration, an increase of fusion percentage was observed with a half-
maximal effect about 10^{-5} M calcium. The changes could not be produced
by magnesium or strontium. On the contrary, Mg^{2+} inhibited release
and fusion. These morphological changes of neurosecretory vesicles
were thus similar to changes demonstrated by Dahl and Gratzl in other
secretory tissues, such as pancreatic islets, liver and medulla.

     It should be remembered that fusion is not exocytosis, but
probably a phenomenon that occurs under the same circumstances as
exocytosis (see e.g. Gratzl et al., 1977). At least there seems to
be a morphological phenomenon related to secretion and induced by
changes in Ca^{2+} concentration which are physiological and specific.
Creutz et al. (1978) have recently isolated a protein from adrenal
medullary cells that causes Ca^{2+}-dependent aggregation of adrenal
medullary granules. Its Ca^{2+} affinity is rather low.

Hypothesis that Ca^{2+}-Binding Regulator Proteins are essential in
the Release Mechanism

     The circumstances in which the increases in the concentration
of free calcium is recognized by the release processes have the
following characteristics among others: Ca^{2+} increases from micro-
molar concentrations in the resting tissue. The concentration of
Mg^{2+} is about 1mM and that of K^+ about a hundred times higher.
It is obvious that recognition of calcium under these circumstances
required binding sites of a high specificity. One should probably
look for proteins with pKd (Ca^{2+}) values in the range 5-8, that is

TABLE I. Ca^{2+} binding proteins from muscle

	M.W	pKd(Ca^{2+})	Capacity (mol Ca^{2+}/ mol prot.)
Myosin	210,000 (15,000,25,000)	4.9	
Ca^{2+} ATPase (65% of SR prot.)	102,000	6.5	2
High affinity CBP (1+11% of SR prot.)	56,000	5.0 (4.0)	1 (25)
Calsequestrin (19% of SR prot.)	44,000	4.3	43
Acidic proteins	20-30,000		
Troponin C	17,846	5.2	2
Parvalbumin	12,000	7	

TABLE II. Ca^{2+} binding proteins isolated from ox neurohypophyses

Source	M.W.	pKd(Ca^{2+})	Capacity (mol Ca^{2+}/ mol prot.)
Soluble fraction (B)	35,000 (14,500)	5	2
"Microsomal" fraction solubilized in deoxy-cholate (C, D)	16,500 (C) 68,000 (D)	5	4
Solubilized in Triton X-100			

The purification was 407 times for the soluble protein fraction. 98 times for the solubilized one.

factors which can be cyclically turned on and off as a second messenger when the calcium concentration of the cell is changed from about 10^{-8} to 10^{-5} M. A number of calcium binding substances from muscle, of which some (with low m.w.) have such a function, are shown in Table I (see also Kretzinger, 1976).

Soluble Calcium-Binding Protein in the Neurohypophysis

We have started a series of studies on isolation of such calcium-binding proteins from ox neurohypophyses. So far we have isolated one (and only one) in a high degree of purity from the soluble fraction (Table II, Russell and Thorn, 1977). It comprised about 1% of the proteins in that fraction. Its unit weight was approximately 15,000. The purification procedure involved ammonium sulphate fractionation, DEAE-cellulose chromatography and gel filtration on Sephadex G-100 and Sephadex G-50, a procedure used previously for isolation of such proteins from whole brain.

Russell, Christensen and Thorn (unpublished observation) have performed amino acid analyses for the purified soluble calcium binding protein from the neurohypophysis. The composition was very similar to that which has been published for cyclic nucleotide phosphodiesterase (and actomyosin ATPase), troponin C-like modulator protein isolated from brain (high in glutamic and aspartic acid, low in tryptophan and half-cystine). It thus seems very likely that it is identical to the calcium dependent modulator protein (CDR) isolated by others from the whole brain.

Membrane-Bound Calcium Binding Proteins in the Neurohypophysis

The membrane-bound calcium binding proteins have not yet been characterized in such detail as the soluble protein. One of them (D) had a molecular weight of approximately 68,000. Because of the very small amount obtained, no further characterization was tried in this study.

Another protein (C) had a molecular weight of about 16,500 and an apparent dissociation constant for calcium of about 1.0×10^{-5} M (binding capacity: 4 mol of calcium per mol of protein). It had different characteristics from the soluble purified protein on gel filtration and PAGE. It would seem difficult to exclude, however, that it might be the same as the soluble one, adsorbed or bound to an inner layer on the cell membrane or on granule membranes. This possibility is especially interesting, since Vincenzi and coworkers (1977) have recently shown that modulator protein in the soluble phase of erythrocytes in the presence of calcium can be bound to plasma membrane and stimulate the calcium dependent ATPase present there. It should also be remembered that Gnegy et al. (1976) have suggested that the modulator protein in rat caudate nucleus can be released from the membranes by a phosphorylation with 3':5'-cAMP dependent protein kinase.

A very small amount of a calcium binding protein (C-1) with a mol. wt. of approximately 10,000 was also found in this fraction.

A fourth calcium binding protein was shown to be present in the fraction obtained when DOC insoluble proteins from the microsomal fraction were solubilized in Triton X-100. This fraction was lost on further attempts at purification.

Probable Complexity of Calcium-Dependent Regulator Protein Systems in Secretory Tissues

Since the discovery of the presence of CDR protein in the brain other factors have been found which modify its function (Wang and Desai, 1977). This of course complicates the picture in terms of our trying to explore a functional role of this protein. The functions of CDR in terms of modification of enzyme activity also seem complicated. It modified cyclic nucleotide phosphodiesterase and actomyosin ATPase (Watterson et al., 1976). It is very interesting that Schulman and Greengard (1978) have recently described CDR as having an important function in mediating Ca^{2+} stimulation of cerebral synaptosomal membrane protein kinase. It may thus be a multifactorial complex, but it is obvious that further work is necessary to clarify these problems.

The cAMP and cGMP systems in the Neurohypophysis

Several groups have seen no effect on release when adding high concentrations of cAMP or dibutyryl cAMP to the incubation medium for isolated neurohypophyseal tissue (for references see Thorn et al., 1978).

The presence of cAMP in the neurohypophysis of rats and an increase of its concentration in rats that have been drinking 2% NaCl solution for 24 h has been demonstrated previously (Ruoff et al., 1976). Poirier et al. (1977) and Bonne et al. (1977) have published results of experiments on presence of adenylate cyclase in plasma membrane and secretory granule fractions from the ox neurohypophysis.

In recent studies (Torp-Pedersen, Treiman and Thorn, submitted for publication) it has been shown that cAMP phosphodiesterase activity is present in plasma membranes from a preparation of isolated nerve endings from bovine neurohypophyses. High activity was also found in high speed supernatants from bovine neurohypophysis homogenates and lysed neurosecretosomes (roughly 40% of the total activity), whereas only negligible activity was associated with a purified secretory granule fraction.

The Ca^{2+} binding protein isolated from ox neurohypophysis might be a regulator in this system.

Possible Role of ATP and of Phosphorylation Processes in the Release

Douglas et al. (1965) and Warberg and Thorn (1969) have shown that metabolic inhibitors abolish or decrease the release of vasopressin in vitro induced by calcium and potassium. There could be several explanations for this fact, one of them that the easily releasable pool or hormone is reduced during the preincubation with the inhibitors. In later experiments it has been shown (Müller et al., 1975), however, that about 30% of the total hormone can be released during prolonged stimulation of the kind involved in the experiments just mentioned. It would thus seem more likely that the inhibitors have interfered with partial processes in the stimulated release involving ATP.

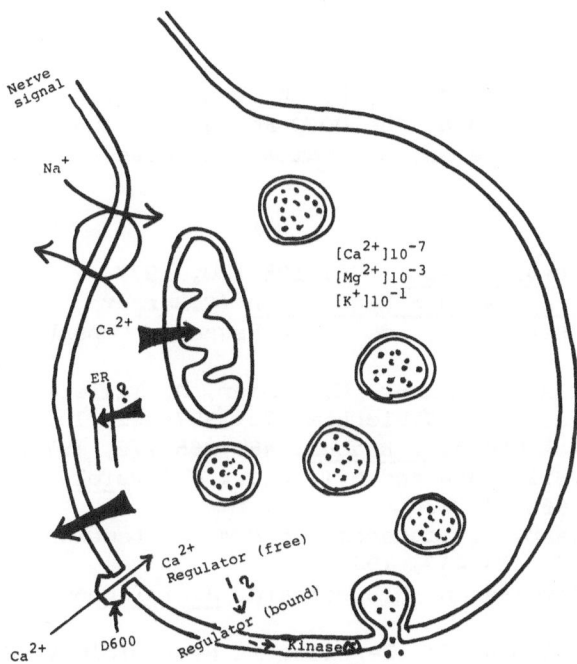

Fig. 2 Sketch of secreting nerve ending with indication of Ca^{2+} transports and Ca^{2+-} induced processes in release.

ATP can cause release of vasopressin from isolated secretory granules (Poisner and Douglas, 1968; Russell and Thorn, 1978). Under these circumstances a phosphorylation of a membrane component was demonstrated in the experiments of Russell and Thorn. In later experiments (Overgaard et al., to be published) no correlation could be demonstrated between release of hormone and ATP metabolism. It might be expected that a considerable part of ATP metabolized goes to other processes than release, so this does not exclude an intimate relation between ATP metabolism and release.

We are currently studying the characteristics of endogenous phosphorylation of proteins in the plasma- and granule-membranes obtained from neurosecretosomes. The distribution of ^{32}P activity, covalently incorporated into the proteins, is followed and quantitated on a high resulution SDS-PAGE system. A complex picture with several phosphorylated bands is obtained, most of them responding to cAMP stimulation. We are presently focusing our attention on effects of Ca^{2+} in physiological concentration range on this system.

Fig. 2 summarized our hypothesis about the role of Ca^{2+} in the mechanism of release of vasopressin.

Acknowledgements

These studies were supported by the Danish Medical Research Council, Nordisk Insulinfond, The NOVO Foundation and the Danish Foundation for the Advancement of Medical Sciences.

References

Andersson, B. Ann. Rev. Physiol. 39: 185-200, 1977.
Bie, P. In: Osmotic and Volume Regulation (C.Barker
  Jørgensen and E. Skadhauge, eds.), Munksgaard, Copenhagen
  pp 260-272, 1978.
Bonne, D., P. Nicolas, M. Lauber, M. Camier, A. Tixier-
  Vidal and P. Cohen. Europ.J.Biochem. 78: 337-342, 1977.
Bridges, T.E. and N.A.Thorn. J.Endocr. 48: 265-276, 1970.
Bridges, T.E., E.W.Hillhouse and M.T.Jones. J.Physiol.
  (Lond.) 260: 746-666, 1976.
Buijs, R.M., D.F.Swaab, J. Dogterom and F.W.van Leeuwen.
  Cell Tiss.Res. 186: 423-433, 1978.
Creutz, C.E., C.J.Pazoles and H.B.Pollard. J.biol.Chem. 253:
  2858-2866, 1978.
deWied, D. and W.H.Gispen. In: Peptides in Neurobiology
  (Gainer, H., ed.), Plenum Press, New York, p. 397, 1977.
Douglas, W.W., A.Ishida and A.M.Poisner. J.Physiol.(Lond.) 181,
  753-759, 1965.
Dreifuss, J.J. and J.S.Kelly. J.Physiol.(Lond.) 220: 87-103, 1972.
Edwards, B.A., M.E.Edwards and N.A.Thorn. Acta Endocr.
  (Kbh.) 72: 417-424, 1973.
Eggena, P. and N.A.Thorn. Acta Endocr. (Kbh.) 65: 442-452, 1970.
Eggena, P. and A.X.Polson. Endocrinology 94: 35-43, 1974.
Gauer, O.H. In: Osmotic and Volume Regulation (C.Barker
  Jørgensen and E. Skadhauge, eds.), Munksgaard, Copenhagen,
  pp 229-242, 1978.

Gnegy, M.E., P.Uzunow and E.Costa. Proc.nat.Acad.Sci. (Wash.) 73: 3887-3890, 1976.

Gratzl, M., G.Dahl, J.T.Russell and N.A.Thorn. Biochim. Biophys.Acta 470: 45-57, 1977.

Halter, J.B., A.P.Goldberg, G.L.Robertson and D.Porter jr. J.clin.Endocr. 44: 609-616, 1977.

Hayward, J.N. Physiol.Rev. 57: 574-658, 1977.

Kandel, E.R. J.Gen.Physiol. 47: 691-717, 1964.

Kendrick,N.C., M.P.Blaustein, R.D.Fried and R.W.Ratzloff. Nature 265: 246-248, 1977.

Kretzinger, R.H. Ann.Rev.Biochem. 45: 239-266, 1976.

Lincoln, D.W., G.Clarke, C.A.Mason and J.J.Dreifuss. In: Neurohypophysis (Moses, A.M. and L.Share, eds.) Karger, Basel, p 101-109, 1977.

Llinas, R. and Heuser, J.E. Neurosci.Res.Progr.Bull.15: 557-685, 1977.

Mathison, R. and K.Lederis. Molec.cell.Endocr. 9: 81-91, 1977.

McKinley, M.J., D.A.Denton and R.S.Weisinger. Brain Res. 141: 89-103, 1978.

Miller, M. and A.M.Moses. In: Neurohypophysis (Moses, A.M. and L.Share, eds.) Karger, Basel, pp. 153-166, 1977.

Müller, J.R., N.A.Thorn and C.Torp-Pedersen. Acta Endocr. 79: 51-59, 1975.

Negoro, H. and R.C.Holland. Brain res. 42: 385-402, 1972.

Peck, J.W. and E.M.Blass. Amer.J.Physiol. 228: 1501-1509, 1975.

Poirier, G., F. Labrie, H.Lemay, A.Duporet, M.Savary and G.Pelletier. Canad.J.Biochem. 55: 555-566, 1977.

Poisner, A.M. and W.W.Douglas. Science 160: 203-204, 1968.

Ruoff, H.J., R.Mathison and K.Lederis. Neuroendocrinology 22: 18-29, 1976.

Russell, J.T. and N.A.Thorn. Acta Endocr. (Kbh.) 76: 471-487, 1974.

Russell, J.P. and N.A.Thorn. Acta physiol.scand. 93: 364-377, 1975.

Russell, J.T. and N.A.Thorn. Biochim.biophys.Acta 491: 398-408, 1977.

Russell, J.T. and N.A.Thorn. Acta Endocr. (Kbh.) 87: 495-506, 1978.

Schulman, H. and P.Greengard. Nature 271: 478-479, 1978.

Share, L. Amer.J.Physiol. 215: 1384-1389, 1968.

Sladek, C.D. and K.M.Knigge. Endocrinology 101: 1834-1838, 1977.

Sofroniew, M.V. and A.Weindl. Endocrinology 102: 334-337, 1978.

Sperk, G. and R.J.Baldessarini. J.Neurochem. 28: 1403-1405, 1977.

Summy-Long, J.Y., L.C.Keil and W.B.Severs. Brain Res. 140: 241-250, 1978.

Terry, L.C. and J.B.Martin. Ann.Rev.Pharmacol.Toxicol 18: 111-125, 1978.

Thorn, N.A. In: Advances in Metabolic Disorders (Luft,
    R. and R.Levine, eds.) vol. 4, Academic Press, New
    York, p 39-73, 1970.
Thorn, N.A., J.T.Russell, C.Torp-Pedersen and M.Treiman.
    Ann.N.Y.Acad.Sci. 307: 618-642, 1978.
Tribollet, E., G.Clarke, J.J.Dreifuss and D.W.Lincoln.
    Brain Res. 142: 69-84, 1978.
Verney, E.B. Proc.roy.Soc.B. 135: 25-106, 1947.
Vinzenzi, F.F. Ann.N.Y.Acad.Sci. 307: in press, 1978.
Wang, J.H. and R.Desai. J.biol.Chem. 252: 4175-4184, 1977.
Warberg, J. and N.A.Thorn. Acta Endocr. (Kbh.) 61: 415-424, 1969.
Watterson, D.M., W.G.Harrelson, jr., P.M.Keller, F.
    Sharief and T.C.Vanaman. J.biol.Chem. 251: 4501-4513, 1976.
Weitzman, R.E., D.A.Fisher, S.Minick, N.Ling and R.
    Guillemin. Endocrinology 101: 1643-1646, 1977.

STUDIES ON THE RELEASE AND DEGRADATION OF HYPOTHALAMIC RELEASING HORMONES BY THE HYPOTHALAMUS AND OTHER CNS AREAS IN VITRO

S. L. Jeffcoate, N. White, G. W. Bennett*, J. A. Edwardson*, E. C. Griffiths[1], R. Forbes[1], and J. A. Kelly[1]
Department of Endocrinology, Chelsea Hospital for Women, London SW3 UK; *Department of Physiology, St. George's Hospital Medical School, London SW17, UK; [1]Department of Physiology, University of Manchester, Manchester, UK.

INTRODUCTION

During the past decade great advances have been made in the study of the hypophysiotrophic functions of the hypothalamus, culminating in the isolation and structural identification of three peptides: TRH, LH-RH and somatostatin (Sandow and König, 1978). Although several other hormonal (e.g. angiotensin) and non-hormonal peptides (e.g. substance P) are also found in the hypothalamus in addition to the remaining hypophysiotrophic hormones as yet unidentified, our studies have been limited to the three 'hypothalamic releasing hormones' whose structure is firmly established.

It is now clear that many peptides, including the so-called hypothalamic hormones, are distributed throughout the central nervous system and even outside it (e.g. somatostatin and TRH in the gastrointestinal tract and TRH in frog skin and retina) and they have actions in the CNS other than their hormonal ones. Evidence is thus accumulating for a neurophysiological role whose significance is still uncertain and many questions remain regarding the regulation of the peptidergic hormones and their integration with other parts of the CNS.

When TRH, LH-RH and, later, somatostatin became available in synthetic form, it was possible to develop sensitive and specific radioimmunoassays for these peptides. Most of our own efforts to apply these assays to biological fluids and to the study of hypothalamic function in vivo have been unrewarding for reasons detailed elsewhere (Jeffcoate 1977, 1978) and the results of such studies remain controversial. These assays have been successfully used in our laboratories to study (a) peptide secretion from isolated nerve

61

terminals/synaptosomes and (b) their degradation by peptidases in
subcellular fractions from neural tissue.

Synaptosomes have been widely used in the investigation of the
metabolic features of pre-synaptic organisation (de Belleroche and
Bradford, 1973) and of the factors controlling the release of peptides
from peptidergic neurons (Edwardson and Bennett, 1977). The location
of the peptides in synapses in specific areas forms one of the three
criteria listed by Reichlin et al (1976) for acceptance of a peptide
as a neurotransmitter. The other two are: there should be specialised
receptor sites for them and there should be local degrading systems
available for their inactivation. Little is known about the specific
receptors for the peptides although receptors for all three have been
identified in the brain, but degradation by peptidases is attracting
increasing attention (Griffiths, 1976; Marks, 1976 for reviews) and
our studies using radioimmunoassays have built up a picture of widespread localised degradation systems for the hypothalamic peptides.

STUDIES ON THE HYPATHALAMUS

The availability of antisera to hypothalamic peptides has
resulted in their identification by immunoassay in crude extracts
of median eminence from several mammalian and submammalian species.
More precise localisation has been achieved by the use of microdissection techniques, followed by immunoassay (e.g. Brownstein et
al, 1974, Kizer et al, 1976) and by the use of immunohistochemistry
(e.g. Fuxe et al, 1976). Each of these approaches has also been
used to localise neurotransmitters within the hypothalamus and is
evidence for an interaction between them and the peptides. Despite
their advantages and the valuable information they generate, these
techniques remain static, giving only a rough indication of control
mechanisms.

Secretion by Hypothalamic Synaptosomes Incubated in Vitro

Synaptosomes isolated from the hypothalamus contain both neurohypophysial and hypothalamic hormones, identifiable by bioassay and
immunoassay (Edwardson and Bennett, 1977) and almost all the immunoreactive TRH and LH-RH is found in synaptosome-containing fractions
when extracts of rat hypothalamus are fractionated on sucrose
gradients.

When the synaptosome fraction is incubated in vitro, a steady
basal release of hypothalamic hormones is observed. Electrical
stimulation causes a marked calcium-dependent increase in this
secretion. The presence of high depolarising concentrations of
potassium ions unexpectedly failed to increase TRH and LH-RH secretion (Bennett et al, 1975). A response in somatostatin secretion

was obtained by Wakabayashi et al (1977) and we have recently confirmed this. Having established the basal functioning of isolated
hypothalamic synaptosomes, we have been studying the effects of
various physiological states and pharmacological treatments on the
secretion of TRH and LH-RH.

  Neurotransmitters   Microdissection and immunofluorescence
studies have shown a close agreement in the distribution of dopamine
(DA) - containing neurones and LH-RH - containing neurones, thus
providing anatomical evidence for an axo-axonic interaction between
them (Fuxe et al, 1976; Kizer et al, 1976). The Karolinska group
consider DA to have an inhibitory effect on LH-RH secretion whereas
others consider the effect of DA to be stimulatory (McCann et al,
1972). The data upon which these ideas are based are essentially
obtained from indirect sources and the synaptosome preparation provides a direct approach to the study of DA and other neurotransmitters
on the secretion of LH-RH and other hypothalamic peptides.

  One finding is that the region of origin of the synaptosomes
may have an important influence on the results. Thus, when sheep
hypothalamus was dissected into 3 regions and synaptosomes prepared
separately, DA stimulated the release of LH-RH and TRH from median
eminence synaptosomes (about 10% of the total net weight of tissue),
but not from those from the periventricular region or the remainder
of the hypothalamus (Bennett et al, 1975). Thus, with 90% of the
tissue unresponsive, it is easy to see how negative results would
be obtained.

  The dose of neurotransmitter is also important. The stimulatory effects we obtained were with DA at concentrations of 10^{-8}M or
less. Using high non-physiological concentrations of 10^{-4}M there is
either no effect, or an inhibition of release. Somatostatin release
is also increased by dopamine (Wakabayashi et al, 1977).

  Apart from dopamine, acetylcholine is the only neurotransmitter
to have a similar distribution to TRH and LH-RH in microdissection
studies (Kizer et al, 1976). It is however without any effect on
their release from synaptosomes (Bennett et al, 1975). The only one
that does is serotonin, which significantly inhibited the release of
TRH (but not LH-RH). In recent, unpublished studies, we have found
an inhibitory effect of the pineal indole, melatonin, on LH-RH
release, an action in line with its well-established anti-
gonadotrophic properties.

  Seasons and cycles   There is some preliminary evidence
that the seasonal and cyclical changes in reproductive function are
reflected in changes in secretion of LH-RH by hypothalamic synaptosomes in vitro (Bennett and Edwardson, 1978). We have observed, for
instance, that the basal secretion of LH-RH by synaptosomes isolated
from sheep hypothalamus in the oestrous season (November to May) is

some six times higher than in the anoestrous period. In the rat
basal release is higher at pro-oestrus than at oestrus. Interestingly, sensitivity to the inhibitory action of gonadal steroids
and cyclic nucleotides is also maximal at pro-oestrus. Simultaneous secretion of TRH in both the sheep and the rat experiments
showed no changes; this is a valuable control parameter.

<u>Gonadal steroids</u>   Feedback regulation of gonadotrophin is at
least partly mediated by the action of gonadal steroids at the neurosecretory nerve terminal. Progesterone or oestradiol-17β (10^{-8}M)
halve the basal secretion of LH-RH from sheep hypothalamic synaptosomes when added <u>in vitro</u> and the effects are additive. LH-RH
secretion from rat hypothalamic synaptosomes is also reduced by
gonadal steroids. Once again, no effects on TRH secretion are seen
(Bennett & Edwardson, 1978).

<u>Cyclic nucleotides</u>   Earlier work by two of us (Edwardson &
Bennett, 1975) indicated that cyclic nucleotides (specifically cyclic
AMP) may be involved in the feedback inhibition of CRF release by
corticosteroids. The effect on LH-RH secretion from hypothalamic
synaptosomes is similar, with an inhibition of secretion by cyclic
AMP, whereas cyclic GMP causes, if anything, a slight increase.
The effects of cyclic AMP are markedly increased by the addition of
very small amounts (10^{-10}M) of progesterone and oestradiol-17β, which
are ineffective themselves. These experiments suggest the possibility of a complex interaction between steroids and cyclic nucleotides
at the level of the synapse in the control of peptide secretion. This
complexity implies that the design and interpretation of experiments
utilising isolated synaptosomes needs careful thought.

Degradation by Peptidases

The possible regulation of hypothalamic peptides by peptidase
degradation is attracting increased attention (see Griffiths, 1976;
Marks, 1976 for recent reviews). Peptidases in extracts of rat and
rabbit hypothalamus rapidly destroy LH-RH and TRH (Griffiths, 1976
for refs) and somatostatin (Griffiths et al, 1977 ab). Initially
the LH-RH studies used bioassays; all our recent work has used the
disappearance of LH-RH, TRH and somatostatin immunoreactivity as
an index of peptidase action. In order to equate this with destruction of the peptide, it is essential to know the structural specificity of the antisera used. For TRH, removal of any part of the
molecule results in a complete loss of immunoreactivity; for LH-RH,
loss of the first 2 or 3 residues from the N terminus is associated
with some retention of immunoreactivity, but it is apparent that
degradation <u>in vitro</u> occurs at the C-terminus (at the terminal amide
and at the 6-7 bond) so that the degradation products are not
measured by this antiserum. Similarly, the somatostatin antiserum

is directly towards the 5-11 region of the peptide, the site at
which degradation is thought to occur (Marks, 1976).

   Subcellular localisation of peptidases   Ultracentrifugation
at 25,000g of hypothalamic homogenates yields two fractions - a
soluble, cytoplasmic fraction and a particulate fraction containing
microsomes and mitochondria. With the exception of TRH inactivation
by rat hypothalamus (but not rabbit), the soluble fraction is more
active, and usually considerably more active, than the particulate
fraction. In general, changes in peptidase activity in the supernatant fraction are reflected by changes in the particulate fraction.
Some of our recent studies using thin-layer chromatography to separate
and identify degradation products have indicated that the enzyme
systems in the two fractions may in fact be different. For TRH, the
soluble enzyme appears to be a deamidase, whereas the particulate
bound enzyme are a pryoglutamyl peptidase cleaving pyroglutamic
acid from histidyl prolineamide and a His-ProNH$_2$ peptidase releasing
ProNH$_2$. For LH-RH, cleavage at the 6-7 bond occurs in both fractions
with further breakdown of the 7-10 fragment. This site of cleavage
is confirmed by the fact that LH-RH analogues with structural alterations at position 6 (D-amino acids) are more resistant than LH-RH to
enzyme action.

   Sex differences in peptidase activity   For both TRH and LH-RH,
peptidase activity is significantly higher in male hypothalamus than
in the female and these differences are seen also in other brain areas
(see below). This may well reflect biochemical differences induced
by sexual differentiation in early life under the influence of androgen. This is supported by the observation that neonatal androgenisation of female rats increases LH-RH peptidase activity in extracts
of hypothalamus (and cerebral cortex) (Griffiths et al, 1976).

   Effect of gonadal steroids   Experiments have been carried
out to test the possibility that the enzymes inactivating LH-RH
could be involved in the steroid feedback mechanisms at the level
of the hypothalamus (Griffiths et al, 1975). In females, ovariectomy
significantly decreased peptidase activity, oestradiol treatment
reversed this effect. Progesterone has the opposite effect, further
decreasing peptidase activity. In the males, orchidectomy decreased,
and testosterone or dihydrotestosterone restored the activity. These
changes which were seen in the soluble fraction only are in line with
the well-established effects of gonadectomy and gonadal steroid injection on gonadotrophin secretion and suggests that these could, in part,
be explained by an action on LH-RH peptidases.

   Age-dependent changes   The changes in TRH and LH-RH inactivation from birth to adulthood have been investigated in the hypothalamus and other brain areas (see below) in the rat. For TRH, enzyme
activity increased progressively in the soluble and particulate
fractions of the hypothalamus up to 15-20 days old and then decreased

to adult levels. For LH-RH, the variations in hypothalamic
peptidases are more complex, enzyme activity being high at birth,
reaching a peak at 5-10 days before falling progressively to their
lowest levels at 35-40 days (the time of puberty) and then rising
again to reach adult levels. Whilst it is possible to associate
these changes with the complex changes in the secretion of gonadal
steroids and gonadotrophins that occur in the rat between the newborn and puberty, the exact role of the LH-RH peptidases in the
control of the development of sexual function remains to be
established.

EXTRA-HYPOTHALAMIC AREAS OF THE CNS

As mentioned in the introduction, evidence is accumulating
for the distribution of LH-RH, TRH (Reichlin et al, 1978) and
somatostatin (Luft et al, 1978; Pimstone and Berelowitz, 1978)
outside the hypothalamus, and this has led us to investigate both
peptide secretion by nerve terminals isolated from different areas
of the CNS, and also peptidase activity in subcellular fractions
from these areas.

Cerebral Cortex

Peptide secretion by synaptosomes   Nerve terminals from the
cerebral cortex of the sheep and rat secrete TRH and somatostatin
(Edwardson & Bennett, 1977, Pimstone and Berelowitz, 1978) and this
is increased by depolarisation with high K^+, an effect which is
calcium-dependent. For TRH, release is stimulated by dopamine,
but not by glycine or somatostatin; this is in contrast to the
brain stem (see below).

Degradation by peptidases   Degradation by soluble and particulate fractions of rat and rabbit cerebral cortex have been shown for
LH-RH, TRH and somatostatin (Griffiths et al, 1975, 1976, 1977ab,
1978). For LH-RH, the same clear sex difference in rat cortex is
seen as in the hypothalamus and neonatal androgenisation in the
female leads to a marked increase to reach male activity levels.
This suggests a possible role of these peptidases in the sexual
differentiation of the brain. Studies from birth to adulthood
have revealed changes in activity, the meaning of which are unclear.
In the particulate fraction there is a sharp rise in activity at 10-15
days in life before a progressive decline to adult levels which may
reflect a more general type of peptidase activity in this fraction.
A similar pattern of development of TRH inactivation is seen, except
that the peak is later - at 25 days. The development of somatostatin
has not yet been studied in detail, but activity is higher in the
soluble than in the particulate fraction and higher in the adult
male than in the female.

Thalamus and Cerebellum

Degradation by peptidases   Peptidase degradation of LH-RH,
TRH and somatostatin is found in subcellular fractions of thalamus
and cerebellum. As with cerebral cortex, there are sex differences
and developmental changes, the importance of which are unclear.
Interestingly, neonatal androgenisation of the female has no effect
on LH-RH peptidases in these areas. In both these areas, LH-RH
degradation is highest in the first 5 days of life, thereafter
declining to adult levels.

Brain Stem

Peptide secretion by synaptosomes   Nerve terminals isolated
from sheep brain stem secrete TRH and somatostatin *in vitro*
(Edwardson & Bennett, 1977). For TRH, dopamine does not stimulate
release; this is in contrast to the effect on hypothalamic and
cortical synaptosomes. On the other hand, glycine (10^{-4}M) and
somatostatin (10^{-8}M), which are without effect on cortical synaptosomes, stimulate TRH release from brain stem synaptosomes.

Effect of TRH on amino acid neurotransmitters   An attempt
has been made to explore the possible interactions between peptidergic neurones and neurones containing putative amino acid neurotransmitters which are abundant in the central nervous system. The effect
of TRH, LH-RH, somatostatin on the K^+ induced release of amino acids
from sheep brain stem synaptosomes have been studied. LH-RH and
somatostatin were inactive. TRH has no effect on basal secretion
but markedly stimulates K^+ induced secretion of glycine, and to a
lesser extent, glutamic acid and aspartic acid.

CONCLUSIONS

Radioimmunoassays have demonstrated the presence of immunoreactive TRH-like and somatostatin-like material throughout the
central nervous system of mammals and throughout the animal king-
dom (Luft et al, 1978; Pimstone and Berelowitz, 1978; Reichlin et
al, 1978). Subcellular localisation has shown the peptides to be
predominantly located in synaptosomes.

The pattern of distribution has been supported and further
illustrated by extensive immunochemical studies, notably by the
group in the Karolinska Institute in Stockholm and by microdissection studies, notably at the National Institutes of Health in the
United States. These studies have explored the geographical correlations between peptidergic neurones and neurotransmitters, in
an attempt to delineate the factors controlling the peptidergic
neurones.

In general, it has been supposed that the neurones contain only one transmitter (a hypothesis associated with the name of Sir Henry Dale); control in the CNS is then mediated via a complex system of polysynaptic pathways and axo-axonic junctions. At each stage, control can either be stimulatory or inhibitory. These complexities make it difficult to draw conclusions from static anatomical data or even from the more dynamic physiological data we have obtained from the synaptosome experiments. The recent histochemical identification of two putative transmitters, somatostatin and noradrenaline, together in peripheral neurones (Luft et al, 1978) has now raised the possibility that this might also be the case in the CNS.

The complex nature of synaptic interactions make difficult the study of factors controlling them and conclusions can only be drawn tentatively from experiments in which synaptosomes are isolated and incubated in vitro. They do, however, provide a tool for investigation of the sites and mechanism of feedback control of peptides in the hypothalamus, in addition to their wider application in neurophysiology. One interesting aspect of our results is the functional difference between synaptosomes from different brain areas and their responses to chemical stimuli.

Peptidase degradation of neural peptides is an essential criterion for acceptance of these substances as neurotransmitters and these results show that it is found throughout the central nervous system. The presence of peptidases does not, of course, necessarily imply the presence of the peptides there, or of a neurotransmitter role. The further study of these peptidases and their activity in different stages of development and different physiological and pathological states should reveal more about their role in the control of the peptidergic neurone.

REFERENCES

Bennett, G. W. and Edwardson, J. A. (1978). Release of immunoreactive LH-RH from hypothalamic nerve-endings; effect of gonadal steroids and cyclic nucleotides. In: Hypothalamic Hormones: Proceedings of Second European Colloquium (D. Gupta and W. Voelter, eds.) Verlag Chemie, Weinheim and New York. (in press)

Bennett, G. W., Edwardson, J. A., Holland, D. T., Jeffcoate, S. L. and White, N. (1975). The release of luteinising hormone-releasing hormone and thyrotrophin-releasing hormone from hypothalamic synaptosomes. Nature (London) 257, 323-325.

Brownstein, M. J., Palkovits, M., Saavedra, J. M., Bassiri, R. M. and Utiger, R. D. (1974). Thyrotrophin-releasing hormone in specific nuclei of rat brain. Science 185, 267-269.

de Bellroche, J. S. and Bradford, H. F. (1973). The synaptosome: an isolated working, neuronal component. In: Progress in Neurology (G. A. Kerkut and J. W. Phillips, eds.) Vol I, pp. 275-298. Pergamon Press, Oxford and New York.

Edwardson, J. A. and Bennett, G. W. (1974). Modulation of corticotrophin-releasing factor release from hypothalamic synaptosomes. Nature (London) 251, 425-427.

Edwardson, J. A. and Bennett, G. W. (1977). Hypothalamic hormones and mechanisms of neuroendocrine integration. In: Biologically active substances: Exploration and exploitation. (D. A. Hems, ed.) J. Wiley and Sons, New York. pp. 281-299.

Fuxe, K., Hökfelt, T., Löfström, A., Johansson, O., Agnati, L., Everitt, B., Goldstein, M., Jeffcoate, S., White, N., Everoth, P., Gustafsson, J. A. and Skett, P. (1976). On the role of neuro-transmitters and hypothalamic hormones and their interactions in hypothalamic and extrahypothalamic control of pituitary function and sexual behaviour. In: Subcellular mechanisms in reproductive neuroendocrinology. (F. Naftolin, K. J. Ryan and I. J. Davies, eds.) Elsevier. Amsterdam, Oxford, New York. pp. 193-246.

Griffiths, E. C. (1976). Peptidase inactivation of hypothalamic releasing hormones. Hormone Research (Basel) 7, 179-191.

Griffiths, E. C., Forbes, R., Jeffcoate, S. L. and Holland, D. T. (1977a). Local degradation of luteinising hormone releasing hormone in the rat central nervous system. Neuroscience Letters 4, 33-37.

Griffiths, E. C., Hooper, K. C., Jeffcoate, S. L. and Holland, D. T. (1975). The effects of gonadectomy and gonadal steroids on the activity of hypothalamic peptidases inactivating LH-RH. Brain Research 88, 384-388.

Griffiths, E. C., Hooper, K. C., Jeffcoate, S. L. and Holland, D. T. (1976). The effect of neonatal androgen on the activity of peptidases in rat brain inactivating LH-RH. Hormone Research (Basel) 7, 218-226.

Griffiths, E. C., Jeffcoate, S. L. and Holland, D. T. (1977b)., Local degradation of somatostatin in the central nervous system. Neuroscience Letters 4, 33-37.

Griffiths, E. C., White, N. and Jeffcoate, S. L. (1978). Local degradation of thyrotrophin-releasing hormone (TRH) in the the central nervous system. Neuroscience Letters (in press).

Jeffcoate, S. L. (1977). Immunoassay of hypothalamic releasing hormones and the regulation of their secretion. In: Proceedings of Fifth Int. Cong. of Endocrinology. Excerpta Medica pp. 181-185.

Jeffcoate, S. L. (1978). Hypophysiotrophic hormones of the hypothalamus. In: Topics in hormone chemistry I (W. R. Butt, ed.) Ellis Horwood, Chichester pp. 13-47.

Kizer, J. S., Palkovits, M., Tappaz, M., Kebabian, J. and Brownstein, M. J., (1976). Distribution of releasing factors, biogenic amines and related enzymes in the bovine medium eminence. Endocrinology 98, 685-695.

Luft, R., Efendic, S. and Hokfelt, T. (1978). Somatostatin - both hormone and neurotransmitter? Diabetologia 14, 1-13.

McCann, S. M., Kalra, P. S., Donoso, A. O., Bishop, W., Schneider, H. P. G., Fawcett, C. P. and Krulich, L. (1973). The role of monoamines in the control of gonadotrophin and prolactin secretion. In: Brain-Endocrine interaction. Median eminence structure and function. (K. M. Knigge, D. E. Scott and W. Weindl, eds.) Karger, Basel pp. 224-235.

Marks, N. (1976). Biodegradation of hormonally active peptides in the central nervous system. In: Subcellular mechanism in reproductive endocrinology. (F. Nuftolin, K. J. Ryan and I. J. Davies, eds.) Elsevier. Amsterdam, Oxford, New York. pp. 129-147.

Pimstone, B. L. and Berelowitz. (1978). Somatostatin-paracrine and neuromodulator peptide in gut and nervous system. South African Med. J. 53, 7-9.

Reichlin, S., Martin, J. B. and Jackson, I. M. D. (1978). Regulation of thyroid-stimulating hormone (TSH) secretion. In: The Endocrine Hypothalamus. (S. L. Jeffcoate and J. S. M. Hutchinson, eds.) Academic Press, London, New York. pp. 230-269.

Reichlin, S., Saperstein, R., Jackson, I. M. D., Boyd, A. E. and Patel, Y. (1976). Hypothalamic hormones. Ann. Rev. Physiol. 38, 389-424.

Sandow, J. and Konig, W. (1978). Chemistry of the hypothalamic hormones. In: The Endocrine Hypothalamus (S. L. Jeffcoate and J. S. M. Hutchinson, eds.) Academic Press, London, New York. pp. 150-211.

Wakabayashi, I., Miyazawa, Y., Kanda, M., Miki, N., Demura, R., Demura, H. and Shizume, K. (1977). Stimulation of immunoreactive somatostatin release from hypothalamic synaptosomes by high K^+ and dopamine. Endocrinol. Japon. 24, 601-604.

THE ROLE OF GTP IN THE COUPLING OF HORMONE RECEPTORS

AND ADENYLATE CYCLASE

M. Rodbell and W. Schlegel

National Institutes of Health

Bethesda, Maryland 20014 U.S.A.

The adenylate cyclase system is one of the most important biological communication systems known. Through this system, a variety of extracellular agents, including hormones and neuro-transmitting agents, regulate the production of the intracellular "messenger," cyclic AMP, that profoundly affects the metabolism, growth, and differentiation of cells responding to these agents. Although it is generally agreed that the receptors for the extra-cellular agents and the enzyme that catalyzes the formation of cyclic AMP from ATP are separate molecules, not understood is the nature of the transduction process whereby binding of the hormones leads to changes in the activity of the enzyme.

A major change in our perception of these systems occurred when it was discovered several years ago that GTP was necessary for the expression of glucagon-activation of adenylate cyclase in hepatic plasma membrane preparations of the enzyme system. This finding, plus those subsequently reported by a number of labora-tories, led to the generally agreed principal that GTP plays a crucial role in hormonal regulation of adenylate cyclase systems and thus in the transduction process. Reviews of the critical studies that led to this conclusion have been published [1,2]. Here we will first give a working hypothesis, based on our studies with the glucagon-sensitive system, which brings into focus the role of what are termed "nucleotide regulatory proteins" in the transduction process. Following the description of this hypothesis, we will present recent data which for the first time gives a structural basis for the transduction process and the role of the nucleotide regulatory proteins in this process.

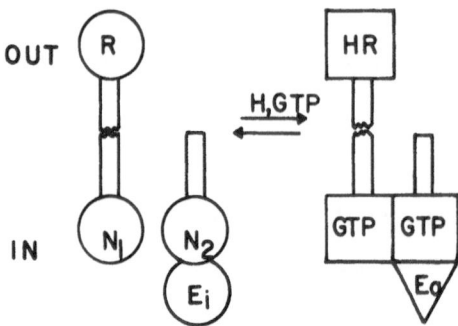

Figure 1. Model for GTP and Hormone (H) action on adenylate
cyclase systems. Shown are four components designated R (for
hormone receptor), N_1 (for nucleotide regulatory component
associated with receptor), N_2 (for nucleotide regulatory component
associated with adenylate cyclase which exists in inactive (E_i)
and activated (E_a). For details, see text.

THE ROLE OF GTP IN THE TRANSDUCTION PROCESS

Our working hypothesis is represented pictorially in Figure 1.
Four fundamental components are depicted. On the outer surface
of the cell membrane is the hormone receptor (R) whereas on the
internal face are two nucleotide regulatory proteins, designated
N_1 and N_2, and adenylate cyclase (E). For the purposes of this
discussion, it matters little whether N_1 and N_2 are separate
molecules from the receptor and enzyme, respectively, although
this is likely the case. The essential feature of the model is
that binding of GTP, derived from the interior of the cell,
brings about the coupling of N_1 and N_2, thus linking functionally
the exterior receptor component to the enzyme system. The hormone,
by preferential binding to the coupled or linked components, acts
in concert with GTP to stabilize the coupled configuration of the
enzyme system. In this case, coupling brings about a change from
an inactive (E_i) to an active (E_a) form of the enzyme. It is
conceivable that other nucleotide regulatory components that
inhibit cyclase activity may be similarly linked to the enzyme
system.

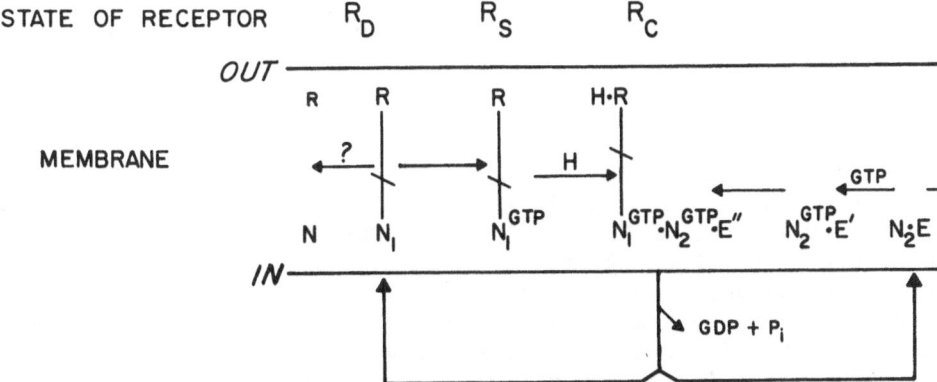

Figure 2. A model for the dynamic actions of GTP and hormones (H) on adenylate cyclase systems. For detailed description, see text.

Figure 2 illustrates a view of the dynamics of the regulatory process. Based on our studies of the kinetic features of the glucagon-sensitive cyclase system and the characteristics of the glucagon receptor[1], the receptor and enzyme can take three fundamental states of binding or cyclase activity.

The receptor, in the absence of GTP binding at the N_1 component binds glucagon by a slow process that leads to an essentially irreversible binding state of the hormone-receptor complex. This tight-binding state of the receptor cannot link to the enzyme system and is termed the desensitized (R_d) state of the receptor. Binding of GTP to N_1 rapidly converts the receptor to two states. One state, designated R_s (for sensitized receptor) is the only form of the receptor that can couple to the enzyme system. It will link to the system in proportion to the concentration of EN_2 present in the membrane. If in excess of the concentration of EN_2, the RN_1.GTP complex will be observed experimentally as a state having a much lower affinity for the hormone than either the desensitized or the coupled states of the receptor.   Thus, when linked to the enzyme system, RN_1.GTP acquires a higher affinity for the hormone. Accordingly, when the system is exposed to GTP, preferential binding of the hormone to R_c, the coupled receptor state, drives the equilibrium toward the receptor-coupled form of the enzyme system.

The enzyme (E) is linked to N_2 which when liganded with GTP, converts the enzyme to a transient activity state, termed E'. This transient state isomerizes into the high V_{max} state, designated E", of the enzyme system. In this high activity state of the enzyme, GTP is hydrolyzed to inert products, $GDP+P_i$, and the nucleotide regulatory components uncouple, causing the system to revert to the ground states of the enzyme and of the receptor. Thus, in this model, continual supply of GTP is required to maintain the system in its

coupled, high activity state. The hormone acts in concert with GTP
to assure rapid transition to the high activity state; paradoxically,
formation of the coupled state leads to hydrolysis of the activating
nucleotide.

This model accounts for the rapid on-off characteristics of the
enzyme system in response to glucagon and GTP. It also accounts for
the fact that analogs of GTP, such as GMP-PNP, that are not hydrolyzed at the terminal phosphate, lead to a persistent high activity
state of the enzyme, irrespective of the linkage of the RN_1 component to the enzyme system. The latter is thus only necessary to
enhance the rate of formation of the activated E" state of the enzyme
whether the nucleotide components are occupied by GTP or by non-
hydrolyzable analogs of GTP.

We believe this model can explain the behavior of a number of
adenylate cyclase systems, including the intensively investigated beta
adrenergic linked adenylate cyclase system (reviewed in ref 2).

The model predicts that the receptor and enzyme cannot interact
within the plane of the membrane unless the nucleotide regulatory
components are occupied by GTP. What are the structures of the
receptor and enzyme that require GTP for coupling and what does GTP
do to these structures that causes integration of the various components into a biologically active configuration? Based on studies
using high-energy radiation analysis of the functional sizes of the
hepatic cyclase system and the glucagon receptor, we believe that we
now have answers to these questions.

TARGET SIZE ANALYSIS OF THE HEPATIC ADENYLATE CYCLASE SYSTEM

Ionizing radiation has been employed to analyze the molecular
size of numerous proteins[4]. The advantage of this procedure is that
the enzyme need not be purified; it relies solely on the analysis of
the loss of biological activity as a function of increasing radiation exposure. Thus, theoretically, it can be employed for analysis
of an enzyme or any other functional process even in crude membrane
preparations. Indeed, it has been used by Metcalfe and associates
in their analysis of the adenylate cyclase system in liver membranes[5]. The theory and its application are relatively straightforward. Exposure to ionizing radiation leads to loss of function
due to destruction of critical structures. The larger the structure,
the more likely it will be "struck" by the ionizing source. Because
the energy source is large (1500 Kcal/mole), function is completely
destroyed; i.e., there are no partially damaged "targets." Therefore, the only remaining activity after radiation exposure is from
units that have escaped ionization. The number of "hits" follows a
Poisson distribution. Accordingly, the surviving biological

activity measured after different exposures decreases as an exponential function of the radiation dose. Since ionization caused by irradiation occurs at random throughout the volume of the sample, the slope of the exponential function relating loss of biological activity to radiation dose is a measure of the volume. From the known average density of proteins and Avagadro's number, a simple formula is used to calculate the molecular weight.

The method has its limitations which, if not appreciated, can lead to misleading interpretations. Since liquid solutions of targets cannot be employed for radiation analysis, the usual procedure is to lyophylize the sample prior to irradiation. However, even with lyophylized preparations, the apparent target size is influenced significantly by the temperature during irradiation, reaching constant values at -110°. A temperature correction factor of 2.8, determined empirically, is used to adjust data obtained at -110°. This permits the use of frozen samples provided that the temperature during irradiation is monitored and maintained. This was important in our studies since lyophilization of plasma membranes resulted in substantial losses of adenylate cyclase activity, particularly when the enzyme was pre-activated by various effector ligands. Because of such losses in activity, we have generally employed frozen samples for radiation analysis.

Two types of procedures were used for analysis of the functional target size of the hepatic enzyme. One was to first irradiate the frozen samples of membranes without prior exposure of the enzyme to activating ligands (GMP-PNP, GTP, fluoride, GTP+glucagon). After rapid thawing, the membranes were immediately assayed for adenylate cyclase activity in the presence of each of the activating ligands. The medium used for assay is the standard medium used in previous studies of the system with the exception that the specific activity of $\alpha-^{32}P$ ATP was at least 50 times higher than usual. The latter was necessary in order to measure cyclase activity with high sensitivity and reliability even with two logs of enzyme decay after irradiation; this also provided the means of testing changes in functional size over a large range of ionizing doses. Irradiation of the samples prior to activation by the various effector ligands gives a measure of the functional size of the enzyme in its basal or ground state of activity. In order to measure the size of the enzyme when it is converted to the high activity state, the membranes were first incubated, under standard cyclase conditions, with the effectors at saturating concentrations. The samples were quick-frozen at liquid nitrogen temperatures and subsequently irradiated with increasing dosage. The thawed samples were then assayed under the same assay conditions used during pretreatment but with high specific activity ATP in the assay medium.

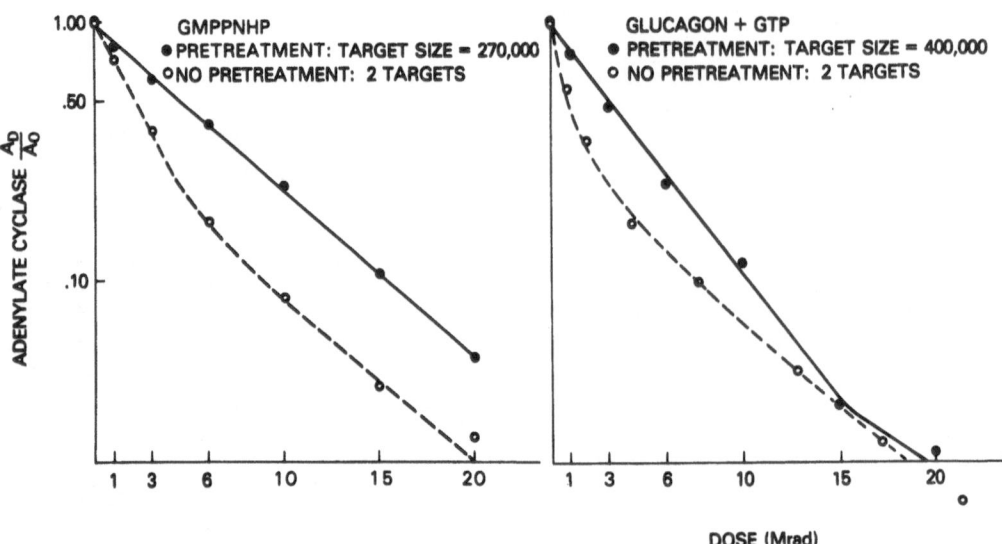

Figure 3. Irradiation inactivation of adenylate cyclase in rat
liver plasma membranes. For the target analysis of the enzyme in
the ground state (No pretreatment), a frozen suspension of plasma
membranes was irradiated at -110°C. After irradiation, the
membranes were thawed and adenylate cyclase activity was measured
by the method of Salomon et al[6] in an incubation mixture containing 10 μM ^{32}P-ATP(1 μCi/nMole), 20 mM MgCl$_2$, 2mM creatinephosphate,
10 U/ml creatine kinase, 20 μM cAMP, 1 mM DTT and 30 mM Tris-HCl
pH 7.5 with either 10^{-5}M GppNHp (left panel) or 10^{-6}M glucagon
plus 10^{-5}M GTP (right panel). Incubations were carried out for 6
minutes at 30°C. For the irradiation inactivation analysis of
the activated states of the enzyme (Pretreatment) the membranes
were incubated for 2 minutes in the assay mixtures described
above but containing only unlabelled ATP; then they were quick
frozen in liquid nitrogen and irradiated frozen at -110°C. After
thawing they were mixed with identical assay mixtures containing
^{32}P-ATP to measure adenylate cyclase activity also for 6 minutes
at 30°C. GppNHp-stimulated activies at zero dose (A$_0$ left panel)
were 9.8 and 13.1 pmoles/mgxmin for no pretreatment and pretreatment;
Glucagon plus GTP stimulated activities at zero dose (A$_0$ right
panel) were 18.4 and 15.6 pmoles/mg for no pretreatment and pretreatment, respectively. The target sizes were calculated as
described in Table I.

     The functional size of the glucagon receptor was also deter-
mined by the radiation procedure. In this case, function was
assayed by the binding of ^{125}I-glucagon to the membranes both
before and after exposure to increasing radiation doses. This

procedure gives the size of the receptor in its unliganded state but is, in fact, the state that yields the tight-binding, or R_d, state of the receptor. Reasonably accurate determinations of the binding function were obtained by measuring either total binding (after subtraction of radioactivity bound in the presence of high concentrations of unlabeled glucagon) or by measuring the amount of labeled hormone displaced from the receptor by the actions of GTP, which converts the receptor to the R_s state. Both measurements gave essentially the same target size of the receptor.

Figure 3 shows representative data of decay curves obtained for adenylate cyclase activity in both the ground state and the activated state using either GMP-PNP (left panel) or glucagon plus GTP (right panel) as activating ligands. Both panels show that the ground state displays non-linear decay curves suggesting that more than one target size of the enzyme pre-existed in the membranes. Assuming two targets, the estimated target sizes (shown in Table I) of the non-preactivated enzyme are $1.2-1.5 \times 10^6$ daltons for the large size and $2.5-3.0 \times 10^5$ for the smaller size of the enzyme. It should be emphasized that the two target sizes were obtained irrespective of the activating ligands used in the final assay medium.

When the enzyme was preactivated with GMP-PNP (or with fluoride ion or GTP, data not shown) a single exponential decay rate was obtained as a function of dose (Figure 1, left panel). Analysis of the slope gave the same size as the small size of the enzyme in its non-pretreated state. Thus, GMP-PNP converted all of the enzyme from a large size to a smaller functional unit having the size of $2.5-3.0 \times 10^5$ daltons. The smaller size, therefore, must represent the activated state of the enzyme system. The high activity state of the enzyme is thus reflected functionally as having one fourth the size of its precursor. This finding suggests that the ground state of the enzyme is an oligomer, possibly a tetramer, whereas the activated state, is the monomer formed by the actions of GMP-PNP.

A significant and reproducible change in the functional size of the enzyme system occurred when it was pre-activated by GTP plus glucagon. As seen in Figure 3 (right panel), again a single exponential was obtained after preactivation but the slope of the decay rate was steeper than that observed with the GMP-PNP-activated enzyme. The size of the GTP-glucagon activated enzyme, recorded in Table I, was $4-5 \times 10^5$ daltons or about 200,000 daltons greater in size than the activated enzyme induced by GMP-PNP. Since glucagon was the only added factor that contributed to this larger size of the enzyme system, we conclude that the increment is due to the linkage of the receptor to the enzyme.

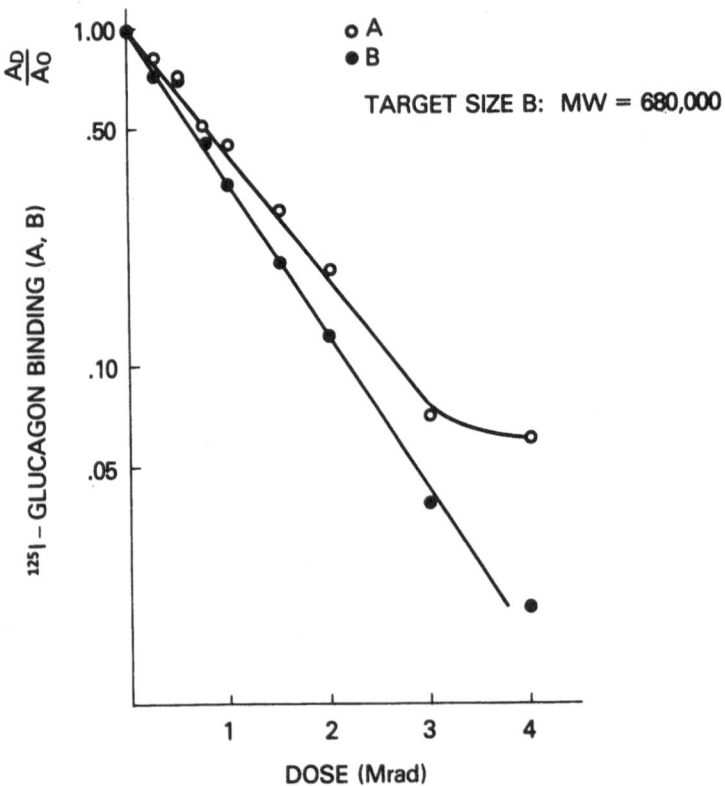

Figure 4. Irradiation inactivation of glucagon receptors in rat
liver plasma membranes. A suspension of rat liver plasma membranes
in 1 mM NaHCO$_3$, pH 7.5, was quick frozen in liquid nitrogen, lyophylized at –20°, and irradiated at 25° with a 13 MeV linear
accelerator. The membranes were re-hydrated with the above
buffer for 30 minutes on ice. Binding of ^{125}I-glucagon was
measured by a filtration assay using 0.5 nM labeled hormone in
the incubation medium. Since GTP causes displacement of a
considerable portion of the bound labeled hormone (in the presence
of unlabeled hormone to minimize re-binding of labeled hormone),
both total binding (A) and GTP-displaceable binding (B) were used
to characterize the target size of the glucagon receptor.

The binding experiment was carried out as follows: The
rehydrated membranes (30–81 µg/ml) were incubated in 20 mM
Tris-HCl, pH 6.9, 0.2% albumin, and 0.5 nM ^{125}I-glucagon for 15
minutes at 30°. Total binding (A) was determined by filtration
of 1 ml aliquots; values were corrected for nonspecific binding
determined by filtration of samples incubated in the presence of
1 µM unlabeled glucagon. After 18 minutes of incubation, GTP was
added to a final concentration of 10 µM together with 1.0 µM
unlabeled glucagon; again 1 ml aliquots were filtered and the
samples counted. GTP-displaceable binding (B) was determined as →

The obvious question is whether the receptor size evaluated from binding studies is commensurate with the increment in size of the functional receptor when it is coupled to the enzyme system. Figure 4 shows a representative experiment of the radiation decay curves obtained for the binding of ^{125}I-glucagon. Both total binding and that displayed by GTP gave approximately the same rates of decay. As recorded in Table I, the estimated size of the receptor in its unliganded R_d state is 6.5-8.0 x 10^5 daltons. Thus the uncoupled R_d state of the receptor is approximately 4 times that of the receptor when it is coupled to the enzyme system. These findings suggest, therefore, that the uncoupled receptor, as is the case of enzyme in its ground state, is an oligomeric structure. Based on the difference in the size of R_d state of the receptor and the functionally coupled form (R_c) induced by the actions of GTP and glucagon, we deduce that the oligomer is a precursor to the coupled form and that GTP converts the oligomeric state of the receptor to a monomer that couples to form the high V_{max} state of the enzyme system in response to hormone+GTP.

Another conclusion that can be drawn from these findings is that the tight-binding form of the receptor requires the oligomeric structure for binding of glucagon to occur. This conclusion is based on the fact that all of the binding measured both before and after irradiation is to the tight binding R_d state of the receptor. If binding to each subunit of 200,000 molecular weight were not dependent on the presence of neighboring subunits in the oligomer, the size of the receptor should have been this value rather than the observed 6.5-8.0 x 10^5 daltons. It would appear, therefore, that binding of glucagon to each subunit of the oligomer leads to conformational changes in the receptor subunits that result in re-arrangement of the subunits, possibly by a concerted mechanism, into a new quaternary state exhibiting a very low dissociation rate constant for the hormone. Positive cooperativity of binding might be expected from these findings. In fact, we have found that binding of glucagon to hepatic membranes yields a Hill coefficient of 1.4-1.6, which is suggestive of a positive-cooperative process.

Figure 4 caption, continued:
the difference in the amount of ^{125}I-glucagon bound in samples containing or lacking GTP during incubation. Specific binding at zero dose was 1.8 pmol/mg protein for total binding and 1.46 pmol/mg for GTP-displaceable binding. The graph shows the ratios of the specific binding at a given dose (A_D) to the specific binding at zero dose (A_o) as a function of dose.

M. RODBELL AND W. SCHLEGEL

TABLE I

RANGE OF TARGET SIZES FOR ADENYLATE CYCLASE AND
THE GLUCAGON RECEPTOR IN HEPATIC PLASMA MEMBRANES

ENZYME			RECEPTOR
Ground State	Activated State		Tight Binding Site
	GppNHp	Glucagon+GTP	
$1.2-1.5 \times 10^6$	$2.5-3 \times 10^5$	$4-5 \times 10^5$	$6.5-8 \times 10^5$
$2.5-3 \times 10^5$			

The sizes of the enzyme and of the receptor were determined
from a series of studies carried out essentially as described in
the legends to Figures 3 and 4. These include studies with samples
irradiated when lyophylized (for the ground state of the enzyme),
frozen(-110°C)samples (always with the preactivated enzyme), and
either lyophylized or frozen samples for the glucagon receptor.
Target size was calculated using the empirical formula derived by
Kepner and Macey[8]. Since the inactivation curves obtained with the
ground state of the enzyme did not follow a single exponential decay
curve, calculations of the sizes of the ground state necessitated
certain assumptions. The critical assumption is that non-linearity
describes two fundamental sizes. This allowed for relatively simple
graphical analysis of the curve. From the shallower slope (smaller
size target) an estimate of size was made by extrapolating the line
to the abscissa. The steeper slope was then adjusted to account for
the contribution of the smaller size component. From the adjusted
slope, the size of the large component was calculated as above.
    A temperature correction factor was required for data obtained
from irradiations carried out at -110°. This was required since
irradiation inactivation studies carried out with several enzymes at
various temperatures showed that target theory, although valid at
any given temperature, is a function of temperature for all enzyme
studies. A correction factor of 2.8 was used in this study; it
was derived from a comparison of the irradiation inactivation of
5-mononucleotidase carried out at 25° in lyophylized hepatic plasma
membranes and irradiation analysis at -110° in lyophylized and
frozen samples.

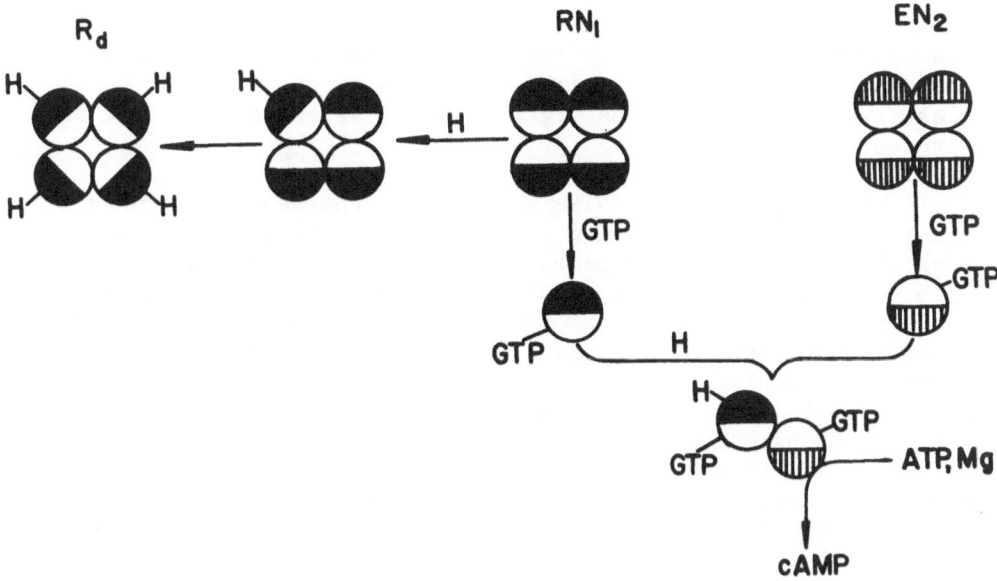

Figure 5. <u>Effects of GTP on the structure of the glucagon receptor
and of adenylate cyclase based on radiation analysis</u>. Circles
represent RN_1 and EN_2 units of receptor and adenylate cyclase,
respectively, described in Figure 1. A tetramer of both the
receptor and enzyme are assumed for the "inactive" oligomeric
structures. Coupling of RN_1 and EN_2 units requires dispersion of
the oligomers to monomers by the actions of GTP on the nucleotide
regulatory components (N_1 and N_2), resulting in the active form
of the enzyme and hydrolysis of GTP, as is depicted in Figure 2.
In the absence of GTP, binding of the hormone to the tetrameric
state of the receptor causes a conformational change in the
receptor subunits, resulting in concerted change of the tetramer
to a new state that displays a slow rate of dissociation of the
hormone from the receptor subunits. This is suggested to represent the desensitized state of the receptor. Binding of GTP to
this state converts it to monomers.

     Figure 5 summarizes the conclusions, presented in schematic
form, drawn from target size analysis. This scheme indicates
that the receptor and enzyme, along with their associated nucleotide regulatory components, are inactive in their oligomeric
states. Activation of the enzyme and coupling of the receptor
derives from an action of GTP at the nucleotide regulatory components which converts the oligomers to the monomeric forms of the
receptor and enzyme. As suggested from kinetic analysis of the
system, uncoupling of the receptor and catalytic components
results from hydrolysis of GTP at the nucleotide regulatory

components, the rate of hydrolysis determining the life-time of
the activated enzyme-receptor complex. In the absence of GTP,
binding of glucagon to the oligomeric state of the receptor
induces the formation of the tight-binding, desensitized state of
the receptor.

In conclusion, target size analysis of the hepatic adenylate
cyclase system has given a new perspective on the structure of
the receptor and enzyme components as well as a basis for the
complex actions of GTP on the enzyme system. Most importantly,
for the first time we need not rely solely on kinetic analysis of
the binding and action of hormones and guanine nucleotides to
decipher the complex actions of these agents. The fact that the
kinetic and structural studies happen to be self-consistent is not
proof for the hypothesis. However, the hypothesis has at least
survived a crucial test. We are currently investigating whether
other hormone-sensitive cyclase systems display similar-sized
structures and the changes in structure induced by GTP and hormone
observed with the glucagon-sensitive system.

As an addendum, a few comments should be noted on the
differences between the hypothesis put forth here and the "mobile
receptor" theory proposed by others[9,10] to explain hormone action
on adenylate cyclase systems. The major premises of the latter
theory are that hormone receptors diffuse freely in the plane of
the membrane and that "coupling" of receptors and enzyme occurs
only when the receptors are liganded by hormones. Accordingly,
the on-off rates of hormone action, according to the mobile
receptor hypothesis, are determined by the binding constants of the
hormones to the coupled receptor-enzyme complex. To some extent,
this theory and the hypothesis presented here bear resemblance.
The essential difference is that GTP, through the nucleotide
regulatory components, controls the coupling process, the activity
of adenylate cyclase, and the conformation or structure of the
receptor required for high affinity binding of the hormone. The
dynamic characteristics of hormone action are determined both by the
binding constants and the rate of breakdown of GTP by the system.

REFERENCES

1. Rodbell, M. (1978) in Molecular Biology and Pharmacology of
    Cyclic Nucleotides (G. Folco and R. Paoletti, Eds.) Elsevier/
    North-Holland Biomedical Press. pp. 1-12.
2. Maguire, M.E., Ross, E.M., and Gilman, A.G. (1977) Adv. Cyclic
    Nucleotide Res. 8, 1-83.
3. Lad, P.M., Welton, A.F., and Rodbell, M. (1977) J. Biol. Chem.
    252, 5942-5946.

4. Kempner, E. and Schlegel, W. (1979) Anal. Biochem. (submitted).
5. Houslay, M.D., Ellory, J.C., Smith, G.A., Hesketh, T.R., Stein, J.M., Warren, G.B., and Metcalfe, J.C. (1977), Biochim Biophys Acta 467, 208–219.
6. Salomon, Y., Londos, C., and Rodbell, M. (1974) Anal. Biochem. 58, 541–548.
7. Lin, M.C., Wright, D.E., Hruby, V.J., and Rodbell, M. (1975) Biochemistry 14, 1559–1563.
8. Kepner, G.R. and Macey, R.I. (1968) Biochim. Biophys. Acta 163, 188–203.
9. De Haen, C. (1976) J. Theor. Biol. 58, 383–400.
10. Cuatrecase, P., Hollenberg, M.D., Chang, K. J., and Bennett, V. (1975) Rec. Prog. Horm. Res. 37, 31–68.

INTERACTIONS BETWEEN HYPOTHALAMIC AND PERIPHERAL HORMONES AT THE ANTERIOR PITUITARY LEVEL

F. LABRIE, J. DROUIN, L. LAGACE, L. FERLAND,
M. BEAULIEU, V. RAYMOND and J. MASSICOTTE

Medical Research Council Group in Molecular
Endocrinology, Le Centre Hospitalier de
l'Université Laval, Quebec G1V 4G2, Canada

Although peripheral hormones have been shown for many years to play a major role in the control of adenohypophyseal activity in man and experimental animals, in vivo approaches could not dissociate between hypothalamic and pituitary sites of action. This area of research was much facilitated by the development of the pituitary cell culture system (Vale et al., 1972; Labrie et al., 1973). In fact, adenohypophyseal cells in primary culture have been extremely useful, not only for assessment of biological activity of analogs of TRH, LHRH and somatostatin (Labrie et al., 1973; Belanger et al., 1974; Labrie et al., 1976a, b) but also for determination of the characteristics of interaction between hypothalamic and peripheral hormones at the adenohypophyseal level (Drouin et al., 1976a, b; Drouin and Labrie, 1976a, b).

Knowing that LHRH stimulates the secretion of both LH and FSH (Borgeat et al., 1972; Labrie et al., 1973), the divergence frequently observed in vivo between the rates of secretion of the two gonadotropins can be best explained by differential effects of gonadal steroids at the pituitary level on the secretion of the two hormones. Emphasis will thus be given on the specific effects of androgens, estrogens and progesterone on basal and LHRH-induced secretion of LH and FSH in anterior pituitary cells in culture. Data describing the effects of "inhibin" of testicular and ovarian origin at the pituitary level on gonadotropin secretion will also be presented.

A combined in vivo and in vitro approach will then be used to study the site of action and the characteristics of interaction of estrogens and thyroid hormone in the control of TSH and PRL secretion. A close correlation will be found between changes of TSH and

PRL responsiveness to TRH and the level of receptors for the neurohormone.

The predominant influence of the hypothalamus on prolactin secretion is inhibitory. Rapidly accumulating evidence suggests that
dopamine (DA) may be the main or even the only inhibitory substance
involved. It thus appeared important to study in detail specificity of the control of prolactin secretion and properties of the adenohypophyseal dopaminergic receptor. Since estrogens are known to
be potent stimulators of prolactin secretion, the interaction of
estrogens with dopaminergic action was then studied at the pituitary level both in vitro and in vivo. Estrogens were found to act
directly at the pituitary level and, more surprisingly, to have
potent antidopaminergic activity on prolactin secretion.

INTERACTIONS BETWEEN LHRH, SEX STEROIDS AND "INHIBIN" IN THE CONTROL OF LH AND FSH SECRETION

Although the influence of the hypothalamus on the secretion of
both gonadotropins is probably exerted exclusively through LHRH, it
is well recognized that gonadal steroids can have a marked influence on LH and FSH secretion.

The recent observation that LHRH can potentiate the LH response to subsequent injection of the neurohormone (Aiyer et al.,
1973; Castro-Vasquez and McCann, 1975; Ferland et al., 1976) illustrates that it is almost impossible to dissociate hypothalamic
and pituitary sites of steroid action under in vivo conditions.
In fact, a stimulatory effect of gonadal steroids on LHRH secretion should lead to an increased LH responsiveness to the neurohormone (in the absence of any direct effect of the steroid at
the pituitary level). The inverse situation should follow the
inhibitory effect of a steroid on LHRH secretion.

This chapter will attempt to summarize pertinent data which
indicate specific and selective effects of androgens, estrogens
and progestins on LH and FSH secretion by a direct action at the
adenohypophyseal level. In support of the proposal first made
by McCullagh (1932), much recent evidence suggests that, in addition to sex steroids, a non-steroidal substance (called inhibin) of testicular origin is involved in specific inhibition of
FSH secretion (Franchimont, 1972; Setchell and Jacks, 1974;
Baker et al., 1976). In order to study the effect of this inhibin-like activity at the adenohypophyseal level, we took advantage of the precision of the pituitary cell system to examine
the effect of porcine follicular fluid or Sertoli cell culture medium on basal as well as on LHRH-induced LH and FSH secretion.

Stimulatory Effects of Estrogens on LH and FSH
Secretion at the Pituitary Level

In vivo data indicate that estrogens can have positive effects
on LH and FSH secretion. These data pertain to the induction of
the afternoon LH surge after administration of estradiol benzoate to
ovariectomized animals (Baldwin et al., 1974), the induction of
early ovulatory LH surges by injection of estrogens at early stages
of the estrous cycle (Everett, 1948) and the abolition of the ovulatory surge of LH at proestrus by administration of an estrogen antiserum (Jewelewicz et al., 1974). These data do not however differentiate between hypothalamic and pituitary sites of positive estrogen action.

As shown in Fig. 1A, preincubation of male rat anterior pituitary cells for 40h in medium containing 1 x 10^{-8}M E_2 increased
the LH responsiveness to LHRH. The LHRH concentration required to
produce a half-maximal stimulation (ED_{50}) of LH release is decreas-

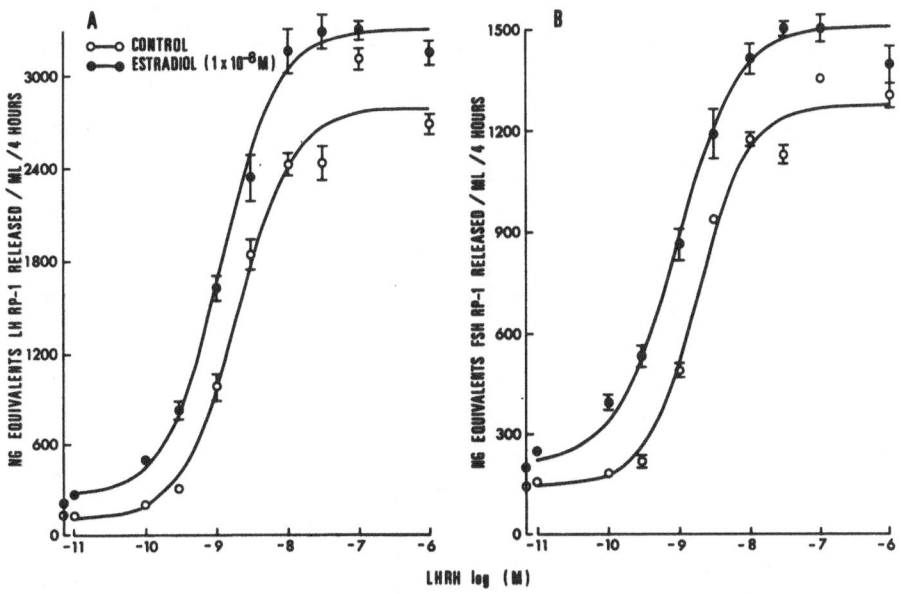

Fig. 1. Effect of increasing concentrations of LHRH on LH (A) and
FSH (B) release by anterior pituitary cells in primary culture preincubated for 40h in the presence of 1 x 10^{-8}M 17β-estradiol (●) or
control medium (o). Anterior pituitary cells were obtained from
adult male rats. The response to LHRH was performed during a 4-h
period after preincubation in the presence or absence of the estrogen. The results are presented as mean ± S.E.M. of data obtained
from triplicate dishes.

ed by E_2 pretreatment from $2.3 \pm 0.03 \times 10^{-9}$ to $1.2 \pm 0.01 \times 10^{-9}$M
($p < 0.01$). It can also be seen in Fig. 1A that preincubation
with E_2 increased the basal LH release from 120 ± 8 ng LH-RP-1 per
ml per 4 h to 205 ± 10 ng per ml per 4 h ($p < 0.01$). Moreover, in
this and similar experiments performed with adenohypophyseal cells
obtained from male and female animals (Drouin et al., 1976a), the
maximal LH response to LHRH is slightly, but not significantly increased. E_2 pretreatment increased both the basal FSH release and
the maximal response of the hormone to LHRH (Fig. 1B). Similar
effects have been previously obtained in female rat anterior pituitary cells (Drouin et al., 1976a).

This stimulatory effect of E_2 at the adenohypophyseal level may
well be, at least partly, responsible for the increased LH and FSH
sensitivity to LHRH observed at proestrus in the rat (Gordon and
Reichlin, 1974; Ferland et al., 1975) and during the preovulatory
period in the human (Nillius and Wide, 1972; Yen et al., 1972).

Opposite Effects of Androgens on LH and FSH Se-
cretion at the Pituitary Level

As illustrated in Fig. 2, pretreatment with 10^{-8}M testosterone

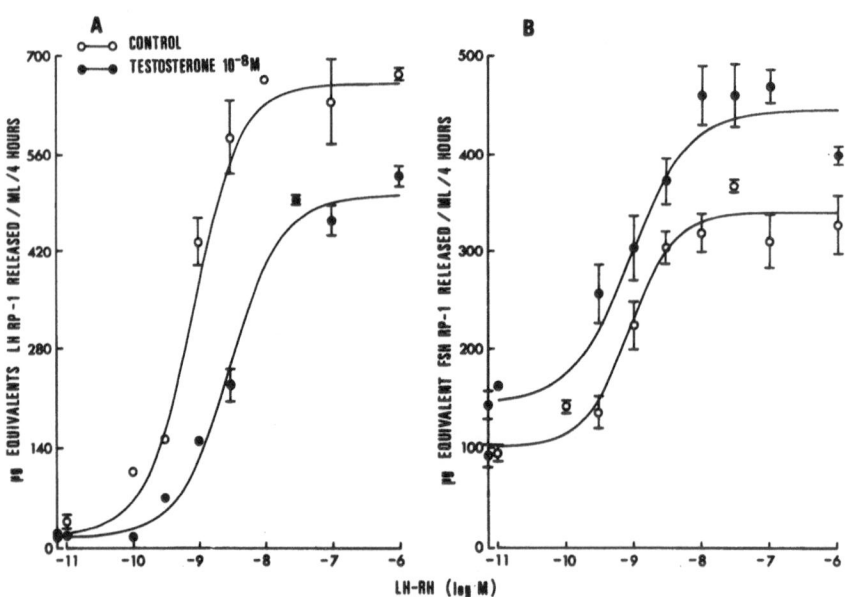

Fig. 2. Effect of increasing concentrations of LHRH on LH (A) and
FSH (B) release by anterior pituitary cells in primary culture
preincubated for about 40h in the presence (o) or absence (o) of
1×10^{-8}M testosterone. Results are expressed as means ± S.E.M.
of triplicate determinations.

led to a marked inhibition of the LH responsiveness to LHRH, the LHRH ED_{50} being increased from 1×10^{-9} to $3 \times 10^{-9}M$ in the presence of the androgen ($p < 0.01$). It can also be seen that the androgen did not affect basal LH release but slightly decreased the maximal response to the neurohormone. In contrast with the LH data, it can be seen that in the same experiment, testosterone did not significantly affect the LHRH ED_{50} ($1 \times 10^{-9}M$) for FSH release. Both the spontaneous and maximal release of FSH were, however, slightly (30-40%) but consistently increased after androgen pretreatment ($p < 0.01$).

These data clearly show that androgens have not only specific but also opposite effects at the pituitary level on LH and FSH secretion. In fact, pretreatment of pituitary cells with androgens can markedly inhibit the LH response to LHRH while the effect on FSH is stimulatory. Qualitatively similar results have been obtained when anterior pituitary cells obtained from male or female rats were used. These findings can offer an explanation for the observations in rat (Swerdloff et al., 1972) and man (Swerdloff and Odell, 1968) of a greater sensitivity of LH than FSH release to the inhibitory action of androgen administration in vivo.

Biphasic Effects of Progesterone on Gonadotropin Secretion at the Pituitary Level

As illustrated in Fig. 3, progesterone alone had no effect on the plasma LH response to LHRH. However, when progesterone was added 48h after the addition of E_2, a completely different response was observed: potentiation of the stimulatory effect of E_2 at short time intervals (4-8h) with progressive inhibition of the LH response to control levels at later time intervals. The effect of progesterone on the FSH response to LHRH was exclusively stimulatory both in the presence or absence of preincubation with E_2. This effect of progesterone on FSH responsiveness was more rapid in the presence of E_2, a maximal effect being reached at 6h while 24h were required in the absence of preincubation with the estrogen.

Since the stimulatory effects of progesterone on LH release in pituitary cells in vitro are small compared to the peak of LH secretion observed in vivo when the steroid is given 2 to 4 days after estrogen priming in ovariectomized rats (Kalra et al., 1973), it is likely that the acute in vivo effect of progesterone is exerted at the level of the central nervous system (Legan and Karsch, 1975). This site of stimulatory action of progesterone is supported by the finding that the progesterone-induced LH surge always occurs in the afternoon irrespective of the time of injection of the steroid (Caligaris et al., 1971). The depressed plasma LH levels found 24h later may well be due to the direct inhibitory effect of progesterone at the pituitary level.

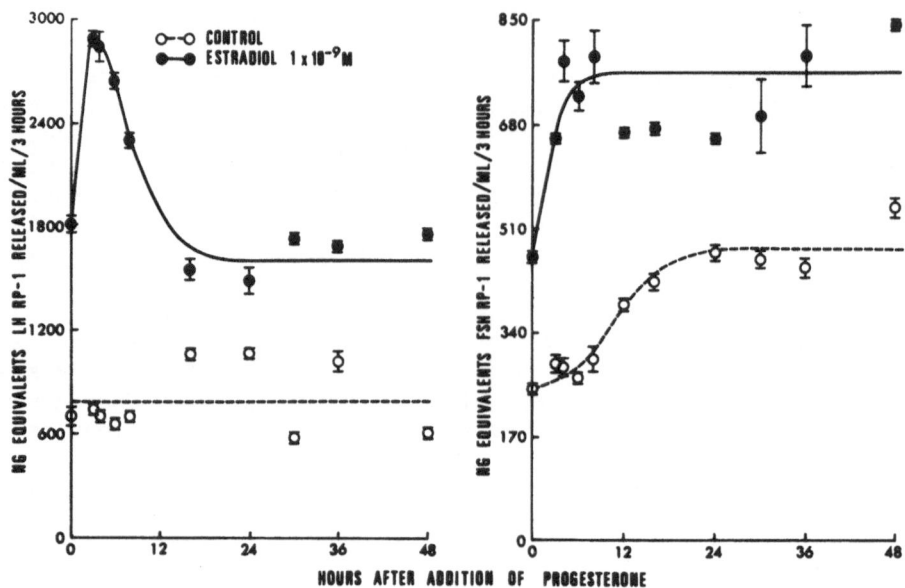

Fig. 3. Time-course of the effect of progesterone on the LH (A) and FSH (B) responses to 0.1 nM LHRH in female rat anterior pituitary cells preincubated for 48 h in the presence (●--●) or absence (o--o) of 1 nM 17β-estradiol. LHRH was present during a 3-h incubation period.

It should be mentioned that while the effects of estrogens and androgens have been studied in detail using anterior pituitary cells obtained from male and female rats and qualitatively similar effects of the two classes of sex steroids were found, the effects of progesterone reported in this review are those obtained using adenohypophyseal cells obtained from female animals only.

Action of "Inhibin" on LH and FSH Secretion at the Pituitary Level

As mentioned earlier, our findings that estrogens and androgens are exclusively stimulatory on FSH secretion provide some support to the concept first proposed by McCullagh (1932) than an inhibitory substance of testicular origin could be involved in the specific control of FSH secretion. As illustrated in Fig. 4, incubation of female rat anterior pituitary cells for 72h in the presence of increasing concentrations of Sertoli cell culture medium (days 5-8 in culture) led to a maximal 40% inhibition of spontaneous FSH release at the highest dilution used (1/8). When Sertoli cell culture medium obtained between days 2 and 5 of culture was used, the inhibitory activity was less potent than with medium obtained between

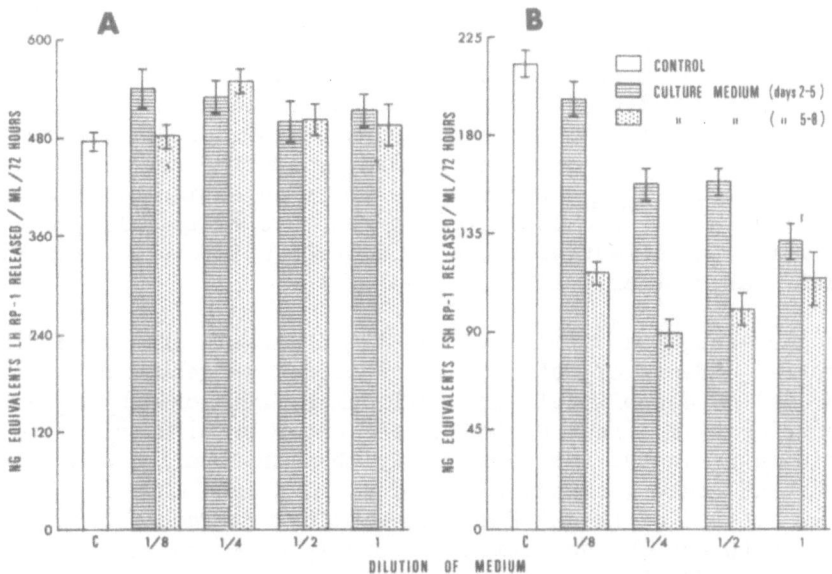

Fig. 4. Effect of increasing concentrations of Sertoli cell culture medium (days 2-5 or 5-8 in culture, 35-day old animals, 1.5 mg protein/7.5 ml culture medium) on basal LH (A) and FSH (B) release during a 72-h incubation with female rat anterior pituitary cells in culture.

days 5 and 8. Specificity of the action of the Sertoli cell culture medium is indicated by the complete absence of its effect on spontaneous LH release (Fig. 4A).

Preincubation of anterior pituitary cells in culture for 72h with increasing concentrations of Sertoli cell culture medium led to a progressive inhibition of LHRH-induced FSH and LH release (data not shown). The inhibitory effect was however more important on FSH than LH release. For example, at the 1/16 dilution of Sertoli cell culture medium, LH and FSH release was inhibited by 25 and 55%, respectively. With undiluted Sertoli cell culture medium, the LH and FSH responses were inhibited by 80 and 95%, respectively. Almost superimposable data were obtained when anterior pituitary cells were incubated with porcine follicular fluid (Lagacé et al., 1978), thus suggesting similar properties of the inhibitory material of testicular and ovarian origin.

The present findings indicate that the in vivo data showing a preferential inhibitory effect of an inhibin-like material on FSH secretion may well be explained by a direct pituitary site of ac-

tion. This inhibitory effect of follicular fluid on basal FSH release is highly specific, no effect on basal LH release being detected at any dose used. Although the inhibitory effect of this
material on LHRH-induced FSH release is observed at lower concentrations than on LH release, it is clear that its inhibitory effect
can be exerted on the secretion of both gonadotropins. This is
analogous to some in vivo experiments where injection of bull or
human seminal plasma led to decreased response of both LH and FSH
to LHRH while the effect on basal gonadotropin levels was restricted to FSH (Baker et al., 1976).

     The present data summarized in Fig. 5 clearly show specific and
differential effects of estrogens, androgens, progesterone and (a)
substance(s) of Sertoli cell and follicular fluid origin on LH and
FSH secretion by a direct action at the anterior pituitary level.
These findings suggest the possibility that the ovarian and testicular "inhibin" could interact with sex steroids and LHRH in the differential control of LH and FSH secretion and explain the changes
of ratio of LH and FSH secretion frequently observed in man (Grumbach et al., 1974; Franchimont et al., 1975) and experimental animals (Ferland et al., 1976).

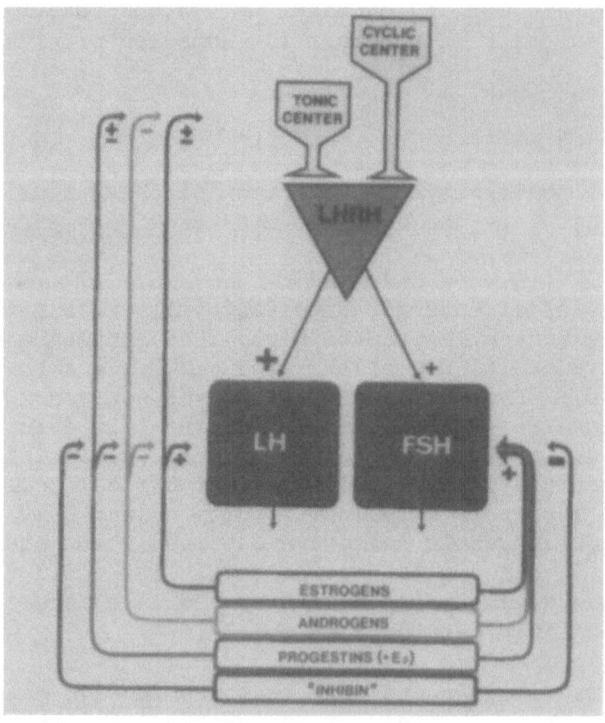

Fig. 5. Schematic representation of the pituitary and/or hypothalamic
sites of sex steroid and "inhibin" action on gonadotropin secretion.

INTERACTIONS BETWEEN ESTROGENS AND THYROID HORMONE
IN THE CONTROL OF TSH SECRETION

Recent studies indicate that the tissue concentration of some
peptide and steroid receptors is not static and that it can be influenced by a variety of hormonal agents. For example, estrogens
can increase prolactin receptor concentrations in liver (Posner et
al., 1975) and induce progesterone receptor levels in uterus (Toft
and O'Malley, 1970).

Since estradiol is well known to increase prolactin secretion
in man (Yen et al., 1974) and rat (Chen and Meites, 1970) and a stimulatory effect of this steroid has also been observed on TSH secretion in the rat (Labrie et al., 1970), it was of interest to study
the possible effect of estrogens and thyroid hormones on the level
of pituitary TRH receptors and to correlate such modifications of
the concentration of TRH receptors with changes in the TSH and
prolactin responses to the neurohormone.

Daily administration of 25 µg of estradiol benzoate to adult
female rats for 9 days led to an approximately 3-fold increase in
the number of pituitary TRH binding sites (De Léan et al., 1977a).
Estrogen treatment was, however, without significant effect on the
TRH-receptor dissociation constant (3.6 ± 0.4 vs 4.1 ± 10^{-8}M). A
significant increase of the level of pituitary TRH binding was observed between 2 and 4 days after the first injection of estradiol
benzoate, with a plateau being reached at 7 days (De Léan et al.,
1977a). Plasma PRL followed a similar pattern although its concentration continued to rise between days 7 and 10. Estrogen administration had no significant effect on basal plasma TSH levels (data
not shown).

Since thyroid hormones are potent inhibitors of the TSH response
to TRH, we next studied the effect of thyroid hormone on TRH binding
in anterior pituitary tissue. The injection of a single large dose
of L-thyroxine (60 µg) to hypothyroid animals led to a progressive
decrease of the number of TRH binding sites to approximately 50% of
control after 24 h, while a further decrease to 40% of control was
measured at 48h (De Léan et al., 1977a).

It was then of interest to study a possible correlation between
changes of adenohypophyseal TRH receptor levels and TSH and prolactin responsiveness to the neurohormone. Clear evidence of the antagonism between estrogens and thyroid hormone is illustrated in Fig.
6. In this experiment, we have studied the effect of treatment for
seven days with increasing doses of L-thyroxine alone or in combination with estradiol benzoate. The subphysiological 0.5 µg dose of
L-thyroxine led to an approximately 100% increase of the TSH response to TRH calculated from the areas under the plasma response curves
after substraction of the basal levels at time zero. Treatment with

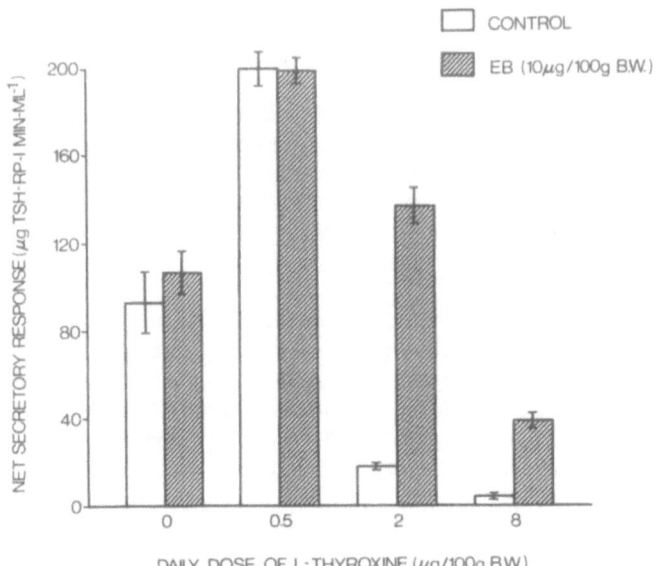

Fig. 6. Effect of treatment with increasing doses of L-thyroxine in
the presence or absence of a fixed dose of estradiol-3-benzoate on
the net TSH secretory responses to TRH in propylthiouracil-hypothy-
roid male rats. Results represent the net areas under the plasma
TSH response curves.

a daily 2 μg dose of L-thyroxine inhibited the response to 20% of
control while a supranormal (8 μg/100 g) dose of the thyroid hormone
blunted the response. Treatment with estradiol benzoate (10 μg/100g)
did not affect the TSH response in control animals or in those treated with a subnormal dose of L-thyroxine (0.5 μg/100 g). However,
the estrogen did elicit a reversal of the inhibitory effects of higher doses of thyroid hormone. As judged by the areas under the curves (Fig. 6),estradiol completely restored the secretory response
inhibited by a physiological dose of L-thyroxine (2 μg/100 g) while
the blunting effect of a higher dose of thyroxine was reversed to 40%
of the control (p < 0.01).

     The present data clearly show that pituitary TRH receptor levels
can be modulated by the stimulatory effect of estrogens and the inhibitory action of thyroid hormone and that changes of TRH receptor
levels are generally accompanied by parallel modifications of the
TSH and PRL responses to the neurohormone. Such a close correlation
observed between changes of the number of TRH binding sites and TSH
and PRL responses to the neurohormone indicate that the availability of TRH receptors could be a rate-limiting step in the action of
TSH in both thyrotrophs and mammotrophs. That at least part of
this effect of estrogens and thyroid hormone is exerted at the pi-

tuitary level has been shown by experiments performed in pituitary
cells in culture (Drouin and Labrie, unpublished data).

The present evidence of a modulation of TRH receptors by estrogens and thyroid hormone may provide some explanation for the clinical observations of the inhibition of both TSH and PRL secretion by
thyroxine in hypothyroid subjects (Snyder et al., 1974), higher TSH
response to TRH in women than in men (Bowers et al., 1971) and reports that oral contraceptives can increase the TSH response to TRH
(Ramey et al., 1975). Moreover, these data clearly show that estrogens and thyroid hormone have opposite effects at the pituitary level on the TSH responsiveness to TRH. The important reversal of
the complete inhibition by L-thyroxine of the TSH response to TRH
by daily injection of 1 µg of estradiol (De Léan et al., 1977a)
indicates that the effect of the estrogen is likely to be of physiological importance. This is also supported by our recent findings
of increased TSH responsiveness to TRH during the afternoon of
proestrus and morning of estrus in four-day cycling rats (De Léan
et al., 1977b).

It should be mentioned that we have recently observed that estrogen treatment in vivo leads also to an increased sensitivity of
the TSH response to TRH with no alteration in the pituitary TSH content. The present findings of parallel changes in pituitary TRH receptor levels and TSH responsiveness to TRH suggest that the number
of TRH receptors are an important site for control of activity of
the TSH-secreting cells (Fig. 7). A similar antagonism between estrogens and thyroid hormones has also been found on prolactin secretion. The stimulatory effect of estrogens is, however, predominant
on prolactin secretion while the inhibitory effect of thyroid hormone is more important on TSH secretion. Although the sensitivity appears somewhat different, both prolactin- and TSH-secreting cells
possess qualitatively similar mechanisms responsive to estrogens
and thyroid hormones.

INTERACTIONS BETWEEN ESTROGENS AND DOPAMINE IN THE
CONTROL OF PROLACTIN SECRETION

Secretion of prolactin from the anterior pituitary gland is
controlled by stimulatory and inhibitory influences of hypothalamic
origin. The predominant influence of the hypothalamus on prolactin
secretion is however inhibitory (Meites et al., 1963). Rapidly accumulating evidence suggests that dopamine (DA) may be the main or
even the only inhibitory substance involved. In fact, destruction
of the dopaminergic cell bodies and terminals resulted in a marked
elevation of plasma prolactin levels (Bishop et al., 1972). Moreover, the prolactin release-inhibiting activity contained in purified hypothalamic extracts could be accounted for by their catecholamine content (Takahara et al., 1974) and preincubation of hypotha-

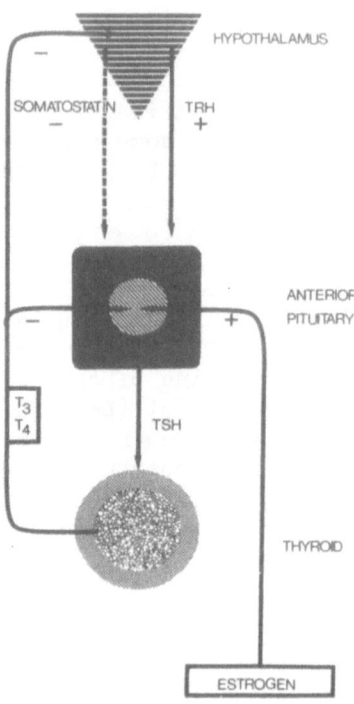

Fig. 7. Schematic representation of the antagonistic effects of 17β-estradiol and thyroid hormone at the anterior pituitary level in the control of TSH secretion and at the hypothalamic level (probably on TRH secretion).

lamic extracts with aluminum oxyde or monoamine oxidase led to a complete loss of prolactin release-inhibiting activity (Shaar and Clemens, 1974). In support of this physiological role of DA in the control of prolactin secretion, DA has recently been measured in portal blood (Ben-Jonathan et al., 1977).

It thus appeared important to study in detail specificity of the control of prolactin secretion and the properties of the adenohypophyseal dopaminergic receptor. Since estrogens are known to be potent stimulators of prolactin secretion, the interaction of estrogens with dopaminergic action was then studied at the pituitary level both in vitro and in vivo.

Specificity of Prolactin Release

Much information about the factors controlling prolactin release could be obtained using rat hemipituitaries (Birge et al., 1970;

MacLeod, 1969; MacLeod and Lehmeyer, 1974; Pelletier et al., 1972; Shaar and Clemens, 1974). However, since adenohypophyseal cells in primary culture proved to be a more precise system to study control of the secretion of many pituitary hormones (Drouin and Labrie, 1976b; Drouin et al., 1976b; Labrie et al., 1973), we have used this model in the present report to investigate in detail the specificity of action of a large series of DA agonists and antagonists on prolactin release. Fig. 8 shows the inhibitory effect of increasing concentrations of various catecholamines and analogs on prolactin release during a 2-hour incubation of rat anterior pituitary cells in primary culture. As measured by the concentrations giving a 50% inhibition of hormone release, the following order of potency was obtained: DA (35 nM) > epinephrine (420 nM) ≥ norepinephrine (540 nm) while the α and β agonists phenylephrine and isoproterenol were without effect up to 10 μM. Apomorphine, the prototype of DA agonists (Anden et al., 1967), was approximately 10 times more potent than DA as inhibitor of prolactin release at an ED_{50} value of 3 nM.

At high concentrations, apomorphine, DA, epinephrine and norepinephrine led to a maximal 85-90% inhibition of basal prolactin release. Serotonin, a neurotransmitter potentially involved in the

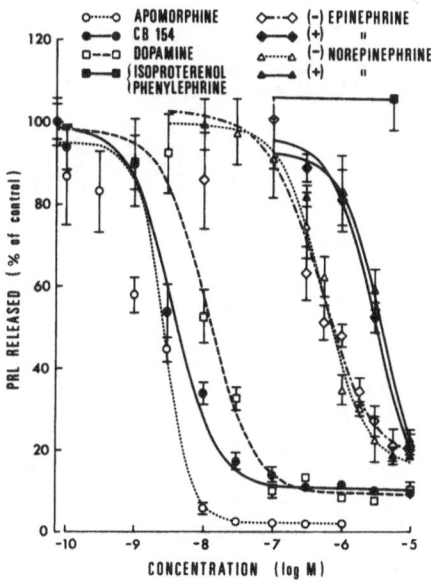

Fig. 8. Effect of increasing concentrations of various catecholamines and analogs on the release of prolactin in anterior pituitary cells in primary culture. Four days after plating, cells were incubated for 3 hours in the presence of the indicated drug concentrations. Data are expressed as means ± S.E.M. of duplicate determinations of triplicate dishes.

control of prolactin secretion in vivo (Kordon et al., 1973; Subramanian and Gala, 1976) was a very weak inhibitor of prolactin release
in cells in culture.   The inhibition of prolactin release by catecholamines showed the expected stereoselectivity since (-) epinephrine and (-)norepinephrine affected prolactin release with a potency
approximately 8 times greater than the corresponding (+) enantiomers.

    In order to further study the specificity of the dopaminergic
control of prolactin release, we took advantage of the precision of
the in vitro system to examine the ability of various dopaminergic
antagonists and their enantiomers to reverse the inhibition of prolactin release induced by DHE (3 nM) or DA (50 nM).   3 nM DHE decreased basal prolactin release by more than 75% (Fig. 9A, B, C).
This inhibitory action of 3 nM DHE could be progressively and completely reversed by the addition of increasing concentrations of
the potent neuroleptic (+) butaclamol (K_D = 0.3 nM) whereas its
clinically inactive enantiomer, (-) butaclamol, had only slight
activity at 10 μM (Fig. 9A).

    In agreement with their relative dopaminergic-blocking activi-
ties in pharmacological tests, α-flupenthixol was approximately 100-

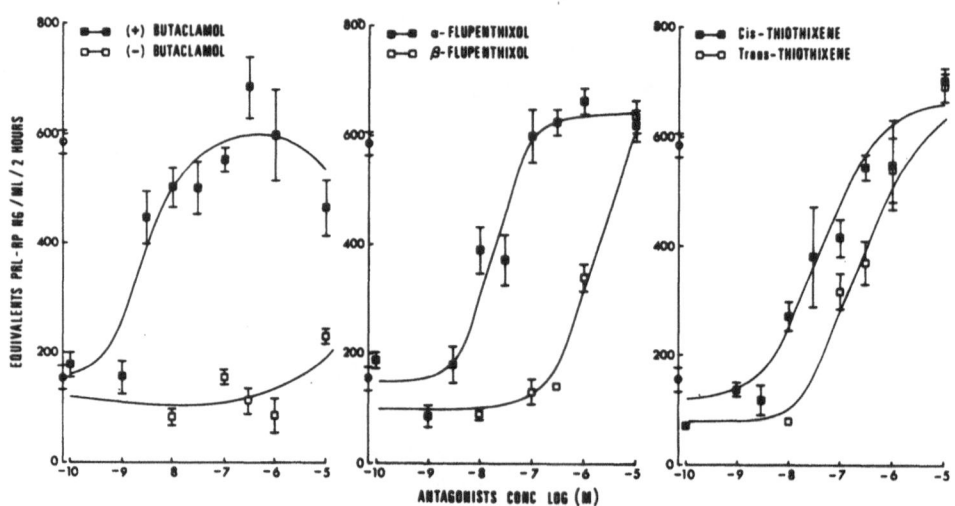

Fig. 9. Ability of various dopaminergic antagonists and their enantiomers to reverse the inhibition of prolactin release by 3 nM dihydroergocornine (DHE) in anterior pituitary cells in primary cul-
culture. (o), basal PRL secretion; (o), PRL secretion in the presence of 3 nM dihydroergocornine. Increasing concentrations of
the indicated compounds were all tested in the presence of 3 nM
dihydroergocornine during a 2-hour incubation.

fold more potent than β-flupenthixol to reverse the DHE-induced inhibition of PRL release (Fig. 9B) whereas cis-thiothixene was more
active than trans-thiothixene by about an order of magnitude (Fig.
9C). Almost identical results were obtained when 50 nM DA was used
to inhibit PRL release (data not shown).

Specificity of the Dopamine Receptor

Binding of [^{3}H]dihydroergocryptine to bovine anterior pituitary membranes displayed a specificity typical of a dopaminergic process (Caron et al., 1978). Agonists competed for [^{3}H]dihydroergo-
cryptine binding with the following order of potency: apomorphine >
dopamine > epinephrine ≥ norepinephrine >> isoproterenol = clonidine.
This relative order of potency closely resembles the potency of these agonists in inhibiting prolactin secretion from rat anterior pituitary. Ergot alkaloids which act as potent dopaminergic agonists
on prolactin secretion are also potent competitors of [^{3}H]dihydro-
ergocryptine binding. These ergot compounds did in fact compete
for binding with a potency which closely parallels their inhibitory
effect on prolactin release in anterior pituitary cells in culture.

The present data indicate that the anterior pituitary gland,
beside its own intrinsic interest, should represent a useful model
for detailed study of the mechanisms of dopaminergic action. In
fact, changes of [^{3}H]dihydroergocryptine binding to the dopaminergic receptor can be correlated with an easily accessible and highly
precise parameter of biological activity, prolactin release in cells
in culture. Such a model of dopamine action has not been previously
available and should be useful for a better understanding of the
mechanisms controlling dopamine receptor-mediated actions.

Antidopaminergic Action of Estradiol at the Pituitary Level

It is well known that estrogens are potent stimulators of prolactin secretion in both man (Frantz et al., 1972) and rat (Chen
and Meites, 1970; De Léan et al., 1977a). These in vivo effects of
estrogens could however be exerted at the hypothalamic and/or pituitary level(s). Since anterior pituitary cells in culture proved to
be an excellent system to study the specificity of action of sex
steroids at the anterior pituitary level (Drouin et al., 1976a, b),
we have used this system, instead of intact pituitaries, to study
the interactions of 17β-estradiol and dopamine on prolactin release.

In agreement with previous data obtained with the dopamine agonist dihydroergocornine (Raymond et al., 1978), it can be seen in
Fig. 10 that the inhibitory effect of CB-154 or both basal (A) and

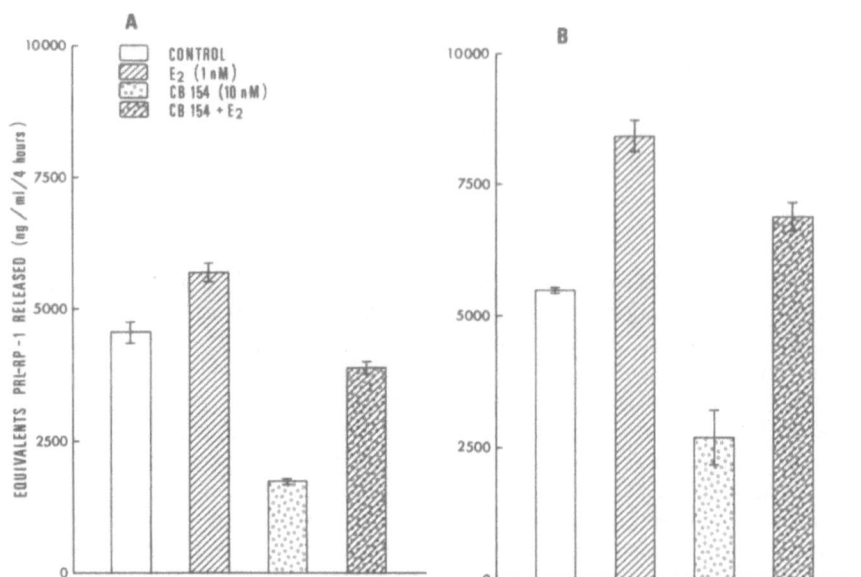

Fig. 10. Effect of 1 nM 17β-estradiol, 10 nM 7-bromoergocryptine (CB-154), a combination of both treatments or the vehicle alone on basal (A) and 10 nM TRH-induced (B) prolactin release in anterior pitui-pituitary cells in culture. Cells were preincubated for 72h in the pre-presence or absence of estradiol before a 4-h incubation in the presence or absence of CB-154 and TRH. Note that estradiol led to an almost complete reversal of the inhibitory effect of CB-154 on prolactin release.

TRH-induced (B) prolactin release, was almost completely reversed by estradiol.

Following our in vitro data showing a potent antidopaminergic effect of estrogens on prolactin secretion, it then became of interest to investigate if such a potent activity of estrogens also occurs under in vivo conditions. The present study was facilitated by our recent findings that the endogenous inhibitory dopaminergic influence on prolactin secretion can be eliminated by administration of opiates, thus making possible study of the effect of exogenous dopaminergic agents without interference by endogenous dopamine.

As illustrated in Fig. 11A, the subcutaneous administration of 100 or 400 ug of dopamine completely prevented the increased plasma prolactin levels following morphine injection in rats ovariectomized two weeks previously. Treatment with estradiol benzoate (20 µg/day) for seven days led to a stimulation of basal plasma prolactin levels (from 14 ± 1 to 56 ± 8 ng/ml) and to a marked increase of the maxi-

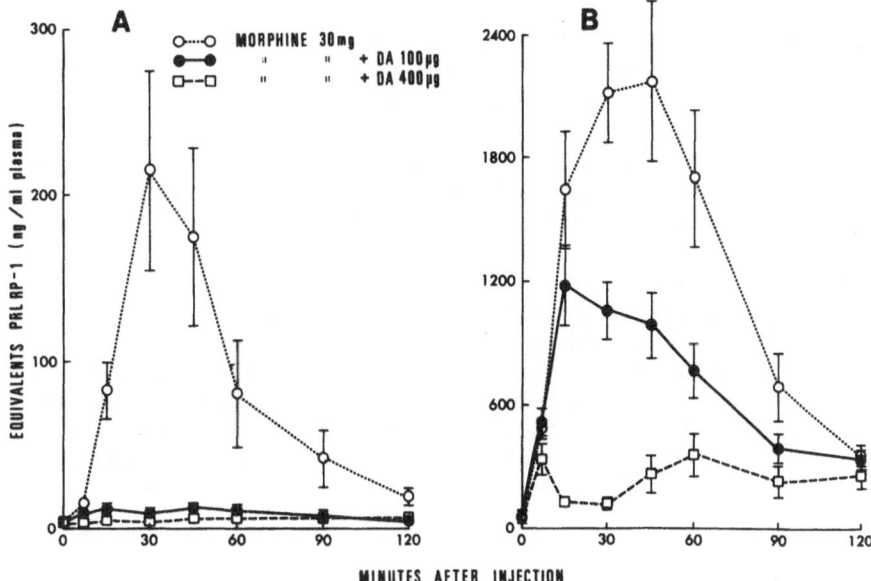

Fig. 11. Effect of estrogen treatment on the inhibitory effect of
dopamine on prolactin release in the female rat. Adult Sprague-
Dawley rats ovariectomized two weeks previously were injected subcutaneously (s.c.) with estradiol benzoate (10 ug, twice a day) for
seven days or the vehicle alone (0.2 ml 0.1% gelatin in 0.9% NaCl)
before insertion of a catheter into the right superior vena cava
under anesthesia (Surital, 50 mg/kg, B.W., i.p.). Two days later,
undisturbed freely-moving animals were injected s.c. with morphine
sulfate (30 mg) alone or in combination with dopamine (100 or 400μg).
Blood samples (0.7 ml) were then withdrawn at the indicated time
intervals for measurement of plasma prolactin concentration. Data
are expressed as mean ± S.E.M. of duplicate determinations of samples obtained from 8-10 animals per group.

mal plasma prolactin response to morphine from 215 ± 60 to 2175 ±
390 ng/ml. The most interesting finding is however that the 100
and 400 μg doses of dopamine which could maintain plasma prolactin
levels at undetectable levels after morphine injection in control
rats led only to 40 and 85% inhibition of prolactin levels, respectively, in animals treated with estrogens.

     These studies clearly demonstrate that estrogens have potent
antidopaminergic activity on prolactin secretion, not only in anterior pituitary cells in culture (Raymond et al., 1978),but also in
vivo, the effect being qualitatively similar in both female and male
animals. As reflected by an increase of the ED_{50} value of dopamine
agonists, the in vitro effect of estrogens was due to a decreased

sensitivity of prolactin release to dopamine action at the anterior
pituitary level. Such findings indicate that higher concentrations
of dopamine in the hypothalamo-hypophyseal portal blood system are
likely to be required to inhibit prolactin secretion under conditions
of high estrogenic influence. The almost complete reversal of the
inhibitory effect of low doses of dopamine by estrogen treatment
clearly indicates an important interaction between estrogens and dopamine at the adenohypophyseal level (Fig. 12).

As mentioned earlier, dopamine appears to be the main factor
of hypothalamic origin involved in the control of prolactin secretion (Labrie et al., 1978; MacLeod and Lehmeyer, 1974). Moreover,
using the dopamine agonist [^{3}H]dihydroergocryptine as tracer and
measurement of prolactin release in pituitary cells in culture, we
have found that the characteristics of binding to the pituitary receptor and the control of prolactin secretion are typically dopaminergic.

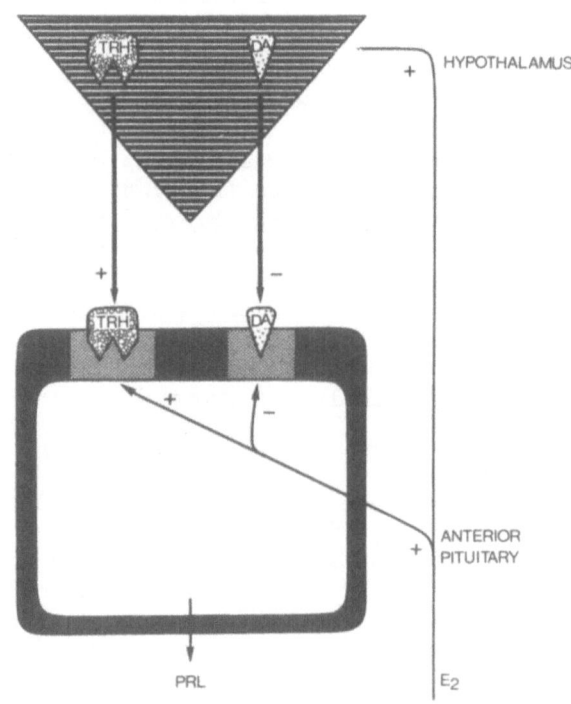

Fig. 12. Schematic representation of the interaction between estrogens, dopamine and TRH in the control of prolactin secretion at the
anterior pituitary level. The stimulatory effect of estrogens on
prolactin secretion appears to be exerted mainly through blockade
of the activity of the dopaminergic receptor.

The role of dopamine in various brain functions has become increasingly evident during the recent years (Anden, 1972; Beiger et al., 1972). Moreover, malfunction of dopaminergic systems can lead to neurological and psychiatric diseases (Snyder et al., 1974; Stevens, 1973). It is thus hoped that knowledge gained about the pituitary dopaminergic system where the effects (changes of prolactin secretion) can be measured with a high degree of precision can help our understanding of less accessible dopaminergic systems in the central nervous system. It is of great interest that treatment with estrogens has recently been found to inhibit the action of dopaminergic agents on circling behavior in rats bearing a unilateral lesion of the nigrostriatal dopaminergic pathways (Bedard et al., 1978) and on striatal acetylcholine accumulation in the rat (Euvrard et al., 1978). Moreover, symptoms of tardive dyskinesia following chronic treatment with neuroleptics have been found to be improved by estrogen treatment in the human (Stevens, 1973).

REFERENCES

(1) Aiyer, MS, Chiappa, SA, Fink, G., and Greig, F. J. Physiol. (London) 234, 81P-82P, 1973.

(2) Anden, NE, Rubenson, A., Fuxe, L., and Hokfelt, T. J. Pharm. Pharmacol. 19, 627-629, 1967.

(3) Anden, N.E. J. Pharmacol. 24, 905-906, 1972.

(4) Baker, H.W.G., Bremmer, W.J., Burger, H.G., De Kretser, D.M., Dulmantis, A., Eddie, L.W., Hudson, B., Keog, E.S., Lee, V.W.K., and Rennie, G.C. Rec. Progr. Horm. Res. 32, 429-472, 1976.

(5) Baldwin, D.M., Ramirez, V.D., and Sawyer, C.H. Fed. Proc. 33, 212, 1974.

(6) Bedard, P., Dankova, J., Boucher, R., and Langelier, P. Can.J. Physiol. Pharmacol. (in press).

(7) Beiger, D., Larochelle, L., and Hormykiewicz, O. Eur. J. Pharmacol. 18, 128-136, 1972.

(8) Belanger, A., Labrie, F., Borgeat, P., Savary, M., Côté, J., Drouin, J. Mol. Cell. Endocrinol. 1, 329-339, 1974.

(9) Ben-Jonathan, N., Oliver, C., Weiner, H.J., Mical, R.S., and Porter, J.C. Endocrinology 100: 452-458, 1977.

(10) Birge, C.A., Jacobs, L.S., Hammer, C.T., and Daughaday, W.H. Endocrinology 86: 120-130, 1970.

(11) Bishop, W., Fawcett, C.P., Krulich, L., and McCann, S.M.
Endocrinology 91, 643-656, 1972.

(12) Borgeat, P., Chavancy, G., Dupont, A., Labrie, F., Arimura,
A., and Schally, A.V. Proc. Natl. Acad. Sci. USA 69, 2677-
2681, 1972.

(13) Bowers, C.Y., Friesen, H.G., Hwang, P., Guyda, H.J., and
Folkers, K. Biochem. Biophys. Res. Commun. 45, 1033-1041,
1971.

(14) Caligaris, L., Astrada, J.J., and Taleisnik, S. Endocrinology
89, 331-337, 1971.

(15) Caron, M.G., Beaulieu, M., Raymond, V., Gagné, B., Drouin, J.,
Lefkowitz, R.J., and Labrie, F. J. Biol. Chem. (in press).

(16) Castro-Vasquez, A., and McCann, S.M. Endocrinology 97, 13-19
(1975).

(17) Chen, C.L., and Meites, J. Endocrinology 86, 503-505, 1970.

(18) De Léan, A., Ferland, L., Drouin, J., Kelly, P.A., and Labrie,
F. Endocrinology 100, 1496-1504, 1977a.

(19) De Léan, A., Garon, M., Kelly, P.A., and Labrie, F. Endocrinology 100, 1505-1510, 1977b.

(20) Drouin, J., and Labrie, F. Endocrinology 98, 1528-1534, 1976a.

(21) Drouin, J., and Labrie, F. Prostaglandins 11, 355-366, 1976b.

(22) Drouin, J., Lagacé, L., and Labrie, F. (1976a) Endocrinology
99, 1477-1481, 1976a.

(23) Drouin, J., De Léan, A., Rainville, D., Lachance, R., and Labrie, F. Endocrinology 98, 514-521, 1976b.

(24) Euvrard, C., Boissier, J.R., Labrie, F., and Raynaud, J.P.
Proc. 7th Int. Congr. Pharmacol., (in press).

(25) Everett, J.W. Endocrinology 43, 389-392, 1948.

(26) Ferland, L., Borgeat, P., Labrie, F., Bernard, J., De Léan, A.,
and Raynaud, J.P. J. Mol. Cell. Endocrinol. 2, 107-115, 1975.

(27) Ferland, L., Drouin, J., and Labrie, F. In Hypothalamus and
Endocrine Functions, pp. 191-209, Ed. Labrie, F., Meites, J.,
and Pelletier, G., Plenum Press, New York, 1976.

(28) Franchimont, P. J. Royal Coll. Physicians (London) 6, 283-298, 1972.

(29) Franchimont, P., Chari, S., and Demoulin, A. J. Reprod. Fertil. 44, 335-350, 1975.

(30) Frantz, A.G., Kleinberg, D.L., and Noel, G.L. Rec. Prog. Horm. Res. 28, 527-590, 1972.

(31) Gordon, J.H., and Reichlin, S. Endocrinology 94, 974-978, 1974.

(32) Grumbach, N.M., Roth, J.C., Kaplan, S.L., and Kelch, P. The control of the onset of puberty, Wiley and Sons, New York, pp. 115-166, 1974.

(33) Jewelewicz, R., Ferin, M., Vande Wiele, R.L., Dyrenfurth, I., and Warren, M. Fertil. Steril. 25, 290-291, 1974.

(34) Kalra, P.S., Fawcett, C.P., Krulich, L., and McCann, S.M. Endocrinology 92, 1256-1268, 1973.

(35) Kordon, C., Blake, C.A., Terkel, J., and Sawyer, C.H. Neuro-endocrinology 13, 213-223, 1973-74.

(36) Labrie, F., Pelletier, G., Raynaud, J.P., Ducommun, P., Delga-do, A., MacIntosh, B., and Fortier, C. Med. Hyg. 28, 266-281, 1970.

(37) Labrie, F., Pelletier, G., Lemay, A., Borgeat, P., Barden, N., Dupont, A., Savary, M., Côté, J., and Boucher, R. In Karo-linska Symposium on Research Methods in Reproductive Endocri-nology (E. Diczfalusy, ed.), Geneva, pp. 301-340, 1973.

(38) Labrie, F., De Léan, A., Drouin, J., Barden, N., Ferland, L., Borgeat, P., Beaulieu, M., and Morin, O. In "Hypothalamus and Endocrine Functions" (F. Labrie, J. Meites, and G. Pel-letier, eds), pp. 147-159, 1976a.

(39) Labrie, F., Pelletier, G., Borgeat, P., Drouin, J., Ferland, L., and Belanger, A. In Frontiers in Neuroendocrinology, vol. 4 (W.F. Ganong, and L. Martini, eds), pp. 63-94, Raven Press, New York, pp.   1976b.

(40) Labrie, F., Beaulieu, M., Caron, M., and Raymond, V. Proc. Int. Congr. on prolactin (C. Robyn and M. Harter, Elsevier-North Holland (in press).

(41) Lagacé, L., Labrie, F., Lorenzen, J., Schwartz, N.B., & Channing, C.P. Submitted.

(42) Legan, S.J., and Karsch, F.J. Endocrinology 96, 571-575, 1975.

(43) MacLeod, R.M. Endocrinology 85, 916-923, 1969.

(44) MacLeod, R.M., and Lehmeyer, J.E. Endocrinology 94, 1077-1085, 1974.

(45) McCullagh, D.R. Science 76, 19-20, 1932.

(46) Meites, J., Nicoll, C.S., and Talwalker, P.K. In Advances in Neuroendocrinology, (Nalbandov, A.V., University of Illinois Press, Urbana, III, pp. 238-288, 1963.

(47) Nillius, S.J., and Wide, L. J. Obstet. Gynaecol. Br. Commonw. 78, 822-827, 1972.

(48) Pelletier, G., Lemay, A., Beraud, G., and Labrie, F. Endocrinology 91, 1355-1371, 1972.

(49) Posner, B.I., Kelly, P.A., and Friesen, H.G. Science 187, 57-59, 1975.

(50) Ramey, J.N., Burrow, G.N., Polak-Wich, R.J., and Donabedian, R.K. J. Clin. Endocrinol. Metab. 40, 712-714, 1975.

(51) Raymond, V., Beaulieu, M., Labrie, F., and Boissier, J.R. Science (in press).

(52) Setchell, B.P., and Jacks, F. J. Endocrinol. 62, 675-676, 1974.

(53) Shaar, C.J., and Clemens, J.A. Endocrinology 95, 1202-1212, 1974.

(54) Snyder, S.H., Banerjee, S.P., Yamamura, H.I., and Greenberg, D. Science 184 1243-1245, 1974.

(55) Stevens, J.R. Arch. Gen. Psychiat. 29, 177-189, 1973.

(56) Subramanian, M.G., and Gala, R.R. Endocrinology 98, 842-848, 1976.

(57) Swerdloff, R.S., and Odell, W.D. Lancet 2, 683-687, 1968.

(58) Swerdloff, R.S., Walsh, P.C., and Odell, W.D. Steroids 20, 13-22, 1972.

(59) Takahara, J., Arimura, A., and Schally, A.V. Endocrinology 96, 462-465.

(60) Toft, D.O., and O'Malley, B.W. Endocrinology 90, 1041-1045, 1970.

(61) Vale, W., Grant, G., Amoss, M., Blackwell, R., and
     Guillemin, R. Endocrinology 91, 562-   1972.

(62) Yen, S.S.C., van den BERG, G., Rebar, R., and Ehara, Y.
     J. Clin. Endocrinol. Metab. 35, 931-934, 1972.

(63) Yen, S.S.C., Ehara, Y., and Siler, T.M. J. Clin. Invest.
     53, 652-655, 1974.

PEPTIDE AND NEUROTRANSMITTER RECEPTORS IN THE BRAIN: REGULATION

BY IONS AND GUANYL NUCLEOTIDES

Solomon H. Snyder

Dept. of Pharmacol., Johns Hopkins University School of

Medicine, Baltimore, Maryland 21205

In recent years attention has focused on the localization of traditional peptide hormones in the brain. In several cases peptides, thought to be confined to the gastrointestinal tract, and associated glands and the pituitary, were reported to occur in the brain. Whether or not all gut peptide hormones exist in the brain as well is by no means clear. Interestingly, peptides first discovered in the brain, such as substance P and enkephalin, have subsequently been shown to occur also in the intestinal tract. Because of their localization in specific neuronal pathways, these peptides have been suggested as neurotransmitters. One way in which the brain peptides resemble conventional neurotransmitters lies in the properties of their receptor sites as defined in biochemical studies.

IONS AND THE OPIATE RECEPTOR

The most striking example of a peptide receptor in the brain was not first recognized as such. While enkephalin peptides are the natural substrates of the opiate receptor, this receptor was first identified as the site at which conventional opiate drugs exert their pharmacological effects. The remarkable localization of opiate receptors to areas of the brain integrating information about pain and emotional perception as well as other functions suggested that the opiate receptor was not an accident of nature. The opioid pentapeptides methionine enkephalin and leucine enkephalin, whose discovery and characterization will not be described here, appear to be the endogenous ligands for opiate receptors. Autoradiographic maps of opiate receptor distribution

throughout the brain closely parallel the localization of
enkephalin containing neurons, though some subtle differences do
exist.

Identifying a receptor site by biochemical assay techniques
and localizing it is a first step toward understanding a neurotransmitter or hormone receptor. Binding studies deal only with
the "recognition" site of the receptor. Somehow recognition information is translated into a second messenger, a change in ion permeability or formation of cyclic nucleotide. Neurophysiologists
have clarified ionic mechanisms of synaptic transmission in certain
cases, such as the role of sodium at nicotinic cholinergic
synapses of the neuromuscular junction and chloride at hyperpolarizing glycine and GABA synapses in the brain. Ionic mechanisms for
other synapses are much less clear, even for such extensively
studied transmitters as norepinephrine and dopamine. While neurophysiological techniques have been of primary importance in early
studies of ionic mechanisms, biochemical investigations may now
have something to offer.

In the case of the opiate receptor, sodium ion appears to
exert selective influences. Low concentrations of sodium ion can
increase the receptor binding of ^{3}H-opiate antagonists while decreasing the binding of agonists (11). The influence of sodium is
selective, since while lithium, whose properties resemble those of
sodium, can reproduce these effects somewhat, other monovalent
cations, such as rubidium, cesium and potassium, fail to discrimi-
nate agonists and antagonists. In part, the augmentation of ^{3}H-
antagonist binding in the presence of sodium is due to the ability
of sodium to cause a dissociation from the receptor of endogenous
enkephalin which has been tightly bound to receptor sites, thus
leaving receptors vacant for the binding of ^{3}H-antagonists (8).
Nonetheless, even when membranes have been preincubated to free
opiate receptors of bound enkephalin, sodium still augments the
binding of ^{3}H-opiate antagonists. These findings suggest that the
opiate receptor may exist in two interconvertible states with respective preferences for binding of agonists and antagonists.

This "sodium" effect on opiate receptor binding has practical
consequences. One can predict whether a drug is an agonist or
antagonist by the effect sodium has on its receptor interactions.
One measures the binding of the drug such as the antagonist ^{3}H-
naloxone in the presence or absence of sodium. Nonradioactive
agonists become substantially less potent in competing for receptor
binding when sodium is added to incubation media, while pure antagonists show no loss in potency. Drugs with both agonist and antagonist properties are affected in an intermediate fashion. Such
mixed agonist-antagonists have important therapeutic properties.
Certain of these agents elicit analgesia, a function of their agonist

effects. But, perhaps because of their antagonist propensities, they are markedly less addicting than conventional opiate agonists. These mixed agonist-antagonists are the most promising of all drugs as relatively addiction-free pain relievers. While it is difficult to identify such agents in screening procedures in intact animals, opiate receptor interactions afford a simple, sensitive, inexpensive yet highly reliable screen.

The influence of sodium on the opiate receptor also has theoretical implications. It suggests that the opiate receptor may exist in two interconvertible states with respective affinities for binding of agonists and antagonists. Sodium apparently binds preferentially to the "antagonist" state. One might speculate that synaptic effects of opiate agonists occur by altering the relative proportions of agonist and antagonist receptors, which changes the tightness of attachment of sodium to the receptor, in turn altering membrane permeability to sodium. This model, based on biochemical data, finds support in recent neurophysiological studies. Zieglgansberger and Bayerl (19) have shown that opiates and opioid peptides inhibit neuronal firing without hyperpolarizing cells. Instead, they apparently block excitatory actions of glutamate and acetylcholine through selective effects at synaptic sodium channels, presumably the sodium sites of the opiate receptor identified in binding studies. Direct evidence of a physiological role of sodium in the actions of opiates has been obtained in the cat nicitating membrane, which responds to opiates in a highly specific fashion. Enero (2) has shown that the effects of opiates on the nicitating membrane are enhanced when one lowers the sodium concentration of the medium, fitting well with predictions based on receptor binding studies.

Opiate receptors are also regulated by divalent cations. Low, physiological concentrations of manganese and magnesium selectively increase the binding of ^{3}H-agonists without augmenting binding of antagonists (9,14). While manganese and magnesium exert this effect, calcium is essentially without influence on receptor binding. A role for endogenous divalent cations eliciting this effect is established by experiments using chelating agents. EGTA, which chelates calcium but not manganese and magnesium, has no effect on receptor binding. By contrast, EDTA, which does chelate magnesium and manganese, selectively decreases receptor binding of ^{3}H-agonists but not ^{3}H-antagonists, an effect opposite to that of the divalent cations themselves. Enero (2) has obtained physiological evidence showing that manganese and magnesium do indeed regulate the normal biological functioning.of the opiate receptor. In the cat nicitating membrane the pharmacological effects of morphine are indeed enhanced by manganese and magnesium selectively. The influences of manganese and magnesium on opiate receptor binding are most marked in the presence of sodium. Thus, these ions potently antagonize

the ability of sodium to decrease agonist binding. This suggests
that normally some interaction takes place between these divalent
cations and sodium at the opiate receptor.

IONIC INFLUENCES ON α-NORADRENERGIC,
β-NORADRENERGIC AND DOPAMINE RECEPTORS

Ionic effects are not restricted to the opiate receptor.
Recently we found that sodium differentiates the binding of [3]H-
agonists and ^{3}H-antagonist to the α-noradrenergic receptors of the
brain (4). As with the opiate receptor this effect is manifested
by sodium and lithium but not by potassium. Also like the opiate
receptor there is an interaction between sodium and divalent
cations. In the presence of sodium, manganese and magnesium selectively enhance receptor binding of ^{3}H-agonists to α-receptors
without altering the binding of ^{3}H-antagonist, ^{3}H-WB-4101. Calcium
is much less active. While manganese can increase agonist binding
to the α-receptor in the absence of sodium, the influence of the
divalent cations is much more pronounced in sodium containing media
(U'Prichard and Snyder, in preparation).

Thus, it is possible that sodium and certain divalent cations
may regulate synaptic activities of norepinephrine at α-receptors
in the brain. Neurophysiological studies have not given a clear
picture of what ionic mechanisms are involved in synaptic transmission of norepinephrine at α-sites, though it is generally felt
that the important constricting influences of norepinephrine on
blood vessels may involve changes in sodium permeability.

In certain ways sodium effects on α-receptors differ from
actions at opiate receptors. While opiate binding sites identified
in these studies are thought to represent a homogeneous population
of receptors which can exist in distinct agonist and antagonist
preferring states, this is not the case with α-sites labeled in our
investigations. The ^{3}H-agonists, clonidine, norepinephrine and
epinephrine, label fully distinct and non-interconverting sites as
compared to the antagonist, ^{3}H-WB-4101 (3). The ergot ^{3}H-dihydro-
ergokryptine (DHE) has considerable affinity for both sites labeled
by ^{3}H-agonists and ^{3}H-antagonists. The numbers of ^{3}H-agonists and
[3]H-antagonist preferring sites in various brain regions and in
various peripheral tissues differs markedly (10,16). In brain
regions enriched in "agonist preferring" sites, the drug specificity
of ^{3}H-DHE binding is that of a ^{3}H-agonist, while in antagonist
preferring regions or tissues ^{3}H-DHE behaves like ^{3}H-WB-4101 (10).

The distinctness of agonist and antagonist preferring α-receptors
in the brain is underlined by the effect of sodium on potencies of
drugs in competing for binding of ^{3}H-agonists and ^{3}H-antagonists.

In the case of the opiate receptor, one can grade drugs on a continuum from pure antagonists to pure agonists by determining the ability of sodium to alter their potency in inhibiting ^{3}H—naloxone binding (11). If ^{3}H—α—agonists and ^{3}H—α—antagonists labeled interconverting states of the same receptor, one could conduct similar experiments at the α—receptor. However, while sodium markedly reduces binding of ^{3}H—agonists to α—receptor sites, it does not alter the potency of the unlabeled agonists in competing for the binding of the antagonist, ^{3}H—WB—4101. Thus, in the case of the α—receptor one cannot speak of ions as regulating the interconversion of two states of the receptor. Instead, sodium specifically decreases the affinity of the agonist preferring receptors for ^{3}H—agonists while having no influence whatsoever on the antagonist preferring receptors.

Sodium also regulates affinities of agonists for β—receptors. Sodium ion decreases the binding of ^{3}H—epinephrine to β—receptors and decreases the potency of agonists in competing for the binding of ^{3}H—antagonists to β—receptors (15). As in the case of opiate and α—receptors this effect of sodium is selective being manifested by sodium and lithium but not by potassium. Both agonists and antagonists appear to bind to one form of the β—receptor. There is little if any evidence for differences in potencies of drugs in competing for binding of ^{3}H—agonists or ^{3}H—antagonists to β—receptors. Thus, both agonists and antagonists appear to label one state of the receptor which does not undergo interconversion.

Dopamine receptors are also regulated by divalent cations. Manganese potently enhances binding of the agonist ^{3}H—apomorphine while magnesium and calcium are less effective (Creese, Usdin and Snyder, in preparation). Except for lowering of nonspecific binding no influences of monovalent cations on dopamine receptors have been detected.

How these ionic effects modulate synaptic transmission at β—noradrenergic and dopamine receptors is far from clear. Both of these receptors are thought to be linked to a neurotransmitter stimulated adenylate cyclase. There has been much speculation as to how synaptic transmission is effected at adenylate cyclase related synapses. One possibility is that the formation of cyclic AMP mediates "metabolic" effects of the receptors while synaptic transmission per se involves a convention change in ion permeability. Alternatively, cyclic AMP could trigger phosphorylation of a membrane protein which in turn alters ion permeability.

A more clear cut relationship of ions and related neurotransmitter receptors is apparent for glycine. Glycine is well estabas a transmitter of postsynaptic inhibition in the spinal cord and brainstem. Glycine hyperpolarizes cells by increasing chloride

permeability. Glycine receptors can be labeled by the binding of
the glycine antagonist, ^{3}H-strychnine (18). The binding of [3]H-
strychnine to glycine receptors is depressed by physiological
concentrations of chloride ion. Moreover, the relative ability of
various anions to inhibit ^{3}H-strychnine binding corresponds closely
to their relative potencies in mimicking the ability of chloride to
reverse inhibitory postsynaptic potentials when injected into target
cells (18). Chloride and other anions also decrease the affinity of
glycine for ^{3}H-strychnine binding sites in proportion to their neurophysiological activities (7). It is well established that chloride
is the ion involved in glycine's synaptic activities. These studies
strongly indicate that the glycine receptors labeled by ^{3}H-strychnine
has associated with them the specific sites which regulate ion
permeability. These findings suggest, but do not prove, that the
ion conductance binding site is part of the same receptor macro-
molecule as the neurotransmitter recognition site.

RECEPTOR REGULATION BY GUANYL NUCLEOTIDES

The classic studies of Rodbell and associates (12) have established a major role for guanyl nucleotides in regulating the function of peptide adenylate cyclase linked receptors. GTP can stimulate the adenylate cyclase directly and can also increase the influence of the appropriate peptide hormone upon cyclase activity. An
abundance of evidence suggests that there might exist distinct GTP
binding sites, one on the adenylate cyclase molecule and one on the
recognition portion of the receptor. A role of guanyl nucleotides
in neurotransmitter linked adenylate cyclase systems is indicated
from the studies of Lefkowitz (5) and Gilman and associates (6).
GTP selectively enhances catecholamine actions at β-receptors in
frog red blood cells and certain tumor cells in culture. Moreover,
GTP influences β-noradrenergic receptor binding. GTP decreases the
affinity of agonists for β-receptor binding sites labeled with [3]H-
antagonists.

How does one relate decreased affinity of an agonist with
enhanced synaptic effectiveness? It is possible that GTP "sensitizes" the receptor to effects of the neurotransmitter. In the
absence of GTP agonists have "too much" affinity for the receptors.
They bind so tightly that they do not dissociate rapidly enough to
facilitate synaptic transmission. Effective synaptic activity can
take place only with a more rapidly dissociating ligand which can
"flip on and off" rapidly.

While it had been thought that GTP effects were restricted to
adenylate cyclase mediated receptors, we have obtained evidence
that such effects may be more universal. Specifically, GTP is a
major regulator of α-receptor binding in the brain, a site at which

TABLE 1   Influence of Guanyl Nucleotides on
Opiate, α–Noradrenergic and Dopamine Receptor Binding

	OPIATE[a]		α–NORADRENERGIC[b]		DOPAMINE[c]	
	+ 100 mM NaCl					
	Ag.	Ant.	Ag.	Ant.	Ag.	Ant.
	^{3}H–DHM	^{3}H–NAL	^{3}H–NE	^{3}H–WB	^{3}H–APO	^{3}H–SPIRO
GTP	18	104	21	101	49	96
GppNHp	40	95	31	103	51	102
GDP	17	88	29	88	57	99
GMP	105	89	97	91	107	92
ATP	90	95	80	90	104	93
ADP	86	75	92	97		
AMP	113	93	96	105		

Values are percent of control binding in the absence of nucleotides. Concentrations of nucleotides employed were 50 μM for opiate,
10 μM of nucleotides for α–noradrenergic and 10 μM for dopamine
receptor.
  a. From S. Childers and S.H. Snyder, in preparation.
  b. From U'Prichard and Snyder (17).
  c. From I. Creese, T. Usdin and S.H. Snyder, in preparation.

Abbreviations are: Ag, Agonist; Ant, Antagonist; DHM, dihydromorphine; NAL, naloxone; NE, norepinephrine; WB, WB-4101; APO, apomorphine and SPIRO, spiroperidol.

the role of cyclic nucleotides is questionable. As had been found
at the β–receptor, at α–sites GTP decreases the binding of ^{3}H–
agonists but not of ^{3}H–antagonists. The effect is selective, since
it is manifested by GTP and GDP but not by GMP, ATP, ADP and AMP
(17) (Table 1). As predicted by theoretical models, GTP decreases
the affinity of agonists by accelerating their dissociation from α–
receptors (Figure 1).

A similar role for GTP has been observed at dopamine
receptors in the brain. GTP decreases the binding of the agonist,
^{3}H–apomorphine, while exerting no effect on the antagonist, ^{3}H–
spiroperidol. This effect is selective, being mediated by GTP and
GDP but not by GMP, ATP, ADP and AMP (Table 1) (Creese, Usdin and
Snyder, in preparation).

Recently Blume (1) observed effects of GTP upon opiate
receptor binding. He found a decrease in binding mediated by GTP
and GDP but not by ADP and AMP. However, he reported marked dis-

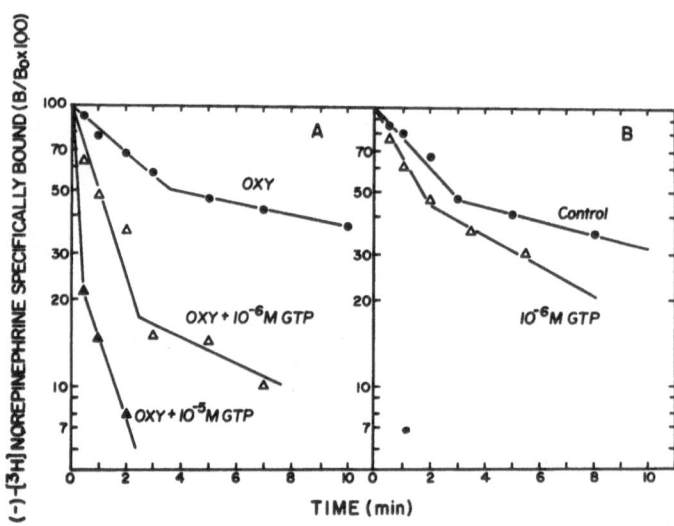

FIG. 1.   Effect of GTP on the time course of dissociation of (-)-
³H-norepinephrine binding to α-receptors. Dissociation of bound
(-)-³H-norepinephrine in calf cortex homogenates was measured at
25°C following incubation with 2 nM (-)-³H-norepinephrine to equilibrium at 25°C. Points shown are those from a single experiment,
performed in triplicate, which was replicated three times.
A, nucleotide added along with oxymetazoline to initiate dissociation. B, nucleotide added at the onset of association.

crepancies between these effects and those reported at other
receptors. If, as predicted by theoretical considerations, GTP
modulates sensitivity of the receptor to neurotransmitter or hormone,
it should decrease affinity of agonists but not of antagonists.
However, Blume reported similar reductions in the binding of both
³H-agonists and ³H-antagonists to opiate receptors. We have also
characterized influences of guanyl nucleotides upon the opiate
receptor and observed rather different effects than those reported
by Blume (Childers and Snyder, in preparation) (Table 1). GTP
selectively decreases the binding of agonists but not antagonists.
The difference between our results and those of Blume relates to
our employment of sodium since the normal environment of biological membranes includes an abundance of sodium. GTP synergizes
with sodium in reducing affinity of agonists but not antagonists.

    How and if opiate effects are related to cyclic nucleotides
is by no means clear. In neuroblastoma-glioma cells in culture
opiates lower concentrations of cyclic AMP via specific receptor
mechanisms (13). Whether such effects occur in mammalian brain
is not at all clear. Thus both in the case of the α-receptor and
the opiate receptor regulation by GTP may not indicate an associ-

ation with adenylate cyclase. Perhaps a GTP site on the receptor maintains it in a physiologically sensitive state. Increases and decreases in sensitivity of a receptor may be an important regulatory mechanism. For instance, when a receptor is bombarded with excessive levels of transmitter or hormone, desensitization may protect the target cell from excessive stimulation.

REFERENCES

1. BLUME, A.J.. Life Sci., in press, 1978.
2. ENERO, M.A. Eur. J. Pharmacol., 45:349-56, 1977.
3. GREENBERG, D.A. and S.H. SNYDER. Mol. Pharmacol., 14:38-49, 1978.
4. GREENBERG, D.A., D.C. U'PRICHARD, P. SHEEHAN and S.H. SNYDER. Brain Res., 140:378-384, 1978.
5. LEFKOWITZ, R.J. and L.T. WILLIAMS. Proc. Natl. Acad. Sci., USA, 74:515-519, 1977.
6. MAGUIRE, M.E., P.M. VAN ARSDALE and A.G. GILMAN. Mol. Pharmacol., 12:335-339, 1976.
7. MÜLLER, W.E. and S.H. SNYDER. Brain Res., in press, 1978.
8. PASTERNAK, G.W., H.A. WILSON and S.H. SNYDER. Mol. Pharmacol., 11:478-484, 1975.
9. PASTERNAK, G.W., A.M. SNOWMAN and S.H. SNYDER. Mol. Pharmacol., 11:735-744.
10. PEROUTKA, S.J., D.A. GREENBERG, D.C. U'PRICHARD and S.H. SNYDER. Mol. Pharmacol., in press, 1978.
11. PERT, C.B. and S.H. SNYDER. Mol. Pharmacol., 10:868-879, 1974.
12. RODBELL, M. and L. LONDOS. Metabolism, 25:1347-1349, 1976.
13. SHARMA, S.K., M. NIRENBERG and W.A. KLEE. Proc. Natl. Acad. Sci., USA, 72:590-594.
14. SIMANTOV, R, A.M. SNOWMAN and S.H. SNYDER. Mol. Pharmacol., 12:977-986.
15. U'PRICHARD, D.C., D.A. BYLUND and S.H. SNYDER. J. Biol. Chem., in press, 1978.
16. U'PRICHARD, D.C., D.A. GREENBERG, P. SHEEHAN and S.H. SNYDER. Brain Res., 138:151-158, 1977.
17. U'PRICHARD, D.C. and S.H. SNYDER. J. Biol. Chem., in press, 1978.
18. YOUNG, A.B. and S.H. SNYDER. Proc. Natl. Acad. Sci., USA, 71:4002-4005.
19. ZIEGLGANSBERGER, W. and H. BAYERL. Brain Res., 115:111-128, 1976.

NEUROPHYSIOLOGY OF HYPOTHALAMIC PEPTIDERGIC NEURONS

L.P. RENAUD, Q.J. PITTMAN and H.W. BLUME

Division of Neurology, Montreal General Hospital

Montreal, Quebec, Canada, H3G 1A4

Recent developments in radioimmunoassay, immunohistochemistry and biochemistry indicate that 'peptidergic' neurons constitute a significant portion of the cells of the central, peripheral and **autonomic nervous systems**. From an historical perspective, central peptidergic neurons have been traditionally associated with the hypothalamus, one of the few areas where they can be identified and examined with electrophysiological techniques. The information provided by these investigations has enhanced our understanding and appreciation of neural participation in the endocrine system, and has provoked other studies related to the role of peptides in interneuronal communication. This chapter will consider these issues as related to the neurophysiology of two groups of accessable putative peptidergic neurons of the hypothalamus i.e. neurohypophyseal and tuberoinfundibular neurons.

THE NEUROHYPOPHYSEAL SYSTEM

Neurosecretory neurons of the supraoptic and paraventricular nuclei are responsible for the synthesis and release of oxytocin and vasopressin(4,9,41,44,95). These nuclei appear to have differentiated from the preoptic nucleus of lower vertebrates(79) where detailed electrophysiological studies in the goldfish provided the initial evidence for the neuronal nature of the neurosecretory neuron(58). In the goldfish preoptic nucleus, and in the supraoptic and paraventricular nuclei of higher species, the identity of neurosecretory neurons has been revealed by the technique of antidromic activation following stimulation in the neurohypophysis (3,7,16,19,34,47,50,53,61,63,70,73,106). Both intracellular and extracellular recordings have indicated that these neurosecretory neurons display the electrical characteristics found in other central neurons, including the ability to generate and conduct orthodromic and antidromic action potentials, and excitatory and inhibitory postsynaptic potentials. With improved recording techniques and innovative approaches, it has been possible to examine in detail the location of neurosecretory neurons, and to investigate possible relationships between their activity patterns and neurosecretion during conditions known to invoke the release of oxytocin and/or vasopressin e.g. parturition, lactation, dehydration, haemorrhage, carotid occlusion, vaginal and uterine distension(3,15,27,31,34,51,63,64,82,104).

Localization of Neurohypophyseal Neurons

Oxytocin and vasopressin have been localized by immunohistochemistry within individual neurons in and around the supraoptic and paraventricular nuclei(2,17,36,55,96,99,102,109). This distribution of peptide-containing neurons is in agreement with the results of retrograde axoplasmic transport studies on the origin of cells of the neurohypophyseal tract(98). Electrophysiological techniques have confirmed this location of neurohypophyseal neurons and have also indicated that these cells are found within the periventricular region and arcuate nucleus (Fig. 1), an area traditionally considered to contain neurons that belong to the tuberoinfundibular system.

Hormone Release and Neuronal Activity

Early anatomic and lesion experiments suggested that the frequency of action potentials of supraoptic and paraventricular neurosecretory neurons was linked to the amount of vasopressin and oxytocin released into the circulation(4,21,44,95). In later work with electrical stimulation in discrete areas of the hypothalamus, or in the isolated neurohypophysis, effective hormone release could be evoked with stimulation frequencies in excess of 30Hz(28,43,45, 57,103), implying that synchronized neuronal activity at this frequency was required for physiological hormone release. Improvements in recording techniques, choices of appropriate stimulation, anaesthesia, selectivity and sensitivity of hormone assays have resulted in a clearer demonstration of the relationship between neurosecretory neuronal activity and hormone release(26,48,49).

Oxytocinergic Neurons. The milk ejection reflex is thought to be regulated by episodic pulses of circulating oxytocin. Single unit recordings from neurosecretory neurons of the supraoptic and paraventricular nuclei in the lactating female rat are in accord

7.2

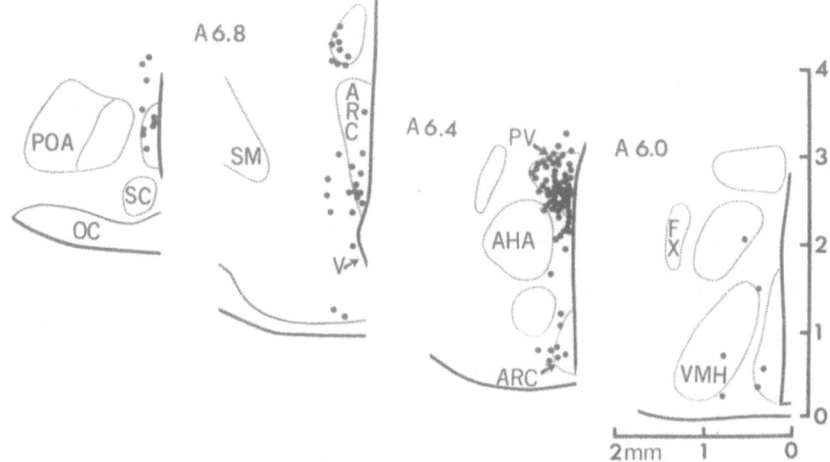

Fig. 1.   The dots superimposed on four schematic coronal sections
of the rat hypothalamus and preoptic area illustrate the location
of a sample population of neurohypophyseal neurons, identified by
antidromic activation from stimulation in the neurohypophysis.
The figure excludes neurohypophyseal neurons of the supraoptic
nucleus.   Note that neurohypophyseal neurons are present in both
the arcuate nucleus (ARC) - periventricular area, and the paraventricular nucleus (PV).
Abbreviations: AHA, anterior hypothalamic area; Fx, fornix; OC,
optic chiasm; POA, preoptic area; SC, suprachiasmatic nucleus; SM,
stria medullaris; V, third ventricle; VMH, ventromedial nucleus.
The illustrated frontal planes are drawn from the rat stereotaxic
atlas of Pellegrino and Cushman (from ref. 14).

with this proposal. Approximately 50% of the neurosecretory
neurons display an irregular or random pattern of activity that is
interrupted every few minutes by an abrupt synchronized neuronal
activation lasting 2-4 secs, characterized by action potential
frequencies in excess of 30 Hz and a subsequent silent interval;
this sequence is followed 12-15 secs later by an increase in intramammary pressure and milk ejection (Fig. 2). A similar rise in
intramammary pressure and milk ejection follows by 10-15 secs the
intravenous injection of synthetic oxytocin or electrical stimulation of the pituitary stalk at 50 Hz(19,63,64,103). In view of
this relationship between milk ejection and neurons that display
synchronized high frequency discharges, this category of neurosecretory neuron has been labeled the 'oxytocinergic' neuron.

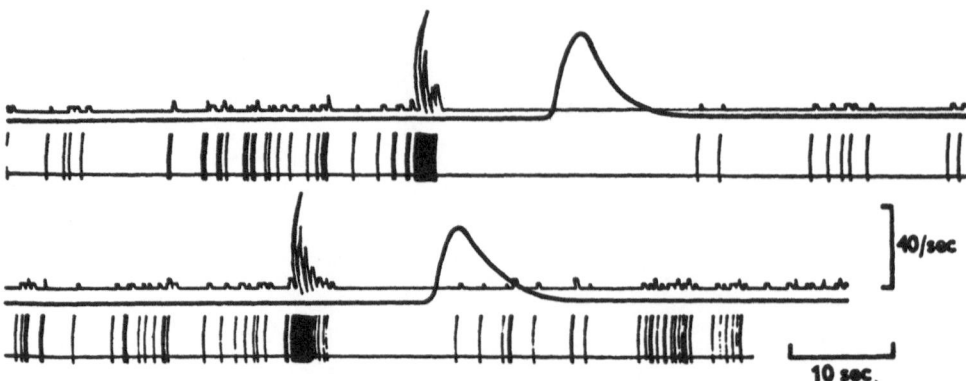

Fig. 2.   Two sets of polygraph records from an 'oxytocinergic' neuron in the rat supraoptic nucleus during suckling to illustrate the nature of the neurosecretory response that precipitates each milk ejection. The bottom trace of each set illustrates neuronal activity, in which each vertical deflection corresponds to a single action potential; the top trace gives an integration of this unit recording where the height of the trace is proportional to frequency. The middle trace is a recording of intramammary pressure. This figure illustrates two responses from this oxytocinergic neuron, characterized by an abrupt change from a random background activity pattern to a marked acceleration in spike frequency and subsequent silent interval, during which there is a rise of about 10 mm Hg in intramammary pressure. Adapted from ref. 64 with permission.

      In the lactating animal vaginal distension also produces an increase in the activity of supraoptic neurosecretory neurons and raises intramammary pressure(31). The magnitude of both the unit activity and intramammary pressure increase is less than that associated with suckling. The specificity of this particular stimulus as a means of identification of oxytocinergic neurons has been questioned(31) since vaginal dilatation produces an increase in the activity of both randomly firing neurons and phasically firing neurons; the latter are usually related to the release of vasopressin (see below).

      Oxytocin is a very potent stimulator of myometrial activity at term(18) and considered to be of importance for both the onset and course of labor. Recordings from paraventricular neurosecretory neurons during labor have revealed an increase in the mean spontaneous firing rate of all neurosecretory neurons, suggesting a tonic release of both neurohypophyseal hormones during labor(15). However, no relationship has been demonstrated between uterine contraction and unit activity.

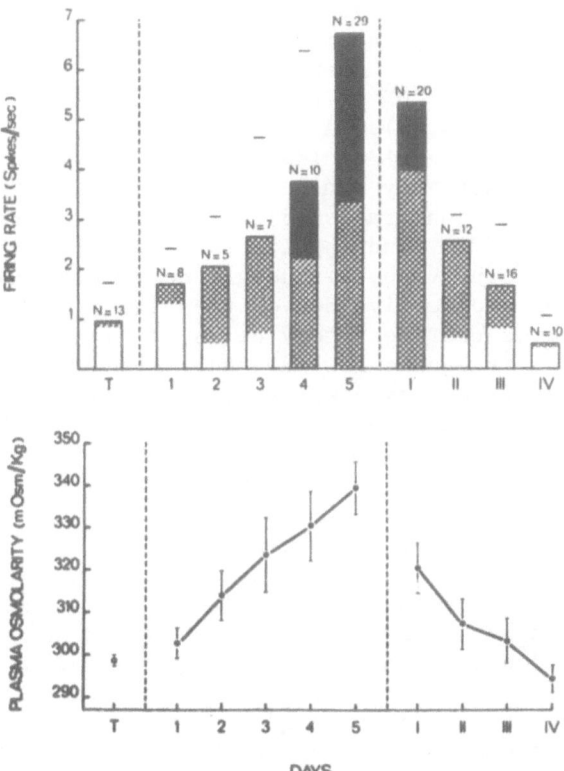

Fig. 3.   Summary of data obtained from supraoptic 'vasopressin-ergic' neurons of the awake monkey during five days of water depri-vation and rehydration. In the upper part of the figure, the height of the blocks represents the mean firing rate of neurons recorded for each day of the experiment. The number of neurons for each day is shown above each block. This figure also illustrates the relative proportion of the different cell types. Note that with dehydration, the percentage of irregularly firing neurons (open area) diminishes, to be replaced by an increase in the percentage of phasic (crosshatched) and continuously firing neurons (black area), a sequence that appears to be reversible upon rehydration. The lower part of the figure illustrates the corresponding changes in plasma osmolarity. Adapted from ref. 3 with permission.

Vasopressinergic Neurons. Early studies on hypothalamic neuro-secretory neurons demonstrated their activation by intracarotid injections of hyperosmolar solutions, and led to the postulate that this activation was related to vasopressin release(20,21,44). It is now known that vasopressin release probably results from an

interaction between a variety of factors including osmotic (e.g. dehydration), volumetric (e.g. haemorrhage) and behavioural (e.g.
drinking) stimuli, and can be correlated with phasic activity
patterns of approximately 20% of the supraoptic and paraventricular
neurosecretory neurons(48,49). A phasic activity pattern is unique
to these cells, and is useful in identifying 'vasopressinergic'
neurons(16). Phasic activity can be provoked by several vasopressin
releasing stimuli i.e. dehydration(3), blood volume depletion(104)
and carotid occlusion(27), and appears to be but one functional
state of activity of the vasopressinergic neuron(3). For example,
under conditions of progressive dehydration the pattern of activity
of certain supraoptic neurosecretory cells of the unanaesthetized
monkey changes from irregular low frequency discharges, to phasic
discharges, to high frequency discharges in concert with increases
in plasma osmolarity; the reverse trend occurs on rehydration
(Fig. 3).

Efferent Connections

     The first electrophysiological evidence that axons of neuro-
hypophyseal neurons have functional connections with other neurons
was the demonstration by Kandel of short latency inhibitory postsynaptic potentials in preoptic neurosecretory neurons following
stimulation of the neurohypophysis(58). Since the inhibitory postsynaptic potentials were evoked by pituitary stimuli limited to the
axons of the neurohypophyseal tract, the presence of axon collaterals in the neurohypophyseal tract was strongly implied. These
central collaterals appear   to participate in recurrent pathways
to the parent neuron, either directly by synapsing with neurosecretory neurons, or indirectly by synapsing with interneurons whose
axons in turn synapse with the neurosecretory neuron (see Fig. 5).
Observations derived from mammalian supraoptic and paraventricular
neurosecretory neurons have yielded similar conclusions(7,29,50,61,
73,74,76). Attempts to locate cells considered to be the inhibitory
interneurons in a recurrent inhibitory pathway have led to conflicting
results. While Barker et al(7) were unable to find electrophysiological evidence of inhibitory interneurons, Koizumi and Yamashita
(61) described Renshaw-like cellular activity in the vicinity of
the supraoptic nucleus following pituitary stalk stimulation, and
suggested that these could be local inhibitory interneurons. Similar
results following stimulation of the neural lobe have been described
by other investigators(73,74).

     In the mammalian nervous system, facilitation as the initial
response to neurohypophyseal stimulation has only been observed in
non-neurosecretory neurons. However, in-vitro studies in the bull-
frog hypothalamo-hypophyseal system have indicated that stimulation
of the neurohypophysis activates both recurrent inhibitory and

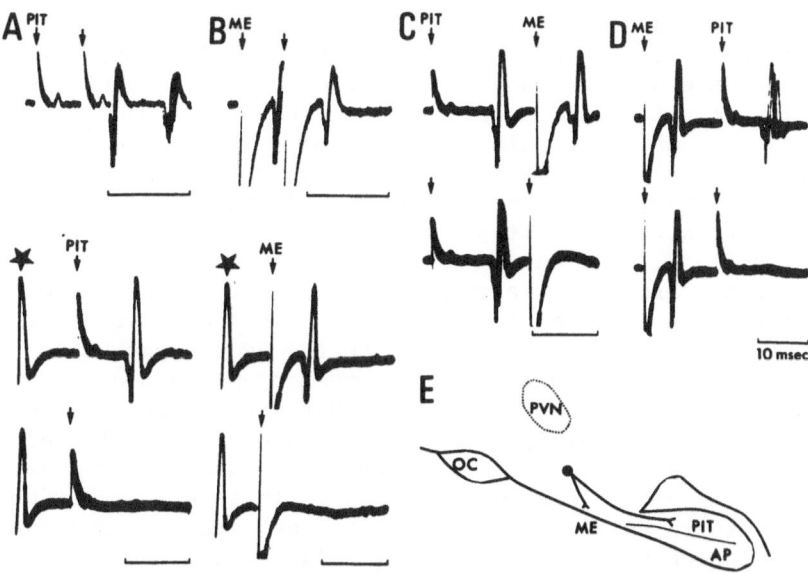

Fig. 4. Superimposed oscilloscope records from a hypothalamic
periventricular neuron to illustrate antidromic activation from
stimulation of both the neurohypophysis (PIT) and median eminence
(ME). In A, B, the upper rows illustrate constant latency responses to two stimuli (arrows) presented to each site; the lower
rows illustrate failure of antidromic activation due to collision
when a suprathreshold stimulus is presented within a critical
period (approximately equal to the antidromic spike latency) after
a spontaneous action potential (star) partially visible at the onset of each sweep. In C, D, the upper traces illustrate antidromic responses to sequential stimulation at each site, while the
lower traces illustrate the failure of the second antidromic response due to collision when the interstimulus interval is
shortened by 1 msec. Interpretation of these observations
suggests a neuron whose axon bifurcates close to the cell soma
and innervates both the median eminence and the neurohypophysis,
as illustrated in E. Negativity is illustrated by an upward deflection. The time calibration is 10 msec in all traces.
Abbreviations: OC, optic chiasm; PVN, paraventricular nucleus;
AP, adenohypophysis.

recurrent facilitatory connections(60). The precise neural network that would account for these observations is uncertain.

    Recent anatomic observations have demonstrated pathways con-
taining vasopressin, oxytocin and their associated neurophysins
in areas extending beyond the neurohypophyseal tract including

amygdala, midbrain, spinal cord and median eminence. This has
raised the question of whether the same neuron might project simultaneously to the neurohypophysis and to one of these other regions.
We have investigated this possibility and have observed that some
neurons in the paraventricular nucleus and periventricular region
demonstrate simultaneous antidromic activation from both the neurohypophysis and median eminence. This would indicate that these
cells send axons to both sites. Two patterns of axon branching
have been defined: one, where the axons to median eminence and
neurohypophysis arise from a common origin near the cell soma (Fig.
4); another where the axon to the median eminence arises as a short
collateral of the main axon coursing to the neurohypophysis. These
findings of a dual innervation of the median eminence and neurohypophysis are contrary to the traditional notion that the tuberoinfundibular and neurohypophyseal systems are anatomically distinct
(100) and may indicate functional interaction in these two systems.
Similar studies were carried out with simultaneous stimulation in
both the neurohypophysis and midbrain region, and have indicated
that certain neurons in the periventricular region, but not in the
paraventricular nucleus, project to both of these areas. Thus,
peptidergic pathways from the paraventricular nucleus to midbrain
regions(17,99) do not appear to arise from neurons that simultaneously project to the neurohypophysis.

Afferent Connections

The supraoptic and paraventricular nuclei have been shown to
receive anatomical connections from many structures, including the
brainstem, amygdala, septum, hippocampus, olfactory tubercle, cortex and mediobasal hypothalamus (62,78,105,108). Some of these
connections are almost certainly involved in reflex mechanisms for
the release of oxytocin and vasopressin, as well as for the release
of these hormones evoked by electrical stimulation in hypothalamic
and extrahypothalamic regions(1,8,10,21,32,52-54,73,97). Some of
these connections have been confirmed by electrophysiological
techniques. Stimulation in the septum, amygdala, midbrain reticular formation and hippocampus evokes short latency changes in
excitability of supraoptic and paraventricular neurosecretory
neurons(61,73,74). Peripheral afferents are also important in the
release of oxytocin and vasopressin(32), presumably through multi-
synaptic pathways that have received less detailed examination.
For example, vagal stimulation produces vasopressin release(32),
but only evokes long latency excitation in supraoptic neurosecretory
neurons(7).

Pharmacology of Neurohypophyseal Neurons

Identification of neurotransmitters involved in the afferent or

recurrent pathways to the supraoptic and paraventricular neurosecretory neurons described above has not been firmly established.
Several investigators have utilized the technique of microiontophoresis to examine the sensitivity of neurosecretory neurons to
putative neurotransmitters (see ref. 59). Particular attention has
been paid to the characteristics of the transmitter involved in the
recurrent inhibitory pathway. The results of these studies will be
reviewed briefly.

The intracarotid injection of acetylcholine and other cholino-
mimetics produces an increase in the activity of supraoptic neurons
and the release of oxytocin and vasopressin(9,48,80). Microiontophoretic application of acetylcholine to neurosecretory neurons
also results in excitation(6,30,72). This effect can be reduced
by atropine, but also displays characteristics suggesting activation of nicotinic receptors(7,30). However, neither iontophoretic nor intravenous atropine administration appears to block
recurrent inhibition in the neurohypophyseal pathway, making acetylcholine an unlikely neurotransmitter in this pathway(13,76).

Norepinephrine is present in both supraoptic and paraventricular nuclei(38). Microiontophoretic application of norepinephrine
to neurosecretory neurons usually results in a decrease in excit-
ability(7,72), an action that is blocked by α-adrenergic antagonists. Recurrent inhibition in the supraoptic nucleus is not
affected by treatment with adrenergic blocking agents or pretreatment with 6-hydroxydopamine(13,76).

Neurosecretory neurons are sensitive to the microiontophoretic application of several amino acids. L-glutamate enhances,
whereas GABA and glycine consistently depress the excitability of
most supraoptic and paraventricular neurosecretory neurons(76).
While strychnine will antagonize glycine evoked responses, and
picrotoxin and bicuculline will antagonize GABA evoked depressions,
only bicuculline has been reported to partially antagonize recurrent
inhibition in the neurohypophyseal pathways(13,76). Therefore it
is possible that GABA could function as a neurotransmitter in this
pathway.

The presence of axon collaterals in neurohypophyseal fibres
(see above) has prompted considerable interest in the possibility
that the neurohypophyseal peptides might act as neurotransmitters
at central synapses in this system(5,13,75). For example, vasopressin has been proposed as a neurotransmitter for the direct
recurrent inhibitory pathway(76). Its application to supraoptic
neurosecretory neurons does produce a decrease in their excitabil-
ity(76). However, evidence that recurrent inhibition is still
present in the vasopressin deficient homozygous Brattleboro rat(33)
argues against a significant role for vasopressin in the hypothalamic recurrent inhibitory pathways. Oxytocin is also an un-

likely neurotransmitter candidate for a direct recurrent inhibitory
pathway, since it is reported to enhance the excitability of neurosecretory neurons in the paraventricular nucleus(70). Nevertheless,
since there are now numerous reports of central neurophysin, vasopressin and oxytocin containing neurons and fibres originating from
the paraventricular nucleus(2,36,55,96,99,102,109) and indications
that these peptides influence behaviour(24,101), one must still
consider that these peptides do have a role in neuronal function.

THE TUBEROINFUNDIBULAR SYSTEM

Neural regulation of adenohypophyseal secretion is considered
to be achieved by hypothalamic tuberoinfundibular neurons, a hetero-
geneous population of parvicellular neurosecretory neurons that
release various factors from their median eminence nerve terminals
into the pituitary portal circulation in response to appropriate
neural stimuli(12,42,44,49,87,94,100). Although the term 'tuberoinfundibular' was initially introduced as an anatomical reference
to neurons located in the mediobasal hypothalamus(100), recent
morphological and electrophysiological findings indicate that axons
from neurons outside this region also project to the median eminence zona externa(65,77,87). For purposes of this presentation,
the 'tuberoinfundibular neuron' is therefore defined as any neuron
that can be demonstrated by electrophysiological criteria (see
below) to project to the median eminence. However, in view of the
evidence presented above for simultaneous innervation of the median
eminence and neurohypophysis (see Fig. 4) this definition is somewhat ambiguous, but will be retained for lack of a better term.
This definition also includes both monoaminergic and peptidergic
neurons with median eminence projections(87). Most monoaminergic
tuberoinfundibular neurons originate from cells located in the
vicinity of the arcuate nucleus(11,38,56).

Localization and Electrical Properties

Electrical stimulation of the <u>surface</u> of the median eminence
near the junction with the pituitary stalk has been utilized by
several investigators to activate terminals of tuberoinfundibular
neurons(65,71,83,92). In the rat, this stimulus produces evidence
for antidromic activation of neurons in both the medial hypothalamus i.e. arcuate, ventromedial, dorsal premammillary and paraventricular nuclei, and the periventricular region and in more rostral
regions i.e. the suprachiasmatic nucleus and medial preoptic area
(65,87). Extracellular records from tuberoinfundibular neurons
and neighboring hypothalamic or other central neurons have shown
no notable differences in their spontaneous, orthodromic or antidromic action potentials. While these neurons are capable of un-
sustained impulse conduction at frequencies greater than 100/sec

impulse conduction in the tuberoinfundibular system is usually under 2.0m/sec(65,83,92). The activity patterns of spontaneously active tuberoinfundibular neurons have been examined in some detail. Some cells discharge randomly, others display activity in the form of bursts of 5-8 action potentials at frequencies up to 300 Hz, and a few of the tuberoinfundibular neurons found in the paraventricular nucleus display phasic activity patterns similar to those noted in vasopressinergic neurons (see above). If these phasic tuberoinfundibular neurons can indeed be associated with vasopressin release in the median eminence, this would be the first indication in the tuberoinfundibular system where activity patterns can differentiate a particular category of tuberoinfundibular neuron.

Efferent Connections

Electrophysiological investigations have indicated that tuberoinfundibular neurons not only innervate the median eminence, but also have central axon collaterals. Some axon collaterals directly or indirectly activate a postsynaptic inhibitory mechanism; with extracellular recordings this is manifested as a transient decrease in excitability following median eminence stimulation(83-85,92,93). Other axon collaterals project to adjacent regions of the hypothalamus i.e. anterior hypothalamic area and paraventricular nucleus, or to extrahypothalamic regions i.e. medial preoptic area, midline thalamic nuclei, amygdala and lateral septum(46,83-85,87,88). Indications for these connections arise from electrophysiological evidence of simultaneous antidromic activation for certain tuberoinfundibular neurons from both the median eminence and one of the areas mentioned above. At present, little is known of the functional significance of these extrahypothalamic axon collaterals. One might postulate that such pathways serve to inform other brain areas of the state of activity in the tuberoinfundibular system. Some of the extrahypothalamic regions that receive these axon collaterals in turn provide afferent fibres to tuberoinfundibular neurons.

Afferent Connections

The results of stimulation and lesion experiments of a number of extrahypothalamic sites including the amygdala, preoptic area, hippocampus and brainstem, have indicated that these regions can influence adenohypophyseal secretion(25,37,67,68,110), presumably through their ability to alter the excitability of different tuberoinfundibular neurons that receive afferent connections from such areas. Recent electrophysiological studies in the rat hypothalamus have supported this proposal. Electrical stimulation in the amygdala, lateral septum, preoptic-anterior hypothalamic area

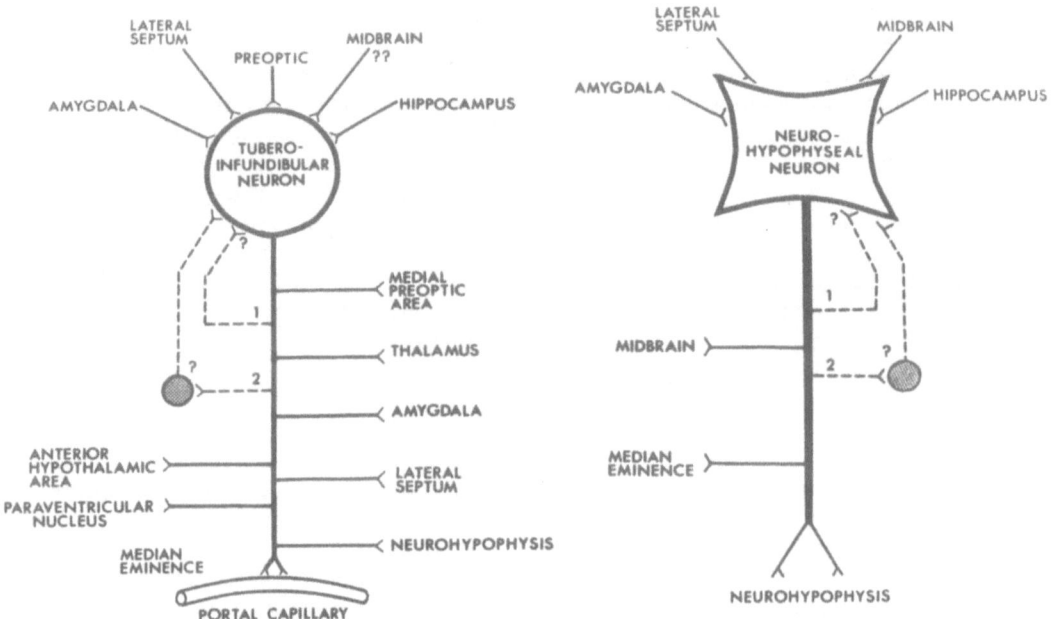

Fig. 5. Summary sketch of known connections of neurons in the tuberoinfundibular system (on the left) and neurohypophyseal system (on the right). Efferent connections are shown in the lower half of each diagram. The heavy vertical lines refer to the main peptidergic axons projecting to the median eminence portal capillaries and neurohypophysis respectively, thus identifying each cell type. Intrahypothalamic axon collaterals in recurrent pathways that either directly or indirectly (through local interneurons) engage the parent neuron, and whose transmitter agent is unknown (?) are depicted by Nos. 1 and 2 in each system. Other axon collaterals extend to the areas indicated. Afferent connections in each system are illustrated at the top of each diagram. Midbrain afferents to the tuberoinfundibular system are still in doubt.

and the dorsal hippocampus has been shown to evoke short latency changes in excitability of tuberoinfundibular neurons in the medio-basal hypothalamus(84-87). The amygdala afferents appear to be topographically organized, and selectively influence tuberoinfun-dibular neurons within the ventromedial nucleus(84). On the other hand, there is little evidence that midbrain reticular formation and periaqueductal gray stimulation affects the excitability of any mediobasal hypothalamic tuberoinfundibular neurons(81). Fig. 5 schematically summarizes these connections.

Pharmacology of Tuberoinfundibular Neurons

While virtually every putative neurotransmitter substance has been considered to modify adenohypophyseal secretion, presumably through an action on the excitability of tuberoinfundibular neurons (39,40,66,91) few investigations have directly examined tuberoinfundibular neuronal excitability during microiontophoretic application of these agents. Of the amino acids, L-glutamate has been shown to increase excitability, whereas GABA produces a decrease in excitability (c.f ref. 71). There is also electrophysiological evidence that GABA may be an inhibitory transmitter in the recurrent inhibitory pathway to tuberoinfundibular neurons(93,107) and in other hypothalamic pathways(87). Tuberoinfundibular neurons have displayed either an increase or a decrease in excitability during microiontophoretic application of dopamine and norepinephrine(71), and catecholamines have been proposed to mediate recurrent facilitation in the tuberoinfundibular system(107). Hypothalamic peptides have also been proposed to mediate or modify neurotransmission at central synapses formed by the axon collaterals of tuberoinfundibular neurons(86). This suggestion is in part based on Dale's postulate(23) that all branches of an axon secrete the same substance(s), and is further supported by evidence that microiontophoretic application of hypothalamic peptides to neurons in the hypothalamus and elsewhere produces prominent changes in their excitability(35,89, 90).

COMMENTS

The introduction of electrophysiology into investigations on central regulation of the endocrine system has significantly increased our knowledge of some hypothalamic pituitary mechanisms. These studies have generally focused on the functions of the magnocellular neurohypophyseal system where it now appears possible to examine neuronal activity patterns and reliably identify specific peptidergic neurons. These studies have also provided a model for the examination of the tuberoinfundibular system. However, the heterogeneity and size of tuberoinfundibular neurons and the lack of identifying stimuli have made studies on this system more difficult. Nevertheless, initial observations have defined the location and connections of tuberoinfundibular neurons, and in certain instances have illustrated a definite topography within these connections. A more detailed study of the activity of tuberoinfundibular neurons may aid in their functional identification, as has been proposed for the phasic tuberoinfundibular neurons in the paraventricular nucleus. Recent observations that individual neurons innervate both the median eminence and neurohypophysis suggest an interaction in the tuberoinfundibular and neurohypophyseal systems.

On the basis of demonstrable connections for axons of both the neurohypophyseal and tuberoinfundibular systems, it would now appear that both types of peptidergic neurons are involved in at least two functions, one related to the pituitary and another related to central neuronal function. Examined in the light of Dale's postulate, any proposal that peptides are important in interneuronal communication as neurotransmitters or neural modulators deserves serious consideration.

ACKNOWLEDGEMENTS

The authors are grateful to the Canadian Medical Research Council and Conseil de la recherche en santé du Québec for financial support. We thank Mrs. M. Walker and Mrs. G. Landrigan for typing the manuscript, and William Ellis and Brian MacKenzie for expert technical assistance.

REFERENCES

1. Anderson, B. and S.M. McCann. Acta physiol. Scand. 35: 191-201, 1955.
2. Antunes, J.L., P.W. Carmel and E.A. Zimmerman. Brain Res. 137: 1-10, 1977.
3. Arnauld, E., B. Dufy and J.D. Vincent. Brain Res. 100: 315-325, 1975.
4. Bargmann, W. and E. Scharrer. Am. Sci. 39: 255-259, 1951.
5. Barker, J.L. In: Peptides in Neurobiology, edited by H. Gainer, New York, Plenum Press. pp. 295-343, 1977.
6. Barker, J.L., J.W. Crayton and R.A. Nicoll. J. Physiol. 218: 19-32, 1971.
7. Barker, J.L., J.W. Crayton and R.A. Nicoll. Brain Res. 33: 353-366, 1971.
8. Beyer, C., L.G. Anguiano and F. Mena. Am. J. Physiol. 200: 625-627, 1961.
9. Bisset, G.W. In: Peptide Hormones, Edited by J.A. Parsons, London, University Park Press, pp. 145-174, 1976.
10. Bisset, G.W., B.J. Clark and M.L. Errington. J. Physiol. 217: 111-131, 1971.
11. Björklund, A., B. Falck, A. Nobin and V. Stenevi. In: Neurosecretion - The Final Neuroendocrine Pathway, edited by F. Knowles and L. Vollrath, New York, Springer-Verlag, pp. 209-222, 1974.
12. Blackwell, R.E., and R. Guillemin. Ann. Rev. Physiol. 35: 357-370, 1973.
13. Bloom, F.E. Neurosci. Res. Program Bull. 10: 122-251, 1972.
14. Blume, H.W., Q.J. Pittman and L.P. Renaud. Brain Res. (in press).
15. Boer, K. and J.W.L. Nolten. J. Endocr. 76: 155-163, 1978.

16. Brimble, M.J. and R.E.J. Dyball. J. Physiol. 271: 253-271, 1977.
17. Buijs, R.M., D.F. Swabb, J. Dogterom and F.W. van Leeuwen. Cell Tiss. Res. 186: 423-433, 1978.
18. Cross, B.A. J. Endocr. 16: 261-276, 1958.
19. Cross, B.A., R.E.J. Dyball, R.G. Dyer, C.W. Jones, D.W. Lincoln, J.F. Morris and B.T. Pickering. Recent Prog. Hormone Res. 31: 243-294, 1975.
20. Cross, B.A. and J.D. Green. J. Physiol. 148: 554-569, 1959.
21. Cross, B.A. and G.W. Harris. J. Endocr. 8: 148-161, 1952.
22. Cross, B.A. and I.A. Silver. Brit. Med. Bull. 22: 254-260, 1966.
23. Dale, H.A. Proc. R. Soc. Med. 318-332, 1935.
24. De Weid, D., B. Bohus and Tj. B. van Wimersma Greidanus. Brain Res. 85: 152-156, 1975.
25. Döcke, F. Neuroendocrinology 14: 345-350, 1974.
26. Dreifuss, J.J. Ann. N.Y. Acad. Sci. 248: 184-201, 1975.
27. Dreifuss, J.J., M.C. Harris and E. Tribollett. J. Physiol. 257: 337-354, 1976.
28. Dreifuss, J.J., I. Kalnins, J.S. Kelly and K.B. Ruf. J. Physiol. 215: 805-817, 1971.
29. Dreifuss, J.J. and J.S. Kelly. J. Physiol. 220: 87-103, 1972.
30. Dreifuss, J.J. and J.S. Kelly. J. Physiol. 220: 105-118, 1972.
31. Dreifuss, J.J., E. Tribollett and A.J. Baertschi. Brain Res. 113: 600-605, 1976.
32. Dyball, R.E.J. Brit. J. Pharmacol. 33: 319-328, 1968.
33. Dyball, R.E.J. J. Endocr. 60: 135-143, 1974.
34. Dyball, R.E.J. and K. Koizumi. J. Physiol. 201: 711-722, 1969.
35. Dyer, R.G. and R.E.J. Dyball. Nature (Lond.) 252: 486-488, 1974.
36. Elde, R. and T. Hökfelt. In: Frontiers in Neuroendocrinology, Vol. 5, edited by W.F. Ganong and L. Martini, New York, Raven Press, pp. 1-33, 1978.
37. Ellendorf, F., J.A. Colombo, C.A. Blake, D.I. Whitmoyer and C.H. Sawyer. Proc. Soc. Exp. Biol. Med. 142: 417-420, 1973.
38. Fuxe, K. and T. Hökfelt. In: Frontiers in Neuroendocrinology, edited by W.F. Ganong and L. Martini, London, Oxford Univ. Press, pp. 47-96, 1969.
39. Ganong, W.F. Life Sci. 15: 1401-1414, 1974.
40. Ganong, W.F. In: Hypothalamic Hormones, edited by M. Motta, F.G. Crossignani and L. Martini. New York, Academic Press, pp. 237-248, 1975.
41. Ginsburg, M. In: Handbook of Experimental Pharmacology, edited by B. Berde, Berlin, Springer-Verlag, pp. 286-371, 1968.

42. Halász, B. In: Frontiers in Neuroendocrinology, edited by W.F. Ganong and L. Martini, London, Oxford University Press, pp. 307-342, 1969.
43. Harris, G.W. Phil. Trans. B. 232: 385-441, 1947.
44. Harris, G.W. Neural Control of Pituitary Gland, London, Edward Arnold, 1955.
45. Harris, G.W., Y. Manabe and K.B. Ruf. J. Physiol. 203: 67-81, 1969.
46. Harris, M.C. and M. Sanghera. Brain Res. 81: 401-411, 1974.
47. Hayward, J.N. J. Physiol. 239: 103-124, 1974.
48. Hayward, J.N. Ann. Rev. Physiol. 37: 191-210, 1975.
49. Hayward, J.N. Physiol. Rev. 57: 574-658, 1977.
50. Hayward, J.N. and D.P. Jennings. J. Physiol. 232: 515-543, 1973.
51. Hayward, J.N. and D.P. Jennings. J. Physiol. 232: 545-572, 1973.
52. Hayward, J.H., K. Murgas, K. Pavasuthipaisit, F.R. Perez-Lopez and M.V. Sofroniew. Neuroendocrinology 23: 61-75, 1977.
53. Hayward, J.N. and W.K. Smith. Arch. Neurol. 9: 171-177, 1963.
54. Hayward, J.N. and J.K. Smith. Amer. J. Physiol. 206: 15-20, 1964.
55. Hökfelt, T., R. Elde, K. Fuxe, O. Johansson, A. Ljungdahl, M. Goldstein, R. Luft. S. Efendic, G. Nilsson, L. Terenius, D. Ganten, S.L. Jeffcoate, J. Rehfeld, S. Said, M. Perez de la Mora, L. Possani, R. Tapia, L. Teran and R. Palacios. In: The Hypothalamus, edited by S. Reichlin, R.J. Baldessarini and J.B. Martin. New York, Raven Press, pp. 69-135, 1977.
56. Hökfelt, T. and K. Fuxe. In: Brain-endocrine Interaction. Median Eminence: Structure and Function, edited by K.M. Knigge, D.E. Scott and A. Weindel. Basel, Karger, pp. 181-223, 1972.
57. Ishida, A. Jap. J. Physiol. 20: 84-96, 1970.
58. Kandel, E.R. J. gen. Physiol. 47: 691-717, 1964.
59. Kelly, J.S. and L.P. Renaud. In: Pharmacology of the Hypothalamus, edited by B. Cox, I. Morris and A. Weston. London, The MacMillan Press (in press) 1978.
60. Koizumi, K., T. Ishikawa and C. McC. Brooks, Brain Res. 63: 408-413, 1973.
61. Koizumi, K. and H. Yamashita. J. Physiol. 221: 683-705, 1972.
62. Léranth, Cs., L. Zaborszky, J. Marton and M. Palkovits. Exp. Brain Res. 22: 509-523, 1975.
63. Lincoln. D.W. and J.B. Wakerley. J. Physiol. 242: 533-554, 1974.
64. Lincoln. D.W. and J.B. Wakerley. J. Physiol. 250: 443-461, 1975.
65. Makara, G.B., M.C. Harris and K.M. Spyer. Brain Res. 40: 283-290, 1972.

66. Makara, G.B. and E. Stark. Neuroendocrinology 16: 178–190, 1974.
67. Martin, J.B. Endocrinology 91: 107–115, 1972.
68. Martin, J.B., J. Kontor and P. Mead. Endocrinology 92: 1354–1361, 1973.
69. Moss, R.L., C.A. Dudley and M.J. Kelly. Neuropharmacology 17: 87–93, 1978.
70. Moss, R.L., R.E.J. Dyball, and B.A. Cross. Exp. Neurol. 34: 95–102, 1972.
71. Moss, R.L., M. Kelly and P. Riskind. Brain Res. 89: 265–277, 1975.
72. Moss, R.L., I. Urban and B.A. Cross. Am. J. Physiol. 223: 310–318, 1972.
73. Negoro, H. and R.C. Holland. Brain Res. 42: 385–402, 1972.
74. Negoro, H. S. Visessuwan and R.C. Holland. Brain Res. 57: 479–483, 1973.
75. Nicoll, R.A. In: Neurotransmitters, Hormones and Receptors: Novel Approaches, edited by J.A. Ferrendelli, B.S. McEwen and S.H. Snyder. Bethesda, Md., Society for Neuroscience, pp. 99–122, 1976.
76. Nicoll, R.A. and J.L. Barker. Brain Res. 35: 501–511, 1971.
77. Palkovits, M., Cs. Léranth, L. Zaborszky and M.J. Brownstein. Brain Res. 136: 339–344, 1977.
78. Palkovits, M. and L. Zaborszky. In: Handbook of the Hypothalamus, edited by P. Morgane and J. Paksepp, New York, Marcel Dekker (in press) 1979.
79. Pickford, G.E. and J.W. Atz. The Physiology of the Pituitary Gland of Fishes. New York, New York Zoological Society.
80. Pickford, M. J. Physiol. 106: 264–270, 1947.
81. Pittman, Q.J., H.W. Blume, R.E. Kearney and L.P. Renaud. Proc. XXVII Int. Cong. Physiol. 12: 525, 1977.
82. Poulain, D.A., J.B. Wakerley and R.E.J. Dyball. Proc. R. Soc. London B., 196: 367–384, 1977.
83. Renaud, L.P. Brain Res. 105: 59–72, 1976.
84. Renaud, L.P. J. Physiol. 260: 237–252, 1976.
85. Renaud, L.P. J. Physiol. 264: 541–564, 1977.
86. Renaud, L.P. In: Approaches to the Cell Biology of Neurons, edited by W.M. Cowan and J.A. Ferrendelli, Bethesda, Md., Society for Neuroscience, pp. 269–290, 1977.
87. Renaud, L.P., H.W. Blume and Q.J. Pittman. In: Frontiers in Neuroendocrinology, Vol. 5, edited by W.F. Ganong and L. Martini, New York, Raven Press, pp. 135–162, 1978.
88. Renaud, L.P., H.W. Blume, Q.J. Pittman, R.E. Kearney and B.W. MacKenzie. Soc. for Neuroscience Abstracts 3: 204, 1977.
89. Renaud, L.P., J.B. Martin and P. Brazeau. Nature 255: 233–235, 1975.
90. Renaud, L.P., J.B. Martin and P. Brazeau. Pharmac. Biochem. Behav. 5: Suppl. I: 171–178, 1976.

91. Rivier, C. and W. Vale. Endocrinology 101: 506–511, 1977.
92. Sawaki, Y. and K. Yagi. J. Physiol. 230: 75–85, 1973.
93. Sawaki, Y. and K. Yagi. J. Physiol. 260, 447–460, 1976.
94. Schally, A.V., A. Arimura and A.J. Kastin. Science 179:
       341–350, 1973.
95. Scharrer, E. and B. Scharrer. Recent Prog. Horm. Res. 10:
       183–240, 1954.
96. Seif, S.M. and A.G. Robinson. Ann. Rev. Physiol. 40: 345–376,
       1978.
97. Sharpless, S.K. and A.B. Rothballer. Amer. J. Physiol. 200:
       909–915, 1961.
98. Sherlock, D.A., P.M. Field and G. Raisman. Brain Res. 88:
       403–414, 1975.
99. Swanson, L.W. Brain Res. 128: 346–353, 1977.
100. Szentágothai, J., B. Flerkő, B. Mess and B. Halász. Hypo-
       thalamic Control of the Anterior Pituitary. Budapest,
       Akademiai Kiado, 1968.
101. Urban, I. and D. De Wied. Pharmac. Biochem. Behav. 8:
       51–59, 1978.
102. Vandesande, F. and K. Dierickx. Cell Tiss. Res. 164:
       153–162, 1975.
103. Wakerley, J.B. and D.W. Lincoln. J. Endocr. 57: 477–493,
       1973.
104. Wakerley, J.B., D.A. Poulain, D.A., R.E.J. Dyball and B.A.
       Cross. Nature 258: 82–84, 1975.
105. Woods, W.H., R.C. Holland and E.W. Powell. Brain Res. 12:
       24–46, 1969.
106. Yagi, K., T. Asuma and K. Matsuda. Science 154: 778–779, 1966.
107. Yagi, K. and Y. Sawaki. Brain Res. 84: 155–159, 1975.
108. Zaborszky, L., Cs. Léranth, G.B. Makara and M. Palkovits.
       Exp. Brain Res. 22: 525–540, 1975.
109. Zimmerman, E.A. In: Frontiers in Neuroendocrinology, edited
       by L. Martini and W.F. Ganong. New York, Raven Press,
       pp. 25–62, 1976.
110. Zolovick, A.J. In: The Neurobiology of the Amygdala, edited
       by B.E. Eleftheriou, New York, Plenum Press, pp. 643–684,
       1972.

THE OPIOID RECEPTORS AND THEIR LIGANDS

Lars Terenius

Department of Pharmacology

University of Uppsala, Uppsala, Sweden

INTRODUCTION

The term opioid is used to describe a morphine-like compound. Typical morphine-like activity includes analgesia, drowsiness, changes in mood and mental clouding (10). Several thousand compounds share these properties of morphine and it is clear that they have a very specific action profile. It is therefore logical to assume that they react with a common, specific receptor. In recent years, however, it has become obvious that subgroups of opioids and opioid receptors must be considered. One particularly interesting subgroup is represented by the endogenous morphine-like substances, the endorphins, which are peptides. It is the purpose of this communication to review studies on opioid receptor heterogeneity.

The structural diversity of compounds with morphine-like activity is quite enormous and knows no parallell in pharmacology, except possibly for the estrogens. Despite the variety in structural elements compatible with analgesic activity, a three dimensional optimum structure has been determined by structure-activity analysis and used to construct a complementary receptor surface (1, 15). However, the extensive structure-activity studies of Portoghese and co-workers as summarized recently (23), show that this approach is not simple. Using stereoisomers of analgesics as receptor probes, they arrived at the conclusion that multiple modes of interaction between an opioid and the receptor or the receptors are possible. Then it is understandable that the opioid receptors will recognize a wide variety of structurally dissimilar compounds. However, the structure-activity analysis does not allow distinction between multiple modes of interaction with different loci on a single receptor or interaction with a family of related receptors.

The chemical work of Portoghese and others has unfortunately not been followed up by extensive pharmacologic analysis. Therefore it is not known whether these differences in mode of interaction with the receptors have any functional consequenses. However, such differences have been observed in a series of substituted benzomorphanes. They have a very complex pharmacology as suggested from their mixed analgesic/anti-opiate activity and they also show dissociation from classical morphine-like compounds in their activity on peripheral tissue responses and on behavior. This led Martin and associates (17, 19) to postulate the existence of several opiate receptors. These results were interpreted in terms of interactions with a minimum number of 3 different receptors (Table 1). The μ-receptor would be the one principally occupied by the vast majority of classical opiates. This site is sensitive to naloxone. The κ- and σ-sites are recognized by a few benzomorphan derivatives. One interesting feature of these sites is their relative insensitivity to naloxone, the μ-receptor being around 10 times more sensitive. Thus, effects on the κ- and σ-receptors can still be observed after blockade of the classical μ-receptor.

TABLE 1. PHARMACOLOGIC PROFILES OF μ, κ AND σ AGONISTS

VARIABLE	AGONIST		
	μ	κ	σ
PULSE RATE	↓	0	↑
PUPILS	↓	↓	↑
RESPIRATORY RATE	↓	0	↑
TEMPERATURE	↓	0	↑
BEHAVIORAL EFFECT	INDIFFERENCE	HYPNOSIS	DELIRIUM

(From ref. 19)

ISORECEPTORS

In principle, receptor multiplicity can depend on differences in primary structure of the receptor molecules or on different stages of conformation of a single receptor molecule. In the former case, the term isoreceptor may be used in analogy with the iso-enzyme concept in biochemistry. Unfortunately, isolation and biochemical identification of receptor molecules have not yet been carried out extensively and therefore there is only indirect evidence for one or the other possibility. Table 2 illustrates the cases where it is likely that one is dealing with true isoreceptors.

The possible existence of isoreceptors is of importance for several reasons. One practical aspect is that it opens a possibility for the medicinal chemist to develop specific receptor agonists or blocking agents. In fact, the subclassification of these receptors has entirely been based on pharmacologic analysis. In several cases, it has led to the development of useful drugs, such as selective β_2-stimulants in asthma therapy or H_2-antagonists in the treatment of peptic ulcer. In the opioid field similar therapeutic applications have not yet been established.

While there seem to be isoreceptors, no one has so far described marked differences in the presynaptic elements for synthesis, storage, or release of a particular neurotransmitter. The genetic coding for the presynaptic build-up may therefore not be coupled to the genetic coding for the receptor elements. If a similar diversity existed in the presynaptic elements, it is likely that this would have been detected during the long history of random screening of neuroactive drugs. The isoreceptor systems may have evolved in parallell with other structural elements of the postsynaptic membrane. For instance, the nicotinic acetylcholine receptor is coupled directly to an ionophore (6), while muscarinic transmission in the superior cervical ganglion seems to be coupled

TABLE 2. ISORECEPTOR SYSTEMS

ADRENERGICS	α, β_1, β_2
CHOLINERGICS	Muscarinic, nicotinic (gangl.) nicotinic (skel. muscle)
HISTAMINERGICS	H_1, H_2
OPIOIDS	μ, κ, σ

to the activation of guanyl cyclase (12). Functionally distinct
from postsynaptic receptors are the presynaptically situated auto-
receptors, which regulate transmitter release (5). It is not yet
known whether structural differences exist between the autoreceptor
and the postsynaptic receptor, in other words, whether they are
isoreceptors, but such differences are possible in the light of
what has been suggested above.

In conclusion, the study of isoreceptors and receptor multiplicity in general gives the basic knowledge for understanding
physiologic mechanisms as well as for therapeutic intervention.

METHODS FOR STUDYING RECEPTOR HETEROGENEITY

This section will briefly describe some experimental approaches
to studies on receptor heterogeneity. The receptor differences discussed so far have been observed in structure-activity studies
usually with different types of blockers, each with selectivity for
one isoreceptor. In the opioid system there are very few known antagonists. The two most frequently used, naloxone and naltrexone,
are almost devoid of agonistic activity and they antagonize almost
every action of opioids. In fact, it is customary to use naloxone
reversal as a criterion for an opioid receptor.

Antagonist reversal can be used to characterize a receptor in
a quantitative way in the so-called pA_2 analysis (26). Here, graded
doses of the antagonist are tested against equiactive doses of
various agonists. If all agonists would act via the same receptor,
each will be affected to the same extent by the antagonist. This
approach has been used to characterize various actions of opiates
and has provided evidence for tissue specific receptor systems (30,
31). Thus, receptors mediating analgesia, respiratory depression,
or lenticular changes differ slightly in naloxone sensitivity.

Over the last 10 years, techniques have been developed for the
direct measurement of receptor affinity by biochemical methods.
These techniques, if properly controlled, may be used to give the
exact affinity constant of a particular drug or endogenous substance for a receptor. Furthermore, they are excellent for the
study of microheterogeneity of a single receptor molecule as induced by conformational transitions, which may occur between the
active, inactive and desensitized stages. In such analysis, it is
customary to use one radioisotope-labelled receptor ligand as the
indicator and various non-labelled substances as competitors. The
affinity of each substance can then be calculated indirectly. It is
very hard to observe binding site heterogeneity with the use of
only one radioindicator, unless differences between the affinities
of the binding sites are several orders of magnitude. A much more
sensitive approach is to use several radioindicators and to com-

pare the affinities of various competitors against each of them. To
compensate for the possible introduction of indicator errors in the
system, pairwise compensated indicator solutions are prepared, each
containing both indicator ligands at identical concentrations but
with alternative ligands labelled (Fig. 1). The competitors will
therefore be tested under identical conditions against both radio-
ligands (35). This method can be used to study receptor selectivity.
Absolute affinities can not be calculated if there is a family of
receptors and only imperfect (= nonselective) receptor indicators
are available.

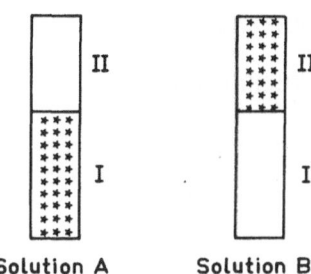

Solution A   Solution B

Fig. 1. Indicator solutions for
receptor selectivity measurements.
Solutions A and B contain two components, I and II, at identical
concentrations. In A, component I
is labelled, in B, component II
is labelled.

RECEPTOR HETEROGENEITY IN THE OPIOID SYSTEM

As indicated above, pharmacologic analysis reveals multiplicity of opioid receptors. One may for instance, investigate if this
is related to the multiplicity of endogenous opioid systems which
has been found to exist both with regard to localization to tissue
and to active peptide (Table 3). There are many examples of the
fact that the same peptides may be present in central and peripheral nervous system and in non-neuronal cells in peripheral tissue (21). The endorphins are no exception but it is probably ex-
ceptional that there are two families of apparently parallelly
acting systems, with enkephalins and β-endorphin as active ligands,
respectively. These systems seem to be truly separate (3, 24, 39).
In most systems studied, enkephalins and β-endorphin show similar
effects. However, the enkephalins are metabolically much more un-
stable in brain or plasma. For instance, despite their receptor
affinities being equal, enkephalin is a very poor analgesic even
if introduced intracerebrally (2, 4, 11). On the other hand, metabolically stable analogues of enkephalins as well as β-endorphin
are potent analgesics (8, 25).

In binding analysis it soon became evident that the peptides
differ from opiate alkaloids in binding selectivity (16, 32). This
can be demonstrated in a test system where various competitors are
tested for affinity against labelled dihydromorphine or a labelled

TABLE 3. MULTIPLE OPIOID SYSTEMS (TENTATIVE LISTING)

SITE	PRINCIPAL PEPTIDE
CNS	ENKEPHALINS
	β-ENDORPHIN
GANGLIA	ENKEPHALINS
ADENOHYPOPHYSIS	β-ENDORPHIN
ADRENAL MEDULLA	ENKEPHALINS
GASTROINTESTINAL TRACT (ENDOCRINE CELLS)	ENKEPHALINS

(See further refs. 9, 14, 27, 33)

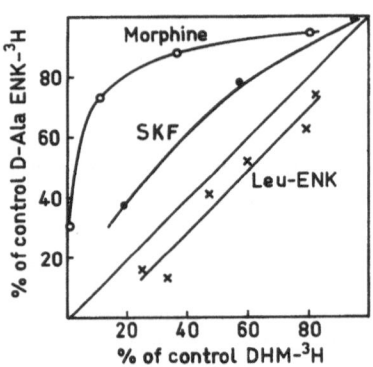

Fig. 2. Receptor selectivity of morphine, SKF-10046 and leu-enkephalin against dihydromorphine-^{3}H (μ-sites) and D-ala^2-leu^5-enkephalin-^{3}H (δ-sites). In this graph, a competitor with no selectivity would fall on the straight line interconnecting the lower left and upper right corners. Curvature upwards to the left indicates preference for μ-sites, downwards to the right, for δ-sites.
The receptor preparation was from rat brain synaptic membranes.

enkephalin (Fig. 2). The graphical procedure allows the detection of very small differences in receptor selectivity (35). A compound with no selectivity for any site should give a straight line between the lower left and upper right corners as shown in the figure. It is then evident that leu-enkephalin shows rather limited selectivity for the enkephalin radioindicator. Morphine, on the other hand, shows marked selectivity for sites labelled by dihydromorphine. For morphine the ratio between the apparent K_D's is of the order 20-30. Other opiates like methadone, levorphanol and etorphine behave as morphine. Interestingly, the same is true for naloxone. Thus, morphine-like agonists (μ-agonists) show marked selectivity for sites which in analogy may be called μ-sites. In doing so, however, it should be realized that there is a long step between binding affinity data and a pharmacologic response, and that such a designation can only be tentative. The peptide sites have been named δ-sites (16), since their relation to the pharmacologic receptors is still uncertain.

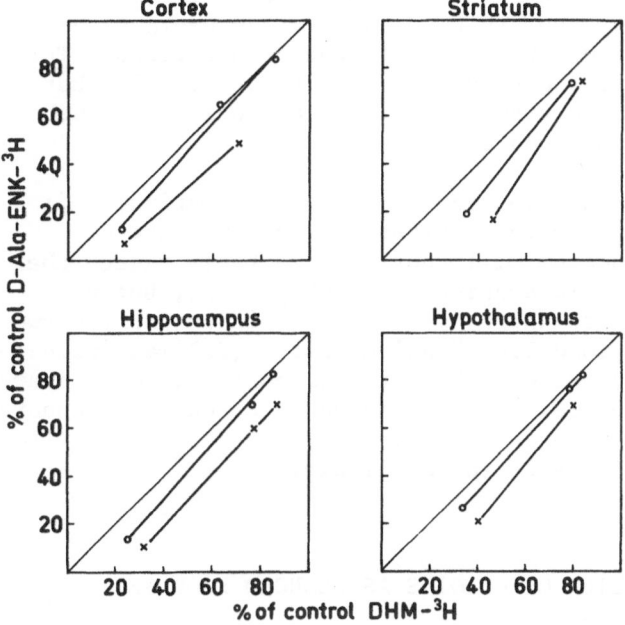

Fig. 3. Selectivity of β-endorphin (circles) and met-enkephalin (crosses) against μ- and δ-sites of receptor preparations from various rat brain areas. See further text to Fig. 2.

One important question is whether the two natural ligands, enkephalin and β-endorphin, differ in selectivity between μ- and δ-sites. Results of such analysis are shown in Fig. 3. Receptor preparations from several CNS areas (also including the diencephalon, medulla oblongata and spinal cord, which are not shown) gave consonant results. The peptides show almost equal affinity for the two sites, maybe β-endorphin is slightly more equal. We have also tested the two enkephalins against each other and have found no differences in selectivity (unpublished). The conclusion from these experiments would be that β-endorphin and enkephalin share very similar sites and in terms of occupation threshold it would not matter which type of receptor they interact with. However, further analysis pointed to a selectivity difference in a system where binding to μ-sites (radioindicator, dihydromorphine) was compared with that to antagonist sites (radioindicator, naltrexone).

Then, β-endorphin shows less selectivity than other tested agonists (unpublished). Our interpretation is that there is probably another set of binding sites with selectivity for β-endorphin, which like the other sites is antagonized by narcotic antagonists. One would have to do direct binding experiments with labelled β-endorphin to settle the matter.

The results of receptor selectivity studies carried out in the author's laboratory are summarized in Table 4. Site μ has been identified with dihydromorphine as the radioindicator, site δ with the natural enkephalins or the synthetic derivative D-ala^2-leu^5-enkephalin and site "antagonist" with naloxone or naltrexone. This is a minimum number of sites observed and it is conceivable that further heterogeneity can be visualized if other indicators such as, for instance β-endorphin would be used.

The proportion of μ- and antagonist-sites, respectively, is shifted by the absence (favouring μ-sites) or the presence of sodium ion. Addition of sodium will therefore markedly reduce the apparent affinity of narcotic agonists and enkephalins, but not that of narcotic antagonists which have equal affinity for the two sites (cf Table 4). This so-called "sodium shift" (22) may indicate that the μ- and antagonist-sites relate to two conformational states of the same receptor molecule, perhaps linked to a sodium ionophore. The evidence for this hypothesis has been summarized recently (29) but no proof for its validity is yet available.

TABLE 4. RELATIVE AFFINITIES OF OPIOIDS AS DEDUCED FROM SELECTIVITY ANALYSIS OF OPIOID RECEPTOR SITES.

SUBSTANCE	SITE μ	SITE ∂	SITE "ANTAGONIST"
ALKALOID AGONISTS	HIGH	LOW	LOW
ENKEPHALINS	HIGH	HIGH	LOW
β-ENDORPHIN	HIGH	HIGH	INTERMEDIATE
ALKALOID ANTAGONISTS	HIGH	LOW	HIGH

Functional correlates for the μ- and δ-site diversity have been sought and found. In the guinea-pig ileum there seems to be about the same proportion of these sites as in brain while in the mouse vas deference there is preponderance of δ-sites (16). Likewise, mouse neuroblastoma x glioma hybrid cells which are sensitive to opioids (28, 37) have mainly δ-sites (38). Is there a connection between the δ-sites and the isoreceptors identified by Martin? Testing of the receptor selectivity of a typical σ-agonist, SKF-10046 (19) in the same binding system (Fig. 2) reveals that it shows less selectivity than morphine-like alkaloids or naloxone. Similar observations have been made by Della Bella and associates (7). Thus, there is some indication that the particular activity profile of the SKF-substance could be related to a particular receptor selectivity profile. In further support of this conjecture is the observation that actions of the SKF-substance require more naloxone for reversal than those of μ-agonists (19). This is exactly what might have been predicted from the selectivity analysis, naloxone being more selective for μ-sites than the SKF- compound (Fig. 2). Therefore there is at least some indirect evidence for a relation between δ-sites and σ-receptors.

The δ-sites of neuroblastoma x glioma cell hybrids seem to be linked to a PGE-sensitive adenyl cyclase (28, 37). Whether this linkage is unique to the system and whether other binding sites or receptor systems are linked to the same cyclase or other effector mechanisms is not known.

One cannot assume automatically that all receptor responses observed on administration of a drug or all the binding sites observed in the test tube, really represent functionally relevant elements. Several years ago, Martin (18) formulated the hypothesis that there may be receptors for which there are no endogenous agonists, the so-called redundancy hypothesis. It could also be that a certain population of receptors shows an unfavourable selectivity profile for the endogenous agonist and it is therefore generally possible only to observe an effect on these receptors if an artificial agonist is introduced into the system. Another situation when such a receptor could be activated is via an unconventional distribution of the endogenous agonist or via the formation of a toxin, a chemically modified agonist. Such a toxin might mimic an artificial agonist. This possibility occurred to us during an investigation of endorphin activity in man under various physiologic and pathologic conditions. For this purpose, we developed a receptor-binding assay for endorphins in cerebrospinal fluid (34). This assay revealed relatively high levels in samples from psychotic patients notably schizophrenics (36). Since some σ-agonists produce psychotomimetic effects, particularly auditory hallucinations, it was postulated that an abnormal endorphin might be produced and be involved in some aspects of schizophrenia. To test this hypo-

TABLE 5. TRIALS OF NALOXONE IN CHRONIC SCHIZOPHRENIA.
A RESPONSE IS CHARACTERIZED BY A TEMPORARY EXTINCTION
OR REDUCTION OF HALLUCINOSIS. DATA ARE COLLECTED FROM
10 DIFFERENT SERIES DESCRIBED IN THE LITERATURE (33).

SERIES	DOSE (MG)	NO. OF RESPONDERS/ NO. OF PATIENTS
1	0.4	4/6
2	0.4	0/7
3	0.4-10	0/14
4	1.2	0/8
5	0.4-1.2	0/12
6	4.0	12/20
7	10	6/9
8	25	7/12
9	0.8	1/10
10	0.4	1/1
		TOTAL 31/99

thesis, naloxone was given to hallucinating schizophrenic patients
(13). The results of this study were partly encouraging and positive results have been observed in other series of selected patients
(Table 5). In fact, it has been claimed very recently that an abnormal β-endorphin with a leucine in position 5 can be isolated
from hemodialyzates of schizophrenics (20). It is not yet known
whether this compound produces psychotomimetic effects and has a
σ-receptor action profile.

CONCLUSIONS

The data summarized here, show that opioid receptors are
heterogenous in test systems both under in vitro and in vivo conditions. Probably some of this heterogeneity should be understood
in terms of isoreceptors, i.e. completely different receptor systems. In that sense the family of opioid receptors is not a special
case but maybe one of the most extensively studied. The endogenous
ligands, the endorphins, show very little selectivity between opioid
agonist sites. No opiate alkaloid which mimics the receptor selectivity of endorphins is yet known. The neuroanatomical and func-

tional correlates to the receptor multiplicity are very incompletely known.

ACKNOWLEDGEMENTS

Work in the author's laboratory is supported by the Swedish Medical Research Council.

REFERENCES

1. Beckett, A.H. and A.F. Casy. J. Pharm. Pharmacol. 6:986-999, 1954.
2. Belluzzi, J.D., N. Grant, V. Garsky, D. Sarantakis, C.D. Wise and L. Stein. Nature (London) 260:625-626, 1976.
3. Bloom, F., E. Battenberg, J. Rossier, N. Ling and R. Guillemin. Proc. Natl. Acad. Sci. USA 75:1591-1595, 1978.
4. Büscher, H.H., R.C. Hill, D. Römer, F. Cardinaux, A. Closse, D. Hauser and J. Pless. Nature (London) 261:423-425, 1976.
5. Carlsson, A. In: Pre- and Postsynaptic Receptors, edited by E. Usdin and W.E. Bunney Jr. New York: Marcel Dekker, 1975.
6. Changeux, J.-P. In: Handbook of Psychopharmacology, vol. 6: Biogenic Amine Receptors, edited by L.L. Iversen, S.D. Iversen and S.H. Snyder. New York: Plenum Press, 1975, p. 235-301.
7. Della Bella, D., F. Casacci and A. Sassi. In: Advances in Biochemical Psychopharmacology, vol. 18, edited by E. Costa and M. Trabucchi. New York: Raven Press, 1978, p. 271-277.
8. Dutta, A.S., J.J. Gormley, C.F. Hayward, J.S. Morley, J.S. Shaw, G.J. Stacey and M.J. Turnbull. Br. J. Pharmacol. 61:481P-482P.
9. Goldstein, A. Science 193:1081-1086, 1976.
10. Goodman, L.S. and A. Gilman (editors). The Pharmacological Basis of Therapeutics (5th ed.). New York: Macmillan Publ. Co., 1975.
11. Gráf, L., J.I. Szekely, A.Z. Ronai, Z. Dunai-Kovacs and S. Bajusz. Nature (London) 263:240-241, 1976.
12. Greengard, P. Nature (London) 260:101-108, 1976.
13. Gunne, L.-M., L. Lindström and L. Terenius. J. Neural Transm. 40:13-19, 1977.
14. Hughes, J. and H.W. Kosterlitz. Br. Med. Bull. 33:157-161, 1977.
15. Lewis, J.W., K.W. Bentley and A. Cowan. Ann. Rev. Pharmacol. 11:241-270, 1971.
16. Lord, J.A.H., A.A. Waterfield, J. Hughes and H.W. Kosterlitz. Nature (London) 267:495-499, 1977.
17. Martin, W.R. Pharmacol. Rev. 19:464-521, 1967.
18. Martin, W.R. Fed. Proc. 29:13-18, 1970.

19. Martin, W.R., C.G. Eades, J.A. Thompson, R.E. Huppler and P.E. Gilberg. J. Pharmacol. Exp. Ther. 197:517-532, 1976.
20. Palmour, R.M., F.R. Ervin, H. Wagemaker and R. Cade. In: Endorphins in Mental Illness, edited by E. Usdin and W.E. Bunney Jr. London: McMillan, 1978, in press.
21. Pearse, A.G.E. Nature (London) 262:92-94, 1976.
22. Pert, C.B. and S.H. Snyder. Mol. Pharmacol. 10:868-879, 1974.
23. Portoghese, P.S. Acc. Chem. Res. 11:21-29, 1978.
24. Rossier, J., T.M. Vargo, S. Minick, N. Ling, F.E. Bloom and R. Guillemin. Proc. Natl. Acad. Sci. USA 74:5162-5165, 1977.
25. Römer, D., H.H. Büscher, R.C. Hill, J. Pless, W. Bauer, F. Cardinaux, A. Closse, D. Hauser and R. Huguenin. Nature (London) 268:547-549, 1977.
26. Schild, H.O. Br. J. Pharmacol. 2:189-206, 1947.
27. Schultzberg, M., T. Hökfelt, J.M. Lundberg, L. Terenius, L.G. Elfvin and R. Elde. Acta physiol. scand., in press, 1978.
28. Sharma, S.K., M. Nirenberg and W.A. Klee. Proc. Natl. Acad. Sci. USA 72:590-599, 1975.
29. Simon, E.J. and J.M. Hiller. Ann. Rev. Pharmacol. 18:371-394, 1978.
30. Smith, A.A., R. Albin and M. Crofford. In: Opiates and Endogenous Opioid Peptides, edited by H.W. Kosterlitz. Amsterdam: Elsevier/North Holland, 1976, p. 289-294.
31. Smits, S.E. and A.E. Takemori. Br. J. Pharmacol. 39:627-638, 1970.
32. Terenius, L. Psychoneuroendocrinology 2:53-58, 1977.
33. Terenius, L. Ann. Rev. Pharmacol. 18:189-204, 1978.
34. Terenius, L. and A. Wahlström. Life Sci. 16:1759-1764, 1975.
35. Terenius, L. and A. Wahlström. Eur. J. Pharmacol. 40:241-248, 1976.
36. Terenius, L., A. Wahlström, L. Lindström and E. Widerlöv. Neuroscience Letters 3:157-162, 1976.
37. Traber, J., K. Fischer, S. Latzin and B. Hamprecht. FEBS Lett. 49:260-263, 1974.
38. Wahlström, A., M. Brandt, L. Moroder, E. Wünsch, G. Lindeberg, U. Ragnarsson, L. Terenius and B. Hamprecht. FEBS Lett. 77:28-32, 1977.
39. Watson, S.J., J.D. Barchas and C.H. Li. Proc. Natl. Acad. Sci. USA 74:5155-5158, 1977.

REGULATION OF THE β-ADRENERGIC RECEPTOR IN THE PINEAL GLAND AND RED CELL MEMBRANES

Julius Axelrod

Laboratory of Clinical Science, NIMH

Bethesda, Maryland

The pineal gland and red cell membranes have served as useful models to study the β-adrenergic receptor and its regulation. The pineal gland has a specific β-adrenergic receptor that is stimulated by noradrenaline released from nerve terminals innervating this gland. Interaction of the pineal β-adrenergic receptor triggers a series of reactions that leads to the synthesis of the pineal specific hormone melatonin. The critical biochemical step that is controlled by the β-adrenergic receptor is the synthesis of the enzyme serotonin N-acetyltransferase. The enzyme forms the precursor for the melatonin forming enzyme, hydroxyindole-O-methyltransferase. The responsiveness of the β-adrenergic receptor and the consequent synthesis of melatonin depends on the prior exposure to catecholamines. Reduced release of noradrenaline causes supersensitivity while increased exposure to catecholamine agonists results in a rapid subsensitivity. Changes in sensitivity is found not only in the β-adrenergic receptor but also in adenylate cyclase, protein kinase and other subcellular events.

Phospholipids are an important component of membranes which influence receptor actions. Two enzymes involved in the biosynthesis of phosphatidylcholine from phosphatidylethanolamine were found in red cell membranes. Stimulation of the β-adrenergic receptor in red cell ghosts caused a marked increase in methylation of phospholipids, membrane fluidity and coupling of the receptor with adenylate cyclase. Stimulation of phospholipid methylation with S-adenosylmethionine increases the number of β-adrenergic receptors in reticulocytes. The interrelationship between β-adrenergic receptor, phospholipid synthesis and translocation and membrane fluidity will be described.

THE β-ADRENERGIC RECEPTOR IN THE PINEAL GLAND

Neurotransmitters and peptide hormones after they are discharged from their storage sites in nerves or endocrine tissues interact with specific post-synaptic receptors on cell surfaces. This initiates a complex series of events that ultimately leads to a specific cellular response. Recently, there has been considerable progress as to how receptors recognize and interact with their specific ligands. However, the biochemical reactions and membrane events that generate complex intracellular reactions are not well understood. During the past several years we have found that the pineal gland can serve as a useful model to study the β-adrenergic receptor. This organ provided a relatively simple system in which the biological consequences of β-adrenergic stimulation has been clarified to some degree. The physiological responses of this gland to noradrenergic stimulation can be readily modified so that the process by which cells can regulate their own responses can be studied. We have also found that the red cell membranes have provided a productive tissue to study intracellular events that regulate the action of β-adrenergic receptors.

The pineal gland is located between two cerebral hemispheres and is innervated by noradrenergic nerves whose cell bodies originate in the superior cervical ganglia (Kappers, 1960). Noradrenaline released from these nerves stimulates the pineal to synthesize melatonin in the gland. This hormone influences reproductive functions and pigmentation. Melatonin is synthesized in the pineal gland by the following steps: tryptophan → 5-hydroxytrytophan → serotonin → N-acetylserotonin → melatonin (Axelrod, 1974). The enzymes catalyzing this biosynthetic pathway are tryptophan hydroxylase, amino acid decarboxylase, serotonin N-acetyltransferase and hydroxyindole O-methyltransferase. The addition of noradrenaline to the pineal gland in organ culture can also stimulate the synthesis of melatonin from tryptophan (Axelrod, et al., 1969). Melatonin synthesis is initiated via a β-adrenergic receptor coupled with adenylate cyclase. Serotonin N-acetyltransferase is the enzyme specifically activated by the β-adrenergic receptor (Klein, et al., 1970). Additions of the β-adrenergic agonist isoproterenol or dibutyryl cyclic AMP stimulates the synthesis of N-acetyltransferase in pineal ʼorgan culture. The β-adrenergic antagonist propranolol or a protein synthesis inhibitor blocks the stimulation of the acetylating enzyme. This indicates that the binding of catecholamine to the β-adrenergic receptor leads to the synthesis of new serotonin N-acetyltransferase molecules. The levels of N-acetyltransferase in the pineal gland shows a marked circadian rhythm, rising at night and falling during the daytime (Klein and Weller, 1970; Deguchi and Axelrod, 1972). During the nighttime there is an increased release of noradrenaline onto the pineal and this causes a 30-fold increase in serotonin N-acetyltransferase

and melatonin formation. Environmental lighting at daytime reduces
the release of noradrenaline to such an extent that the induction
of N-acetyltransferase is suppressed. The acetylating enzyme in
the pineal can be increased by an injection of isoproterenol during
the daytime when the enzyme is very low (Deguchi and Axelrod, 1972).

REGULATION OF SENSITIVITY OF THE β-ADRENERGIC RECEPTOR IN THE PINEAL

Development of supersensitivity, after denervation or long
term reduction of the β-adrenergic stimulation is a well-known phenomena. Until recently little was known as to how such a change in
responsitivity of the receptor occurs. The fact that the nerve impulses flow and release of noradrenaline onto the pineal can be easily
changed by placing the rat in light or darkness proved to be an excellent experimental tool to study super- and subsensitivity of the
β-adrenergic receptor. Denervation of the pineal gland resulted in
a marked increase to responsiveness to the induction of N-acetyltransferase or elevation of cyclic AMP after exposure of the pineal
gland to isoproterenol in organ culture (Deguchi and Axelrod, 1973).
Continuous lighting which reduces the release of noradrenaline, also
elicited supersensitivity to stimulation of cyclic AMP and N-acetyltransferase by isoproterenol. Conversely the repeated injections
of isoproterenol in a rat led to a diminished response with respect
to N-acetyltransferase induction and cyclic AMP elevation when the
pineal was removed and treated with the isoproterenol in organ
culture. These results demonstrated that the responsiveness of the
pineal gland is dependent on the degree of previous exposure of the
β-adrenergic receptor to the neurotransmitter noradrenaline. The
changes in response occurred rapidly within hours.

In a series of studies we have found that there are multiple
sites involved in the regulation of sensitivity of the β-adrenergic
receptor. These sites are adenylate cyclase, cyclic AMP-dependent
protein kinase, phosphodiesterase and protein synthesizing mechanisms
for N-acetyltransferase. The triggering step in the sequence of
events in the formation of the pineal hormone, melatonin, is the
binding of the neurotransmitter noradrenaline to the β-adrenergic
receptor on the cell surface of the pineal cell.

Recently, ligands have been found that specifically bind to the
β-adrenergic receptor. One such ligand ^{3}H-dihydroalprenolol binds to
the β-adrenergic receptor of many cells with the characteristics consistent with the known properties of this receptor (Lefkowitz, et
al., 1974; Zatz et al., 1976). The binding of ^{3}H-dihydroalprenolol
is saturable, stereospecific, rapid and reversible. The order of
potency of agonist or antagonist in competing for the binding
sites is similar to their potency in stimulating or inhibiting
adenylate cyclase activity. In pineal glands in which supersensi-

tivity was produced by exposure to light there was more binding of the ^{3}H-dihydroalprenolol than in subsensitive pineal cell membranes obtained from rats placed in darkness (Kababian, et al., 1975). A prior injection of isoproterenol produced a subsensitivity to the induction of N-acetyltransferase and also caused reduction in a number of available binding sites. The increase and decrease in a number of receptor sites were not dependent on new protein synthesis nor was there a change in the conformation of the available binding sites for agonist or antagonist. Changes in response of adenylate cyclase to isoproterenol correlated with changes in binding sites under similar conditions in super- and subsensitivity. The injection of isoproterenol produced a rapid decrease (within 1/2 hour) in the binding of ^{3}H-dihydroalprenolol and to the stimulation of cyclic AMP synthesis by isoproterenol.

During the onset of darkness there is a fifty-fold elevation of serotonin N-acetyltransferase and melatonin in the rat pineal. These marked changes in responsivity cannot be only accounted for by the increased release of noradrenaline or by the elevation of the number of binding sites of the β-adrenergic receptor. There must be other cellular changes beyond the β-adrenergic receptor. Adenylate cyclase of pineal glands made supersensitive by exposure to light were much more responsive to stimulation by sodium fluoride (Kebabian, et al., 1975) or cholera toxin (Zatz, 1977). Both of these compounds bypass the β-adrenergic receptor and act directly on the adenylate cyclase. These results indicate that the reduced release of noradrenaline when rats are exposed to light results in a increase in responsivity to adenylate cyclase as well as an increase in the number of receptor sites on the pineal cell surface.

Changes in adenylate cyclase also affects the synthesis and accumulation of cyclic AMP. In pineal glands from rats exposed to light there is an increased accumulation of the cyclic AMP after stimulation (Kebabian, et al., 1975). Dibutyryl cyclic AMP, a compound that bypasses the receptor adenylate cyclase complex, is more effective in inducing N-acetyltransferase in a supersensitive pineal (light exposed rats) than a subsensitive pineal (dark exposed rats) (Romero and Axelrod, 1975). This suggests that there are intracellular factors as well as the membrane receptor-adenylate cyclase complex that are changed in super- and subsensitivity.

The induction of N-acetyltransferase activity by cyclic AMP is presumably acting via a protein kinase. Marked differences in protein kinase activity were found in supersensitive and subsensitive pineals (Zatz and O'Dea, 1976). After exposure to light there was almost twice as much protein kinase activity after isoproterenol stimulation both in basal activity and after the addition of cyclic AMP. All of these findings indicate that there are multiple sites in the regulation of the sensitivity to β-adrenergic complex.

These are β-adrenergic binding sites, adenylate cyclase, cyclic
AMP dependent protein kinase. Each step appears to be regulated
by and reflects an amplification of the preceding step. The entire
cascade is initiated by changes in the rate of neurotransmitter
release from sympathetic nerve endings.

MEMBRANE LIPID METHYLATION AND THE β-ADRENERGIC RECEPTOR

In view of the rapid changes in responsivity of the β-
adrenergic receptors to variations in transmitter release without
synthesizing new receptors, we suspected that there were changes
in the properties of the cell membrane in which these receptors are
embedded. It now appears that the methylation of membrane lipids
are an important factor in the actions of the β-adrenergic receptor.
The discovery of the role of membrane phospholipids in the regulation of the β-adrenergic receptor came about in an unexpected manner.
In studying the effect of protein carboxymethylase in exocytoses in
the adrenal gland, we found that lipids in this gland were actively
methylated when Mg^{++} and S-adenosylmethionine were present. This
prompted a search for the Mg^{++}-dependent enzyme responsible for
lipid methylation. Two enzymes were identified and characterized
in the adrenal medulla microsomes that catalyzed the methylation of
phosphatidylethanolamine to phosphatidylcholine (Hirata, et al.,
1978). These enzymes had markedly different properties and were
shown to be critical in the regulation of the activity of the β-
adrenergic receptor. The first enzyme methylated phosphatidylethanolamine to form phosphatidyl-N-monomethylethanolamine. The
enzyme had an absolute requirement for Mg^{++} and a low Km for S-
adenosylmethionine. The second enzyme catalyzes the two successive
methylations of phosphatidyl-N-monomethylethanolamine to form
phosphatidylcholine. The latter enzyme did not require Mg^{++} and
had a high Km for S-adenosylmethionine.

The second enzyme could be easily removed from membranes
while the second was not readily solubilized. This suggested an
asymmetric arrangement for these enzymes in membranes as had been
previously found with their substrates phosphatidylethanolamine
and phosphatidylcholine.

To study the asymmetry of the lipid methyltransferases and
its possible relationship to the β-adrenergic receptor we used
red cell membranes (Hirata and Axelrod, 1978). Red cells can be
easily made into ghosts and also turned inside-out. These cells
have β-adrenergic receptors and the membrane responsivity of
these receptors can increase in immature reticulocytes.

The two enzymes that convert phosphatidylethanolamine to
phosphatidylcholine were found to be present in high concentrations

in red cell membranes. Using red cell ghosts that were right side
out and inside out, these enzymes were found to be asymmetrically
arranged in the membrane. The first enzyme faced the cytoplasmic
side and the second was located on the external surface of the
membrane. Phospholipids in red cells were found to be translocated
from the inner to the outside surface of the membrane. This
transmembrane movement of phospholipids was rapid, occurring within
two minutes. The methylation and translocation of phosphatidylethanolamine facing the cytoplasmic side of the membrane to phosphatidylcholine on the outer membrane takes place by a stepwise
methylation within the membrane. Mg^{++} plays a critical role in this
process. Changes in transmembrane movement of phospholipids has a
profound effect on fluidity and lateral movement of proteins.

     Once the nature of the enzymatic synthesis of phospholipids
and its translocation within membranes was characterized, we were
in a position to study the relationship between β-adrenergic receptor, membrane phospholipid synthesis and fluidity. For these
studies rat reticulocyte ghosts were used since the immature red
cells have considerably more β-adrenergic receptor sites than
mature erythrocytes. When reticulocyte ghosts were preloaded with
^{3}H-S-adenosylmethionine and then exposed to the β-adrenergic
agonist isoproterenol there was a marked increased synthesis of
methylated phospholipids. When the reticulocyte was exposed to
isoproterenol the synthesis of phosphatidylcholine in membranes
was increased more than five-fold in one hour. Experiments using
low concentrations of S-adenosylmethionine to mainly stimulate
the methylation of phosphatidylethanolamine and high concentrations
of S-adenosylmethionine necessary to activate the second methyl-
transferase, isoproterenol acted on both enzymes. Several
additional experiments demonstrated that the increase in membrane
phospholipid synthesis was specifically stimulated by the β-
adrenergic receptor (Hirata, Strittmatter and Axelrod, unpublished).
The increase of lipid methylation was stereospecific, D-isopro-
terenol was not active. Propranolol blocked the effect of isoproterenol and the elevation of phosphatidylcholine synthesis by
catecholamine agonists showed the same order of potency as for the
activation of adenylate cyclase: isoproterenol > noradrenaline
   > adrenaline. The α-adrenergic blocking agent pentolamine
showed no effect. When the adenylate cyclase was stimulated
directly with cholera toxin or NaF there was no increase in
phospholipid methylation indicating that the receptor but not
adenylate cyclase was involved. Comparative studies with reticulocytes and matured red cell ghosts indicated a direct correlation
between number of β-adrenergic sites and an increase in phospholipid methylation induced by isoproterenol. These experiments
suggested that the β-adrenergic receptor on the outer membrane in
some manner suppressed the activity of the phosphatidylcholine
synthesizing enzyme also present on the outer surface of the

membrane. Once the β-adrenergic receptor was occupied by an agonist it appears that the methyltransferases are derepressed so that their activity is increased.

The increased synthesis of phospholipids could change the fluidity of the membrane so that the lateral movement of receptors are more rapid (Hirata and Axelrod, submitted). Using a membrane marker to measure fluidity of red cell membranes diphenyl 1,3,5 hexatrien, it was found that when membrane phospholipid methylation was increased (by the introduction of S-adenosylmethionine) so was the fluidity. A direct relationship between elevation of membrane fluidity and the formation of phosphatidyl-N-monomethylethanolamine was found. These observations suggest that stimulation of the β-adrenergic receptor increases phospholipid methylation which in turn makes the membrane more fluid. The change in fluidity would then speed the lateral movement of the β-adrenergic receptor and increase the chance of the receptor to couple with adenylate cyclase. Additional experiments showed that if methylation of membrane phosphatidylethanolamine was increased with S-adenosyl-methionine, adenylate cyclase activity was also increased. When methylation of phospholipids was prevented by S-adenosylhomo-cysteine the coupling of the β-adrenergic receptor with adenylate cyclase was blocked. Stimulation of methylation of phospholipids in reticulocyte membranes resulted in an increase in the number of β-adrenergic sites available for binding with radioactive ligands. Whether neuropeptides and other hormones acting on specific membrane receptors stimulate phosphatidylcholine synthesis, membrane fluidity, and receptor availability remains to be established.

REFERENCES

Axelrod, J.: Science 184, 1341-1348, 1974.
Axelrod, J., Shein, H.M., and Wurtman, R.J.: Proc. Natl. Acad. Sci. USA 62, 544-549, 1969.
Deguchi, T., and Axelrod, J.: Proc. Natl. Acad. Sci. USA 69, 2547-2550, 1972.
Deguchi, T., and Axelrod, J.: Proc. Natl. Acad. Sci. USA 70, 2411-2414, 1973.
Hirata, F., and Axelrod, J.: Proc. Natl. Acad. Sci. USA, 75, 1718-1721, 1978.
Hirata, F., Viveros, O.H., Diliberto, Jr., E.J., and Axelrod, J.: Proc. Natl. Acad. Sci. USA, 1978, in press.
Kappers, J.A.: Z. Zellforsch. Mikrosk. Anat. 52, 163-215, 1960.
Kebabian, J.W., Zatz, M., Romero, J.A., and Axelrod, J.: Proc. Natl. Acad. Sci. USA 72, 3735-3739, 1975.
Klein, D.C., Berg, G.R., and Weller, J.L.: Science 168, 979-980, 1970.
Klein, D.C., and Weller, J.L.: Science 169, 1093-1095, 1970.

Lefkowitz, R.J., Mukherjee, C., Coverstone, M., and Caron, M.C.:
    Biochem. Biophys. Res. Commun. 60, 703-709, 1974.
Romero, J.A., and Axelrod, J.: Proc. Natl. Acad. Sci. USA 72,
    1661-1665, 1975.
Zatz, M.: Life Sci. 21, 1267-1276, 1977.
Zatz, M., Kebabian, J.W., Romero, J.A., Lefkowitz, R.J., and
    Axelrod, J.: J. Pharmacol. Exp. Ther. 196, 714-722, 1976.
Zatz, M., and O'Dea, R.F.: J. Cyclic Nucleotide Res. 2, 427-439,
    1976.

CYCLIC NUCLEOTIDE AND PROTEIN PHOSPHORYLATION MECHANISMS IN THE

CENTRAL NERVOUS SYSTEM

Paul Greengard

Department of Pharmacology, Yale University School of

Medicine, New Haven, Connecticut   06510   U.S.A.

A. Cyclic AMP, Cyclic GMP and Neuronal Function

In the past few years, a beginning has been made in understanding the molecular basis by which the activity of nerve cells
is regulated. As in the case with hormones acting on non-neuronal
tissues (1), there is now evidence that cyclic AMP may act as an
intracellular mediator for the actions of certain neurotransmitters
on nerve cells (for recent reviews, see 2-8). There is also evidence that certain of the effects of some other neurotransmitters
on their target cells may be mediated by cyclic GMP (2-8). In
addition, there is now evidence that receptor-second messenger
systems can be utilized for two types of modulation of synaptic
transmission (Figure 1). Thus, the postsynaptic cell of the
receptor-second messenger modulatory synapse can be either the presynaptic or the postsynaptic cell of the synapse being modulated.
In other words, the receptor-second messenger system can be utilized
for either presynaptic (Figure 1A) or postsynaptic (Figure 1B) modulation of synaptic transmission. In addition to mediating the
effects of modulator neurons on the permeability properties of
input and output neurons, cyclic nucleotides appear to have many
other roles in neuronal function (4,5,8).

The demonstration of the existence of neurotransmitter-sensitive adenylate cyclases in nervous tissue, their regional and subcellular distribution, and the properties of these enzymes suggest
a role for cyclic AMP in the physiology of synaptic transmission.
For each of the neurotransmitters listed in Table 1, an adenylate
cyclase activated by low concentrations of the neurotransmitter in
question has been found in nervous tissue. Moreover, compounds

A) Presynaptic modulation

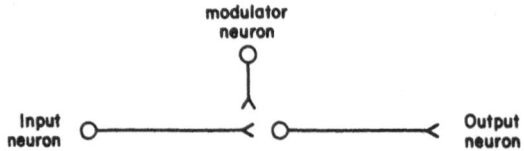

B) Postsynaptic modulation

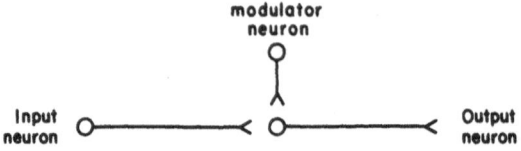

FIGURE 1. Two types of modulation of synaptic transmission in
which cyclic AMP has been implicated as a second messenger.
(A) Presynaptic modulation. In this type of modulation, the postsynaptic cell of the receptor-second messenger synapse is in turn
a presynaptic terminal, and is labeled "input neuron." (B) Postsynaptic modulation. In this type of modulation, the postsynaptic
cell of the receptor-second messenger synapse is also the postsynaptic cell for another synapse, and is labeled "output neuron."
(Taken from reference 8.)

which act as physiological agonists (i.e. mimic the physiological
actions) of each of these neurotransmitters stimulate the appropriate adenylate cyclase. Conversely, those substances which act
as antagonists of the physiological effects of these neurotransmitters in general block the activation by the neurotransmitter of
the particular adenylate cyclase. Representative antagonists for
these neurotransmitters are also included in Table 1. For the

TABLE 1

Neurotransmitters Associated With Cyclic AMP System

Neurotransmitter	Antagonist
Dopamine	Chlorpromazine
Serotonin	LSD
Norepinephrine (β)	Propranolol
Histamine (H$_2$)	Metiamide
Octopamine	Phentolamine

(Taken from reference 8.)

group of neurotransmitters listed in Table 1, the close similarity between the properties of the neurotransmitter receptor, as characterized in physiological experiments, and the properties of the neurotransmitter-sensitive adenylate cyclase, using a variety of physiological and pharmacological agonists and antagonists, suggest an intimate association between neurotransmitter receptors and adenylate cyclase molecules.

Table 2 lists those neurotransmitters that cause increases in cyclic GMP in certain target cells. Each of these neurotransmitters is capable of causing increases in cyclic GMP in postsynaptic cells under conditions in which they produce a physiological response. In addition, substances that mimic the physiological effects of these neurotransmitters also mimic the neurotransmitter in causing increases in cyclic GMP in postsynaptic cells. Conversely, substances that act as antagonists of the physiological effects of these neurotransmitters block the increase in cyclic GMP brought about by the appropriate neurotransmitter in the postsynaptic cells. Representative antagonists for these neurotransmitters are included in Table 2. The available data suggest that cyclic GMP mediates some effects, and modulates other effects, of neurotransmitters on postsynaptic cells. The possible relationship of the neurotransmitter-induced increase in cyclic GMP to the physiological response of the postsynaptic cells is discussed in detail elsewhere (8).

TABLE 2

Neurotransmitters Associated With Cyclic GMP System

Neurotransmitter	Antagonist
Acetylcholine (muscarinic)	Atropine
Histamine (H_1)	Diphenhydramine
Norepinephrine (α)	Phentolamine
Glutamate	?

(Taken from reference 8.)

Studies of cyclic nucleotides in several neuronal preparations, including the superior cervical ganglion (8) and the mammalian cerebellum (2), leave little doubt that synaptic transmission under physiological conditions can result in an increase in the cyclic nucleotide content of certain postsynaptic neurons. The precise role(s) of the cyclic nucleotides formed in the postsynaptic cells, during physiological activity, is less well established. We have proposed that cyclic nucleotides formed in neurons in response to neurotransmitter action may have multiple roles including, but by no means limited to, the generation of certain slow postsynaptic potentials (4,8,9). Other possible roles for cyclic AMP in nervous tissue include regulation of microtubule function and of neurotransmitter biosynthesis. The increase in cyclic AMP during increased impulse flow may also serve to mobilize carbohydrate (10) and lipid reserves, and thus help to meet the additional energy requirements associated with elevated functional activity in nervous tissue. In certain invertebrate nerve networks, an increase in cyclic AMP mediated by serotonin has been shown to be responsible for the increased neurotransmitter release that accompanies behavioral sensitization (11). It is also of interest that, in various model systems, cyclic AMP may act at a transcriptional level to regulate the de novo synthesis of a number of proteins with specific functional roles (see 8 for review). In the case of the nervous system, it is possible that elevation of the cyclic AMP content of postsynaptic cells in response to increased presynaptic impulse flow may result in altered synthesis of certain proteins having specific functional roles in synaptic physiology. This might constitute one component of the mechanism by which long-term information storage (i.e., learning) and retrieval (i.e., memory) can occur. Finally, it is probably important to view these several synaptic and nonsynaptic actions of cyclic AMP not so much in isolation but rather as part of an integrating system.

B. Cyclic AMP-Dependent Protein Phosphorylation

I would like now to discuss a mechanism by which cyclic AMP may mediate the effects of neurotransmitters on postsynaptic cells. The protein kinase hypothesis provides a theoretical mechanism by which one simple molecule, cyclic AMP, could achieve a diversity of biological responses (12). According to this hypothesis, cyclic AMP directly regulates the activity of a single class of enzymes, namely protein kinases. The specificity of the response to the cyclic AMP then resides in the nature of the protein kinases, and especially in the nature of the substrate proteins for the protein kinases in the various tissues.

An increasing amount of evidence supports the protein kinase hypothesis. Thus, numerous examples have now been found (for reviews, see 4,7,8,13-16) of systems in which the effects of cyclic AMP on various metabolic and physiological responses appear to be mediated through protein phosphorylation. At the present time, it seems likely that in eukaryotic organisms most, and conceivably all, of the metabolic and physiological effects of cyclic AMP will prove to be mediated through regulation of protein phosphorylation.

The literature on the subject of protein phosphorylation, even when restricted to the nervous system, is now vast. In this presentation, I shall review one study of this subject from our own laboratory (17-21). In view of the apparent importance of cyclic AMP-dependent protein phosphorylation in the nervous system, my colleagues and I have searched for substrate proteins for cyclic AMP-dependent protein kinase in nervous tissue. The ability of endogenous protein kinase in a synaptic membrane fraction from the rat caudate nucleus to phosphorylate endogenous substrate proteins in this fraction is demonstrated by the experiment shown in Figure 2. As shown by the protein staining pattern in this Figure, a large number of distinct bands were found upon gel electrophoresis of the solubilized proteins from this synaptic membrane fraction. Of these many bands, the phosphorylation of three was markedly stimulated by cyclic AMP. We refer to these three bands as Proteins Ia, Ib and II. Recently, it has been possible to resolve Protein II into two bands (20), one of which has been identified as the regulatory subunit of a protein kinase present in synaptic membranes (19); this regulatory subunit is phosphorylated by the catalytic subunit in an autophosphorylation reaction (19). Proteins Ia plus

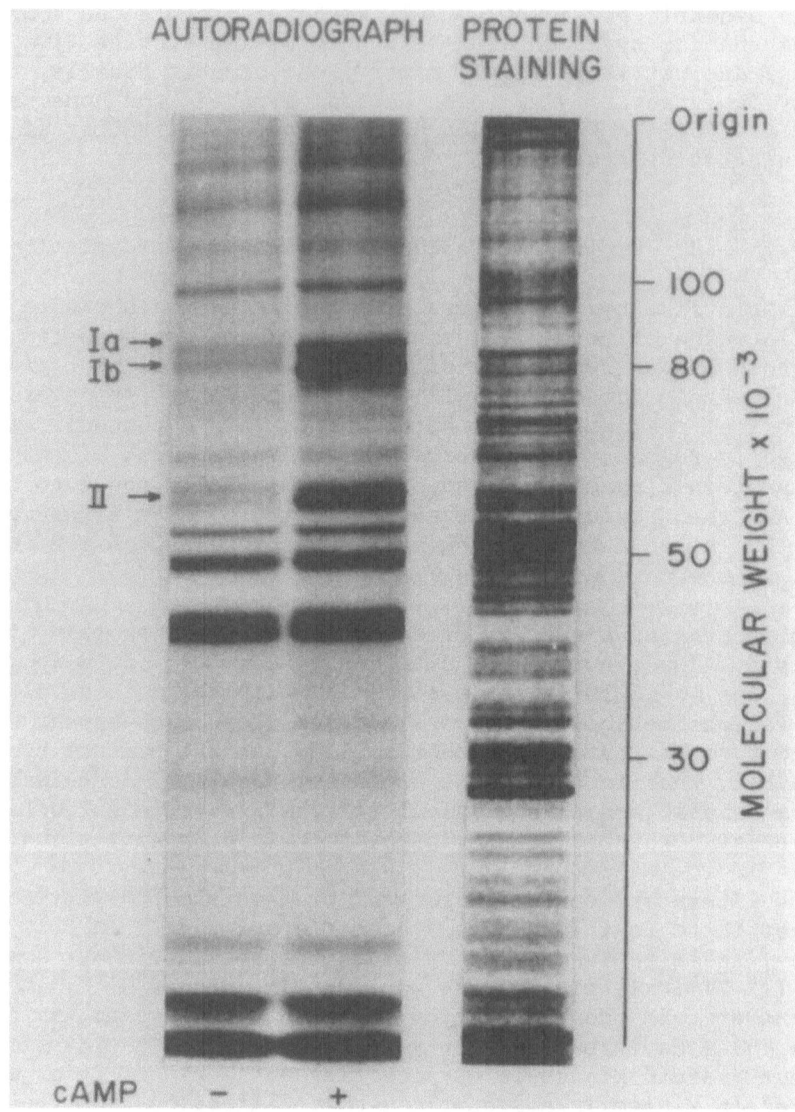

FIGURE 2. Effect of cyclic AMP on endogenous protein phosphorylation in a synaptic membrane fraction from rat caudate nucleus. The synaptic membrane fraction was incubated with 7 μM [γ-^{32}P]ATP for 15 seconds at 30°C, in the absence (−) or presence (+) of 10 μM cyclic AMP. An aliquot of the reaction product was then subjected to sodium dodecyl sulfate polyacrylamide gel electrophoresis. The separated proteins were located in the gel by standard procedures of protein staining. Autoradiography was then carried out to determine those protein bands into which radioactive phosphate had been incorporated. (Modified from reference 21.)

TABLE 3

Some Biological Properties of Protein I

1. Present only in nervous system

2. Within nervous system, present only in neurons

3. Within neurons, present only in synaptic membranes
    and synaptic vesicles

4. Appears simultaneously with synapse formation during
    development

5. Substrate for specific membrane-bound protein kinase
    and protein phosphatase

(Taken from reference 8.)

Ib, collectively designated Protein I, are present only in nervous
tissue (18). In view of the possibility that Protein I might be
involved in the physiology of synaptic transmission, we have investigated its biological and biochemical properties. The results
of those studies indicate that Protein I has a number of unique
properties.

Some of the biological properties of Protein I are summarized
in Table 3. Not only does Protein I appear to be confined to the
nervous system (17,18), but within the nervous system Protein I
appears to be present only in neurons (21). This latter conclusion
is based on observations that injection, into brain, of kainic acid,
a substance which selectively destroys neurons (22-24), causes an
extensive depletion of several brain phosphoproteins, including
Proteins Ia and Ib. Subcellular fractionation studies carried out
by Dr. Tetsufumi Ueda, in collaboration with the research group of
Philip Siekevitz at Rockefeller University, indicate that, within
neurons, Proteins Ia and Ib are present in high concentration in
synaptic membrane fractions, and in particularly high concentration
in fractions enriched in synaptic vesicles and postsynaptic densities (25). The concentration of Proteins Ia and Ib in postsynaptic densities is much higher than that in the synaptic membrane
fractions from which the postsynaptic densities are prepared; therefore, it is possible that the content of Proteins Ia and Ib in synaptic membrane fractions may be attributable to the postsynaptic
densities present in these fractions. In agreement with the results

TABLE 4

Some Chemical Properties of Protein I Purified From Bovine Brain

		Protein IA	Protein IB
1.	Molar proportion	1	2
2.	Molecular weight	86,000	80,000
3.	Isoelectric point	10.3	10.2
4.	Stokes radius	59 Å	59 Å
5.	Shape	highly elongated	highly elongated
6.	Amino acid composition	rich in glycine and proline	rich in glycine and proline
7.	Other structural features	a globular region (phosphorylated) and an elongated collagenous tail (not phosphorylated)	

(Taken from reference 8.)

of subcellular fractionation studies, immunocytochemical studies also indicate that Protein I is associated with synaptic vesicles and postsynaptic densities (26). Correlation between the time course of synapse formation and the appearance of Proteins Ia and Ib in the brains of developing rats and guinea pigs has provided further evidence that Proteins Ia and Ib are associated with synaptic structures (27). Protein I serves as a substrate for a specific Type II membrane-bound cyclic AMP-dependent protein kinase (17,20,28) and a specific membrane-bound phosphoprotein phosphatase (19).

Recently, alterations in the phosphorylation of Proteins Ia and Ib have been observed in intact brain slices and in whole animals in response to various chemical agents. For instance, several central nervous system depressants, including pentobarbitol, chloral hydrate and urethane, administered to mice intraperitoneally, decrease the state of phosphorylation of Protein I and, conversely, the convulsant drug pentylenetetrazole increases the state of phosphorylation of Protein I, in whole brains of mice in vivo (29). These results indicate that the state of phosphorylation of Protein I, a protein confined to the synaptic region of neurons, can be altered by changes in neuronal activity, supporting the possibility that it may play an important role in the physiology of the synapse.

Biochemical studies of Protein I indicate that this molecule is unusual from a chemical point of view. Protein I is present in the brains of all mammalian species examined. Using bovine brain, in order to have large amounts of starting material, Protein I has been purified several thousand-fold to apparent homogeneity. Some of the unusual chemical properties of purified Protein I (18) are summarized in Table 4.

C. Regulation of Protein Phosphorylation by Calcium, Cyclic GMP and Steroid Hormones

Finally, I would like to discuss protein phosphorylation in a broader context than that of its association with the cyclic AMP system. Some of the studies in our laboratory in the past few years were designed to test the possibility that phosphorylated proteins may play a much wider role in biological regulation than one limited only to mediating the effects of cyclic AMP. The results obtained thus far suggest that phosphorylated proteins probably mediate not only the effects of cyclic AMP and those regulatory substances that act through cyclic AMP, but also certain of the effects of a variety of other regulatory agents as well. Thus, it seems that protein phosphorylation may be a final common pathway for the actions of many types of biological regulatory agents. This concept is illustrated schematically in Figure 3. Experimental data suggest that all of the regulatory agents listed in Figure 3 achieve at least certain of their effects through protein phosphorylation. This subject was reviewed in some detail recently (7,8), and therefore I shall limit my discussion to brief comments about three of the regulatory substances, namely Ca^{2+}, cyclic GMP and the steroid hormones, which have been found to affect the phosphorylation of specific proteins in their target tissues.

1. Regulation by Ca^{2+} of protein phosphorylation

It seemed possible that one or more of the known effects of calcium in the presynaptic nerve terminal, such as regulation of neurotransmitter synthesis or release, might be mediated or modulated through the phosphorylation of specific proteins. For this reason, Bruce Krueger and Javier Forn studied the effect of calcium influx on protein phosphorylation in intact synaptosomes. They found (30) that agents known to increase Ca^{2+} transport across the plasma membrane of nerve terminals altered the phosphorylation of specific endogenous proteins in intact synaptosomes from rat brain. Synaptosome preparations, preincubated in vitro with $^{32}P_i$, incorporated ^{32}P into a variety of specific proteins. Veratridine and high (60 mM) K^+, which increase Ca^{2+} transport across membranes,

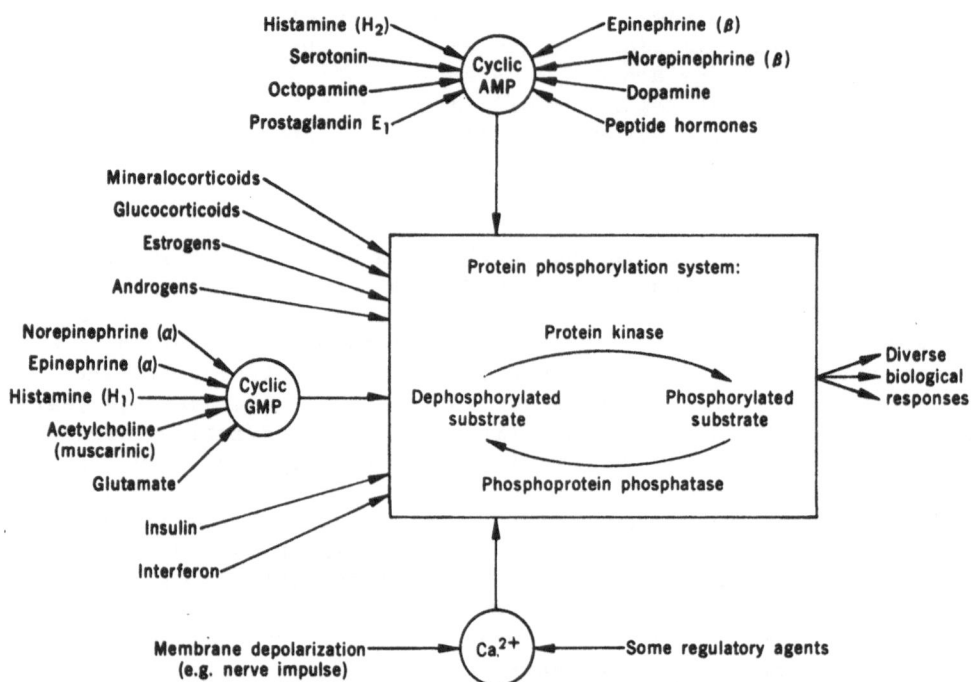

FIGURE 3. Schematic diagram of postulated role played by protein phosphorylation in mediating some of the biological effects of a variety of regulatory agents. The diagram gives examples of regulatory agents some of whose effects may be mediated through regulation of the phosphorylation of specific proteins, and is not intended to be complete. In addition to cyclic AMP and a variety of neurotransmitters and hormones whose effects are mediated through cyclic AMP, these regulatory agents include cyclic GMP, a variety of neurotransmitters and hormones whose effects are mediated through cyclic GMP, Ca^{2+}, and agents whose effects are mediated through Ca^{2+}, as well as several classes of steroid hormones, insulin, and interferon. For brevity, the numerous peptide hormones whose effects are known to be mediated through cyclic AMP, and the various regulatory agents believed to act through translocation of Ca^{2+}, are not listed individually. It seems likely that some, but not necessarily all, of the biological responses elicited by any given one of the regulatory agents shown are mediated through the protein phosphorylation system; for simplicity, pathways from regulatory agent to biological response that do not involve protein phosphorylation are not shown. (Taken from reference 7.)

through a mechanism involving membrane depolarization, as well as the calcium ionophore A23187, each markedly altered the incorporation of ^{32}P into several specific proteins as determined by sodium dodecyl sulfate-polyacrylamide gel electrophoresis and autoradiography. All three agents failed to alter protein phosphorylation in calcium-free medium. The data suggested that conditions which cause an accumulation of calcium by synaptosomes lead to a calcium-dependent increase in phosphorylation of specific endogenous proteins. The results support the possibility that these phosphoproteins may be involved in the regulation of certain calcium-dependent nerve terminal functions such as neurotransmitter synthesis and release.

The mechanism by which calcium entry stimulates the phosphorylation of synaptosomal proteins has been partially clarified by experiments of Howard Schulman using lysed synaptosomes (31). He found that stimulation by calcium of protein phosphorylation required the presence, along with calcium, of a heat-stable calcium-binding protein. The results indicated that regulation by calcium of calcium-dependent protein kinase activity in nervous tissue is mediated by the same calcium-binding protein that regulates cyclic nucleotide phosphodiesterase (32,33) and adenylate cyclase (34) activities.

These studies of nervous tissue demonstrated that calcium is a regulatory agent some of whose effects may be mediated by protein phosphorylation. In addition, recent studies indicate that many other tissues possess a protein kinase that requires both calcium and the same calcium-binding protein for activity, suggesting that this enzyme may be of general importance as a mediator of the regulatory actions of calcium (35).

2. Regulation by cyclic GMP of protein phosphorylation.

Several lines of evidence indicate that cyclic GMP plays an important role as an intracellular regulatory substance (7,36). Moreover, there are some highly suggestive findings concerning the molecular basis for its actions in various biological systems. A number of years ago, Dr. J. F. Kuo and I undertook a study of the mechanism by which cyclic GMP might exert its biological effects. By analogy with cyclic AMP-dependent protein kinase, it seemed reasonable to postulate the existence of cyclic GMP-dependent protein kinase that might mediate the biological effects of cyclic GMP. Through the use of column chromatography, it was possible to demonstrate a cyclic GMP-dependent protein kinase in lobster muscle and to separate it from a cyclic AMP-dependent protein kinase present in the same tissue (37). Subsequently, cyclic GMP-dependent protein kinases have been demonstrated in a variety of other invertebrate and vertebrate tissues (for review, see reference 8). How-

ever, in contrast to the situation with the cyclic AMP-dependent
protein kinases, where many substrates, some with known and some
with unknown function, have been found (see 7,8,13-16 for reviews),
efforts to find endogenous proteins that are specific substrates
for these cyclic GMP-dependent protein kinases proved difficult.
Nevertheless, endogenous substrates for cyclic GMP-dependent protein
kinases have been identified in three locations, namely in membranes
of smooth muscle (38), in membranes of intestinal brush border epi-
thelium (39) and in cerebellar cytosol (40). An experiment with the
cerebellum is illustrated in Figure 4. The results of this and
other experiments have led us to conclude that the 23,000 dalton
protein shown in Figure 4 is a natural substrate for the cyclic GMP-
dependent protein kinase that is also present in cerebellar cytosol.
We are currently attempting to purify and characterize this soluble
23,000 dalton substrate protein for the cyclic GMP-dependent protein
kinase from mammalian cerebellum. Hopefully, studies of the biological and chemical properties of this molecule will contribute to
an understanding of the role of cyclic GMP and of cyclic GMP-dependent
protein phosphorylation in nervous tissue.

3. Regulation by steroid hormones of protein phosphorylation.

        Steroid hormones have multiple actions on their target tissues.
Many of these actions are either synergistic with, or antagonistic
to, the effects of those hormones whose actions are known to be
mediated through cyclic AMP. Although steroid hormones have been
found to affect cyclic AMP levels and the enzymes (adenylate cyclase
and phosphodiesterase) involved in cyclic AMP metabolism in a variety
of tissues, only a few aspects of steroid-hormone action can be
accounted for by the effects of these hormones on cyclic AMP levels
(41). It is of significance, therefore, that representatives of all
of the major classes of steroid hormones, including the mineralo-
corticoid, glucocorticoid, estrogen, and androgen classes, affect
protein phosphorylation systems in their target tissues (42,43).
In every instance, administration in vivo of the appropriate steroid
hormone affects the autophosphorylation of the regulatory subunit
of a cyclic AMP-dependent protein kinase by the catalytic subunit
of this enzyme. The effect is specific for the respective target
tissue, occurs with low doses of the steroid hormones, can be
demonstrated within 1 hour of steroid hormone administration, and
is associated with alterations in the level of cyclic AMP-dependent
protein kinase activity (43,44). The detailed mechanism by which
the steroid hormones affect the autophosphorylation of the regulatory subunit of cyclic AMP-dependent protein kinase is not yet
understood. However, the effect of the steroids is indirect and
requires de novo protein synthesis, in contrast to the direct effect
of cyclic AMP on protein kinases. The effect of steroid hormones
and of cyclic AMP on the same cyclic AMP-dependent protein kinase

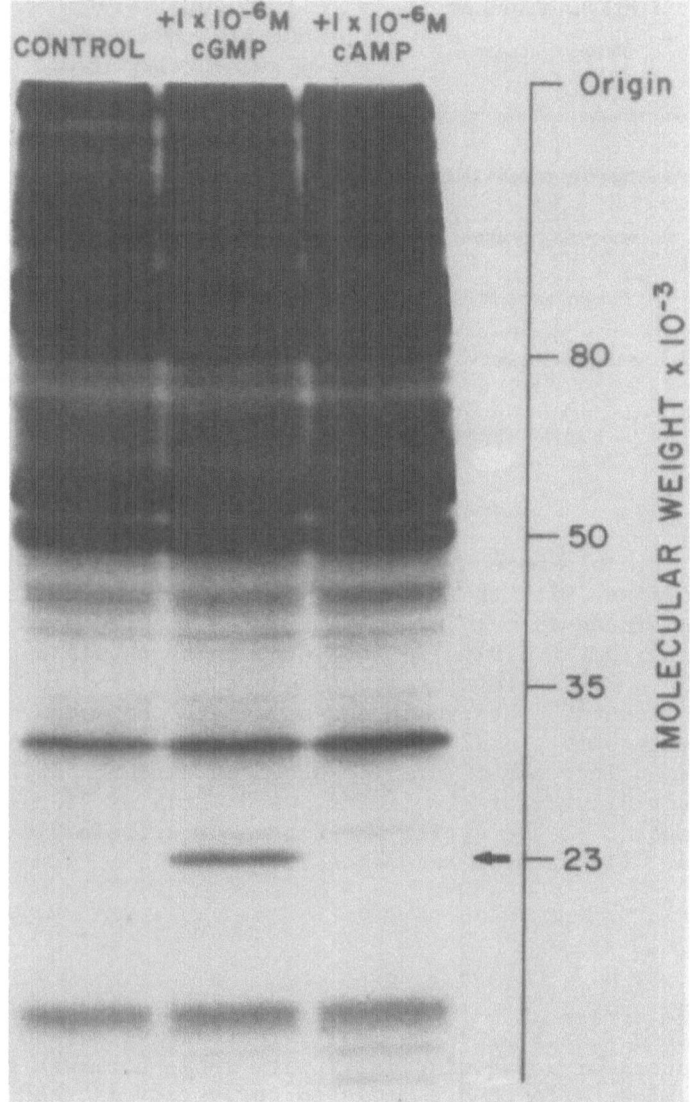

FIGURE 4. Cyclic GMP-dependent endogenous protein phosphorylation
in cytosol from rabbit cerebellum. The cytosol fraction was passed
over a Sephadex G-25 column to remove endogenous ATP and cyclic nucleotides and was then incubated with $[\gamma-{}^{32}P]$ATP for 1.5 minutes at 30°C,
in the absence or presence of cyclic GMP or cyclic AMP as indicated.
The entire sample was then subjected to sodium dodecyl sulfate-polyacrylamide gel electrophoresis and autoradiography. Cyclic GMP, but
not cyclic AMP, stimulated the phosphorylation of a protein with an
apparent molecular weight of 23,000 daltons, indicated by the arrow.
(Modified from reference 40.)

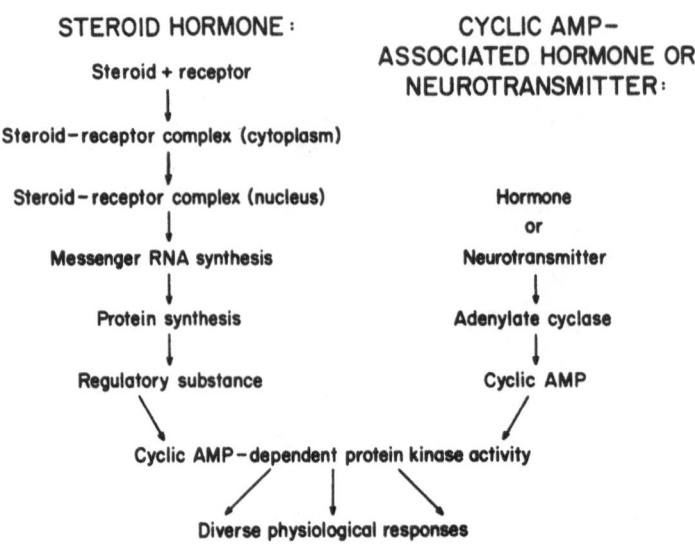

FIGURE 5. A possible molecular basis for the biological interactions
of steroid hormones with those hormones and neurotransmitters whose
effects are mediated through cyclic AMP. The steroid hormones,
through a mechanism involving de novo protein synthesis but not involving cyclic AMP, regulate cyclic AMP-dependent protein kinases in
their target tissues. The cyclic AMP-associated hormones and neurotransmitters, through a mechanism not involving de novo protein synthesis, cause an increase in cyclic AMP which directly affects protein kinase activity. For simplicity, mechanisms of steroid hormone
action independent of the cyclic AMP-dependent protein kinase system
are not shown. (Taken from reference 7.)

(Figure 5) may provide a molecular basis for the ability of the
steroid hormones to act synergistically (permissive action of the
steroid hormones) or antagonistically with those hormones and neurotransmitters whose effects are mediated through cyclic AMP (43,44).

    In conclusion, a variety of experimental data, summarized in
the latter part of this article, and in greater detail elsewhere
(7,8), indicate that protein phosphorylation represents a final
common pathway in biological regulation. Of relevance to this
Symposium, several regulatory agents, including cyclic AMP, cyclic
GMP, calcium and the steroid hormones have been shown to affect the
state of phosphorylation of specific proteins in the nervous system.
Clarification of the detailed manner in which these phosphoproteins
may regulate function in the nervous system is an exciting area for
future research.

REFERENCES

1. Robison, G. A., Butcher, R. W., and Sutherland, E. W., Cyclic AMP (Academic Press, New York) 1971.
2. Bloom, F. E., Reviews of Physiology, Biochemistry, and Experimental Pharmacology (Springer, Berlin) 1975.
3. Daly, J. W., in Handbook of Psychopharmacology, 5 (Iversen, L. L., Iversen, S. H., and Snyder, S. H., eds.) pp. 47-130 (Plenum Press, New York) 1975.
4. Greengard, P., Nature, 260, 101-108 (1976).
5. Beam, K. G., and Greengard, P., Cold Spring Harbor Symposia on Quantitative Biology, The Synapse, Vol. 40, 157-168 (1976).
6. Nathanson, J. A., Physiol. Rev., 57, 157-256 (1977).
7. Greengard, P., Science, 199, 146-152 (1978).
8. Greengard, P., Cyclic Nucleotides, Phosphorylated Proteins and Neuronal Function (Raven Press, New York) 1978.
9. Kanof, P. D., Ueda, T., Uno, I., and Greengard, P., in Approaches to the Cell Biology of Neurons, Vol. II (Cowan, W. M., and Ferrendelli, J. A., eds.) pp. 399-434 (Society for Neuroscience Publishers, Baltimore) 1977.
10. Edwards, C., Nahorski, S. R., and Rogers, K. J., J. Neurochem., 22, 565-572 (1974).
11. Brunelli, M., Castellucci, V., and Kandel, E. R., Science, 194, 1178-1181 (1976).
12. Kuo, J. F., and Greengard, P., Proc. Nat. Acad. Sci. U.S.A., 64, 1349-1355 (1969).
13. Krebs, E. G., Curr. Top. Cell Regul., 5, 99-133 (1972).
14. Langan, T. A., Adv. Cyclic Nucl. Res., 3, 99-154 (1973).
15. Rubin, C. S., and Rosen, O. M., Annu. Rev. Biochem., 44, 831-885 (1975).
16. Nimmo, H. G., and Cohen, P., Adv. Cyclic Nucl. Res., 8, 145-266 (1977).
17. Ueda, T., Maeno, H., and Greengard, P., J. Biol. Chem., 248, 8295-8305 (1973).
18. Ueda, T., and Greengard, P., J. Biol. Chem., 252, 5155-5163 (1977).
19. Uno, I., Ueda, T., and Greengard, P., Arch. Biochem. Biophys., 183, 480-489 (1977).
20. Lohmann, S. M., Walter, U., Sieghart, W., and Greengard, P., manuscript in preparation.
21. Sieghart, W., Forn, J., Schwarcz, R., Coyle, J. T., and Greengard, P., submitted for publication (1978).
22. Coyle, J. T., and Schwarcz, R., Nature, 263, 244-246 (1976).
23. McGeer, E. G., and McGeer, P. L., Nature, 263, 517-519 (1976).
24. Hattori, T., and McGeer, E. G., Brain Res., 129, 174-180 (1977).
25. Cohen, R. S., Blumberg, F., Berzins, K., Siekevitz, P., Ueda, T., and Greengard, P., manuscript in preparation.
26. Bloom, F. E., Battenberg, E., Ueda, T., and Greengard, P., manuscript in preparation.

27. Lohmann, S. M., Ueda, T., and Greengard, P., Proc. Nat. Acad.
     Sci. U.S.A.., in press (1978).
28. Uno, I., Ueda, T., and Greengard, P., J. Biol. Chem., 252,
     5164-5174 (1977).
29. Strombom, U., Forn, J., and Greengard, P., manuscript in prepa-
     ration.
30. Krueger, B. K., Forn, J., and Greengard, P., J. Biol. Chem.,
     252, 2764-2773 (1977).
31. Schulman, H., and Greengard, P., Nature, 271, 478-479 (1978).
32. Cheung, W. Y., Biochem. Biophys. Res. Commun., 38, 533-538 (1970).
33. Kakiuchi, S., and Yamazaki, R., Biochem. Biophys. Res. Commun.,
     41, 1104-1110 (1970).
34. Brostrom, C. O., Huang, Y.-C., Breckenridge, B. McL., and Wolff,
     D. J., Proc. Nat. Acad. Sci., U.S.A., 72, 64-68 (1975).
35. Schulman, H., and Greengard, P., manuscript in preparation.
36. Goldberg, N. D., O'Dea, R. G., and Haddox, M. K., Adv. Cyclic
     Nucl. Res., 3, 155-223 (1973).
37. Kuo, J. F., and Greengard, P., J. Biol. Chem., 245, 2493-2498
     (1970).
38. Casnellie, J. E., and Greengard, P., Proc. Nat. Acad. Sci.
     U.S.A., 71, 1891-1895 (1974).
39. DeJonge, H. R., Nature, 262, 590-593 (1976).
40. Schlichter, D. J., Casnellie, J. E., and Greengard, P., Nature,
     273, 61-62 (1978).
41. Thompson, E. B., and Lippman, M. E., Metabolism, 23, 159-202
     (1974).
42. Liu, A. Y.-C., and Greengard, P., Proc. Nat. Acad. Sci. U.S.A.,
     71, 3869-3873 (1974).
43. Liu, A. Y.-C., and Greengard, P., Proc. Nat. Acad. Sci. U.S.A.,
     73, 568-572 (1976).
44. Liu, A. Y.-C., Uno, I., Walter, U., and Greengard, P., manuscript
     in preparation.

CONTRASTING PRINCIPLES OF SYNAPTIC PHYSIOLOGY:

PEPTIDERGIC AND NON-PEPTIDERGIC NEURONS

Floyd E. Bloom

Arthur Vining Davis Center for Behavioral

Neurobiology, The Salk Institute, La Jolla, CA 92037

INTRODUCTION

A major unresolved theoretical issue in Neurobiology is the nature of the information encoded into chemical neurotransmitters. One aspect of that central issue is the question of whether the nervous system uses an exclusive set of chemical commands to regulate and respond to the endocrine system. Alternatively, the chemical signals of neurons and endocrine glands may simply be part of a general chemical vocabulary by which neurons communicate with cells.

Currently, synaptic pharmacology is besieged by an onslaught of new transmitter-like chemicals discovered at rates far faster than the sites and mechanisms of operation for these substances can be determined. These newly recognized substances, mainly peptides (see 3,4,53,54,55) pose an important challenge for the emerging logic of recognized synaptic operations: are these substances acting in ways generally analogous to amino acids and small amines or do they represent one or more additional classes of chemical operations by which nerve cells communicate? Although data to settle this important conceptual issue are still lacking, I have attempted here and in related essays (8,11,13,14) to develop some arguments by which such communicative events can be evaluated and compared.

Too many or too few?

When the list of putative neurotransmitter chemicals was relatively short, many neuroscientists tacitly assumed that transmitter molecules were functionally equivalent elements differing only by virtue of whether the action they produced was excitation or inhibition. Armed with that generalization, the transmitter for a given circuit did not really require identification provided that the qualitative sign for the inter-cellular operation could be defined. With the logic operations restricted to excitation and inhibition, theoretical constructs of neuronal ensembles to explain information processing appeared only to be matters of circuitry. In this way a neuron could be viewed as a highly complex logic gate with the properties of a living transistor. However, such a view fails to encompass many highly developed biochemical properties of the nerve cell and in particular the biochemical specialization required to synthesize, store, release and respond to a specific neurotransmitter. Let us consider three hypothetical "explanations" for the diversity of neurotransmitter substances.

First, the transmitter could serve to identify the cell or cell population emitting the signal. If so, the number of transmitters in a given nervous system should bear a relationship to neuron number. However, if the response of a cell to a particular signal were to retain meaning as to the signal origin, and if there are multiple possible origins, additional features would be required. Each signalling molecule would also need an intracellular "symbolization" (78) in which each surface receptor system would be keyed to (for example) the production of a specific metabolic intermediate or the translocation of a particular ion. The specific intracellular perturbation would thus mediate the cell surface receptor's recognition of the agonist to which that receptor was on the alert. As currently conceived, all synaptic transmitters change ion permeability or metabolite concentrations. The significance of such actions has been thought to lie only in the resultant change in membrane polarization or ion shunting, rather than in the specific molecular mechanism by which the change is effected. The number of ions (Na^+, K^+, Cl^-, Ca^{++}) and intracellular metabolites (cyclic AMP, cyclic GMP) available to symbolize responses to individual transmitters is far smaller than the number of potential transmitters. Therefore, individual excitatory and inhibitory transmitters could not symbolize connectivity arrangements unless a given target cell received only one afferent system for each available intracellular symbol. A contrasting view is that cells accept signals blindly according to independent rules of connectivity, and are regulated by the intracellular effects of these signals oblivious of the origin or the molecular structure of the transmitter.

A second hypothetical explanation would be that the chemical output of a functional center (i.e., a nucleus, brain region, or any homotypic population of cells) might also be symbolized by a single transmitter. Thus, cells of the nucleus locus ceruleus are noradrenergic, the raphe nuclei cells are serotonergic, spinal motoneurons are cholinergic and cerebellar Purkinje cells are GABA-ergic. In this case, the number of transmitter molecule species is related to the number of cell classes but not necessarily to the number of cells. Pearse (53) has suggested that all cells secreting polypeptide hormones in gut, endocrine, and neural tissues belong to an embryologically homogeneous super-class of cells to which he has given the acronym "APUD" (for amine producing, precursor uptake, and decarboxylation; 53,54). In order to incorporate under this banner the numerous possible candidates, Pearse further has suggested (53) that in some cases the amine-producing capacity is lost as the cell becomes specialized for peptide-production. Erspamer (see 7) has pointed out the amino acid sequence homologies within 5 families of non-mammalian peptides; substance P, gastrin, and cholecystokinin (59) may be the first known representatives of these peptides in mammals.

A common embryologic origin of all cells which secrete peptides or amines could explain their simultaneous existence in some sympathetic ganglia (36) without clarifying the functional principle which selects the transmitter to be employed. Bunge and associates (17) suggest this choice of transmitter may only be finally specified by the target cells when the sympathetic innervation arrives there. The conservation of homologous cell lines across separated organ systems such as the "APUD" cells and the conservation of homologous amino acid sequences within peptides found across the phylogenetic scale suggest that these cells and peptides may be important specific regulatory signals. The potential value of this view is immediately diminished because several distinct nuclei presumably employ the same transmitter and yet share no obvious functional or spatial qualities to predict the targets which they innervate. For example, substance P, and somatostatin, are each found within some groups of primary sensory neurons of the dorsal root ganglia, and enkephalin as well as neurotensin are found in dorsal horn interneurons. However, these substances also occur within other groups of cells which have no known sensory function (34,36,38).

A third hypothetical view of transmitters would be that the same transmitter is released by all those cells needed to execute individual steps leading to fundamental biobehavioral operations such as feeding, sleeping, mating, defending, or learning or even more detailed macro-functions such as body temperature or blood glucose regulation. Execution of this type of function might then require connections which coordinate the outputs of several chemically- coded nuclei. Theoretical proposals to explain the functions of specific peptides on this operational level of analysis (4,15) frequently employ this argument: the presence of the same peptides in multiple sites (e.g. brain, gut, and endocrines) implies that the peptide can purposefully integrate all three sets of tissues leading to a concerted multicellular action. Despite the appeal of the "one transmitter-one behavioral function" concept, pharmacological approaches directed against neuronal cell populations containing the same monoamine transmitter have yet to provide compelling evidence that any complex behavioral function is so simply encoded. On the other hand, recent work on peptides related to drinking (57) and pain perception (35) may prove more successful. However, what function could be subserved by gastrin neurons in the cerebral cortex and gastrin-secreting cells in gut to account for the spatial distribution of this peptide? The answer to this biological riddle may well lie in better understanding of the central representation of visceral monitoring neurons and their ontogeny.

Future experimental work may well serve to strengthen any of the three general views of transmitters mentioned above. As one example, the list of intracellular symbols could be enlarged considerably if cyclic nucleotide-dependent protein kinases were compartmentalized to the vicinity of certain synapses; in this case, changes in the phosphorylation state of unique protein substrates could symbolize that synapse's action by means of the target protein's altered functional properties (28). Similarly, more incisive operational definitions of the "purpose" of certain behaviors may also eventually provide general functional abstractions which will clarify the integral functions of widely separated cells sharing the same peptide or amine. Thus, cells secreting angiotensin (57) or vasopressin (77) could be seen not as signals for drinking or blood pressure elevation but rather as signals to control cell volume (49) or extracellular fluid movement. At the same time, alternative views of peptides and amines may still be needed to provide different frameworks for interpretation of the forthcoming data.

Neurohormones, Neuromodulators and Neurotransmitters

Many recent papers deal with the thorny semantic question of the terms neurotransmitter, neurohormone and neuromodulator. As man-made linguistic terms, they can--in the words of Humpty-Dumpty--mean anything the user wishes. However, for neuroscientists attempting to probe for fundamental principles of interneuronal communication, it would appear worthwhile to dwell a moment on the matter.

Peptide-secreting cells of the hypothalamo-neurohypophysial circuit were once conceptually segregated as "neuro-secretory" cells, a form both fish and fowl, which secreted hormone broadly via the blood stream in response to conventional neuronal signals (61). However, these neurosecretory neurons are now known to make central inter-neuronal connections which are presumably mediated by the same posterior pituitary peptide "hormone" (53,77). Other peptide-containing central neurons also constitute unique and discrete interneuronal circuits (34,37) with no evidence that their secretion is primarily directed into vascular or diffuse extracellular spaces. Thus, all neuronally-released peptides are probably not exclusively broadly acting "neurohormones", and all "neurohormones" are probably neither exclusively peptides nor diffusely acting.

The criteria often applied to the identification of neurotransmitter substances rely mainly on demonstrations that a substance contained within a neuron is secreted by that neuron at its synaptic junctions in order to transmit information to its post-synaptic cells. Thus, on this level of description, given a defined functional effect of neuron A on neuron B, if a substance contained within A produces the A effect on B, then that substance is a neurotransmitter.

However, several authors find attributes of peptides which to them force the creation of a new class of molecular messenger to which they have given the term neuromodulator (5,25); others (20) would add the monoamines to this category. When Florey (23) first used the term modulator substance he included any substance of "cellular and non-synaptic origin which affects the excitability of nerve cells and represents a normal link in the regulatory mechanisms that govern the performance of the nervous system." In his view modulators like CO_2, or NH_3, altered the responsiveness of neurons to transsynaptic actions of pre-synaptic neurons. Florey also postulated that substances released from non-synaptic sites would be expected to act for longer periods of time than substances released from synapses, but, interestingly he also indicated his view that the mechanism by which the modulation was mediated might well be accomplished through the same types of membrane permeability changes produced by transmitters. These properties would all fit with Frederickson's (25) use of the term modulator, based mainly on the abnormally long time course exhibited by some peptides thus viewed as modulators. Assuming that the time course of a peptide may reflect other biological constraints (see below), the basic distinction between the Florey "modulator" (23) and the more general term neurotransmitter rests squarely on the issue of the synaptic or non-synaptic nature of the release and response sites for a given substance. These issues in turn rest upon the techniques used to define "synapses" structurally (41) and those used to define receptivity without directly examining the consequences (9) of neurally released signals. In my view, present data are fully compatible with the idea that in brain release and response sites for peptides and monoamines are "synaptic."

The puzzle deepens, however, when cells in tissues other than brain also secrete the same molecule secreted by a neuron at its synapses. Thus catecholamines are secreted by the adrenal medulla as an endocrine hormone, by cells of the pancreatic islets and other sites as paracrine hormones, and by cells of the sympathetic ganglion and noradrenergic brain nuclei as transmitters (10,32,67,68,69,70,74,83). Similarly peptides such as β-endorphin contained within central mammalian neurons (12) are released as endocrine hormones from the posterior and anterior pituitary (12) and others such as somatostatin may be released as transmitters from the peripheral autonomic nervous systems (35,36). Although no active re-uptake system for conservation of secreted neuropeptides has yet been described, all other aspects of the neurobiology of peptide-secreting neurons are analogous to other chemically identified neurons. However, an important issue of functional difference between neurotransmitters and hormones does arise when the sensitivity of the target cell receptors are examined.

Cells respond to circulating (i.e., endocrine) catecholamines and peptide hormones in the concentration range of M^{-8} or less. Estimates of the concentrations of acetylcholine and amino acids secreted into synaptic clefts (i.e., neurotransmitters) are in the range of M^{-3} or more. If these data are viewed in the light of emerging concepts of receptor auto-regulation (39), it can be proposed that receptors to a substance released from considerable distance must have higher sensitivity to their messenger than those in which the material is released from only a few hundred angstroms away, as in synaptic junctions. The latter lower sensitivity could also be viewed as tonic functional desensitization of synaptic receptors resulting from the high concentrations estimated to be released within the synaptic clefts. If this hypothesis were correct, the receptor affinities of peptides released at paracrine and putative synaptic sites would also be expected to be much lower than affinities of receptors for the same substances acting as hormones. Furthermore, if the scale of receptor affinity is used as the operational discrimination between an action of a neurotransmitter (at low sensitivity receptors) and a hormone (at high sensitivity receptors) the data obtained from studies of brain receptors (22,79,80,81) indicate that affinities to monoamines and peptides are virtually identical with the properties of "real" transmitters like GABA (6).

The term neuromodulator is also being used (5,85) in quite a different way specifically to describe actions of opioid and other peptides which alter "synaptic receptor-coupled conductances without direct activation of such conductances." (5) This definition derives from studies in which enkephalins reduce the response of spinal cord neurons in vivo (85) or in vitro (5) to iontophoretic pulses of the excitant amino acid glutamate, but do not themselves produce any change in membrane conductance. Similar "modulatory" effects have been reported for the effects of thyrotropin releasing hormone and Substance P (54) on perfused frog spinal cords. While this operational definition of neuro modulation is certainly precise enough to carry distinction and grounds for future tests, the choice of the test cell system is clearly the critical issue. Firstly, if the cells tested do not normally receive such peptidic messages, the responses seen may have little physiologic relevance. Secondly, Zieglgansberger and Champagnat (84) have recently demonstrated that the location of the glutamate receptor on distal dendrites of spinal motor neurons results in conductance changes which are not detectable in the perikaryon. Given this differential topography of receptors and receptor-coupled conductance changes, peptides could also be conceived as acting on synaptic receptors which are remote from the site of intracellular impalements to alter the effects of other afferent transmitters onto the same regions of target cells by changes in membrane properties which have escaped detection. At any rate in the case of the central catecholamine systems studied by our group (see below) changes in membrane potential and conductance do occur and, therefore, under this definition (5), the term neuromodulator could not be extended to that system.

Domains of Synaptic Operations

A simple interpretation of the preceeding survey is that the chemical nature of a neurotransmitter (i.e., amino acid, amine or peptide) does not per se provide insight into the functional role mediated by the neurons which secrete it. If that generalization is tenable, then, what other discriminative features of the system may be more useful for this purpose?

On two domains, space and time, the operations of all neurons can be charted for comparative analysis. The spatial domain of a neuron is the total target cell area to which that neuron sends information. Similarly, the temporal domain is the time course of the neuron's effects on its targets.

Let us now ask whether the spatial and temporal domains of chemically-characterized neuronal circuits provide any hints to the nature of the operations such circuits may perform.

Present concepts of the structural organization of the brain rely heavily on two principles of connectivity. In the primary sensory and motor pathways, the prevailing principle has been the classical concept of hierarchical relationships in which the transmission of information is highly sequential and specific (62). Under this concept, primary receptors transmit to primary relay neurons and on upwards to the ultimate sensors in the cortex; for motor output systems the reverse would be true, progressing downwards from the cortex to the final common motor output. While parallel processing routes may be added into these hierarchies for internal refinement of control, the classical concept of the hierarchical or "throughput" system is that destruction of one link incapacitates the chain. No transmitter molecule for any such throughput system, up or down has as yet been identified.

The second widely applied principle of neuronal connectivity is that of the local circuit neuron (58): typically a small, frequently unipolar neuron whose efferent processes bear the morphology of dendrites. The basic feature of a local circuit neuron is that its connections are established exclusively within the local vicinity in which the cell body is found. Such small interneurons can exert significant control over information flowing through that locale, and may do so through "dendritic" release of their transmitter and without action potentials (58). In some cases of inhibitory local circuits, the amino acids gamma-aminobutyrate (GABA) or glycine (GLY) have been implicated as the transmitter (54).

When GABA or GLY have been examined, their receptors appear coupled to mechanisms which increase membrane conductance to ions whose equilibrium potentials lie near or below resting membrane potentials and frequently produce hyperpolarizations (54). At other sites in the brain, other presumptive amino acid-mediated excitatory and inhibitory synapses also appear to operate through increased membrane conductance to specific ions yielding depolarizations or hyperpolarizations, respectively. These systems generally operate over slightly longer times and greater spaces than local circuit neurons (53,54).

However, by comparison, the monoaminergic neurons operate over much larger spatial domains and even more prolonged temporal epochs. The dopamine-containing cell systems alone cover an extremely wide range from the ultra-short systems of the retina (21) and olfactory bulbs (37) to the longer, more highly arborized meso-cortical systems (44,46,47,51). Within the spatial domain, the serotonergic systems seem to be as divergent and extensive as the noradrenergic systems, but are perhaps somewhat more succinct in the temporal display of their synaptic actions (2,82,32,64,67,68,74). Cholinergic systems may also cover a broad spatial domain, but neither their circuitry nor their synaptic time spans of action within the CNS are as yet specifiable (43).

Those neurons that contain biologically active peptides, as determined immunohistochemically (12,33,34,77), do not fit as readily into this two-dimensional domain map as they might if all such neurons represented functionally equivalent units of a single coherent operational class: some presumptive peptidergic cells are small interneurons, such as the enkephalin-and neurotensin-containing cells (35,37,79,80), while others cover significantly broader spatial domains, such as those cells in the CNS and peripheral nervous system which are immunoreactive to antibodies against somatostatin, substance P, β -endorphin (12) or luteinizing hormone releasing factor (33). Unfortunately, there are as yet no data with which to specify for any of these peptidergic systems the duration of the synaptic actions they may mediate. Data from studies applying the peptides exogenously to test systems suggests that their durations of action under these conditions may be longer than the effects of simple amino acids (53,54,60).

Other Domains

Neuronal systems differ from each other in more ways than can be expressed simply in terms of the spatial and temporal domains over which they operate. At least one additional domain can be approached in this analysis and that domain I have termed tentatively, as "energy" (14). This third domain of neuronal operation defines the functional properties, i.e., mechanisms and consequences, of the synaptic operations of that given neuron. The dimensions of the putative energy shifts associated with a specifiable synaptic action are still to be calculated. Indeed "energy" as such may not even be the correct quality in which to quantitate this domain. However, I use this term to distinguish transmitters producing passive membrane responses (i.e., permitting responses to pre-existing electrochemical gradients) from transmitters producing active responses on the membrane and other segments of the target cell (e.g., the activation or inactivation of an enzyme or an ion-exchange carrier process). Along this energy "dimension" we can obtain further separation between the passive response operations mediated by simple amino acids and some amines (like nicotinic cholinergic actions, see 83) and those mediated by beta-noradrenergic and dopaminergic actions (11,69,52) in which one consequence is the activation of adenylate cyclase (19,52) While some peptide hormones act on their peripheral targets through activation of adenylate cyclase, there is as yet no compelling evidence now to ascribe this sort of "energy" transduction in mammals to centrally active peptides.

A three dimensional domain map for chemically coded neural circuits could provide an integrated comparison of their different functional properties. In order to lend greater substance to this scheme, let us consider briefly a chemically-defined system with apparent integrative capacities: the central noradrenergic systems studied in our laboratory (10,11,24,26,30,31,32,65-69,71-74,76). Others (16,18,40,45,48,63,83) have offered additional biologic examples by which the nuances and unsolved problems of similar chemical integrative processes are becoming illuminated.

Space, Time and Energy Operations

The noradrenergic projection systems fit neither into the classical hierarchical or through-put systems (62) nor into the local neuron principles of organization (58). Rather, noradrenergic projections follow highly divergent trajectories in which a few norepinephrine-containing neurons reach out to target cells in many distant regions of the brain bearing no obvious direct functional relationship to each other. Although locus coeruleus fibers do innervate wide regions of the brain, the innervation patterns of the axons are neither random nor all pervasive; within laminated cortical structures (44) and within designated thalamic and hypothalamic nuclei (47), noradrenergic fibers show characteristic branching patterns and organized innervation densities (46). In cerebral, cerebellar, and hippocampal cortices a major portion of the input seems to be directed towards the through-put output cells as targets (11,51).

Within the group of noradrenergic projections which have been tested physiologically, operations on the time domain indicate the effects to have long latencies (30-50 msec or more) and long durations (300-600 msec or more). If the "effects" of the pathway are to be judged on the basis of changes in target cell firing patterns or transmembrane properties, the source cells must be experimentally stimulated sufficiently to determine the quality and quantity of the effect. However, the locus coeruleus appears to exert a potent intranuclear feedback inhibition (1,29) which would tend to dampen the changes in firing imposed on these neurons by their afferent connections. Although locus coeruleus neurons can fire in short rapid bursts under certain behavioral conditions (see 11) the long duration, high frequency stimulus trains generally used to activate some behavioral effects of this nucleus may well have been highly unphysiological (66,68,75).

On the third domain, "energy", the effects of the coeruleo-cerebellar, and coeruleo-hippocampal projections, and the iontophoretic simulations of the effects of coeruleo-cortical projections (76) all adhere closely to the interpretation that the action of neurally released noradrenaline is mediated by beta adrenergic receptors coupled to adenylate cyclase, and hence according to the "second messenger" scheme of Sutherland (11,19,52; see also 27). The effects as judged by firing rates and transmembrane effects are overtly inhibitory, with hyperpolarizations accompanied by increased membrane resistance However, when the effects are examined on other aspects of target cell functioning in different experimental contexts (24,26,64,67,68) the effects of the locus coeruleus appear to fit better the designation of "biasing " or "enabling" (see 38) than they do simple "inhibition." The biasing or enabling function means only that in the epoch over which noradrenergic receptors are active, certain chemical messages received through other receptors can be enhanced or weighted (see 11,51). Such enabling effects can be regarded as predictable outcomes of the observed changes in membrane properties (83).

Elsewhere (11) I have suggested that the combination of electrophysiological and biochemical changes produced in target cells by the noradrenergic fibers should be considered together as a holistic set of responsive changes. Although more easily observed, the electrophysiologic events which result from activation of this system (hyperpolarization, increased membrane impedance) are in this holistic view (11) simply another index of the altered state of the target cells resulting from this form of neurotransmission. For example, (see 11,69) hyperpolarizing changes in transmembrane potential accompany the response of many non-neural cells to hormones and neurotransmitters which activate cyclic nucleotide synthesis. In the heart, catecholamines not only increase the force and frequency of cardiac contractions, but also activate lipolysis and glycogenolysis to provide the cardiac muscle with increased substrates for energy metabolism. Taken as a whole, the electrophysiologic shifts in the properties of the target cell membrane and the concomitant shifts in intracellular metabolism could provide a cell state specific for an altered mode of information processing. On one hand, these effects could be considered "modulatory" in the sense that adenylate cyclase-coupled noradrenergic receptors have altered information transmitted by other afferents for processing by the common target cell. However, the argument again returns to semantics. If the act of modulating is the physiologic function of locus coeruleus neurons, then surely their activity transmits that "modulate" or "enable" message.

Documentation of the holistic view of neurotransmitter function will require better knowledge of when noradrenergic circuits are naturally called into action, and of the purposes of those actions in the biologic functions of the target cells. Based on what is documentable now, however, this system would appear ideally suited to integrate across both time and space, and to shift the metabolic "gears" of the cells to which that message is transmitted.

Finally, let us assume, at least temporarily, that many classes of chemically-coupled transductive systems exist to express the effects of transmitter receptors, and that these couplings can be ion- or substrate-specific, that they can be dependent or independent variables of energy production or Ca++ translocations, and that they can operate actively or passively over a wide range of transmitter specific time periods. Where in this array of conceivable actions do we place the growing list of systems apparently mediated by peptides (57,42), some of which may co-exist within cells previously thought to be monoaminergic (36)? In my view, there is not now any reason to assume a priori that peptides must transmit messages which are qualitatively different in mechanism than those mediated by either amino acids or monoamines. I prefer the concept that cells employ a variety of specific molecules to communicate with their targets and to achieve operational responses across space and time.

I view the neural control of the endocrine system as a splendid example of the generality of intracellular chemical communication. Subsequent data are to be expected on the evolutionary significance of individual transmitter families and the functions of the neurons which secrete them, to be supplemented by quantitative accounts of the metabolic "cost" and "gain" to the transmitting cell of encoding its message in one or another molecular form, as well as the means to protect active sites on agonists aimed at distant receptors coupled to specific effector mechanisms. All these factors must eventually be brought into the solution of the cryptogram we have considered here. Hopefully, future Nobel Symposia will then be able to clarify intercellular operations and then to epitomize the roles played by chemical signals across their own unique temporal and spatial domains in brain, gut and endocrine tissues.

References

1. Aghajanian, G.K., J.M. Cedarbaum, and R.Y. Wang. Brain Res. 136: 570, 1977.
2. Aghajanian, G.K. and R.Y. Wang. In: Psychopharmacology: A Generation of Progress., edited by M.A. Lipton, A. diMascio, K. Killam. New York: Raven Press, 1978, p. 171.
3. Barker, J.L. In: Peptides in Neurobiology, edited by H. Gamer. New York: Plenum Press, 1977, p.295.
4. Barker, J.L. and T.G. Smith, Jr. Neurosci. Symp. 2:340, 1977.
5. Barker, J.L., J.H. Neale, T.G. Smith, Jr., and R.L.

Macdonald. Science 199: 1451, 1978.

6. Bennett, J.P., Jr. and S.H. Snyder. J.Biol.Chem. 251:
    7423, 1976.
7. Bertaccini, G. Pharmacol. Rev. 28: 127, 1976.
8. Bloom, F.E. Brain Res. 62: 299, 1973.
9. Bloom, F.E. Life Sci. 14: 1819, 1974a.
10. Bloom, F.E. In: The Neurosciences: Third Study Program,
    edited by F.O. Schmitt. Massachusetts: MIT Press, 1974b,
    p. 989.
11. Bloom, F.E. Rev. Physiol. Biochem. Pharmacol. 74: 1, 1975.
12. Bloom, F.E., J. Rossier, E.L.F. Battenberg, A. Bayon, E.
    French, S.J. Henriksen, G.R. Siggins, N. Ling, and R.
    Guillemin. In: Endorphins in Mental Health Research,
    edited by E. Usdin, 1978, in press.
13. Bloom, F.E. Biosystems 8: 179, 1977b.
14. Bloom, F.E. In: Neurosciences--A Fourth Study Program.
    edited by F.O. Schmitt and F.G. Worden. Boston: MIT Press,
    1978 (in press).
15. Brown, M., J. Rivier, and W. Vale. Abstr. Int. Soc. Psycho-
    neuroendocrinology, 1977.
16. Brunelli, M., Castellucci, V. and E.R. Kandel. Science 194:
    1178, 1976.
17. Bunge, R., M. Johnson, C.D. Ross. Science 199: 1409, 1978.
18. Castellucci, V. and E.R. Kandel. Aplysia. Science 194:
    1176, 1976.
19. Daly, J. In: Cyclic Nucleotides in the Nervous System.
    New York: Plenum Press, 1977.
20. Dismukes, K. Nature 269: 557, 1977.
21. Dowling, J.E., B. Ehinger, and W.L. Hedden. Invest.
    Ophthalmol. 15: 916, 1975.
22. Enna, S.J. and Snyder, S.H. Brain Res. 115: 174, 1976.
23. Florey, E. Fed. Proc. 26: 1164, 1967.
24. Foote, S., R. Freedman, and A.P. Oliver. Brain Res. 86:
    229, 1975.
25. Frederickson, R.C.A. Life Sci. 21: 23, 1977.
26. Freedman, R., B.J. Hoffer, D.J. Woodward, and D. Puro. Exp.
    Neurol. 55: 269, 1977.
27. Gahwiler, B.H. Nature 259: 483, 1976.
28. Greengard, P. Science 199: 146, 1978
29. Guyenet, P.G. and G.K. Aghajanian. Brain Res. 136: 178, 1977.
30. Hoffer, B.J., G.R. Siggins, and F.E. Bloom. Science 166:
    1418, 1969.
31. Hoffer, B.J., G.R. Siggins, and F.E. Bloom. Brain Res. 25:
    523, 1971.
32. Hoffer, B.J., G.R. Siggins, A.P. Oliver, and F.E. Bloom.
    J. Pharmacol. Expt. Ther. 184: 553, 1973.
33. Hoffman, G.E., K.M. Knigge, J.A. Moynihan, V. Melnyk, and
    A. Arimura. Neuroscience 3: 219, 1978.
34. Hokfelt, T., R. Elde, O. Johansson, A. Ljungdahl, A.
    Schutzberg, and K. Fuxe et al. In: Psychopharmacology--

A Generation of Progress, edited by A. diMascio, M.
Lipton and K. Killam. New York: Raven Press, 1977, in
press.

35. Hokfelt, T., L.G. Elfvin, R. Elde, M. Schultzberg, M.
Goldstein, and R. Luft. Proc. Nat. Acad. Sci. 74:3587,
1977.

36. Hokfelt, T., A. Ljungdahl, L. Terenius, R. Elde, and G.
Nilsson. Proc. Nat. Acad. Sci. U.S.A. 74: 3081, 1977.

37. Hokfelt, T., N. Halasz, A. Ljungdahl, O. Johansson, M.
Goldstein, and D. Park. Neurosci. Lett. 1: 85, 1975.

38. Hore, J., J. Meyer-Lohmann, and V.B. Brooks. Science 195:
584, 1977.

39. Kahn, C.R. J. Cell Biol. 70: 261, 1976.

40. Kandel, E.R. In: Cellular Basis of Behavior. edited by W.H.
Freeman. San Francisco: W.H. Freeman, 1976.

41. Koda, L.Y., J.A. Schulman, and F.E. Bloom. Brain Res. 145:
140, 1978.

42. Kosterlitz, H.W., J. Hughes, J.A.H. Lord, and A.A.
Waterfield. Neurosci. Symp. 2: 291, 1977.

43. Krnjevic, K. In: Handbook Psychopharmacology, edited by
L.L. Iversen, S.D. Iversen, and S.H. Snyder. New York:
Plenum, 1975, 6, 97.

44. Levitt, P. and R.Y. Moore. Brain Res., 1978 in press.

45. Libet, B., H. Kobayashi, and T. Tanaka. Nature 258:
155, 1975.

46. Lindvall, O. and A.A. Bjorklund. Acta Physiol. Scand. Suppl.
412: 1, 1974.

47. Lindvall, O., A. Bjorklund, A. Nobin, and U. Stenevi. J.
Comp. Neurol. 154: 317, 1974.

48. Macdonald, R.L. and P.G. Nelson. Science 199: 1449, 1978.

49. Macknight, A.D.C. and A. Leaf. Physiol. Rev. 57: 510, 1977.

50. Malvin, R.L., D. Mouw, and A.J. Vander. Science 197: 171,
1977.

51. Moore, R.Y. and F.E. Bloom. Ann. Rev. Neurosci. 1: 129, 1978.

52. Nathanson, J. Physiol. Rev. 57: 157, 1977.

53. Nicoll, R.A. Neurosci. Symp. 1: 99, 1976.

54. Nicoll, R.A. In: Psychopharmacology--A Generation of
Progress edited by M. Lipton, A. diMascio, K. Killam.
New York: Raven Press, 1978, p. 103.

55. Pearse, A.G.E. Nature, 262: 92, 1976.

56. Pearse, A.G.E. and T. Takor. Clin. Endocrin. Suppl. 5: 229,
1976.

57. Phillips, M.I., D. Felix, W.E. Hoffman, and Ganten. Neurosci.
Symp. 2: 308, 1977.

58. Rakic, P. Neurosci. Res. Progr. Bull. 13: 293, 1975.

59. Rehfeld, J.F. Nature 271: 771-773, 1978.

60. Renaud, L. J. Physiol. 264: 541, 1977.

61. Scharrer, B. J. Neuro-Visc. Rel. Suppl. IX: 1, 1969.

62. Schmitt, F.O., P. Dev, and B.H. Smith. Science 193: 114,
1976.

63. Schulman, J.A. and F.F. Weight. Science 194: 1437, 1976.
64. Segal, M. Exp. Brain Res. 29: 553-565, 1977.
65. Segal, M. and F.E. Bloom. Brain Res. 72: 79, 1974a.
66. Segal, M. and F.E. Bloom. Brain Res. 72: 99, 1974b.
67. Segal, M. and F.E. Bloom. Brain Res. 107: 499, 1976a.
68. Segal, M. and F.E. Bloom. Brain Res. 107: 513, 1976b.
69. Siggins, G.R. In: Psychopharmacology: A Generation of
    Progress, edited by M.A. Lipton, A. diMascio, and K.
    Killam. New York: Raven Press, 1978, p. 143.
70. Siggins, G.R., D.L. Gruol, A.L. Padjen, and D.S. Forman.
    Nature 270: 263, 1977.
71. Siggins, G.R., B.J. Hoffer, and F.E. Bloom. Science 165:
    1018, 1969.
72. Siggins, G.R., B.J. Hoffer, and F.E. Bloom. Brain Res. 25:
    535, 1971a.
73. Siggins, G.R., A.P. Oliver, B.J. Hoffer, and F.E. Bloom.
    Science 171: 192, 1971b.
74. Siggins, G.R., B.J. Hoffer, A.P. Oliver, and F.E. Bloom.
    Nature 233: 481, 1971c.
75. Stein, L. Nebr. Symp. Motiv. 22: 113, 1975.
76. Stone, T.W. and D.A. Taylor. J. Physiol. 266: 523, 1977.
77. Swanson, L. Brain Res. 128: 346, 1977.
78. Tomkins, G. Science 189: 760, 1975.
79. Uhl, G., J. Bennett, and S. Snyder. Brain Res. 130:
    299, 1977.
80. Uhl, G., M. Kuhar, and S. Snyder. Proc. Nat. Acad. Sci.
    U.S.A. 74: 4059, 1977.
81. U'Prichard, D.C. and S.H. Snyder. Life Sci. 20: 527,
    1977.
82. Wang, R.Y. and G.K. Aghajanian. Brain Res. 120: 85, 1977.
83. Weight, F.F. In: The Neurosciences--Third Study Program.
    edited by F.O. Schmitt, F.G. Worden. Cambridge: MIT Press,
    1974, p. 929.
84. Zieglgansberger, W. and J. Champagnat. Brain Res., 1978,
    in press.
85. Zieglgansberger, W. and J.P. Fry. In: Opiates and
    Endogenous Opioid Peptides, edited by H. Kosterlitz.
    Amsterdam: Elsevier/North Holland Biomedical Press, 1976,
    p. 231.

REGULATION OF NEUROPEPTIDE RELEASE IN RAT BRAIN

Leslie L. Iversen & Thomas M. Jessell

MRC Neurochemical Pharmacology Unit
Department of Pharmacology
University of Cambridge, England

The discovery of the hypophysiotropic peptides marked a large step forward in our understanding of the central regulation of pituitary function. Little is known, however, of the mechanisms which control the release of these substances from peptidergic neurones. The hypothalamic peptidergic neurones receive numerous synaptic inputs, involving various of the conventional amine and amino acid transmitters. In addition it seems likely that local chemical regulatory mechanisms may operate at the level of the presynaptic peptidergic nerve terminals.

Presynaptic regulation of this type has been most thoroughly studied for adrenergic neurones elsewhere in the nervous system. Thus, the release of noradrenaline from nerve terminals in the peripheral sympathetic nervous system and from neurones in the brain is subject to presynaptic regulation by a variety of chemical factors. Released noradrenaline may inhibit its own further release by acting on presynaptic receptors of a pharmacological type similar to classical α-adrenoceptors (Starke, 1977; Langer, 1977). In addition, noradrenaline release may be inhibited by presynaptic receptors which respond to acetylcholine, dopamine, β-adrenoceptor agonists, angiotensin II and opiate drugs and enkephalins (for review see Starke, 1977; Langer, 1977; Westfall, 1977). The noradrenergic terminals in different tissues may respond selectively to some but not all of the above influences, suggesting that the pattern of distribution of presynaptic regulatory receptors differes in different organs.

Studies on the release of hypophysiotropic peptides from hypothalamic synaptosomes have suggested that similar presynaptic regulatory mechanisms may play an important role in controlling peptide release (Edwardson & Bennett, 1974). It is clear that peptidergic neurones exist in many regions of mammalian CNS outside the hypothalamus (for review see Hughes, 1978). Our own recent studies of the factors influencing the release of the neuropeptide substance P from nerve terminals in CNS also indicate the existence of presynaptic regulatory mechanisms in this category of peptidergic neurone. The present chapter will review these findings, and describe the application of a simple in vitro model for studies of the release of this and other brain peptides from defined regions of mammalian CNS.

RELEASE OF SUBSTANCE P FROM RAT BRAIN IN VITRO

Substantia Nigra

Within the mammalian CNS the highest concentrations of the undecapeptide substance P are found in the substantia nigra (Brownstein et al, 1976; Kanazawa & Jessell, 1976). Using a radioimmunoassay based on the procedure described by Powell et al (1973) we find that the content of substance P in rat substantia nigra is 1735 $\pm$ 106 pmol/g wet weight (mean $\pm$ SEM, n=4). Immunohistochemical studies indicate that the peptide is concentrated in a dense fibre and terminal network in both the pars reticulata and pars compacta (Hökfelt et al, 1975; Cuello & Kanazawa, 1978). The cell bodies of the substance P-containing nerve terminals in the substantia nigra appear to originate in the corpus striatum and globus pallidus (Kanazawa et al, 1977; Emson et al, 1978a). The high density of substance P terminals in this brain area thus make it a suitable model tissue for studies of the release of the peptide from CNS neurones (Jessell, 1978).

Release studies were performed using slices of rat substantia nigra tissue carefully dissected from 300μm thick sections of fresh chilled brain with the aid of a dissecting microscope. Although the tissue is highly enriched in the peptide, the average wet weight of nigral tissue pooled from both sides of the rat brain is only approximately 6mg. Nigral tissue from two rat brains, however, provided a sufficient amount of substance P to give easily measurable rates of release. The nigral tissue was cross-chopped at 200μm intervals, and the fragments were placed in a small superfusion chamber (volume 350μl) fitted with 50μm mesh nylon mesh filters at each end to prevent loss of tissue. The tissue was then superfused by pumping Krebs-bicarbonate solution at 37°C through the chamber at a rate of 375μl/min. Superfusate samples were collected at 1 min intervals and the substance P content in each sample, and in extracted nigral tissue recovered at the end of the experiment, was determined by radioimmunoassay. To protect against loss of the peptide by metabolism or by adsorption the superfusion solution contained bovine serum albumin (0.5% w/v). This in vitro procedure can be applied in principle to studies of the release of any putative neurotransmitter from defined small regions of brain. The apparatus has four channels which are run in parallel, and superfusate samples are collected by use of an automatic fraction collector. The relatively small dimensions and amount of tissue used, together with a rapid superfusion rate are designed to ensure that released material is rapidly removed to mimize the possibility of local degradation of the released substance in the tissue.

Using this procedure, it was found that the basal efflux of substance P from rat substantia nigra remained constant (after the first few minutes of superfusion) for more than 30 min. The amounts of substance P in spontaneous efflux samples were small (approximately 8 fmol substance P/mg tissue/min) but easily measurable by the radioimmunoassay procedure, which allowed measurements of as little as 5-8 fmol/sample. The rate of spontaneous release of peptide represented approximately 0.5% of tissue stores per minute.

Raising the potassium concentration in the superfusing medium to 47mM for 2 min (Fig. 1A) evoked approximately a 5-fold increase in the rate of substance P efflux. The rate of release returned to basal levels soon after restoration of normal potassium concentration. The potassium-evoked release of substance P seems to represent a valid model system for studying the stimulus-secretion coupling mechanism in substance P neurones. In common with other transmitter release systems, the

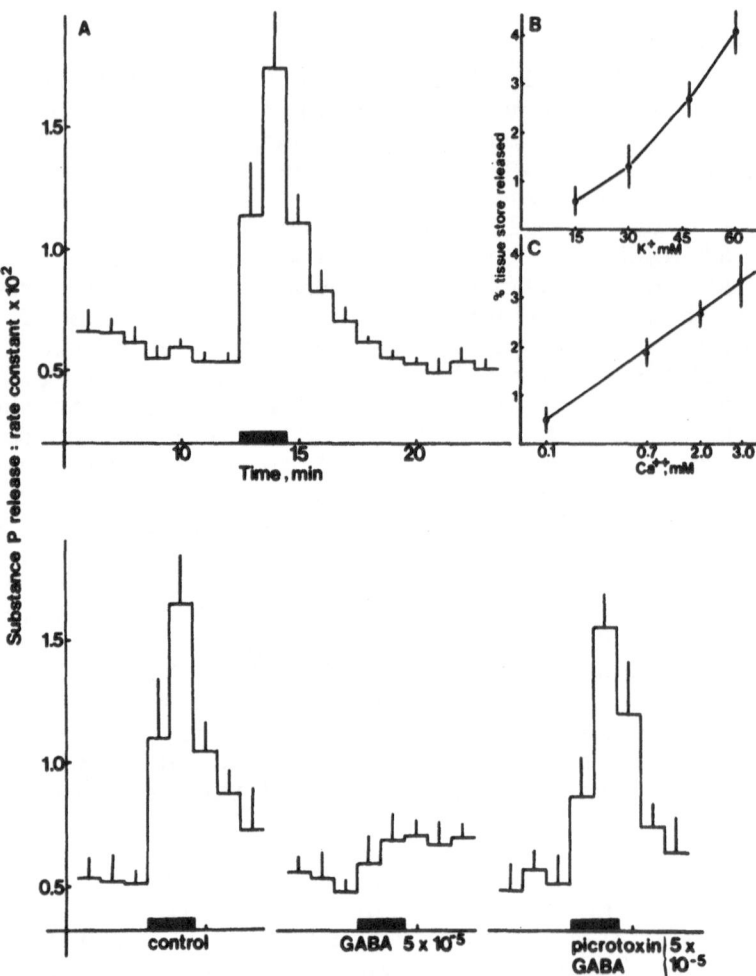

Fig. 1. Release of substance P from superfused slices of rat
substantia nigra.
A. Potassium evoked-release of immunoreactive substance P from
    slices of rat substantia nigra. (▬): K^+ 47mM. Release is
    calculated as the fractional efflux rate constant, and
    expressed as the percentage of tissue stores released per
    minute. Spontaneous efflux represents approximately 8 f mol
    mg^{-1}. min^{-1}. Each point is the mean $\pm$ SEM of at least 8
    determinations.
B. Dose dependency of potassium-evoked release of substance P.
    Each point is the mean $\pm$ SEM of 4-6 determinations.
C. Effect of Ca^{++} concentration of release of substance P evoked
    by 47mM K^+. Each point is the mean $\pm$ SEM of 4-6 determinations.
D. Inhibition of the potassium-evoked release of substance P
    by GABA, and reversal by picrotoxin. (▬): K^+ 47mM.
    Each point is the mean $\pm$ SEM of 6 determinations.

potassium evoked release of substance P is completely
dependent on the presence of calcium ions in the
external medium (Fig. 1C). The effects of potassium
could also be mimicked by another chemical depolarising
stimulus, veratridine (50μM) (Jessell, 1978). The
releasing effects of veratridine were again calcium-
dependent, and could also be completely blocked by
addition of the sodium channel blocker tetrodotoxin
(0.5μM) to the perfusing medium, (Jessell, 1978).

In an attempt to investigate the factors controlling the release of substance P, the effect of other
neurotransmitters present in the substantia nigra were
examined. Addition of dopamine (50μM) (Reubi et al,
1977), or morphine (10μM) or the opioid peptide
β-endorphin (0.5μM) (Jessell & Iversen, 1977) had no
effect on either the spontaneous or potassium-evoked
efflux of substance P. the addition of GABA (10μM) for
7 min prior to the potassium pulse, however, significantly inhibited the release normally evoked by high
potassium. Increasing the GABA concentration to 50μM
produced a much greater inhibition of potassium-evoked
release, without affecting spontaneous release (Fig. 1,
lower part). This effect of GABA appears to involve
GABA receptors, since the inhibitory effects of GABA on
substance P release were reversed by the GABA antagonist
drugs picrotoxin (Fig. 1) or bicuculline (50μM), which
by themselves had no effects on spontaneous or potassium-
evoked release. Furthermore, the potent GABA agonist
drug muscimol (50μM) mimicked the inhibitory effects of
GABA on potassium-evoked release of substance P.

These results show that the peptide substance P is
released from nerve terminals in rat substantia nigra
by a calcium-dependent mechanism, supporting the view
that this peptide may normally be released as a neurotransmitter from such neurones. Similar results, using
brain synaptosome preparations, have been reported by
Schenker et al (1976) and by Lembeck et al (1977). The
inhibitory effects of GABA on the evoked release of
substance P from substantia nigra suggest that GABA,
which is present in high concentrations in this brain
region, may regulate substance P release from striato-
nigral afferent fibres by a local effect on GABA
receptors located on substance P-containing presynaptic
nerve terminals.

Spinal Trigeminal Nucleus

In addition to its presence within intrinsic
neuronal pathways in brain, substance P is also concentrated in the terminals of some primary afferent
neurones in the spinal cord and in sensory nuclei of
cranial nerves (Hökfelt et al, 1975; Cuello & Kanazawa,
1978). The spinal division of the trigeminal nerve
nucleus in rat brain stem contains a high density of
substance P-containing sensory terminals. Radioimmunoassay of this region indicates that it contains the
second highest substance P concentration of any CNS
region in rat (1396 $\pm$ 153 pmol/g wet weight).

Immunohistochemical studies indicate that the
spinal division of the trigeminal nerve nucleus contains
a high density of substance P in nerve terminals in the
superficial substantia gelatinosa, which represents a
rostral extension of the dorsal horn of spinal cord.
The substance P-containing nerve terminals disappear
after section of the trigeminal nerve or surgical removal of the Gasserian ganglion (Cuello et al, 1978a).
This readily accessible brain region thus represents a
suitable model for studies of the release of substance
P from primary sensory terminals.

Substance P release was examined, using slices of
spinal trigeminal nerve nucleus (pooled tissue from two
rat brains, approximately 20 mg wet weight), as described above. In these experiments the aminopeptidase
inhibitor bacitracin (20μM) was added to the perfusion
fluid in order to inhibit metabolic degradation of the
released peptide (Jessell & Iversen, 1977). The rate of
spontaneous efflux, as in the substantia nigra, represented only a small proportion of total tissue stores
(0.2-0.4% per minute), and again exposure to high
potassium elicited a marked increase in substance P
release, which was calcium-dependent (Fig. 2A,G).

Because opiate receptors and the endogenous opioid
peptide enkephalin are known to be highly concentrated
in the region of primary sensory nerve terminals in
substantia gelatinosa of spinal cord and trigeminal
nerve nucleus (Atweh & Kuhar, 1977; Simantov et al,
1977; Elde et al, 1977), we examined the effects of
opiate drugs and endorphins on substance P release from
this area. Addition of morphine to the superfusing
medium 7 min prior to a potassium pulse did not affect
the spontaneous efflux of substance P, but produced a
concentration-dependent inhibition of the evoked

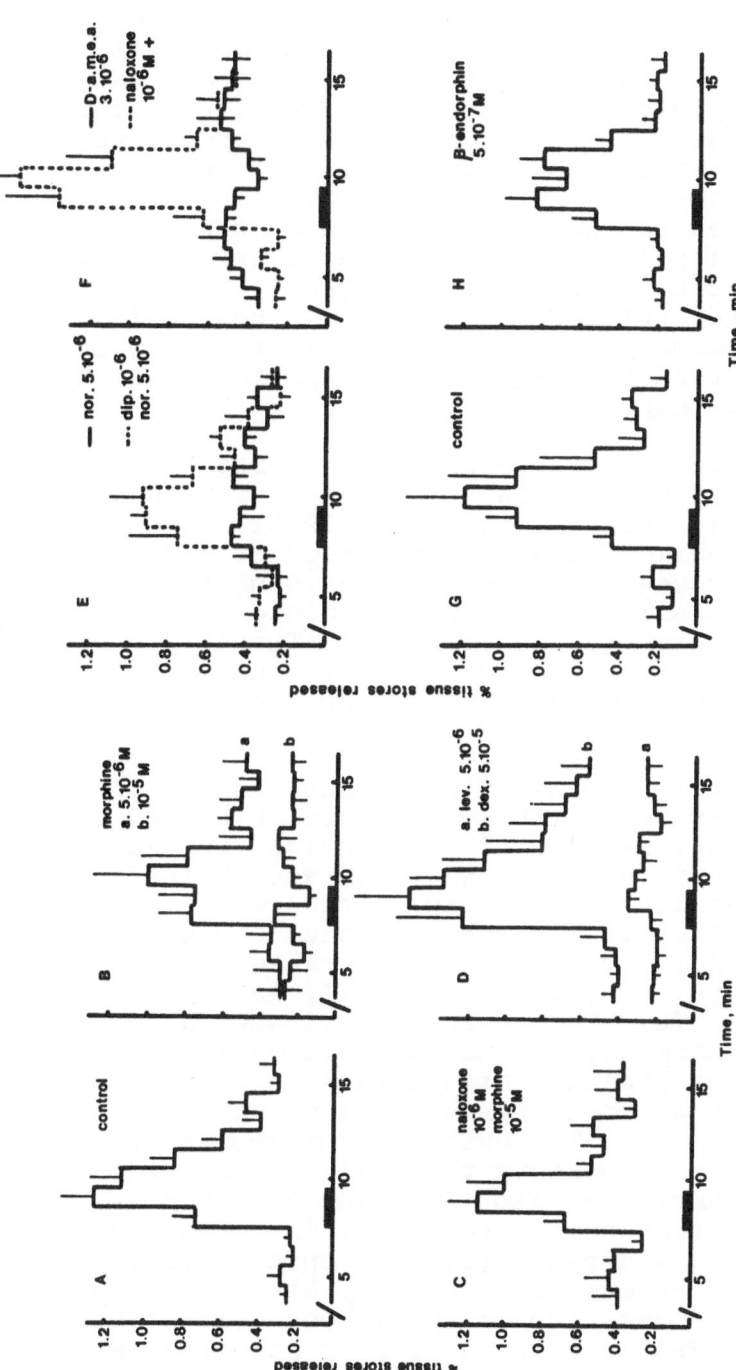

Fig. 2. Effect of opiates on the release of substance P from superfused slices of rat trigeminal
nucleus. A. Potassium evoked release of substance P (details as Fig. 1). B. Inhibition of
potassium-evoked release of substance P by morphine. C. Naloxone reversal of morphine induced
inhibition. D. Effect of levorphanol (lev) and dextrorphan on potassium-evoked substance P
release. E. Inhibition by normorphine and reversal by diprenorphine. F. Inhibition by
D-[ala]2-Met-enkephalin amide (D-a.m.e.a.) and reversal by naloxone. G.H. Inhibition by
(porcine) β-endorphin.

release of peptide (Fig. 2B). Substance P release was
almost completely suppressed at a morphine concentration of 10μM). Continuous superfusion with the opiate
antagonist drug naloxone (1μM) did not affect either
spontaneous or potassium-evoked release of substance P,
but almost completely abolished the inhibitory effects
of 10μM morphine (Fig. 2C). The opiate response was
stereospecific in that superfusion of rat spinal
trigeminal slices with levorphanol (5μM) inhibited the
potassium-evoked release of substance P, whereas the
pharmacologically inactive enantiomer dextrorphan was
inactive at a concentration of 50μM (Fig. 2D). The
potassium-evoked release of substance P could also be
inhibited by normorphine (5μM) and this effect was
partially reversed by simultaneous superfusion with the
antagonist drug diprenorphine (1μM) (Fig. 2E).

A metabolically stable analogue of met-enkephalin,
(D-Ala)[2]-methionine-enkephalin amide – which is a
potent analgesic agent (Pert et al, 1976) – was also
tested. This peptide at a concentration of 3μM almost
completely supressed potassium-evoked substance P
release, and this inhibitory effect could again be
reversed by naloxone (1μM) (Fig. 2F). The larger opioid
peptide β-endorphin was also effective in supressing
potassium-evoked release of substance P (0.5-2.0μM)
(Fig. 2H).

The selective localization of substance P in small
diameter sensory nerve fibres and terminals, and the
calcium-dependent release of the peptide from such
terminals support the view that substance P functions as
a sensory transmitter. There is, furthermore, considerable evidence that in its sensory function substance P
may be particularly concerned with transmission in pain
pathways (for review see Cuello et al, 1978;
Wolstencroft, 1978). The inhibition of substance P
release by opiates and by opioid peptides might be
explained by the presence of presynaptic opiate receptors on primary afferent nerve terminals which contain
substance P (Fig. 3). The existence of presynaptic
opiate receptors on sensory terminals had already been
suggested by Lamotte et al (1976) who observed a
decreased number of opiate receptors in dorsal horn
after rhizotomy. A direct effect of opiates in supressing substance P release from primary sensory neurones
in organ culture has also recently been demonstrated by
Mudge et al (1978). The inhibition of substance P
release by opiates might explain the analgesic actions
of such drugs at the spinal cord level (Yaksh & Rudy,

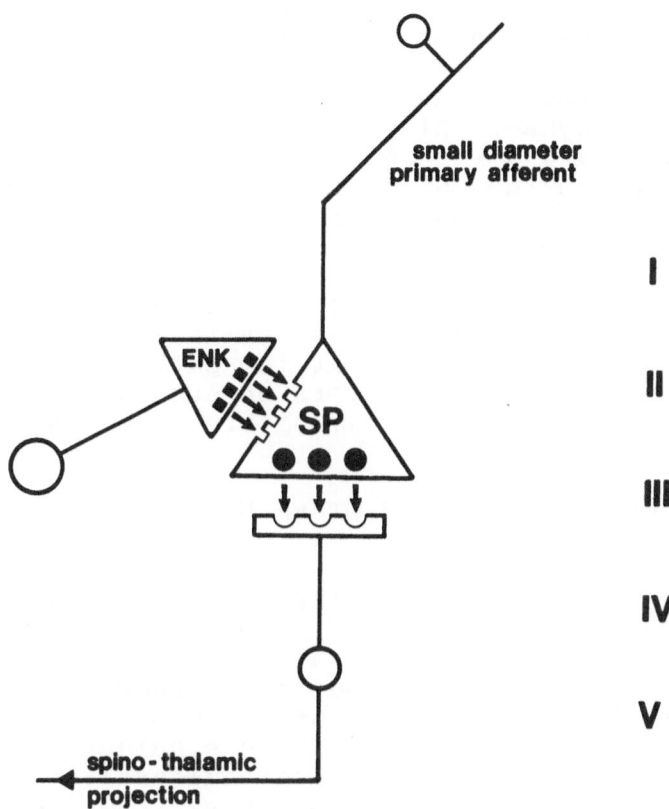

Fig. 3.   Schematic representation of a possible mechanism for opiate-induced suppression of SP release. SP is shown localised within the terminal of a small diameter afferent fibre which forms an excitatory axodendritic synapse with the process of a spinal cord neurone originating in lamina IV or V and projecting rostrally. A local enkephalin-containing inhibitory interneurone (ENK), confined to laminae II and III forms a presynaptic contact on the terminal of the primary afferent. Opiate receptor sites are depicted presynaptically on the primary afferent terminal. Numbers on the right refer to the laminae of Rexed.

1976), and suggests a possible role for enkephalins in the substantia gelatinosa in inhibiting the transmission of stimuli in pain pathways (Fig. 3).

The inhibition of substance P release by presynaptic opiate receptors is also consistent with other studies which have shown that opiates can act presynaptically to inhibit the release of a number of other transmitters. Thus opiate receptors are known to be located presynaptically on peripheral cholinergic (Paton, 1957) and adrenergic (Hughes et al, 1975) neurones, and in the CNS on cholinergic, noradrenergic and dopaminergic neurones (Jhamandas et al, 1971; Loh et al, 1976; Taube et al, 1976; Pollard et al, 1977).

The inability of morphine or β-endorphin to inhibit substance P release from nerve terminals in rat substantia nigra suggests that the pattern of distribution of presynaptic receptors on substance P-containing nerve terminals varies among brain regions.

IN VITRO RELEASE OF OTHER NEUROPEPTIDES

Enkephalin

Recent immunohistochemical studies have shown that the enkephalins are localised in neurones and nerve terminals in various regions of the CNS (Elde et al, 1977; Simantov et al, 1977). These studies, together with the results of detailed regional analyses of enkephalin content by radioimmunoassay (Hong et al, 1977; Kobayashi et al, 1978) have indicated that the rat globus pallidus contains a particularly high density of enkephalin-containing nerve terminals. The enkephalin concentration in globus pallidus is 6-8 times higher than in any other brain region. The results of lesion studies suggest that the cell bodies of the enkephalin neurones with terminals in globus pallidus are localized in the adjacent caudate nucleus and putamen (Cuello & Paxinos, 1978). The globus pallidus thus represents a suitable tissue for in vitro studies of enkephalin release from nerve terminals in mammalian brain (Iversen et al, 1978a).

Slices of rat globus pallidus were superfused in vitro as described above. The superfusion medium contained bovine serum albumin (0.1% w/v) and bacitracin

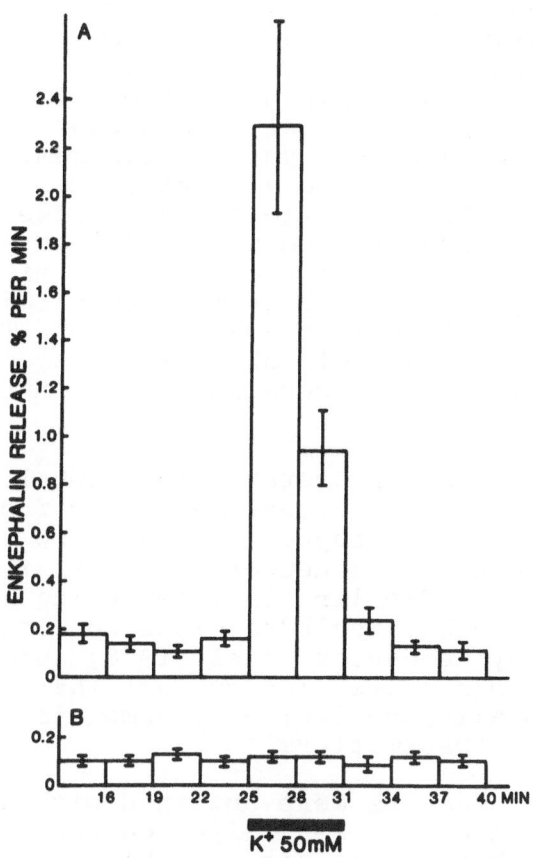

Fig. 4. Release of enkephalin immunoreactivity from slices of
rat globus pallidus superfused with normal (a) or calcium-free
medium (b). The slices of pallidal tissue (about 30 mg wet weight)
were superfused at a rate of approximately 300μl min^{-1}, and after
an initial wash period of 13 min, samples of the superfusate were
collected at 3 min intervals for subsequent radioimmunoassay.

(30μg/ml) to protect the released peptide. The spontaneous efflux of enkephalin-like immunoreactivity was
low (approximately 0.2% of tissue stores per minute)
and was increased more than 10-fold during exposure to
high potassium (50mM) (Fig. 4A). In a 6 min period of
exposure to high potassium there was significantly less
release of enkephalin during the latter half of this
period than initially (Fig. 4) and release rapidly
returned to basal levels after removal of the high
potassium stimulus. A single potassium stimulus released almost 10% of the total tissue stores of enkephalin, and a second potassium stimulus administered
9 min after the first evoked a significantly smaller
release of enkephalin than the first. Unlike the other
neuropeptides studied, the total tissue content of
enkephalin decreased markedly during a 40 min superfusion in vitro, with a loss of more than 60% of the
initial content in freshly prepared slices. The
potassium-evoked release of enkephalin was completely
supressed if calcium ions were omitted from the superfusion fluid (Fig. 4B).

These results demonstrate that enkephalin can be
released by a calcium-dependent mechanism from nerve
terminals in rat brain, supporting the view that it may
normally be released as a neurotransmitter from nerve
terminals in brain. Similar findings, using striatal
slices (which probably included globus pallidus) have
been described by Henderson et al (1978) and Osborne
et al (1978). Our results also suggest that the tissue
stores of enkephalin are labile by comparison with other
amine or peptide transmitters.

The initial results were obtained with a radioimmunoassay based on antisera directed towards leu-
enkephalin, so that it was not possible to distinguish
between leu- and met-enkephalin in the tissue extracts
or release samples. Subsequent experiments have made
use of both leu- and met-enkephalin radioimmunoassay
procedures, together with separation of the two peptides
by high pressure liquid chromatography (Bajon et al,
1978). The results of these experiments indicate that
both leu- and met-enkephalin are released from rat
globus pallidus in response to high potassium. Furthermore, the ratio of the two pentapeptides in the release
samples is similar to that found in the tissue, after
corrections are made for the more rapid degradation of
released met-enkephalin following its release. Despite
the use of bovine serum albumin and bacitracin to

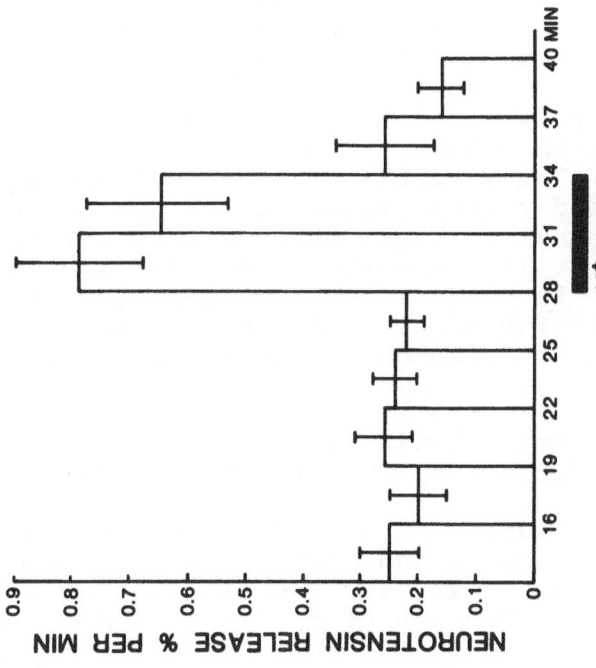

Fig. 6. Potassium-evoked release of neurotensin from slices of rat hypothalamus.
Duplicate 100μl aliquots of superfusate
samples were used for neurotensin measurements using a radioimmunoassay method.

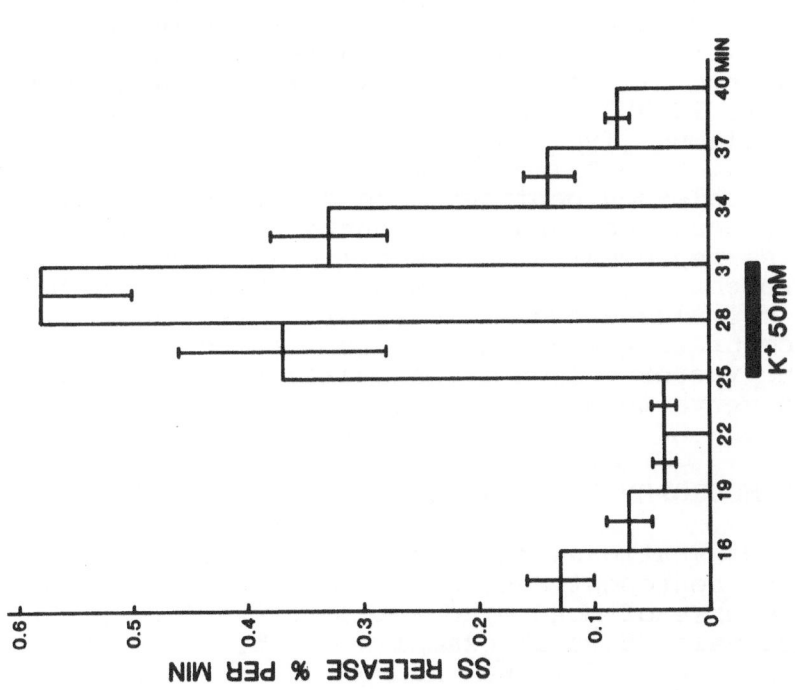

Fig. 5. Potassium-evoked release of somatostatin
from rat amygdala slices. Slices of amygdala were
superfused with a modified Krebs bicarbonate
solution at 37°C, at a rate of approximately
300μl min^{-1}.

protect released enkephalins, experiments in which
trace amounts of tritiated enkephalins were added to
the superfusion medium indicated that 63% of met-enkephalin and 23% of leu-enkephalin are degraded during a
single passage over the tissue slices. The tissue
stores of met-enkephalin also appear to be more labile
than those of leu-enkephalin, since met-enkephalin is
lost more rapidly from tissue stores during superfusion
than leu-enkephalin. These observations strongly support
a transmitter role for both enkephalins in rat globus
pallidus. The in vitro release system also offers a
valuable model for studies of the influence of other
transmitters and drugs on enkephalin release.

Somatostatin, Neurotensin and
Vasoactive Intestinal Peptide

These peptides have been described recently as
occurring in neurones and nerve terminals in mammalian
brain (Kobayashi et al, 1977; Brownstein et al, 1975;
Uhl et al, 1977; Larsson et al, 1976; Elde et al, 1978).
Preliminary studies on the release of these neuropeptides from brain slices, using the procedure described
above, have shown that a calcium-dependent release of
all three peptides can be demonstrated in response to a
high potassium stimulus. Somatostatin release was
demonstrated from slices of rat hypothalamus and amygdala
(Iversen et al, 1978b); neurotensin release from slices
of rat hypothalamus (Iversen et al, 1978b) and VIP
release from rat hypothalamus (Emson et al, 1978b). In
each case the released peptides were detected by radioimmunoassay procedures, and the amounts present in
efflux samples represented only small proportions of
the total present in the tissue stores. The ability to
measure the stimulus-evoked release of these neuropeptides from brain tissue again offers a model for
examining the factors which may normally regulate their
release from nerve terminals in the brain. (Figs. 5 & 6)

EFFECTS OF NEUROPEPTIDES ON RELEASE OF MONOAMINES

Just as conventional neurotransmitters may influence
the release of neuropeptides, so neuropeptides can also
affect the release of monoamine transmitters from presynaptic terminals. Several examples of this type of
interaction are known for the endorphins, where presynaptic opiate receptors are known to exist on the

terminals of cholinergic and adrenergic cells (see
above). We have recently described a similar action of
substance P on dopamine and 5-HT release in slices of
rat substantia nigra in vitro (Reubi et al, 1978).
Substance P and the related peptide eledoisin caused an
increase in the release of newly synthesized 5-HT
(Fig. 7) and dopamine. Many other interactions of this
type probably remain to be discovered.

CONCLUSIONS

The neuropeptides in all probability represent a
new family of chemical messenger substances in mammalian CNS, of which the hypophysiotropic peptides represent only one specialized category. The release of
peptides from peptidergic neurones may be regulated by
presynaptic mechanisms, as in the case of the better
known monoamine neurotransmitters. The development of
simple in vitro models for studying the stimulus-evoked
release of neuropeptides from nerve terminals in defined
brain regions provides a means of examining such local
regulatory mechanisms, and the results obtained with
substance P suggest that in different brain regions
GABA and enkephalins may play important roles in

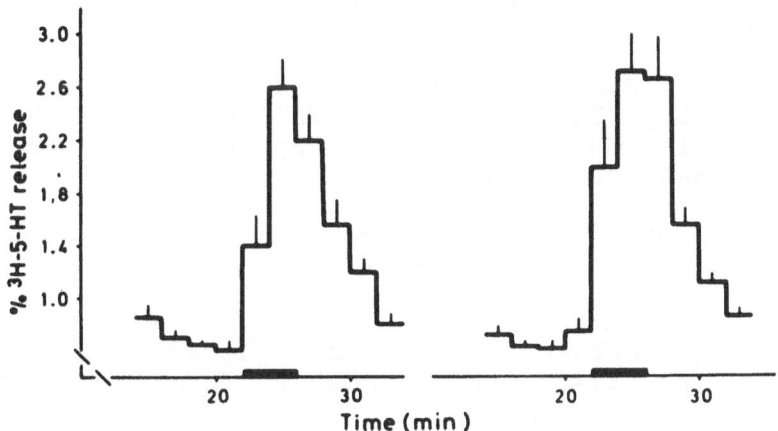

Fig. 7. Effect of substance P and eledoisin on newly synthesized
^{3}H-5-HT release from slices of rat substantia nigra. Left:
Release evoked by a 4 min pulse of 50μM substance P (black bar)
(mean $\pm$ SEM n=6). Right: Release evoked by a 4 min pulse of 50μM
eledoisin (mean $\pm$ SEM n=4). Efflux is expressed as percent of
tissues stores of ^{3}H-5-HT released per min.

controlling the release of this excitatory transmitter candidate.

REFERENCES

ATWEH, S.F. & KUHAR, M.J. Brain Research, 129, 1-12, 1977

BAJON, A., ROSSIER, J., MAUSS, A., BLOOM, F.E., IVERSEN, L.L. & GUILLEMIN, R. Proc. Nat. Acad. Sci. US., (in press), 1978

BROWNSTEIN, M., ARIMURA, A., SATO, H., SCHALLY, A.V. & KIZER, J.S. Endocrinology, 96, 1456-1461, 1975

BROWNSTEIN, M.J., MROZ, E.A., KIZER, J.S., PALKOVITS, M. & LEEMAN, S.E. Brain Research, 116, 299-305, 1976

CUELLO, A.C. & KANAZAWA, I. J. Comp. Neur. 178, 129-156, 1978

CUELLO, A.C. & PAXINOS, G. Nature, 271, 178-180, 1978

CUELLO, A.C. del FIACCO, M. & PAXINOS, G. Brain Research (in press) 1978a

CUELLO, A.C., EMSON, P., del FIACCO, M., GALE, J., IVERSEN, L.L., JESSELL, T.M., KANAZAWA, I., PAXINOS, G. & QUIK, M. In "Centrally Acting Peptides", pp. 135-155, ed. J. Hughes, Macmillan, London, 1978b

ELDE, R., HOKFELT, T., JOHANSSON, O. & TERENIUS, L. Neuroscience, 1, 349-355, 1977

ELDE, R., HOKFELT, T., JOHANSSON, O., LJUNGDAHL, A., NILSSON, G. & JEFFCOATE, S.L. In "Centrally Acting Peptides", pp. 17-35, ed. J. Hughes, Macmillan, London, 1978

EDWARDSON, J.A. & BENNETT, G.W. Nature, 251, 425-427, 1974

EMSON, P.C., JESSELL, T.M., PAXINOS, G. & CUELLO, A.C. Brain Research (in press) 1978a

EMSON, P.C., FAHRENKRUG, J., SCHAFFALITZKY DE MUCKADELL, O.B., JESSELL, T.M. & IVERSEN, L.L. Brain Research, 143, 174-178, 1978b

HENDERSON, G., HUGHES, J. & KOSTERLITZ, H.W. Nature, 271, 677-679, 1978

HOKFELT, T., KELLERTH, J.O., NILSSON, G. & PERNOW, B. Brain Research, 100, 235-252, 1975

HONG, J.S., YANG, H.Y.T., RACAGNI, G. & COSTA, E. Brain Research, 122, 541-544, 1977

HUGHES, J. (Ed) "Centrally Acting Peptides", Macmillan, London, 1978

HUGHES, J., KOSTERLITZ, H.W. & LESLIE, F.M. Brit. J. Pharmac. 53, 371-381, 1975

IVERSEN, L.L., IVERSEN, S.D., BLOOM, F.E., VARGO, T. & GUILLEMIN, R. Nature, 271, 679-681, 1978a

IVERSEN, L.L., IVERSEN, S.D., BLOOM, F.E., DOUGLAS, C., BROWN, M. & VALE, W. Nature, 273, 161-163, 1978b

JESSELL, T.M. Brain Research (in press), 1978

JESSELL, T.M. & IVERSEN, L.L. Nature, 268, 549-551, 1977

JHAMANDAS, K., PHILLIS, J.W. & PINSKY, C. Brit. J. Pharmac., 43, 53-66, 1971

KANAZAWA, I. & JESSELL, T.M. Brain Research, 117, 362-367, 1976

KANAZAWA, I., EMSON, P.C. & CUELLO, A.C. Brain Research, 119, 447-453, 1977

KOBAYASHI, R.M., BROWN, M. & VALE, W. Brain Research, 126, 584-588, 1977

KOBAYASHI, R.M., PALKOVITS, M., MILLER, R.J., CHANG, K.J. & CUATRECASAS, P. Life Sciences, 22, 527-530, 1978

LAMOTTE, C., PERT, B. & SNYDER, S.H. Brain Research, 112, 407-412, 1976

LANGER, S.Z. Brit. J. Pharmac., 60, 481-497, 1977

LARSSON, L.I., FAHRENKRUG, J., SCHAFFALITZKY DE
MUCKADELL, O., SUNDLER, F., HAKANSON, R. & REHFELD, J.F.
Proc. Nat. Acad. Sci. US., 73, 3197-3200, 1976

LEMBECK, F., MAYER, N. & SCHINDLER, G. N.S. Archiv.
Pharmacol., 301, 17-22, 1977

LOH, H.H., BRASE, D.A., SAMPATH-KHANNA, S., MAR, J.B.
LEONG WAY, E., & LI, C.H. Nature, 264, 567-568, 1976

MUDGE, A.W., LEEMAN, S.E. & FISCHBACH, G.D. (submitted)

OSBORNE, H., HOLLT, V. & HERZ, A. Eur. J. Pharmacol.,
48, 219-221, 1978

PATON, W.D.M. Brit. J. Pharmac., 12, 119-127, 1957

PERT, C.B., PERT, A., CHANG, J.K. & FONG, B.T.W.
Science, 194, 330-332, 1976

POLLARD, H., LLORENS-CORTES, C. & SCHWARTZ, J.C. Nature,
268, 745-747, 1977

POWELL, D., LEEMAN, S.E., TREGEAR, G.W., NIALL, H.D. &
POTTS, J.T. Nature, 241, 252-254, 1973

REUBI, J.C., IVERSEN, L.L. & JESSELL, T.M. Nature, 268,
652-654, 1977

REUBI, J.C., EMSON, P.C., JESSELL, T.M. & IVERSEN, L.L.
N.S. Archiv. Pharmacol. (in press)

SCHENKER, C., MROZ, E.A. & LEEMAN, S.E. Nature, 264,
790-792, 1976

SIMANTOV, R., KUHAR, M., UHL, G. & SNYDER, S.H. Proc.
Nat. Acad. Sci. US., 74, 2167-2171, 1977

STARKE, K. Rev. Physiol. Biochem., Pharmacol., 77,
1-124, 1977

TAUBE, H.D., BOROWSKI, E., ENDO, T. & STARKE, K. Eur. J.
Pharmacol., 38, 377-380, 1976

UHL, G.R., KUHAR, M.J. & SNYDER, S.H. Proc. Nat. Acad.
Sci. US., 74, 4059-4063, 1977

WESTFALL, T. Physiol. Rev., 57, 660-728, 1977

WOLSTENCROFT, J.H. In "Chemical Communication Within the Nervous System and its Disturbance in Disease", pp. 135-160, ed. A. Taylor & M.T. Jones, Pergamon Press, Oxford, 1978

YAKSH, T.L. & RUDY, T. Science, 192, 1357-1358, 1976

STUDIES ON INTERACTIONS OF EPINEPHRINE NEURONAL SYSTEMS WITH

OTHER NEURONAL SYSTEMS

M. Goldstein, A. Sauter, F. Hata, J.Y. Lew, Y. Baba,
J. Engel, K. Fuxe and T. Hokfelt
Neurochemistry Laboratories, Department of Psychiatry,
New York University Medical Center, New York, New York
and Department of Histology, Karolinska Institute,
Stockholm, Sweden

INTRODUCTION

Immunohistochemical techniques have proven to be useful
in the identification and mapping of various monoaminergic
neuronal systems. Phenylethanolamine-N-methyl transferase (PNMT),
the enzyme which catalyzes the formation of epinephrine (E)
(E forming enzyme; EC 2.1.1.28) has been localized in specific
brain regions, providing the first morphological evidence for
the existence of E neurons in the brain. The PNMT containing
cell bodies have so far been located in two cell clusters, C_1
and C_2 cell groups of the reticular formation in the medulla
oblongata. The PNMT positive terminals are mainly found in certain visceral afferent or efferent nuclei of the brain and spinal
cord, in the ventral periventricular gray of lower brain stem
and in certain nuclei of the hypothalamus (1,2). The adrenergic
innervation pattern suggests that central E neurons may be implicated in various functions of the CNS (e.g., neurohormone
secretion, temperature regulation, food and water intake, modulation of blood pressure, respiration, regulation of sleep and
wakefulness and certain types of behavior). Recent morphological
and biochemical data suggests that interactions among different
neurotransmitter systems occurs at various levels. In this
presentation we will describe some of the properties of the E
neuronal systems and its interaction with the norepinephrine (NE)
neuronal system following exposure of animals to stressful
stimuli. We will also present data on the possible involvement
of E neuronal system in the development of genetic hypertension.
Finally, we will discuss some possible interactions of the amine
neuronal systems with substance P neuronal systems.

METHODS AND MATERIALS

The PNMT activity was assayed by previously described procedure (3) using phenylethanolamine as a substrate. The method
was rendered by substituting H^3-methyl-S-adenosyl methionine
(SAM) for C^{14}-SAM (4). Epinephrine and norepinephrine were
assayed by a radioenzymatic procedure (5). Rats brains were removed from the cranium immediately after decapitation. Frontal
sections (500 μm) were cut from the level of the nucleus nervi
facialis to the pyramidal decussation. Five slices were made;
the most caudal slice was designated "a" and the most rostral
"e" (Fig. 1). The selected regions were then punched out
(1.0 mm in diameter) under the dissecting microscope.

Locus coeruleus, hypothalamus, and cerebral cortex were
dissected as previously described (6). The PNMT inhibitor SKF
64139, 7,8-dichloro-1,2,3,4 tetrahydroisoquinoline HCl was
kindly given to us by Dr. R.G. Pendleton from SKF, Philadelphia,
U.S.A.

The Effect of Stress on E and NE Neuronal Systems

Several studies have shown that the metabolism of NE in the
CNS is changed by exposure of animals to stressful stimuli (7,8).
However, the participation of the E neuronal systems in the stress
reaction was not previously investigated. We have therefore
studied the effects of electric footshock on E levels in specific
regions of the CNS (6). It is evident from the results presented
in Table 1 that exposure of animals to footshock stress resulted
in a significant lowering of E levels in analyzed regions of the
CNS. The E levels were not significantly decreased by electric
footshock in the C_2 rostral region of the medulla oblongata, an
area which has a high density of PNMT containing cell bodies
(1,2). The data in Table 1 also shows that treatment of animals
with the PNMT inhibitor SKF 64139 (9) results in a significant
lowering of E levels in the CNS (10). However, the E levels were
significantly lower in animals pre-treated with the PNMT inhibitor and then exposed to footshock stress than in animals either
exposed to footshock alone or in non-stressed animals given the
PNMT inhibitor. These results demonstrate that E participates
in the stress reaction, namely it produces a decrease in the
steady state levels of central E and an increase in the turnover
of central E.

As expected, electric footshock stress induces a significant
lowering of NE levels in the analyzed regions of the CNS (Table 2).
It is noteworthy that the NE levels were not significantly lower
in the C_2 rostral region of the medulla oblongata and in the
frontal cortex (11). PNMT inhibition had no significant effect
on central NE levels. However, animals pre-treated with the
PNMT inhibitor and then exposed to stress showed significantly

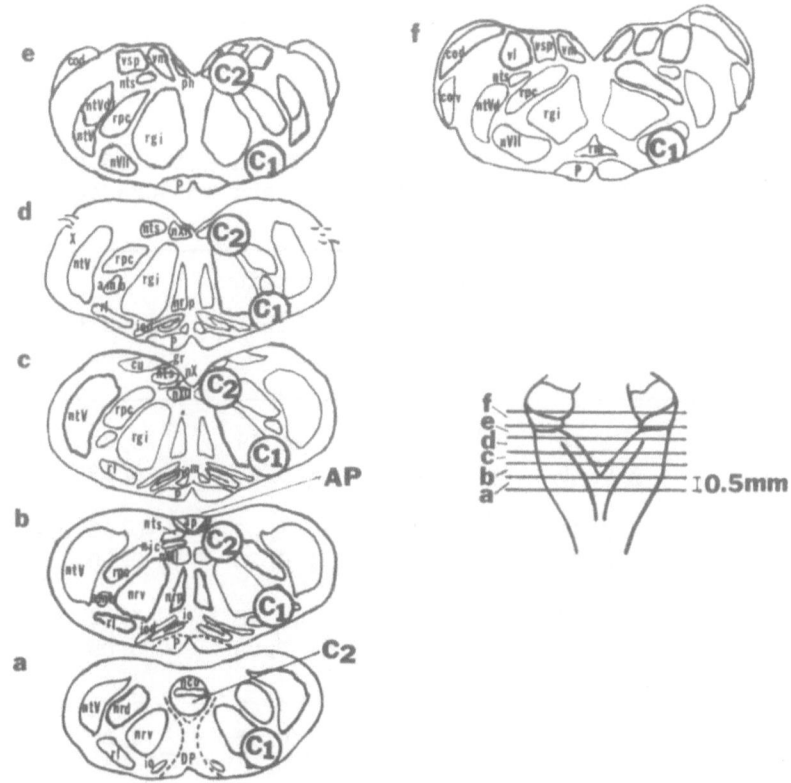

Figure 1: Schematic illustration of the areas dissected as C_1
          and C_2 regions.
          P 8.0 mm (a) is the most caudal level and P 5.0 mm (e)
          is the most rostral level.
          Abbreviations were used as previously reported (12).

greater depletions of NE than untreated animals exposed to stress
(Table 2). This effect was particularly impressive in cerebral
cortex, where in the untreated animals stress did not reduce NE
levels, but in the SKF 64139 treated animals stress reduced the
NE levels significantly. These results further support the
hypothesis that E terminals which innervate locus coeruleus might
be inhibitory (13). The enhanced lowering of NE elicited by
stress in animals pre-treated with a PNMT inhibitor might be due
to a decreased inhibitory input of E neurons. Since SKF 64139
at higher concentration inhibits also monoamine oxidase (R. G.
Pendleton, private communication) we cannot rule out at present
the possibility that this PNMT inhibitor elicits the NE lowering
following stress by some other mechanisms which are not related
to the lowering of E stores in the CNS.

Table 1

Effect of PNMT Inhibition on E Levels in Selected Regions
of the CNS in Non-Stressed (Controls) and in Stressed Rats

Brain Region	ng/10 mg Protein			
	Controls	Stress	Inhibitor	Inhibitor + Stress
C_1-ros	1.74 ± 0.15	1.40 a ± 0.10	1.11 a ± 0.08	0.70 c,e ± 0.09
C_1-cau	2.37 ± 0.21	2.03 b ± 0.05	1.43 a ± 0.20	0.83 c,e ± 0.10
C_2-ros	1.46 ± 0.20	1.33 ± 0.15	0.90 a ± 0.05	0.64 c,e ± 0.05
C_2-cau	3.62 ± 0.22	2.67 a ± 0.40	2.47 a ± 0.15	2.04 d,f ± 0.30
HT	5.94 ± 0.40	4.12 a ± 0.40	3.88 a ± 0.28	3.27 c,f ± 0.30
LC	1.89 ± 0.32	1.41 b ± 0.16	1.33 a ± 0.10	0.88 c,e ± 0.08

ros: rostral; cau: caudal; HT: hypothalamus; LC: locus
coeruleus

The results are the means from 6 animals ± S.D.

Statistically significant different from:

- control: a) $p \leq 0.01$ b) $p \leq 0.05$

- stress: c) $p \leq 0.01$ d) $p \leq 0.05$

- inhibitor: e) $p \leq 0.01$ f) $p < 0.05$

The dissection of brain regions was carried out as
previously described (6).

Table 2

Effect of PNMT Inhibition on NE Levels in Selected Regions of
the CNS in Non-Stressed (Controls) and in Stressed Rats

Brain Region	ng/10 mg Protein			
	Controls	Stress	Inhibitor	Inhibitor + Stress
C_1-ros	52.4 ± 2.7	37.4 a ± 1.8	50.9 ± 4.3	26.2 b,c ± 1.5
C_1-cau	61.1 ± 1.3	51.7 a ± 2.0	62.3 ± 3.1	31.3 b,c ± 4.2
C_2-ros.	35.3 ± 2.5	31.0 ± 2.5	33.7 ± 2.6	21.8 b,c ± 2.1
C_2-cau	102.2 + 7.9	61.4 a + 1.8	95.9 + 5.4	40.8 b,c + 6.2
HT	125.1 +10.7	89.4 a + 7.6	124.2 + 4.2	76.0 b,c + 6.7
LC	48.8 + 2.7	37.8 a + 2.1	51.1 + 1.1	29.4 b,c + 2.5
FC	11.4 + 0.9	11.8 + 1.5	11.2 ± 1.8	5.8 b,c + 0.6

ros: rostral; cau: caudal; HT: hypothalamus; LC: locus
coeruleus; FC: frontal cortex

The results are the mean from 6 animals ± S.D.

Statistically significant different from:
    control   a) $p \leq 0.01$
    stress   b) $p < 0.01$
    inhibitor   c) $p \leq 0.01$

The dissection of brain regions was carried out as
previously described (6).

Genetic Variation in Central PNMT and E Levels in
Various Strains of Rats Including the Spontaneously
Hypertensive Rat (SH-rats)

The immunofluorescence studies (1, 2) have shown that PNMT
containing neurons are localized in brain stem nuclei which are
known to exert control over the cardiovascular system. Further
support for the idea that central E may be involved in the control of blood pressure was obtained from pharmacological (14)
and biochemical studies (15, 16). PNMT activity was found to be
elevated in the C_2 and C_1 regions, while E levels were found to
be elevated in the C_2 but not in the C_1 region of the medulla
oblongata in 4-6 week old SH-rats when compared with normotensive
Wistar Kyoto rats (WK-rats) (15, 16). Since there is a significant genetic variation in the level of catecholamine biosynthetic
enzymes in various inbred rat strains (17) we have investigated
the distribution of PNMT activity and E in the medulla oblongata
of SH-rats and in two normotensive strains namely in WK-rats and
in regular Wistar rats. The results in Table 3 show that in all
three analyzed strains the PNMT activity increases progressively
from the caudal to the rostral parts in the C_1 and C_2 regions of
the medulla oblongata. These findings are compatible with the
immunofluorescence results (1, 2) which show that the density of
PNMT containing cell bodies is higher in the C_1 and C_2 rostral
regions while the density of the PNMT containing dendrites and
terminals is higher in the corresponding caudal regions (1, 2).
The data in Table 3 also shows that the PNMT activity in all
parts of the C_1 and C_2 regions is higher in the Wistar rats than
in the WK-rats or SH-rats. In the C_2 region the enzyme activity
of SH-rats is significantly higher than in the WK-rats in all
analyzed parts with the exception of the first caudal part
(section a, Fig. 1). However, in the C_1 region the PNMT activity
of the SH-rats is not significantly different from WK-rats with
the exception of the middle part of this structure (section c,
Fig. 1) where the enzyme activity is higher in the SH-rats than
in the WK-rats.

The results presented in Table 4 show the E levels in the
caudal-rostral parts of the C_1 and C_2 regions in SH-rats and in
the two normotensive strains. The E levels are in all parts of
the C_1 and C_2 regions of the medulla oblongata significantly
higher in the Wistar rats than in the WK-rats or SH-rats. The E
levels in the SH-rats occur in higher concentrations than in the
WK-rats only in the mediocaudal parts of the C_2 region (section
a and b) and C_1 region (section b). The E levels in the SH rats
were found to be also higher than in the WK-rats in the area
postrema. However, the dissected area postrema (section b, Fig. 1)
probably contains some medial aspects of the nucleus tractus
solitarii complex and the E may not necessarily be localized in
the area postrema.

Table 3

Distribution of PNMT Activity in the
Medulla Oblongata in Different Strains of Rats

Brain Section	C_1 Region			C_2 Region		
	Wistar	WK-rats	SH-rats	Wistar	WK-rats	SH-rats
a	16.9 ± 1.2	9.7[a] ± 0.7	10.6[a] ± 0.6	18.9 ± 1.3	6.50[a] ± 0.5	6.0[a] ± 0.4
b	18.3 ± 1.1	7.3[a] ± 0.4	6.8[a] ± 0.4	36.9 ± 1.8	6.9[a] ± 0.5	9.6[a,c] ± 0.7
c	118.6 ± 8.2	50.9[a] ± 3.6	65.0[a,c] ± 4.5	59.9 ± 4.5	23.6[b] ± 1.2	38.6[b,d] ± 1.8
d	136.3 ± 8.5	67.6[a] ± 3.8	69.7[a] ± 4.0	102.8 ± 6.1	70.8[b] ± 5.0	89.5[b,c] ± 6.5
e	164.7 ± 12.5	92.5[a] ± 7.5	91.7[a] ± 8.0	170.0 ± 13.0	99.7[a] ± 7.5	126.7[b,c] ± 10.5

The results are the means from at least six experiments ± S.D.

The animals in all experiments were 4-6 weeks old.

The enzyme activity is expressed in pmoles of formed product per mg protein per 1 hr.

Differs from Wistar rats: a) $p \leq 0.001$
                              b) $p \leq 0.01$

Differs from WK rats: c) $p < 0.01$
                              d) $p \leq 0.001$

     The results of our study indicate that there is a signifi-
cant genetic variation in the levels of brain PNMT activity and
E. It appears that at least two separate components influence
brain PNMT and E levels. One genetic component is not related to
blood pressure levels and another related to blood pressure
levels. The genetic component which is not related to blood
pressure levels has a more pronounced influence on brain PNMT and
E levels.

Table 4

Distribution of E Levels in the Medulla Oblongata
in Different Strains of Rats

pg/mg protein

Brain Section	C_1 Region			C_2 Region		
	Wistar	WK-rats	SH-rats	Wistar	WK-rats	SH-rats
a	227.0 ± 20	137.5[a] ± 14	156.6[a] ± 12	365.0 ± 22	242.0[a] ± 18	274.5[a] ± 18
b	275.0 ± 21	134.5[a] ± 11	164.0[a,b] ± 11	453.0 ± 26	230.5[a] ± 14	305.0[a,b] ± 24
c	306.0 ± 35	211.5[a] ± 16	234.5[a] ± 15	475.0 ± 31	276.0[a] ± 17	345.5[a,c] ± 21
d	370.5 ± 31	283.0[a] ± 22	295.5[a] ± 24	495.0 ± 35	335.5[a] ± 29	323.0[a] ± 27
e	315.0 ± 26	210.5[a] ± 16	203.0[a] ± 16	330.5 ± 27	228.5[a] ± 19	235.0[a] ± 20

The results are the means from at least six experiments ± S.D.
The animals in all experiments were 4-6 weeks old.
Differs from Wistar rats: a) $p \leq 0.01$
Differs from WK rats: b) $p \leq 0.01$
                                     c) $p \leq 0.05$

     To elucidate the role of specific central adrenergic neurons
in the regulation of blood pressure the effects of PNMT inhibitors
on blood pressure were investigated. The systemic administration of the PNMT inhibitor SKF 7698 (15) or of SKF 64139 (10)
results in a lowering of blood pressure. Recently, we found that
intraventricular (18) or intracerebral administration of SKF 64139
results in a marked elevation of blood pressure (J. Engel, Y. Baba,
M. Goldstein, unpublished data). In conscious rats prepared for
intra-arterial recording of blood pressure bilateral administration of 100 μg SKF 64139 into the medial parts of the C_2 region
results in a marked elevation of blood pressure for a period of

30-45 min. The intraventricular injection of E causes a dose
related fall in blood pressure (19) and these results suggest that
E neuronal systems may have an inhibitory influence on central
regulation of blood pressure. However, intracerebral administration of a structural analog of SKF 64139, namely SKF 72223
which is not a PNMT inhibitor (Pendelton, private communication)
also elevates the blood pressure. This finding raises some doubts
whether the elevation of blood pressure by SKF 64139 is related
to its inhibition of the E forming enzyme. It should be pointed
out that SKF 64139 and SKF 72223 have some structural resemblance
with papevarine, which is also an isoquinoline derivative. Since
papevarine is a known vasodilator it is possible that following
systemic administration of the SKF compounds the blood pressure
is decreased due to vasodilation. The SKF compounds may also have
some antidopaminergic activity. In preliminary studies we have
found that they reduce the apomorphine induced rotation in rats
with 6OH-DA lesions of the substantia nigra. However, the SKF
compounds displace very poorly the binding of ^{3}H-spiroperidol to
striatal membranes. (J. Engel and M. Goldstein, unpublished data)

Blockade of α-adrenergic Receptors and Alterations in Central NE
and E Levels

It was postulated that the blockade of the hypotensive and
respiratory actions of clonidine by piperoxane and yohimbine might
be due to the blockade of central E receptors (13). The spontaneous firing of locus coeruleus neurons was found to be inhibited
by NE and E and this effect is blocked by piperoxane (20). These
studies suggest that piperoxane may block the central adrenoreceptors. We have therefore investigated the effects of piperoxane
alone or in combination with a PNMT inhibitor on central E and NE
levels.

The results presented in Table 5 show that piperoxane alone
doesn't significantly change the NE or E levels in the analyzed
regions of the brain. As expected the E but not the NE levels were
significantly decreased in animals treated with the PNMT inhibitor
(NE levels were even slightly increased). The E levels were lower
in animals pre-treated with the PNMT inhibitor and then treated
with piperoxane than in animals treated with each of these drugs
alone. It is noteworthy that piperoxane causes a depletion of NE
levels in animals pre-treated with the PNMT inhibitor. These results suggest that piperoxane may block central E receptors which
could lead to a feedback mediated activation of E neurons. The
piperoxane elicited lowering of NE following PNMT inhibition could
be explained if one assumes that the E containing neurons which
probably innervate the NE cell bodies of the locus coeruleus exert
a tonic inhibition on the NE neurons. The blockade of the central
adrenergic receptors coupled with a lowering of E levels may result in an accelerated activation of NE neurons.

Table 5

Effect of Piperoxane Alone or in Combination with the PNMT
Inhibitor SKF 64139 on Brain E and NE Levels

	Brain Region					
	Hypothalamus		C_1-Rostral		C_2-Rostral	
Treatment	NE	E	NE	E	NE	E
	ng/g tissue					
Controls	2397 ± 180	58.1 ± 4.0	884 ± 75.0	29.1 ± 2.5	1130 ± 90.0	22.5 ± 2.0
SKF 64139	2524 ± 220	44.7 ± 3.5	1074 ± 90.0	19.7 ± 1.5	1440 ± 115.0	16.2 ± 1.5
Piperoxane	2880 ± 300	57.5 ± 4.5	948 ± 89.5	29.1 ± 3.0	1265 ± 95.5	25.0 ± 2.0
SKF 64139 + Piperoxane	1709[b] ± 160	26.4[a, c] ± 2.8	574[b] ± 43.5	10.6[a, c] ± 0.8	913 ± 75.5	10.8[a, c] ± 0.75

The animals were treated with 20 mg/kg of SKF 64139 (i.p.) and 1
hr later with 60 mg/kg (i.p.) of piperoxane. The animals were
sacrificed 1 hr after administration of piperoxane and the brain
regions were dissected as previously described (6).

The results are the means from at least four experiments ± S.D.

Statistically signficant different from:

Control: a) $p \leq 0.01$ b) $p \leq 0.05$

SKF 64139   c) $p \leq 0.01$
or piperoxane

DISCUSSION

The distribution of PNMT seems to parallel the distribution
of E, suggesting that this neurotransmitter is formed and stored
in the PNMT containing neurons. It remains to be elucidated whether
other N-methyl phenylethanolamines besides E are localized in the
PNMT containing neurons. The biochemical analysis of the PNMT
and E distribution indicates that the enzyme levels are higher
in the cell bodies while the E levels are probably higher in the
terminals. The lowering of brain E levels by Dopamine β hydroxylase and PNMT inhibitors (10) provides further evidence that E is
formed in the brain from norepinephrine.

The data obtained from the stress experiments and from the
studies with α-adrenergic blocking agents indicates that E neuronal
systems may have a tonic inhibitory influence on the NE neuronal
systems. The inhibitory influence of the E neuronal systems on
the NE neuronal systems could be of importance in the central
regulation of blood pressure. The findings that PNMT activity and
E levels are higher in the C_2 regions of Wistar rats than in SH-
rats further support the idea that E exerts a vasodepressor effect
on regulation of blood pressure. However, the lack of specific
PNMT inhibitors makes it difficult to assess the role of central E
in the maintenance of normal blood pressure levels.

Evidence has been recently accumulated that peptidergic neuronal systems and especially Substance P (SP-containing neurons)
(21, 22) interact with monoaminergic neuronal systems. Immunohistochemical studies have revealed that virtually all catecholamines
containing cell bodies are surrounded by SP-positive nerve endings
(21). The NE and E cell bodies in the C_2 region, the NE cell bodies
in the locus ceruleus, as well as the DA cell bodies in the substantia nigra are innervated by SP-containing neurons (21). The
findings that intracerebral administration of SP increases the
turnover of NE and DA (23) provided the first biochemical evidence
for the interactions between SP and monoaminergic systems.

Since the regions containing E cell bodies, namely the C_2 and
C_1 regions of the medulla oblongata, are also densely innervated
with SP-containing fibers we have investigated the effects on antiserum to SP on brain PNMT activity. Treatment with antiserum to
SP produces a small decrease (approximately 25%) of PNMT activity
in the medial C_1 and C_2 regions of the medulla oblongata. Although
this data is still of preliminary nature and further studies are
required to confirm this observation, nevertheless, one is tempted
to suggest that SP may have an excitatory influence on E neurons.

Antibodies directed towards putative peptide transmitters or
catecholamine synthesizing enzymes are mainly used for mapping out
neuronal systems. It appeared to us that these antibodies could
perhaps also be used for monitoring postsynaptic functional changes.
We have therefore investigated the effects of antiserum to SP on
the binding characteristics of ^{3}H-spiroperidol to striatal membrane

sites (24). Preliminary results show that the specific binding of
^{3}H-spiroperidol is higher (approximately 15-20%) in striatal
membranes of rats treated with antiserum to SP than in the corresponding controls. Scatchard plots reveal that in the striatum
of anti SP treated animals there is an increase in the total number
of binding sites. (Treated: binding sites = 39.7 pmole per gram
tissue; controls: binding sites = 28.8 pmoles per gram tissue.)
It is noteworthy that chronic treatment with neuroleptics as well
as degeneration of DA neurons induced by 6OH-DA elicits also an
increase in DA receptor binding sites (24,25). In analogy the
increase in DA receptor binding sites elicited by treatment with
antiserum to SP may be due to the diminished SP mediated excitatory
input which leads to a decrease activity in the nigro-striatal DA
neurons. If indeed antiserum to SP interferes with DA transmission
then one could not exclude the possibility that some mental and
neurological disorders may be of auto-immune nature.

SUMMARY

The interactions of the E neuronal systems with the NE neuronal systems were investigated in animals exposed to stressful
stimuli or in animals treated with the α-adrenoreceptor blocking
agent piperoxane. Exposure of animals to stress results in a
significant lowering of NE and E levels in specific brain regions.
The stress induced lowering of NE levels was enhanced by pre-treatment of the animals with a PNMT inhibitor. The blockade of α-
adrenoreceptors with piperoxane did not change significantly the
NE and E levels in the CNS. However, in animals pre-treated with
a PNMT inhibitor piperoxane elicited a significant lowering of NE.
These results suggest that upon activation of the adrenergic neuronal systems the E neuronal system may have an inhibitory influence on the NE neuronal systems.

The distribution of PNMT activity and of E was investigated
in specific regions of the medulla oblongata in SH-rats and in two
normotensive strains, namely in WK-rats and in regular Wistar rats.
The PNMT activity and E levels are higher in the C_1 and C_2 regions
of the medulla oblongata of the Wistar rats than in the WK-rats or
SH-rats. The PNMT activity and the E levels are higher in the
SH-rats than in the WK-rats in the mediocaudal parts of the C_1 and
C_2 regions of the medulla oblongata. It appears that two separate
components influence brain PNMT and E level; one genetic component
not related to blood pressure levels and another related to blood
pressure levels.

References

1. Hokfelt, T., Fuxe, K., Goldstein, M. and Johansson, O.
   Acta Physiol. Scand. 89: 286-288 (1973).

2. Hokfelt, T., Fuxe, K., Goldstein, M. and Johansson, O. Brain Research 66: 235-251 (1974).

3. Deguchi, T. and Barchas, J.D. J. Biological Chem. 246: 3175-3181 (1971).

4. Saavedra, J.M., Palkovits, M., Brownstein, M.J. and Axelrod, J. Nature 248: 695 (1974).

5. Da Prada, M. and Zurcher, G. Life Sciences 19: 1161-1174 (1976).

6. Sauter, A.M., Baba, Y., Stone, E.A. and Goldstein, M. Brain Research 144: 415-419 (1978).

7. Bliss, E.J., Ailion, J. and Zwanaiger, J. J. Pharm. Exp. Therap. 164: 122-134 (1968).

8. Corrodi, H., Fuxe, K., and Hokfelt, T. Life Sciences 7: 107-112 (1968).

9. Pendelton, R.G., Kaiser, C. and Gessner, G. J. Pharm. Exp. Therap. 197: 623-632 (1976).

10. Sauter, A.M., Lew, J.Y., Baba, Y. and Goldstein, M. Life Sciences 21: 261-266 (1977).

11. Thierry, A.M., Tassin, J.P., Blanc, G. and Glowinski, J. Nature 263: 242-244 (1976).

12. Goldstein, M., Lew, J.Y., Matsumoto, Y., Hokfelt, T. and Fuxe, K. In: Psychopharmacology: A Generation of Progress (Eds. M.A. Lipton, A. DiMascio and K.F. Killam) Raven Press, New York, pp 261-269 (1978).

13. Bolme, P., Corrodi, H., Fuxe, K., Hokfelt, T., Lidbrink, P. and Goldstein, M. Eur. J. Pharmacol. 28: 89-94 (1974).

14. Bolme, P. and Fuxe, K. In: Central Action of Drugs in Blood Pressure Regulation (Eds. D.S. Davies and J.L. Reid) University Park Press, Baltimore, pp 61-62 (1975).

15. Saavedra, J.M., Grobecker, H. and Axelrod, J. Science 191: 483-484 (1975).

16. Wijnen, H.J.L.M., Versteeg, D.H.G., Palkovits, M. and DeJong, W. Brain Research 135: 180-185 (1977).

17. Nagaoka, A. and Lovenberg, W. Eur. J. Pharmacol. 43: 297-306 (1977).

18. Fuxe, K., Bolme, P., Jonsson, G., Schwarcz, R., Goldstein, M. and Hokfelt, T. To be presented at Second European Neurosciences Meeting (in Press).

19. Borkowski, K.R. and Finch L. Proceedings of the B.P.S.: 130P (1977).

20. Cedarbaum, J.M. and Aghajanian, G.K. Eur. J. Pharmacol. 44: 375-385 (1977).

21. Hokfelt, T., Elde, R., Johansson, O., Ljungdahl, A., Schultzberg, N., Fuxe, K., Goldstein, M., Nilsson, G., Pernow, B., Terenius, L., Ganten, D., Jeffcote, F.L., Richfeld, J. and Fiad, S., The Distribution of Peptide Containing Neurons in the CNS. In: Psychopharmacology - A Generation of Progress Eds. M.A. Lipton, K.F. Killam and A. diMasio. John Wiley and Sons, New York, in press.

22. Mroz, E.A., Brownstein, M.J. and Leeman, S.E., Brain Research 113: 597-599 (1976).

23. Magnusson, T., Carlsson, A., Fisher, G.H., Chang, D. and Folkers, K., J. Neural Trans. 38: 89-93 (1976).
24. Burt, D.R., Creese, I., and Snyder, S.H., Science, 196: 326-327 (1977).
25. Creese, I., Burt, D.R., and Snyder, S.H. Science 197: 596-598 (1977).

Supported by NIMH and NSF grants.

EFFECT OF PEPTIDES ON BRAIN MONOAMINES AND ON GROSS BEHAVIOUR

Arvid Carlsson, J.A. Garcia-Sevilla and Tor Magnusson

Department of Pharmacology, University of Göteborg

400 33 Göteborg, Sweden

INTRODUCTION

The discovery of a large number of endogenous peptides with possible neurotransmitter functions in various brain regions (see Hökfelt, This Symposium) raises the question as to the role of these neuropeptides in brain physiology as well as in neuro-psychiatric disorders. The neuropeptides do not readily pass through the blood-brain barrier and thus it is doubtful to what extent systemic administration of the neuropeptides will help to elucidate their functions in the CNS. In the future we may expect that synthetic neuropeptide analogues capable of penetrating this barrier will become available. In fact, such development is already underway. Meanwhile we must resort to local application procedures. In the present investigation we have studied the effects of intra-cerebroventricular injections of several neuropeptides as well as of synthetic analogues. In view of the apparently close relation-ship between the neuropeptides and the monoamines (see Hökfelt, this Symposium) we have studied the effects of the neuropeptides on the synthesis and utilization of the brain monoamines. In addi-tion, observations on gross behaviour and locomotor activity were recorded.

METHODS

Male Sprague-Dawley rats weighing 220-270 g were used. A poly-ethylene cannula was implanted into each lateral ventricle as de-scribed earlier (5). The experiments were performed two days after the operation when the gross behaviour of the animals was normal.

The diverse neuropeptides were injected intracerebroventricularly (i.c.v.) into the right ventricle, followed within a minute by the same dose into the left ventricle (10 μl peptide solution followed by 5 μl saline into each ventricle). The brain monoamine synthesis was studied by measuring the accumulation of 3,4-dihydroxyphenyl-alanine (DOPA) and 5-hydroxytryptophan (5-HTP) during 30 min following inhibition of the aromatic L-amino acid decarboxylase by a supramaximal dose of 3-hydroxybenzylhydrazine (NSD 1015, 100 mg/kg i.p.) (3). The utilization of dopamine (DA), noradrenaline (NA) and 5-hydroxytryptamine (5-HT) was determined by the disappearance of the amines after synthesis inhibition by benserazide (800 mg/kg i.p.) (1). In some experiments the locomotor activity was measured using the Animex II electronic motility meter (LKB-Farad Ltf., Stockholm) which charts the movements as they occur.

The following neuropeptides were studied: synthetic human β-endorphin (gift from Prof. C.H. Li, University of California, San Francisco); [Leu^5]-enkephalin, [Met^5]-enkephalin and [D-Ala^2]-methionine-enkephalin amide (DALA) (Serva Feinbiochemica, Heidelberg); [D-Ala^2, MePhe^4, Met(O)^5-ol]-enkephalin (FK 33-824) (gift from Dr. Roemer, Sandoz Ltd., Basle); substance P (SP) and SP-analogues, luteinizing hormone-releasing hormone (LH-RH), neurotensin (NT) and [Gln^4]-NT, thyrotropin-releasing hormone (TRH) and [DOPA^2]-TRH (gift from Prof. K. Folkers, Institute for Biochemical Research, The University of Texas at Austin, Austin, Texas); cyclic somatostatin, [Thi^2]-TRH, melanostatin (MIF) and pGlu-Leu-Gly-NH-C$_2$H$_5$ (gift from Dr. H. Sievertsson, AB Kabi, Stockholm).

RESULTS

Morphine and Endorphins

The main findings of these experiments are summarized in Table 1. Morphine (3-30 mg/kg i.p., 1 h before sacrifice) enhanced the DOPA formation, but not that of 5-HTP, in the brain of rats treated with the aromatic L-amino acid decarboxylase inhibitor NSD 1015 (5). This effect was clearly dose dependent in the dopamine-rich areas, i.e. the limbic forebrain and the corpus striatum. However, in the noradrenaline-predominant parts of the brain only the highest dose of morphine significantly increased the DOPA formation. When the time interval for morphine (30 mg/kg i.p.) was increased to 2 h, 5-HTP formation as well as tyrosine and tryptophan concentrations were increased in all brain regions. On the other hand, naloxone (10 mg/kg i.p.), a well known opiate antagonist, decreased the DOPA formation in the dopamine-rich areas but not in the noradrenaline-predominant areas. In addition, naloxone also decreased slightly the 5-HTP formation in the corpus striatum. These effects were still apparent in a lower dose (1 mg/kg i.p.) and after a longer time interval (90 min) (5).

Table 1. Effects of Endogenous Peptides and Synthetic Analogues on the Formation of DOPA and 5-HTP, on the Levels of Their Respective Precursors in Brain, and on Gross Behaviour in Rats.

Peptide (or morphine)	Dose range (μg/rat)	Effects on				Gross behaviour[c]
		DOPA	5-HTP	Tyr	Try	
Morphine	3-30 mg/kg i.p.	$+$[a]	$+$[a]	$+$	$+$	$-$ $-$ Catatonia
β-Endorphin	2.5-20	$+$	$+$[a]	$+$	$+$	$-$ $-$ Catatonia
[Leu^5]-Enkephalin	500	$+$	$+$[a]	$+$	$+$	$-$ $-$ Catatonia
[Met^5]-Enkephalin	500	$+$	$+$[a]	$+$	$+$	$-$ Catatonia
DALA	4-250	$+$	$+$	0	$+$	$-$ $-$ Catatonia
FK 33-824	0.03-1	$+$	$+$	0	$(+)$	$-$ $-$ Catatonia
Substance P (SP)	50-120	$+$	0	0	0	$+$ Rotation
[Ile^7]-SP	50	0	0	0	0	$(+)$
[Des-Met^{11}]-SP	50	0	0	0	0	0
[D-Leu^8, D-Phe^9]-SP	50	$+$ $+$	0	0	$(+)$[a]	$+$ $+$ Rotation
[D-Phe^7, D-Phe^8]-SP$_{5-11}$	25	$+$ $+$	0	0	$(+)$	$+$ $+$ Rotation
[Lys^5, D-Leu^8, D-Phe^9]-SP$_{5-11}$	12.5-50	$+$ $+$	$+$	$+$	$(+)$	$+$ $+$ Rotation
[Gly$^{6, 7 \text{ or } 8}$]-SP$_{5-11}$	50	$+$[a]	0	0	0	$+$
SP$_{5-11}$	50	$+$	0	$(+)$	0	$+$ $+$ Rotation
[Gly$^{10 \text{ or } 11}$]-SP$_{5-11}$	50	0	0	0	0	$(+)$
pGlu-His-Pro-NH$_2$ (TRH)	500	/	/	/	/	0 Tremor
pGlu-Thi-Pro-NH$_2$	50-1000	$+$	$+$	$+$	$+$	$+$ $+$ Wet dog sh.
pGlu-DOPA-Pro-NH$_2$	1000	?[b]	0	$+$	0	$+$ Wet dog sh.
Pro-Leu-Gly-NH$_2$ (MIF)	50-1000	$+$[a]	$+$[a]	0	0	0
pGlu-Leu-Gly-NH-C$_2$H$_5$	50-1000	$+$[a]	0	$(+)$	0	$+$
Neurotensin (NT)	50-200	$+$	0	$+$	$+$	$-$ $-$ Loss of tone
[Gln^4]-NT	50-200	$+$	$+$	$+$	$+$	$-$ $-$ Loss of tone
Somatostatin	5-20	$+$	$+$[a]	0	0	$+$ $+$ Rotation
LH-RH	50-200	$+$	0	0	0	$\pm$ Rotation

a) Regional differences in activity.

b) High DOPA level, probably derived from administered peptide.

c) Only dominating feature of behavioural pattern is indicated. + and − indicate increased and decreased motor activity, respectively. 0 indicates no change in motor activity, ± periods of increase and decrease. Number of symbols indicate degree of change. Symbol in parenthesis: slight or inconsistent change.

Similarly to morphine, the i.c.v. injections of β-endorphin (2.5-20 μg/rat) also increased the DOPA formation in all brain regions studied (Fig. 1) (5). After a single dose (5 μg/rat) of β-endorphin the maximum increases (45-100 %) in DOPA accumulation were reached at 1-2 h. The increase in DOPA formation induced by β-endorphin (5 μg/rat) was antagonized by naloxone (10 mg/kg s.c., 1 min before β-endorphin). The 5-HTP formation was significantly increased (50-80 %) in all brain regions except hemispheres 1-2 h after the i.c.v. injections of β-endorphin (5 μg/rat). β-Endorphin (2.5-20 μg/rat) did not change the tyrosine or tryptophan concentrations 40 min after its i.c.v. administration. However, when this time interval was increased up to 2 h, there were for both precursors significant increases (30-60 %) in all brain regions (5).

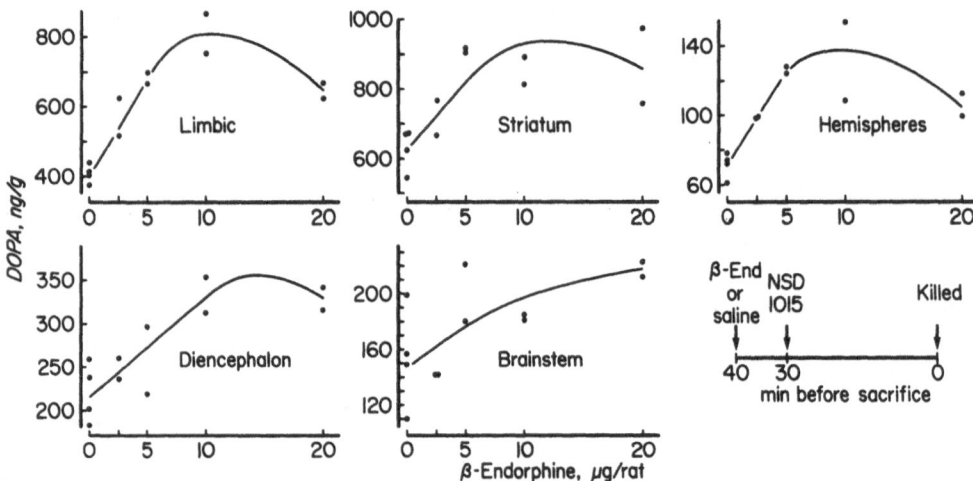

Fig. 1. Dose-response curves for the effect of β-endorphin on DOPA formation in rat brain regions. β-Endorphin was injected intracerebroventricularly (i.c.v.) 10 min before NSD 1015 (100 mg/kg i.p.) and the rats were killed after another 30 min. Control rats received i.c.v. the same volume of saline 10 min before the NSD 1015. Each point represents 2 pooled brain parts. (From ref. (5))

Like β-endorphin, the two enkephalins, [Leu⁵]- and [Met⁵]-enkephalin, also increased the DOPA formation in all brain regions but at much higher dosage level (500 μg/rat i.c.v.) (Table 1). The increases induced by [Leu⁵]-enkephalin (35-65 %) were significantly higher than those of [Met⁵]-enkephalin (10-30 %). Naloxone (10 mg/kg s.c.) antagonized these increases in all brain regions, except in the lower brain stem for the [Met⁵]-enkephalin. The two opioid peptides markedly increased the brain tyrosine concentrations, [Met⁵]-enkephalin being more active than [Leu⁵]-enkephalin. Surprisingly this effect was not blocked, or only partially, by naloxone. Similarly, the increases in the formation of 5-HTP and in the tryptophan concentrations were at most partially blocked by naloxone.

The two synthetic enkephalin analogues DALA and the FK 33-824 compound showed a similar biochemical pattern but with the expected differences in potency (Table 1) (9,10). Thus, the i.c.v. administration of DALA (4-256 μg/rat) enhanced the DOPA formation in a dose-dependent manner and by a naloxone-sensitive mechanism (Fig. 2). Similar results were obtained after the i.c.v. injections of the FK 33-824 compound (0.01-1 μg/rat). The only difference was in potency, FK 33-824 being about 100 times more potent than DALA. Both synthetic analogues increased the formation of 5-HTP and the concentrations of tryptophan by a naloxone-sensitive mechanism. In contrast to the enkephalins, neither DALA nor FK 33-824 increased the brain tyrosine concentrations.

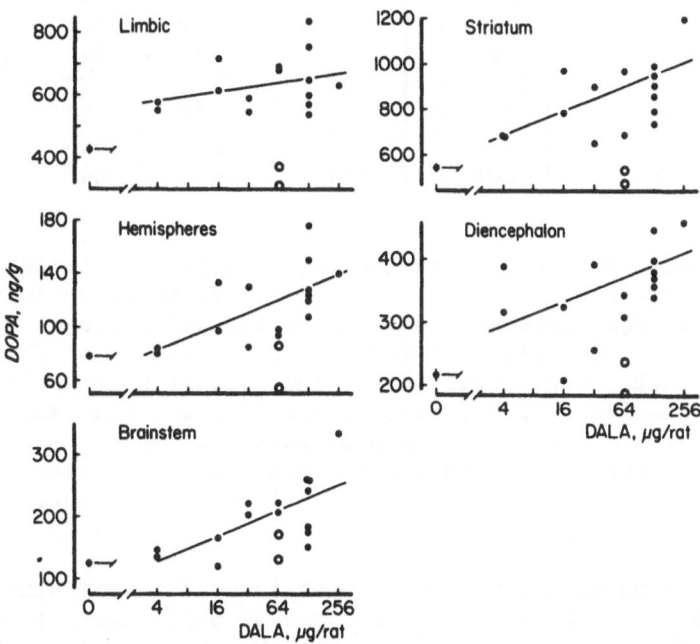

Fig. 2. Dose-response curves for the effect of DALA on DOPA formation in rat brain regions. DALA was injected i.c.v. 10 min before NSD 1015 (100 mg/kg i.p.) and the rats were killed after another 30 min. Control rats received i.c.v. the same volume of saline 10 min before the NSD 1015. Each point represent 2 pooled brain parts. Open symbols: rats pretreated with naloxone, 10 mg/kg s.c., 30 min before DALA.

After inhibition of monoamine synthesis by benserazide, the enkephalins significantly accelerated the disappearance of DA and had a similar though small and statistically nonsignificant effect on NA. [Leu^5]-enkephalin increased the utilization of DA stronger than did [Met^5]-enkephalin (p<0.025). For both peptides the significantly accelerated disappearance of 5-HT was accompanied by simultaneous increases in 5-HIAA brain levels (Table 2). Similar results were obtained with the two synthetic analogues keeping in mind the differences in potency. The FK 33-824 compound did not accelerate the disappearance of 5-HT but increased the 5-HIAA brain levels (Table 2).

Table 2. Effect of Peptides on Monoamines and 5-Hydroxyindoleacetic Acid (5-HIAA) Levels in Whole Rat Brains after Synthesis Inhibition

Peptide	Dose, μg/rat	DA	NA	5-HT	5-HIAA
[Leu^5]-enkephalin	500+200	66†††	82	85††	124
[Met^5]-enkephalin	500+200	82†	81	89†	149†
DALA	50+ 20	75†††	76††	87†	106
FK 33-824	1+0.4	78††	82†	104	120†
Substance P	100+ 40	89††	73†	79†	–
Neurotensin (NT)	200+ 80	73†††	64††	82††	150††
[Gln^4]-NT	200+ 80	77†††	72†	89†	121
Somatostatin	20+ 8	65††	69	84†	172††

All rats received benserazide (800 mg/kg i.p.) and immediately afterwards the peptides were injected i.c.v. in doses indicated (half of the dose into each lateral ventricle) followed within 25 min by a second injection (40 % of the first dose) (half of the dose into each lateral ventricle). The animals were killed 60 min after benserazide. Control rats received i.c.v. the same volume of saline after pretreatment with benserazide as above. Values are in per cent of controls; shown are the means of 4-5 determinations. Significances: † p<0.05; †† p<0.01; ††† p<0.001 (Student's t-test).

Both morphine and the endorphins induced a very similar behavioral pattern in the rat, but with marked differences in potency for the different compounds (Table 1). Morphine (3-30 mg/kg i.p.) induced a loss of both the corneal reflex and the tail-pinch reflex. The lower dose of morphine (3 mg/kg) increased the activity of the rats (sniffing, chewing and locomotion) and the higher dose (30 mg/kg) induced clear catatonia. Rats treated with 10 and 30 mg/kg i.p. of the opiate antagonist naloxone appeared slightly sedated and after the highest dose (100 mg/kg) some of the animals were slightly cataleptic. The i.c.v. injection of β-endorphin (5-20 μg/rat) caused a loss of both the corneal and the eyelid reflex-

es, exophtalmus, analgesia (loss of tail-pinch reflex) and induced
salivation and catatonia in less than 10 min. This behaviour was
long-lasting. In naloxone-pretreated rats none of these effects
appeared. After the i.c.v. administration of the enkephalins (500
µg/rat) or its synthetic analogues DALA (4-250 µg/rat) and   FK
33-824 (0.003-1 µg/rat) the rats showed the same behavioural pattern. The differences in potency were very striking; even a very
low dose of the FK 33-824 (3 ng/rat) clearly induced a loss of the
corneal reflex, exophtalmus, piloerection and cataleptic postures
in some of the rats. Naloxone was able not only to prevent but also
to reverse within 2 min all the manifestations of a fully developed
behaviour induced by a large dose of DALA (250 µg/rat i.c.v.).

Whether or not this behavioural pattern is correlated with the
endorphin-induced changes in brain monoamines remains to be clarified.

Substance P (SP) and Analogues

The data on substance P, parts of which are summarized in
Tables 1 and 2, have been obtained in collaboration with Professor
Karl Folkers and his group, The University of Texas at Austin (4,
8 and unpublished data).

The i.c.v. injection of SP (50 µg/rat) stimulated DOPA formation significantly in all brain regions (Fig. 3). This effect was
stronger in the NA-rich areas, i.e. the hemispheres, the diencephalon and the lower brain stem (35-50 %) than in the DA-predomi-
nant parts of the brain, i.e. the limbic forebrain and the corpus
striatum (16-17 %). The increases in DOPA formation induced by SP
were blocked in all brain parts except lower brain stem by the administration of naloxone (10 mg/kg s.c.) 2 min before the i.c.v.
injection of SP (Fig. 3). Naloxone alone (10 mg/kg s.c., 42 min
before sacrifice) had no effect on DOPA formation in any brain
region (data not shown). SP (50 µg/rat) did not modify the 5-HTP
accumulation or the brain concentrations of tyrosine or tryptophan.
With higher doses (105 or 132 µg/rat) SP caused within 30 min a
slight decrease in DA and NA levels in the whole brain but no significant change in 5-HT levels (4). After inhibition of monoamine
synthesis by benserazide, SP significantly accelerated the disappearance of DA, NA and 5-HT (Table 2) (8).

A total of 30 structural analogues of SP were evaluated in the
rat with respect to the action on brain monoamine synthesis as well
as on behaviour. The main results of a selected number of these SP-
analogues are summarized in Table 1. The decapeptide [Des-Met[11]]-
SP was inactive both on behaviour and on monoamine synthesis. The
analogue [Ile[7]]-SP was inactive biochemically but still moderately

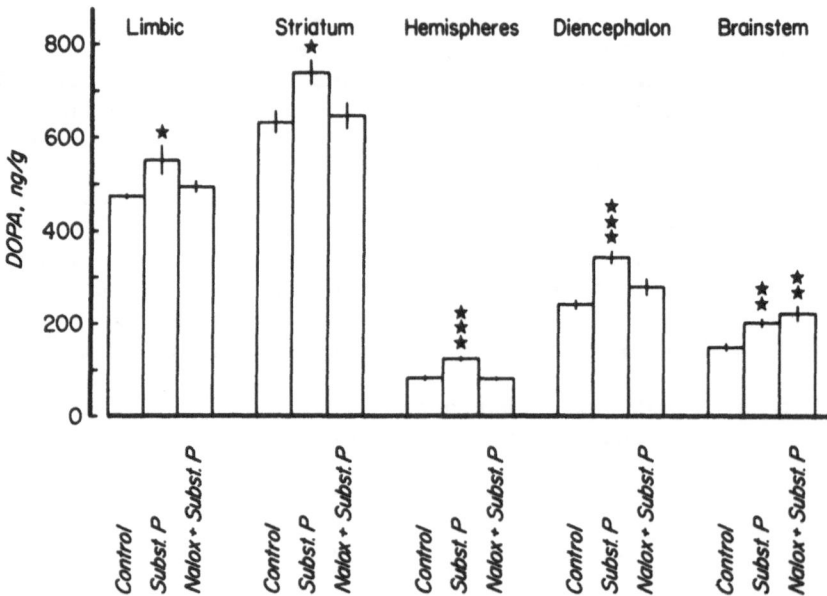

Fig. 3. Reversal by naloxone of the substance P-induced increase
in DOPA formation in rat brain regions. Substance P (50 μg/rat)
was injected i.c.v. 10 min before NSD 1015 (100 mg/kg i.p.) and
the rats were killed after another 30 min. Control rats received
i.c.v. the same volume of saline 10 min before the NSD 1015. Naloxone (10 mg/kg) was given subcutaneously 2 min before substance P.
Bars represent means ± s.e.m. of 4-8 experiments, each comprising
the pooled brain parts of 2 rats.
   ★ p<0.05; ★★ p<0.01; ★★★ p<0.001   (Student's t-test).

increased locomotor activity. These results indicate that the
[Met[11]]-NH$_2$ terminal end of the SP molecule is essential for biological activity, and that the [Phe[7]] residue also plays an important role. The most potent peptides studied in modifying both
behaviour and brain monoamine synthesis were the analogue [D-Leu[8],
D-Phe[9]]-SP and the heptapeptides [D-Phe[7], D-Phe[8]]-SP$_{5-11}$ and [Lys[5],
D-Leu[8], D-Phe[9]]-SP$_{5-11}$. The reduction of the length of the SP molecule and the replacements in positions 7 and 8 with D-enantiomers
of the naturally occurring amino acids produced an analogue,
[D-Phe[7], D-Phe[8]]-SP$_{5-11}$, with a much stronger and more long-lasting
activity than the parent compound. With the other two analogues,
the above mentioned chemical modifications also induced an increase
in activity. Like SP, these three analogues had a stronger effect
on DOPA formation in the NA-rich areas (65-140 %) than in the DA-
predominant parts of the brain (55-100 %). [Lys[5], D-Leu[8], D-Phe[9]] -
SP$_{5-11}$ was the only peptide which in contrast to SP increased the
formation of 5-HTP after NSD 1015. The shortening of the SP molecule and the replacement in positions 6, 7, 8, 10 or 11 with gly-

cine produced an interesting group of peptides with different activities. While [Gly^{6,7} ^{or} ^{8}]-SP$_{5-11}$ did not remarkable modify the activity of SP, SP$_{5-11}$ itself (with [Gly^9]) increased the behavioral activity of the parent compound without any further important change in brain monoamine synthesis. Finally, [Gly^{10}]-SP$_{5-11}$ and [Gly^{11}]-SP$_{5-11}$ were inactive on the formation of DOPA but still moderately increased locomotor activity. These two peptides behaved like [Ile^7]-SP (Table 1).

As mentioned, the behavioural profiles of SP and SP-analogues were different. The i.c.v. injection of SP (50-120 μg/rat) induced in most cases rotation to the left or to the right (contralateral to the site of injection), initially horizontally but later, especially after large doses, along the length axis of the body ("barrel rotation"). The rotatory movements as a rule lasted for several minutes and were followed by a longer period of general excitation with incessant locomotion round the cage. Also vasodilatation (red ears and paws) and strong salivation were present. [Des-Met^{11}]-SP was completely inactive; [Ile^7]-SP, [Gly^{10}]-SP$_{5-11}$ and [Gly^{11}]-SP$_{5-11}$ induced a moderate SP-like behavioural pattern; [Gly^{6,7} ^{or} ^{8}]-SP$_{5-11}$ behaved as SP; and [D-Phe^7, D-Phe^8]-SP$_{5-11}$, [Lys^5, D-Leu^8, D-Phe^9]-SP$_{5-11}$, SP$_{5-11}$ and [D-Leu^8, D-Phe^9]-SP induced an increased SP-like behavioural pattern. The last mentioned analogue induced a very strong change in the behaviour of the rats with a complete loss of the muscle tone after long-lasting barrel rotation. As mentioned earlier, naloxone blocked the SP-induced DOPA formation in several rat brain regions. However, none of the behavioural effects induced by SP were antagonized by naloxone. On the contrary, the increase in motor activity induced by SP was potentiated by naloxone (Fig. 4).

These results show that after chemical manipulation of the SP molecule it is possible to obtain great variations in activity and to dissociate to some extent the behavioural effects from those on brain monoamine synthesis.

Other Peptides

Table 1 summarizes the main results of this section. The i.c.v injections of the TRH-analogue pGlu-Thi-Pro-NH$_2$ (1 mg/rat) significantly increased the accumulation of DOPA and 5-HTP in the various regions of rats treated with NSD 1015 (unpublished work in collaboration with Dr. H. Sievertsson and his group). The brain tyrosine and tryptophan concentrations were also increased. On the other hand, the i.c.v. administration of the tripeptide [DOPA^2]-TRH (1 mg/rat) markedly increased the normally almost undetectable DOPA levels in the various brain regions of otherwise untreated rats (Table 3; unpublished work in collaboration with Prof. Karl Folkers

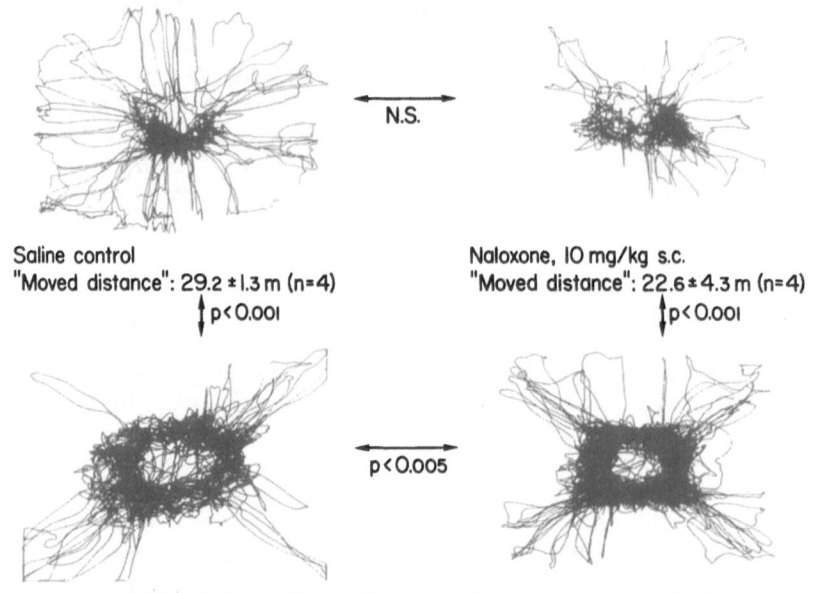

Saline control
"Moved distance": 29.2 ± 1.3 m (n=4)
↕ p< 0.001

Naloxone, 10 mg/kg s.c.
"Moved distance": 22.6 ± 4.3 m (n=4)
↕ p< 0.001

N.S.

p< 0.005

Substance P, 50 μg/rat i.c.v. Rec. 0-10 min
"Moved distance": 55.5 ± 4.0 m (n=4)

Naloxone + Substance P. Interval 2 min
"Moved distance": 76.4 ± 3.7 m (n=4)

Fig. 4. Graphic representation of the effects of substance P, naloxone, or naloxone plus substance P on the motor activity of a rat
using the Animex II electronic motility meter (LKB-Farad, S-126 34
Hägersten-Stockholm, Sweden). Saline and substance P (50 μg/rat)
were injected i.c.v., naloxone (10 mg/kg) was given s.c. 2 min before the i.c.v. administration of saline or substance P. All dia-
grams represent 10 min of recording (the first 10 min after the
i.c.v. injection). Lines within the central portion of the figure
represent horizontal movements on the cage floor, while those outside the central portion represent rearings on the walls of the
cage. Three additional experiments gave similar results. The "moved
distance" during the 10 min was calculated for all 4 experiments
from the cage size (350 mm by 200 mm) and the number and size of
the movements. Shown are the means ± s.e.m. Statistical significances were calculated according to one way analysis of variance
followed by t-test. Note: Increased locomotion but reduced rearing
after SP.

and his group). These high DOPA concentrations were probably derived from the administered peptide. Furthermore, there was a simultaneous increase in the levels of DA and NA in the brain. The
distribution of newly formed catecholamines differed from that obtained after the i.c.v. injection of an equimolar dose of L-DOPA;
the peptide gave rise to much higher levels, especially of NA, in
the NA-predominated hemisphere portion (Table 3).

The tripeptide Pro-Leu-Gly-NH$_2$ (MIF) (1 mg/rat i.c.v.) significantly increased the DOPA formation in all brain regions, except the hemispheres, and that of 5-HTP in the limbic forebrain and corpus striatum. The MIF-analogue pGlu-Leu-Gly-NH-C_2H_5 (1 mg/rat) increased the accumulation of DOPA in the same brain regions but not the formation of 5-HTP (unpublished work in collaboration with Dr. H. Sievertsson and his group).

Table 3. Concentrations of DOPA, Dopamine and Noradrenaline in Rat Brain Regions after Administration of [DOPA^2]-TRH or L-DOPA, 40 min before Death. Shown are the means ± s.e.m.

	Limb	Stri	Hem	Dien	Stem
	DOPA, ng/g				
Control	< 10	< 10	< 10	< 10	< 10
[DOPA^2]-TRH (n=4) 1 mg/rat i.c.v.	7025 ± 613	2353 ± 438	16339 ± 3414	4341 ± 595	5462 ± 567
[DOPA^2]-TRH (n=2) 1.85 mg/rat s.c.	46 ± 12	32 ± 3	57 ± 2	16 ± 8	31 ± 1
L-DOPA (n=2) 0.488 mg/rat i.c.v.	15487 ± 2843	3831 ± 701	30083 ± 2204	19540 ± 1217	37184 ± 2923
	Dopamine, per cent of control				
[DOPA^2]-TRH (n=4) 1 mg/rat i.c.v.	322 ± 21	161 ± 17	3633 ± 600	1300 ± 119	3659 ± 464
[DOPA^2]-TRH (n=2) 1.85 mg/rat s.c.	96 ± 1	110 ± 6	4 12	97 ± 25	110 ± 9
L-DOPA (n=2) 0.488 mg/rat i.c.v.	341 ± 171	180 ± 20	2528 ± 352	2207 ± 133	5552 ± 529
	Noradrenaline, per cent of control				
[DOPA^2]-TRH (n=2) 1 mg/rat i.c.v.	98 ± 10	90 ± 2	250 ± 20	111 ± 2	127 ± 8
L-DOPA (n=2) 0.488 mg/rat i.c.v.	85 ± 6	81 ± 2	125 ± 12	88 ± 1	115 ± 2

The i.c.v. injections of both NT and [Gln^4]-NT (50-200 μg/rat) increased the formation of DOPA in several brain regions in a dose-dependent manner (unpublished data in collaboration with Professor Karl Folkers and his group). In the corpus striatum, a DA-rich

area, [Gln⁴]-NT was about twice as active as NT in increasing the
formation of DOPA. NT was completely ineffective regarding the formation of 5-HTP, while [Gln⁴]-NT consistently increased the accumulation of this 5-HT precursor (Fig. 5). The two neuropeptides also
increased the tyrosine and tryptophan concentrations in all brain
regions. After inhibition of monoamine synthesis by benserazide,
both NT and [Gln⁴]-NT significantly accelerated the disappearance
of DA, NA, as well as 5-HT which was accompanied by a simultaneous
increase in 5-HIAA brain concentrations (Table 2).

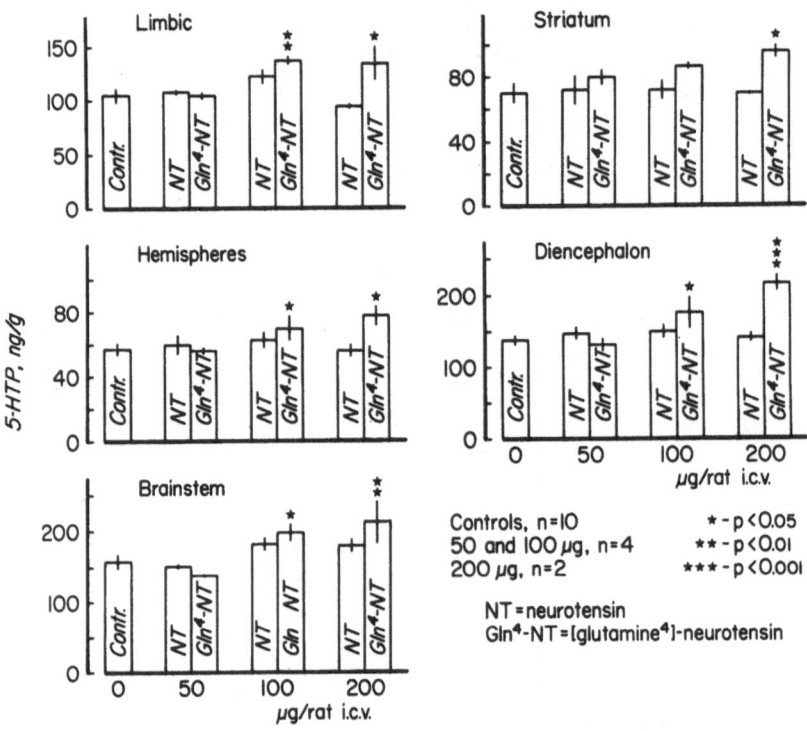

Fig. 5. Effect of various doses of NT and [Gln⁴]-NT on 5-HTP formation in rat brain regions. The peptides were injected i.c.v.
10 min before NSD 1015 (100 mg/kg i.p.) and the rats were killed
after another 30 min. Control rats received i.c.v. the same volume
of saline 10 min before the NSD 1015. Each experiment comprises
the pooled brain parts of 2 rats. Stars above bars refer to comparison with saline-injected control rats. For the statistics one
way analysis of variance followed by t-test was used.

        Somatostatin (5-20 μg/rat i.c.v.) also increased the formation
of DOPA and 5-HTP in a dose-related manner in several brain regions. No effect was seen on tyrosine and tryptophan concentra-

tions. After inhibition of monoamine synthesis by benserazide, somatostatin significantly accelerated the disappearance of DA and also had a small, statistically insignificant, effect on NA. The significantly accelerated disappearance of 5-HT was accompanied by a significant increase in 5-HIAA brain levels (Table 2) (6).

The decapeptide LH-RH (50-200 µg/rat i.c.v.) moderately increased the accumulation of DOPA, but not of 5-HTP, in the various brain regions. No effect was seen on tyrosine and tryptophan concentrations.

The behavioural profiles induced by the diverse hypothalamic releasing hormones and analogues were different. The i.c.v. injection of TRH (500 µg/rat) led to intense muscle tremor and pilo-erection. Its analogue pGlu-Thi-Pro-NH$_2$ (100-1000 µg/rat i.c.v.) induced, in addition to the mentioned picture, a clear increase in locomotor activity with repeated wet dog shake behaviour. The TRH-analogue pGlu-DOPA-Pro-NH$_2$ (1 mg/rat i.c.v.) induced a peculiar behaviour where a short period of increased locomotor activity with stereotypies was followed by a long lasting fixed squirrel-like posture (the rat sits on its hindlegs in an upright position) with very frequent wet dog shakes. This behavioural pattern was completely different from that induced by an equimolar dose of L-DOPA (488 µg/rat i.c.v.) where, after a very short lasting increase in activity, the rats showed an apparently normal behaviour. The tripeptide MIF did not induce any clear change in behaviour, whereas its analogue pGlu-Leu-Gly-NH-C$_2$H$_5$ (1 mg/rat i.c.v.) increased locomotor activity. The i.c.v. injection of somatostatin (5-20 µg/rat) induced stereotype behaviour, increased locomotor activity, strong salivation and long lasting barrel rotation (6). LH-RH (50-200 µg/rat i.c.v.) produced intermittent episodes of sedation (decreased locomotor activity, low muscle tone, reduced corneal reflex) and hyperactivity (increased locomotor activity, barrel rotation and convulsions).

Finally, the tridecapeptide NT and its analogue [Gln^4]-NT induced a very complex picture where the rats exhibited reduced muscle tone, a long lasting loss of both the righting and corneal reflexes, rotational behaviour and convulsions.

COMMENTS

The present observations indicate that intracerebroventricularly injected neuropeptides are capable of inducing behavioural patterns characteristic for each type of neuropeptide. High structural specificity is illustrated by several instances of a drastic change in activity after a minor or moderate modification of the chemical structure. In favour of specificity is also the ability of naloxone to prevent or reverse the characteristic behavioural pattern induced by opioid peptides, whereas the behavioural effects of substance P were not prevented but actually enhanced by the opiate-receptor antagonist.

The doses of the neuropeptides required for behavioural and biochemical effects are large compared to the amounts normally present in brain. This may be explained by relatively poor penetration from the CSF into the brain tissue in conjunction with high peptidase activity. The importance of the latter is underlined by the considerable increase in potency obtained by analogues presumably resistant to peptidase activity. In the opioid peptide series this increase amounted to several orders of magnitude, the profile of activity remaining largely unaffected. In fact, the most potent opioid peptide analogue, FK 33-824, proved active also after systemic administration in agreement with published data (10), although the dose had to be increased considerably. A very interesting development of neuropeptide pharmacology can thus be anticipated.

A striking example of the importance of peptidases is afforded by [DOPA^2]-TRH, which was found to induce a rapid accumulation of DOPA, DA and NA in brain. The regional distribution of catecholamine formation differed from that induced by DOPA administered by the same intracerebroventricular route, suggesting that the distribution of peptidase activity in the brain may contribute to the profile of activity. This possibility opens up interesting perspectives.

The observations on monoamine synthesis and utilization indicate a close relationship between neuropeptides and monoaminergic systems. The data suggest that a large number of neuropeptides are capable of stimulating monoamine turnover, presumably by increasing the firing of monoaminergic neurons. However, a closer examination of these relationships requires a more detailed analysis of the synaptic connections between peptidergic and monoaminergic neurons as well as their possible co-existence within one and the same neuron (see Hökfelt, This Symposium).

The relationship between stimulation of monoaminergic systems and behavioural effects of neuropeptides appears to be complex. For example, the stimulation of motor activity by substance P is prevented by reserpine and α-methyltyrosine, suggesting involvement of catecholaminergic systems, whereas the rotational activity persisted indicating that the rotational movements occur independently of monoaminergic activity (2).

SUMMARY

Several neuropeptides as well as synthetic analogues were injected intracerebroventricularly to conscious rats and effects on the synthesis and utilization of monoamines in the brain and on gross behaviour and motor activity were recorded.

Different groups of peptides induced characteristic behavioral patterns, the specificity of which was underlined by the influence of even small changes in chemical structure. Naloxone prevented the effects of opioid peptides but rather enhanced the motor stimulation induced by substance P.

Several peptides stimulated the turnover of monoamines in the brain.

The importance of peptidase activity was demonstrated by the rapid formation of DOPA, dopamine and noradrenaline from $[DOPA^2]$-TRH.

ACKNOWLEDGEMENTS

This study was supported by grant No. 00155 from the Swedish Medical Research Council. J.A. Garcia-Sevilla was supported by a research grant from the Juan March Foundation, Madrid, Spain.

REFERENCES

1. Andén, N.-E., K. Fuxe and T. Hökfelt. Eur. J. Pharmac. 1:226, 1967.
2. Carlsson, A. In: M.A. Lipton, A. DiMascio and K.F. Killam (Eds.), Psychopharmacology: A Generation of Progress. New York: Raven Press, 1978, p. 1057.
3. Carlsson A., J.N. Davis, W. Kehr, M. Lindqvist and C.V. Atack. Naunyn-Schmiedeberg's Arch. Pharmac. 275: 153, 1972.
4. Carlsson, A., T. Magnusson, G.H. Fisher, D. Chang and K. Folkers. In: U.S. von Euler and B. Pernow (Eds.), Substance P. New York: Raven Press, 1977, p. 201.
5. Garcia-Sevilla, J.A., L. Ahtee, T. Magnusson and A. Carlsson. J. Pharm. Pharmac. (in press).
6. Garcia-Sevilla, J.A., T. Magnusson and A. Carlsson. Brain Research (in press).
7. Hökfelt, T., L.G. Elfvin, R. Elde, M. Schultzberg, M. Goldstein and R. Luft. Pro. Natl. Acad. Sci. USA 74: 3587, 1977.
8. Magnusson, T., A. Carlsson, G.H. Fisher, D. Chang and K. Folkers. J. Neural Trans. 38: 89, 1976.

9. Pert, C.B., A. Pert, J.-K. Chang and B.T.W. Fong. Science 194: 330, 1976.
10. Roemer, D., H.H. Buescher, R.C. Hill, J. Pless, W. Bauer, F. Cardinaux, A. Closse, D. Hauser and R. Huguenin. Nature 268: 547, 1977.

CURRENT APPROACHES TO STEROID HORMONE-CELL INTERACTIONS

Etienne-Emile BAULIEU

Inserm U33, Faculté de Médecine, Université Paris-Sud

Lab. Hormones, 94270-Bicêtre, France

In this paper, we give a summary of the findings which have led to the present understanding of steroid hormone receptors (Parts I, II and III), and we discuss a new physiological approach to receptor pharmacology, based on differential hormone binding properties and on the cellular plurality of receptors (Part IV), and a new frontier in steroid hormone action, the plasma membrane (Part V).

I. Specific steroid hormone binding proteins in target cells are "receptors".

The estradiol receptor (ER) has been a "pioneer" not only in the field of steroid receptors but also with respect to the receptors of other hormones and neuro-transmitters. The concept of a receptor was originally defined by pharmacologists who, however, only studied them at the phenomenological level. It was not until high specific activity ^{3}H-estradiol was available that endocrinologists and biochemists were able to study the physico-chemical interactions.

The natural estrogen, estradiol, is retained in its target organs (1). This basic observation was explained by the intrinsic properties of an intracellular protein component of target cells when a strict specificity, of obvious biological significance, was demonstrated by in vitro binding experiments with soluble extracts (2,3). Two techniques were instrumental in this : (i) gradient ultracentrifugation of low salt cytosol extracts permitted the isolation of "8S" ^{3}H-hormone complexes (4) ; (ii) the rational use of adsorbents (charcoal in the first place), according to the differential dissociation principle (5), allowed quantitative binding studies (2). The steroid receptors show highest affinity for the most biologically active hormones with K_{Deq} values on the order of O.1 nM, which relate to the circulating concentrations of these hormones. The very slow dissociation rates of the hormone-receptor complexes (6,7) are compatible with the long retention observed in vivo (1).

On the basis of these properties, despite only circumstantial evidence that these intracellular binding proteins actually operate in mediating hormone action (8), they were called "receptors". They are certainly distinct from the other steroid binding proteins found in plasma, testicular fluid, liver, prostate, etc., whose binding properties do not correlate with the biological activities of their ligands (8). For instance, human Sex Steroid Binding Plasma Protein (SBP) (9) which binds estradiol tightly binds testosterone with greater affinity, in contrast to ER which has very weak affinity for androgens. Conversely, diethylstilbestrol (DES), an active synthetic non steroidal estrogen, binds to ER but hardly at all to SBP or rat α-fetoprotein (α-FP), also called Estradiol Binding Protein (10). Recent findings with antibodies to ER (11), α-FP (12) and SBP (13) confirm that the intracellular estradiol receptor is distinct from the plasma proteins binding estradiol, and show that the estradiol receptors of different organs in different mammals have common antigenic determinants (14).

Receptors have been found for each of the five physiologically well-defined steroid hormones. Besides ER there are : androgen receptor (AR), first described in rat ventral prostate (15-17) ; progesterone receptor(PR), initially found in guinea pig uterus (18) and in chick oviduct (19) ; glucocorticosteroid receptor (GR), first observed in thymus cells (20) and mineralocorticosteroid receptor (MR), first detected in toad bladder (21) and then studied in rat kidney (22). In all cases, ligand affinities correspond to the biological activities of most agonists, and sometimes of antagonists. Receptor concentrations in responsive cells are on the order of 10^4/cell, with values varying between 10^3 and 5.10^4. The receptors for a given hormone seem physico-chemically very similar, regardless of the target organ and animal species.

For each natural hormone, binding experiments have essentially indicated a single category of binding sites of high affinity (K_{Deq} = O.1 nM) - no cooperative binding has been indicated. No work based on purification, antigenicity or binding activity has substantiated the notion of receptor heterogeneity. The frequently referred to "heterogeneity" of receptor, which corresponds mainly to differences in size or extractibility, may therefore artifactual (due to aggregation, proteolysis). The significance of the "insoluble" nuclear receptor (23) and the reason why a fraction of progesterone binding sites do not translocate to the nucleus (24) are still obscure. Possibly R can display distinct interactions with several structures of the same cell. Moreover, R may interact selec-

tively with different acceptors and/or effectors in different cells
and this could be the basis for the same hormone giving rise to different effects via the same receptor.

The structures of the steroid receptors are not known, despite
much progress in affinity chromatography techniques (25-27). We only
know that they are proteins (MW approx. 70,000 in the case of the
calf uterus ER binding units), antigenic whether extracted from the
cytosol (11) or nucleus (14)), and that partial proteolysis removes
a portion of the protein without changing the hormone binding characteristics (27-28 bis).

II. Steroid hormone receptor complexes act in the cell nucleus.

Steroid hormone receptor complexes concentrate in the nucleus
of target cells, where presumably they trigger the main hormonal
effect(s) by modifying gene transcription.

It has been known for a long time that many responses elicited
by steroid hormones implicate changes in protein and RNA synthesis
(29). Recent evidence demonstrates an accumulation of specific
m-RNA(s) when specific protein synthesis is increased (30,31). That
receptors are implicated in these events has been suggested by two
series of observations.

Firstly, in most systems the receptor is cytoplasmic in the
absence of hormone and translocates to the nucleus with the bound
steroid after exposure to the hormone (32,33). This change in subcellular distribution apparently reflects "activation" of the receptor - a steroid-dependent phenomenon increasing its affinity for
polyanions and DNA in particular (34,35)$^{\circ}$. Autoradiography studies
have shown the preferential localization of steroids in target cell
nuclei (36) and, more recently, a selective association of receptor
with solubilized chromatin elements has been described (37). However, it will be seen later that nuclear transfer of the receptor
after ligand binding does not necessarily lead to further action.
Thus there is some distinction between acceptor and effector sites
of R, as depicted in Fig. 1. In addition translocation of the receptor is probably not a prerequisite for hormone action in all cases
(i.e. when receptor is already located in the nucleus in absence of
hormones (38,39) or when previously translocated by antihormone
(see Part IV)).

Secondly, increased incorporation of radioactive-RNA precursors
has been observed when nuclei of target cells have been exposed to

---

$^{\circ}$ The further fate of hormone receptor complexes in the living cells
is not known. "Deactivation" of R or alteration by proteolysis are
amongst the possibilities that have been proposed.

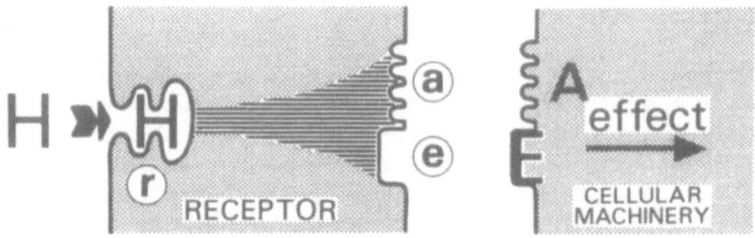

Figure 1 : Phenomenological portrait of steroid hormone (H) receptor (R)
    r is the binding site of H, and can stand for "receptor",
"receiving", "regulatory". r is hormone specific, binds reversibly
and is the last interaction of hormone with the cellular machinery
before action is triggered (this would not be the case if r was the
binding site of a transport protein).
    The receptor includes a coupling mechanism (transduction) bet-
ween r and site(s) a and/or e.
    The effect takes place at the level of the effector structure
E of the cell machinery. The receptor has a site e (for "effect",
"effector", "executive") interacting with E, either catalyzing a
chemical reaction (R is then an allosteric enzyme) or simply binding to E (in that case, sites e and a may be confounded). In both
cases, some specificity should be present to explain that different
effects are promoted by different steroid-receptor complexes in the
same cell. A is a (another) part of the cellular machinery called
Acceptor to which the a site of the receptor can bind. A corresponds
operationally to the nuclear positioning of the hormone-receptor
complexes. The affinity of the receptor for A is unknown, and the number of A sites may be large, and while they may not all be identical,
their affinity is probably relatively weak (106). It is possible, but
not demonstrated, that a small number of high affinity interaction
sites are hidden by a larger number of weaker affinity sites (107),
whether many proteins (as in sex hormone induced cell growth) or only
a few (as when aldosterone acts on sodium transport) are increased.
In other words, the relationship between receptor binding in the nucleus and the size of the hormone responses' is unknown.
    The effect of an antihormone binding to the r site may result
in changes in the conformation of the r site, consequently modifying
transduction to the a and/or e sites, and thus abolishing hormone
effect.
    We have not indicated in this figure the possibility of another
site on the receptor molecule for antagonistic ligands (108, 89),
nor sites possibly involved in the inactivation of the receptors,
as during progesterone action (92).
    Data of several laboratories (30, 52, 109-113) are integrated
in this scheme.

cytoplasmic extracts containing hormone-receptor (40). This observation gave some evidence for the direct participation of the receptor (hormone alone has no effect), although knowledge of eukaryotic transcription at that time was, and still is, very primitive. So far no unequivocal evidence has been obtained which demonstrates that the addition of steroid receptor to a chromatin preparation can result in a change (for instance an increase in specific m-RNA) mimicking the response observed in the intact cell (41). Indeed all steroid responses are poorly understood mechanistically. For instance, inhibition of protein synthesis has no effect on the increase in viral RNA by glucocorticosteroid in GR cells (42), in contrast to the induced early augmentation in conalbumin and ovalbumin mRNAs (43). Moreover, the increase in conalbumin (and m-RNA$_{con}$) occurs very early after estradiol administration, and much later after progesterone, whereas the increase in ovalbumin (and m-RNA$_{ov}$) due to estradiol occurs later than that due to progesterone, but before the effect of the latter on conalbumin (31, 44). These kinetics, observed in the same chick oviduct tubular cells, cannot be explained by current knowledge, and the concept of a "simple" interaction of the receptor(s) with some segment(s) of the DNA encoding a regulatory signal for transcription, therefore seems less tenable.

III. Steroid hormone receptor concentrations vary in target cells and this plays an essential role in hormonal receptivity.

Steroid hormone receptor concentrations can vary under hormonal, developmental, genetic, pathological and pharmacological circumstances. These variations in spite of the limited information on structure and molecular mechanism of activity of receptors, are important in the understanding of all hormonally controlled processes. Endocrinology should therefore include the study of not only hormone but receptor variations.

Hormone-induced variations were the first to be described in detail, and the study of the uterine progesterone receptor during the estrus cycle led to discovery of the double regulatory control of steroid receptor concentrations (45). A positive control is readily observed when, under physiological and pharmacological conditions, progesterone receptor synthesis is increased by estradiol (18, 19) - a result probably explaining the priming effect of estrogen on progesterone action. Such a positive control is more difficult to analyse when it is the hormone itself which increases the concentration of its receptor, as for estradiol and androgens. Inversely, progesterone exerts a negative control by accelerating the decay of its available receptor binding sites (45)-an "inactivation" process which is not fully understood at the molecular level. "Down regulation", similar to the progesterone effect, then has been described for many polypeptide hormone receptors (46) and has attracted much attention since it has been correlated functionally with the desensitization of target cells and physically with an "internalization" process.

Positive and negative controls of receptor concentration are not the same according to the cell type, state of development, hormonal, metabolic and functional circumstances. Such a complexity is obvious when one studies the interplay between progesterone and estradiol actions, since besides the above-mentioned data, it has been observed that progesterone may increase estradiol receptor concentrations directly or decrease the estradiol-dependent increase in estradiol receptor (47-49), etc. A remarkable sequence of variations is found in the human endometrium during the normal menstrual cycle (50) (Fig. 2). Subcellular distribution and changes in concentration vary greatly but apparently follow the same rules as established for laboratory animal model systems. Incidently, the concentration of estradiol receptor and its control are not identical in two neighbouring tissues of the uterus : endometrium and myometrium (47, 51, 52).

Changes in receptor concentration accompany not only functional variations but also ontogenic development. Timely programed changes in androgen receptivity of the prostatic (53) or mammary (54) buds are not under hormonal control, whereas other developmental changes are hormone-dependent, for example the peak of estrogen sensitivity (55) and ER concentration (56) in the rat uterus coincide with plasma estrogen increase at 10 days. However, the developmental curves of ER concentrations in other organs, such as the hypothalamus and pituitary, do not follow the same pattern (57).

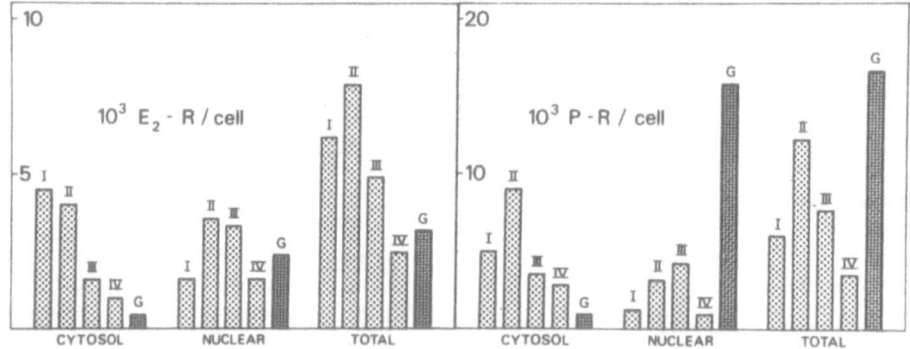

Figure 2 : ER and PR in the cytoplasm and the nuclei of human endometrium during the normal menstrual cycle divided into four periods with respect to the Lh peak (day O) ; I : before -5, II : between -5 and O, III : between O and 5 and IV : after +5. G are pregnancies (6-8 weeks) (after 50, 63 and U. Verma, unpublished). Total = cytosol + nuclear.

A clearcut genetic variation is observed in the tfm mutants
(androgen receptor negative) who display the feminine phenotype
even if testosterone levels are normal (58). In vitro genetic studies of lymphoid cells have confirmed that in the absence of receptor, glucorticosteroids show no killing effect (59). They also demonstrated gene dosage and gave an occasion to correlate hormone sensitivity and receptor concentration : when more receptor is present,
less hormone is needed (60).

The variation in steroid receptor concentrations is also important in clinical situations. In human breast (61, 62) and endometrial (62,63) cancers, one observes a wide range of concentrations
for the estradiol and the progesterone receptors, unrelated to the
known histological and etiological particularities of these tumors.
Correlation between responses to endocrine treatment and a selected
threshold concentration of ER have been proposed (61), although
possible cellular heterogeneity makes interpretation and prognosis
difficult. Nonetheless it was clearly shown that in the absence of
detectable receptor or at concentrations below the selected
threshold, the tumors show no hormonal response.

Remarkably, whatever the cause of the changes in receptor concentration, whether functional, ontogenic, genetic, pathologic or
pharmacological (as seen in the next chapter), the binding specificity and affinity of the receptors has always been the same.
This implies that one is dealing with an apparently purely quantitative
regulatory system. However, this relatively simple approach does not
explain all observed effects and the relationship between the
concentration of hormone-receptor complexes and the extent of the
response depends on the one considered. In the chick oviduct, for
instance, half maximal induction of ovalbumin necessitates approximately twice the hormone (whether estradiol or progesterone) needed
to obtain half maximal conalbumin increase (64). According to the
dose, estradiol and progesterone can be agonistic or antagonistic
for the same parameter (ovalbumin synthesis) and under certain circumstances, progesterone may be simultaneously synergic to estradiol
for ovalbumin and conalbumin synthesis and antiestrogenic with respect to the increase in DNA polymerase (65). Several hypothetical
mechanisms could be responsible for these situations. For instance
different numbers of the same receptor may be engaged in the hormone regulation of different gene products or there may be as yet unknown intermediary step(s) between the receptor and the measured
responses.

In conclusion, presence of the receptor is necessary for steroid hormone responses and its concentration is directly implicated
in the extent of the effect obtained. The concentration of nuclear
hormone receptor complexes and its persistancy is related to the
extent of the effect obtained (66, 68). Refinements of these notions
should take into account the stereochemical and kinetic parameters

of the interactions between hormones and receptors brought to light
by binding studies (discussed further on), and the post-receptor
elements of the cellular machinery involved in the response (69).

IV. <u>Physiological pharmacology of natural and synthetic hormones</u>
<u>and antihormones, based on the binding properties, plurality</u>
<u>and changes in concentration of receptors.</u>

   <u>Binding properties and plurality of receptor.</u>
   <u>A</u> given hormone may interact with <u>several</u> receptors. It was
originally demonstrated that estradiol interacts with AR (70), with
a K_{Deq} approx. 3 nM while estradiol binds to ER with K_{Deq} 0.1 nM.
ER and AR were found together in cloned cells and when estradiol is
added, it apparently exerts a strong negative signal at the AR level,
overriding the androgen effect (71, 72). Conversely, the binding of
androgens to ER, although difficult to demonstrate physically because
of very weak affinity, is well documented, since testosterone and DHT
at pharmacological doses provoke estrogenic activity (73). This is
also the case for usually neglected metabolites such as 3β-androsta-
nediol (P. Robel and J. Hochman, unpublished). Interpretation of the
observed superimposition of autoradiography distributions of AR and
ER in several parts of the brain (74, 75), already difficult because
of binding properties, may be further complicated by the known local
conversion of androgens to estrogens (76).

   Progesterone provides another example since, besides its own
receptor, it binds to AR, GR and MR (62). Progesterone is a weak androgen in prostatic cells (and moreover a progesterone derivative,
cyproterone acetate, is an antiandrogen), and is also an antigluco-
corticosteroid in hepatoma and lymphoid cells, and an antimineralo-
corticosteroid in kidney. This suggests that a natural hormone such
as progesterone may be active in cells through different receptors
at the same time. This is the case in MCF 7 human mammary cancer
cells where progesterone can find 3 direct chanels for receptor
interaction with the cell machinery (PR, AR and GR), plus, probably,
a fourth indirect one via the regulation of ER concentration (77).
In fact it is conceivable that a steroid can be active in a cell even
in the absence of its own receptor. Such theoretical considerations indicate the difficulty of predicting the therapeutic value
of a given hormone drug, even in the case of "natural" steroids.

   If several different receptors can bind the same hormone, it
follows that <u>a given receptor can bind several hormones</u>. The heterologous hormone will exert either an agonistic effect with respect to the receptors own steroid, or a different effect, possibly
antagonist no prediction can be made from binding properties alone.

   The preceding notions are applicable to metabolites and synthetic analogs.

Hormone metabolites. In the ventral prostate, androstanolone
(DHT) formed in this tissue by the 5α-reductase enzyme, is more active than testosterone (78) and is bound preferentially by AR (79,
80). In levator ani or skeletal muscles (81), testosterone is the
natural androgen ligand of AR. The AR is the same in all organs,
but it would seem that the higher affinity interaction is necessary
at a given moment in development, as indicated by the incomplete
development of male external genitalia, prostate and facial hair
in genetic 5α-reductase deficiency (82). In the central nervous system, it is possible that estrogens, which are testosterone metabolites (76), may mediate via ER (and not AR) some functions previously
attributed to testosterone itself (masculinisation of the hypothalamus, sexual behaviour) while testosterone or its reduced metabolites are involved in gonadotropin control via AR. Catechol estrogens,
formed in abundance in the brain (83), and binding to ER (84), may
also have a regulatory role of their own (eventually antagonistic
(85)).

Synthetic analogs. Many steroidal and non-steroidal synthetic
analogs are used in therapeutics. The earliest known, DES, has remarkable therapeutic effects in prostatic cancer (86), and also adverse
action in offspring exposed in utero (87). It is not known whether
DES activities are related only to its estrogenicity. Even if the
structure (analyzed by X-ray) is compatible with an estradiol-like
interaction with ER (88), differences exist which may, as for all
ligands which are not the natural one, provoke conformational changes in the receptor and lead to "abnormal" effects. Moreover, unlike
estradiol, DES does not bind to AR, and this may cause an imbalance
not observed with other estrogens also binding to AR. In a way, diethylstilbestrol is "purer" estrogen than the natural hormone, just
as dexamethasone is a "purer" glucocorticosteroid than cortisol since
it interferes with MR to a negligible extent (89). Synthetic analogs
may show binding kinetics and/or a metabolism very different from
the natural hormones. These parameters influence their actions, agonist or antagonist in nature, whose occurrence and extent are related to the lenght of time spent by the active hormone-receptor
complexes in the nucleus. Certain synthetic compounds can, by their
binding specificity and slow dissociation rate from receptors, be
conveniently used as tags for labelling receptors (90) and eventually
constitute selective therapeutic agents.

The differential binding of hormones and analogs also applies
to plasma proteins. Consequently the latter may influence availability
of most natural steroids to the receptor. Conversely synthetic steroids, binding very little to plasma proteins, acquire more potency,
as in the case of DES, 11β-methoxy-estradiol, R 5020 (a progesterone
analog), dexamethasone, etc (18). The differential distribution of
plasma proteins in different tissues may be important in explaining
differences in the uptake and activities of natural and synthetic

hormones (eg. corticosterone and dexamethasone in different brain regions (75)).

Changes in receptor concentration.
As demonstrated previously for the PR increase by estradiol, these may also be important pharmacologically. A new example has been obtained in endometrial cancer which is sensitive to progestagen therapy, and where we find PR in all cases (63). Since the treatment becomes inefficient with time and the progesterone-like compounds "inactivate" the PR (45), it is of interest to note that tamoxifen (a triphenylethylene derivative with antiestrogenic properties) can increase significantly PR concentration in the cancerous tissue in vivo, even though the compound is not a strong estrogen by other criteria (63). We have also observed a selective inhibition of the ER content of L-cells cultured in the presence of androgen (72). Whether receptor concentration can be selectively modulated by other drugs than hormonal steroids and related compounds is still unknown. It has been reported that perphenazine decreases estradiol receptor concentration in the brain (91).

Antihormones.
The principles applied to hormones also apply to antihormones, which antagonize hormone action at the target cell level. Again we can envisage two types of mechanisms.

Decreases in receptor concentration should diminish hormone activity. Based on this and the fact that progesterone "inactivates" its own receptor, we have proposed a method for "mid cycle" contraception in which a ligand having little or no progestational activity prematurely inactivates PR, which is highly increased at this period of the cycle (92). In the future, drugs decreasing selectively the concentration of hormone receptors may be of great use.

Nevertheless at the present time, one should simply envisage the antihormones as showing competitive binding to the receptor site of the hormone which they antagonize. The simplest and best antagonist would be a compound binding with enough affinity to exclude efficiently the hormone, a metabolism slow enough that it is available for a long period, and, last but not least, a structure such that its interaction with the receptor binding site does not lead to any hormonal agonistic effect. In other words, a "pure" antihormone of great activity. Such requirements have been met by tamoxifen, an antiestrogen, for which no estrogenic effect but a complete estrogen antagonism have been demonstrated in the chick oviduct system (93), thus providing a model for understanding antihormone action. Preliminary indications show that its rate of association with the receptor is relatively slow, the rate of dissociation rapid and the overall affinity rather high (C. Geynet, unpublished). Its slow metabolism is responsible for its lengthy availability to the cell. After binding to the receptor, "activation" takes place in the sense that there is transfer of tamoxifen-receptor complexes to the nucleus, where they stay for a long time while no estrogenic activity occurs. Ten mg of tamoxifen are required to abolish the effects of 1 mg of estradiol (actually a very high dose difficult to antagonize). Tamoxifen, when given after estrogen has been administered, stops hormone action (94), and conversely estradiol given to an animal receiving tamoxifen can trigger estrogenic effects (E. Muvihill and R. Palmiter, personal communication). Several important conclusions can be derived from these observations. The first is that hormone or antihormone action is not critically dependent on cytoplasmic receptor concentration, since tamoxifen and estradiol are active even under circumstances where the cytoplasmic receptor is low and most of the receptor is located in the nucleus. The second is that the cytoplasmic receptor is not necessarily required to allow a ligand to enter the nucleus. Thus the data suggest that the receptor is decisive for "interpreting" the molecular structure of the ligand, since the estradiol-receptor complexes are active and the tamoxifen-receptor complexes are inactive in terms of estrogenic activity, and that the importance of the receptor is not in terms of an intracellular transport proteins.

Why in the past has the anti-steroid hormone effect been hypothetically linked to the limiting availability of cytoplasmic receptor (95) ? Probably because the antihormone blocks ER synthesis, which depends itself on estrogen action. The very fact that tamoxifen is an antagonist in the chick system when injected at the same time as estradiol, indicates that its effect is not mediated by the decrease in cytoplasmic receptor, but rather the decrease in the receptor may secondarily have an antihormone consequence as indicated above. Indeed, the simplest hypothesis is that tamoxifen is an allosteric ligand for the receptor giving the wrong conformation in terms of estrogen action. It also follows that the notion of "acceptor" sites, experimentally responsible for the binding of the "activated" receptor in the nucleus, does not help very much in the understanding of hormone and antihormone activities.

Although tamoxifen has provided us with a relatively clearcut example, the problem is not often as simple with other so-called antihormones. Even tamoxifen is very difficult to understand in systems other than chick oviduct since it (and triphenyl ethylene congeners) displays mixed agonist and antagonist properties (95). In the central nervous system (in the rat), the antihormone (nafoxidin) is antiestrogenic for one behavioural test (lordosis) and estrogenic for another (eating) (96). We have seen that in endometrial cancer, tamoxifen represents a "partial" estrogen giving potentially increased receptivity to progestational agents (63).

Although details of the integrated interactions between ligand, receptor and effector structure(s) will probably remain hidden for a long time, it is already apparent that the studies of ligand-receptor binding kinetics and of the availability of ligand to the target cells can help in the selection of better drugs. It is known that "weak" agonists of low affinity, due to fast dissociation rates, display antagonist effects when administered together with the agonist. However repeated administration will lead to full agonist activity, as demonstrated in the case of estriol vs estradiol (97). Thus, with well chosen doses and an appropriate rhythm of administration, one can obtain hormone or antihormone activity as a consequence of the dynamics of binding, the metabolic characteristics, and/or conformational changes of the receptor.

In conclusion the basic mechanisms for understanding hormone and antihormone effects are reasonably well understood. This is not to say that they are not difficult to apply in devising new drugs, as indicated by binding studies now being regarded as only the first of several screening tests. It is our belief, however that this task is readily accomplishable.

V. Interaction of steroids at the surface membrane level reinitiates meiotic division in Xenopus laevis oocyte.

Recently, we have been interested in the reinitiation of meiosis in Xenopus laevis oocytes, which can be studied in progesterone-containing medium (98-100). The observed effects include the breakdown of the germinal vesicle (nucleus) and an increased accumulation of specific proteins. Microinjection of progesterone into oocytes does not provoke meiosis, and protein changes can take place even in enucleated oocytes. Thus, the progesterone effects appear to be neither dependent on an intracellular receptor (actually undetectable in our laboratories) nor on transcription.

Progesterone and many other steroids (Table II) are active at concentrations of 10 nM-1 μM, which are somewhat higher than the level at which intracellular receptors interact with steroid hormones[ooo]. The stereochemical diversity of active steroids is impossible to reconcile with the characteristics of steroid hormone-receptor interactions in somatic cells, but nonetheless some degree of specificity is apparent. We conclude therefore that there is a locus in oocytes showing a different steroid binding specificity to that of classical receptors. Recent experiments (102) suggest that decisive interaction of progesterone with the oocyte takes place at the membrane surface. A progesterone analog (androsta-4-ene-3-one-17β-

_____

[ooo] Physiologically progesterone is released from surrounding follicle cells, and probably establishes a relatively high steroid concentration. Such a "local" distribution of steroids is also known to occur when testosterone is released from Leydig cells to the seminiferous tubules in testis and when adreno-cortical glucocorticosteroids reach cells of the medulla.

TABLE

STEROIDS REINITIATING MEIOSIS IN XENOPUS LAEVIS OOCYTES IN VITRO (100).

AGONISTS

PROGESTERONE	The most active natural hormone.
R 5020	A synthetic progestagen.
DEOXYCORTICOSTERONE, CORTISOL, TESTOSTERONE	) Hormones non-physiologically re-) lated to maturation.
19-NORTESTOSTERONE	A synthetic androgen.
PREGNENOLONE$^+$, ANDROSTENEDIONE^{++}	Prehormones, metabolizable to progesterone$^+$ or testosterone^{++}.
5α or β-DIHYDROPROGESTERONE 5α or β-DIHYDROTESTOSTERONE	) Metabolites not metabolizable) back to their respective hormone.

ANTAGONISTS

(WITH LITTLE OR NO AGONIST EFFECT)

16α-METHYL-, OR ETHYL-, OR ETHYNYL-19-NORTESTOSTERONE	)) Synthetic derivatives of the) 19-nortestosterone series.
R 2341	)
OESTRADIOL, 8-ISO-OESTRADIOL	Natural and synthetic estrogens.

NEITHER AGONIST NOR ANTAGONIST

13β-PROPYL-CORTISOL, R 2323	Synthetic derivatives with 2C or 3C-13β side-chain.
DIETHYLSTILBESTROL	Synthetic non steroidal estrogen.
CHOLESTEROL, DEOXYCHOLIC ACID	C27 and C25 compounds.

carboxy-acid), when linked to a soluble polymer (amino-propyl-poly-
ethylene-oxide of MW 20,000) , which cannot enter the cell, is
able to promote oocyte meiosis. Controls, including radioactive derivatives, indicated that this could not be attributed to release
of the steroid and/or entry of any hormonal material into the oocytes. Other experiments showed that progesterone action can be mimicked by drugs showing, at millimolar concentration, a "membrane sta-
bilizing effect" or "local anaesthetic activity" (such as proprano-
lone, dibucaine, chlorpromazine, etc.). These compounds displace
membrane-bound Ca^{2+} and/or change cellular Ca^{2+} distribution in many different systems. Meiotic maturation, can also be provoked by

ionophore 23187 (if Ca^{2+} or Mg^{2+} is present), specific Ca^{2+} ionto-
phoresis or lanthanum, whilst the progesterone effect is suppressed
when oocytes are injected with Egta (review in 100). These data support the hypothesis that calcium ions may be the second messenger
for a steroid signal at the membrane level.

In addition to the specific problem of meiosis, these results
may encourage the study of a new type of subcellular target for steroid hormones in somatic cells. Sex steroids, such as androgens and
estrogens, are known to be cell division and/or differentiation
promoting factòrs but so far no mechanistic studies have been made.
Specific interaction of estradiol at the surface of uterine cells has
been described (102). The specificity of steroid structures used in
anaesthesia (103) should be studied further. Interaction at the surface may or may not be related to entry of the steroid into the cell
(104), and may or may not be related to rapid changes in membrane
permeability and/or polarization.

This first instance of steroid hormone action initiated at the
cell surface level without involving the nucleus, does not deny that
steroid hormones act through the decisive contribution of intracellular receptor at the nuclear and, presumably transcriptional, level
in most somatic target cells (30, 31, 52, 89). Interaction of the
hormone at another cellular level may well be complementary to action
in the nucleus (this could be the case for the early effect of androgens upon initiation factor eIF-2 of protein synthesis) (105), or
occasionally be an alternative for some aspects of steroid action
(cell division ?). That a hormone may have several different molecular targets and mechanisms of action in a given responsive cell is
not at all impossible. In regard to this, it is interesting to note
that the possibility of internalization and intracellular activity
of polypeptide hormones is of current interest in several laboratories.

VI. Conclusions.

The existence of specific intracellular hormone receptors in
somatic target cells and their mediation of hormone action via gene
transcription is now well established. Further understanding of steroid hormone mechanisms of action, at the molecular level, will bene-
fit from purification of the respective receptors and immunological
studies with specific receptor antibodies. However, this approach is
likely to be hampered by the current limited understanding of chromatin interactions and gene functioning, and progress in this field
is therefore difficult to estimate.

The observations that several hormones may bind to one receptor,
that several receptors can interact with the same hormone, that
several receptors may be present in the same cell and that receptor
concentrations vary under different conditions, have led to the de-

velopment of a new physiological approach to pharmacology. This approach is readily applicable to synthetic agonists and antagonists (antihormones) and should permit progress in both applied (ie. diagnosis and therapy) and theoretical terms.

The recent observation that steroid hormones can act at the cell surface of amphibian oocytes and bring about metabolic changes without nuclear intervention, opens up a new field of investigation. It remains to be established if this target exists in somatic cells and whether it is an alternative or parallel site for certain hormone activities.

Thus, the field of steroid hormone-cell interactions is expanding and we should see much progress made in the next few years, despite the above-mentioned limitations to classical approaches.

References

1. Jensen, E.V. and Jacobson, H.I., Rec. Progr. Hormone Res., 18: 387, 1962.

2. Baulieu, E.E., Alberga, A. and Jung, I., C.R. Acad. Sci. Paris, 265: 354, 1967.

3. Toft, D., Shyamala, G. and Gorski, J., Proc. Nat. Acad. Sci. USA, 57: 1740, 1967.

4. Toft, D. and Gorski, J., Proc. Nat. Acad. Sci. USA, 55: 1740, 1967.

5. Milgrom, E. and Baulieu, E.E., Biochim. Biophys. Acta, 194: 602, 1969.

6. Best-Belpomme, M., Fries, J. and Erdos, T., Eur. J. Biochem., 17: 425, 1970.

7. Truong, H. and Baulieu, E.E., Biochim. Biophys. Acta, 237: 167, 1971.

8. Baulieu, E.E. In "Endocrinology" (Scow, R.O., ed.), Excerpta Medica, Amsterdam, International Congress Series n°273, pp. 30-62, 1973.

9. Mercier-Bodard, C., Alfsen, A. and Baulieu, E.E., Acta Endocrinol. suppl. 147, 64: 204, 1970.

10. Raynaud, J.P., Mercier-Bodard, C. and Baulieu, E.E., Steroids, 18: 767, 1971.

11. Fox, L.L., Redeuilh, G., Baskevitch, P., Baulieu, E.E. and Richard-Foy, H., Febs Letters, 63: 71, 1976.

12. Radanyi, C., Mercier-Bodard, C., Secco-Millet, C., Baulieu, E.E. and Richard-Foy, H., Proc. Nat. Acad. Sci. USA, 74: 2269, 1977.

13. Renoir, J.M., Fox, L.L., Baulieu, E.E. and Mercier-Bodard, C., Febs Letters, 75: 83, 1977.

14. Greene, G.L., Closs, L.E., Fleming, H., De Sombre, E.R. and Jensen, E.V., Proc. Nat. Acad. Sci. USA, 74: 3681, 1977.

15. Fang, S., Anderson, K.M. and Liao, S., J. Biol. Chem., 244: 6584, 1969.

16. Mainwaring, W.I.P., J. Endocrinol., 45: 531, 1969.

17. Baulieu, E.E. and Jung, I., Biochem. Biophys. Res. Commun., 38: 599, 1970.

18. Milgrom, E., Atger, M. and Baulieu, E.E., Steroids, 16: 741, 1970.

19. O'Malley, B.W., Sherman, M.R. and Toft, D.O., Proc. Nat. Acad. Sci. USA, 67: 501, 1970.

20. Munck, A. and Wira, C. In "Advances in biosciences" (Raspé, G., ed.), Pergamon Press Vieweg, Oxford, Vol. 7, p. 301, 1971.

21. Sharp, G.W.G., Kowack, C.L. and Leaf, A., J. Clin. Invest., 45: 450, 1966.

22. Swaneck, G.E., Chu, L.H. and Edelman, I.S., J. Biol. Chem., 245: 5382, 1970.

23. Lebeau, M.C., Massol, N. and Baulieu, E.E., Eur. J. Biochem., 36: 294, 1973.

24. Mester, J. and Baulieu, E.E., Eur. J. Biochem., 72: 405, 1977.

25. Sica, V., Nola, E., Parikh, I., Puca, G.A. and Cuatrecasas, P., Nature New Biol., 244: 36, 1973.

26. Kuhn, R.W., Schrader, W.T., Smith, R.G. and O'Malley, B.W., J. Biol. Chem., 250: 4220, 1975.

27. Truong, H., Geynet, C., Millet, C., Soulignac, O., Bucourt, R., Vignau, M., Torelli, V. and Baulieu, E.E., Febs Letters, 35: 289, 1973.

28. Erdos, T., Biochem. Biophys. Res. Commun., 32: 338, 1968.

28.bis. Wrange, O. and Gustafsson, J.A., J. Biol. Chem., 253: 856, 1978.

29. Mueller, G.C., Herranen, A.M. and Jervell, K.F., Rec. Progr. Hormone Res., 14: 95, 1958.

30. Means, A.R. and O'Malley, B.W., Science, 183: 610, 1974.

31. Schimke, R.T., Pennequin, P., Robins, D. and McKnight, G.S. In "Hormones and Cell Regulation" (Dumont, J. and Nunez, J., eds.), European Symposium, North Holland Publishing Company, Amsterdam, Vol. 1, pp. 209-221, 1977.

32. Jensen, E.V., Suzuki, T., Kawashima, T., Stumpf, W.E., Jungblut, P.W. and DeSombre, E.R., Proc. Nat. Acad. Sci. USA, 59: 632, 1968.

33. Gorski, J., Toft, D.O., Shyamala, G., Smith, D. and Notides, A., Rec. Progr. Hormone Res., 24: 45, 1968.

34. Higgins, S.J., Rousseau, G.G., Baxter, J.D. and Tomkins, G.M., J. Biol. Chem., 248: 5873, 1973.

35. Milgrom, E., Atger, M. and Baulieu, E.E., Biochemistry, 12: 5198, 1973.

36. Edelman, I.S., Bororoch, R. and Porter, G.A., Proc. Nat. Acad. Sci. USA, 50: 1169, 1963.

37. Massol, N., Lebeau, M.C. and Baulieu, E.E., Nucl. Acids Res., 5: 723, 1978.

38. Mester, J. and Baulieu, E.E., Biochim. Biophys. Acta, 261: 236, 1972.

39. Zava, D.T. and McGuire, W.L., J. Biol. Chem., 252: 3709, 1977.

40. Raynaud-Jammet, C. and Baulieu, E.E., C.R. Acad. Sci. Paris, 268: 3211, 1969.

41. Schwartz, R.J., Kuhn, R.W., Buller, R.E., Schrader, W.T. and O'Malley, B.W., J. Biol. Chem., 251: 5166, 1976.

42. Ringold, G.M., Yamamoto, K.R., Bishop, J.M. and Varmus, H.E., Proc. Nat. Acad. Sci. USA, 74: 2879, 1977.

43. McKnight, G.S., Cell, 14: 403, 1978.

44. Palmiter, R.D., Moore, P.B., Muvihill, E.R. and Emtage, S., Cell, 8: 557, 1976.

45. Milgrom, E., Luu Thi, M., Atger, M. and Baulieu, E.E., J. Biol. Chem., 248: 6366, 1973.

46. Kahn, C.R., J. Cell Biol., 70: 261, 1976.

47. Mester, J., Martel, D., Psychoyos, A. and Baulieu, E.E., Nature, 250: 776, 1974.

48. Hsueh, A.J., Peck, E.J. and Clark, J.H., Nature, 254: 337, 1975.

49. Tseng, L. and Gurpide, E., J. Clin. Endocrinol. Metab., 46: 635, 1978.

50. Bayard, F., Damilano, S., Robel, P. and Baulieu, E.E., J. Clin. Endocrinol. Metab., 46: 635, 1978.

51. Alberga, A. and Baulieu, E.E., C.R. Acad. Sci. Paris, 261: 5226, 1965.

52. Jensen, E.V. and DeSombre, E.R., Ann. Rev. Biochem., 41: 203, 1972.

53. Jost, A. In "The Harvey Lectures", Academic Press, New York, Series 55, p. 201, 1961.

54. Kratochwil, K. and Schwartz, P., Proc. Nat. Acad. Sci. USA, 73: 4041, 1976.

55. LeGoascogne, C. and Baulieu, E.E., Biologie Cellulaire, 30: 195, 1977.

56. Gorski, J., Sarff, M. and Clark, J. In "Advances in biosciences" (Raspé, G., ed.), Pergamon Press Vieweg, Oxford, p. 5, 1971.

57. Raynaud, J.P. and Moguilewsky, M. In "Système nerveux, Activité sexuelle et reproduction" (Soulairac, A., Gautray, J.P., Rousseau, J.P. and Cohen, J., eds.), Masson et Cie, Paris, p. 85, 1976.

58. Gehring, U., Tomkins, G.M. and Ohno, S., Nature New Biol., 232: 106, 1971.

59. Sibley, C.H. and Tomkins, G.M., Cell, 2: 221, 1974.

60. Bourgeois, S. and Newby, R., Cell, 11: 423, 1977.

61. Jensen, E.V., Block, G.E., Smith, S., Kyser, K. and DeSombre E.R., Nat. Cancer Inst. Monogr., 34: 55, 1971.

62. McGuire, W.L., Raynaud, J.P. and Baulieu, E.E. In "Progesterone receptors in normal and neoplastic tissues ; Progress in Cancer Research and Therapy" Raven Press, New York, Vol. 4., 1977.

63. Levy, C., Wolff, J.P., Nicolas, J.C., Robel, P. and Baulieu, E.E., in preparation.

64. Palmiter, R.D., Muvihill, E.R., McKnight, G.S. and Senear, A.W., Symposium on chromatin, Cold Spring Harbor, 1978, in press.

65. Sutherland, R.L., Lebeau, M.C., Schmelck, P.H. and Baulieu, E.E., Febs Letters, 79: 253, 1977.

66. Anderson, J.N., Peck, Jr., E.J. and Clark, J.H., Endocrinology, 95: 174, 1974.

67. Katzenellenbogen, B.S. and Gorski, J., J. Biol. Chem., 247: 1229, 1972.

68. Sutherland, R.L., Mester, J. and Baulieu, E.E. In "Hormones and cell regulation" (Dumont, J. and Nunez, J., eds.), North Holland Publishing Company, Amsterdam, p. 31, 1977.

69. Croce, C.M., Koprowski, H. and Litwack, G., Nature, 249: 839, 1974.

70. Jung-Testas, I. and Baulieu, E.E., C.R. Acad. Sci. Paris, 279: 671, 1974.

71. Feyel-Cabanes, T., Secchi, J., Robel, P. and Baulieu, E.E., Cancer Res., 1978, in press.

72. Jung-Testas, I. and Baulieu, E.E., 1978, in preparation.

73. Rochefort, H. and Garcia, M., Steroids, 28: 549, 1976 ; Garcia, M. and Rochefort, H., Steroids, 29: 11, 1977.

74. Stumpf, W.E. and Sar, M., Fed. Proceed., 36: 1973, 1977.

75. McEwen, B.S., Scientific American, 235: 48, 1976.

76. Naftolin, F., Ryan, K.J., Davies, I.J., Reddy, V.V., Flores, F., Petro, Z., Kuhn, M., White, R.J., Takaoka, Y. and Wolin, L., Rec. Progr. Hormone Res., 31: 295, 1975.

77. Horwitz, K.B., Costlow, M.E. and McGuire, W.L., Steroids, 26: 785, 1975.

78. Baulieu, E.E., Lasnitzki, I. and Robel, P., Nature, 219: 1155, 1968.

79. Bruchovsky, N. and Wilson, J.D., J. Biol. Chem., 243: 2012, 1968.

80. Anderson, K.M. and Liao, S., Nature, 219: 277, 1968.

81. Jung, I. and Baulieu, E.E., Nature New Biol., 237: 24, 1972.

82. Imperato-McGinley, J., Guerreo, L., Gautier, T. and Peterson, R.E., Science, 186: 1213, 1974.

83. Paul, S.M. and Axelrod, J., Science, 197: 659, 1977.

84. Davies, I.J., Naftolin, F., Ryan, K.F., Fishman, J. and Siu, J., Endocrinology, 97: 554, 1975.

85. Paul, S. and Skolnick, P., Nature, 266: 559, 1977.

86. Huggins, C., Science, 156: 1050, 1967.

87. Bibbo, M., Gill, W.B., Azizi, F., Blough, R., Fang, V.S., Rosenfield, R.L., Schumacher, G.F., Sleeper, K., Sonek, M.G. and Wied, G.L., J. Obst. Gynecol., 49: 1, 1977.

88. Hospital, M., Busetta, M., Bucourt, R., Weintraub, H. and Baulieu, E.E., Mol. Pharmacol., 8: 438, 1972.

89. Baulieu, E.E., Atger, M., Best-Belpomme, M., Corvol, P., Courvalin, J.C., Mester, J., Milgrom, E., Robel, P., Rochefort, H. and DeCatalogne, D., Vitamins and Hormones, 33: 649, 1975.

90. Raynaud, J.P., Ojasoo, T., Bouton, M.M. and Philibert, D., Cancer Res., 1978, in press.

91. Ginsburg, M., Maclusky, N.J., Morris, I.D. and Thomas, P.J., Brit. J. Pharmacol., 59: 397, 1977.

92. Baulieu, E.E., Eur. J. Obst. Gynecol. Reprod. Biol., 4: 161, 1975.

93. Sutherland, R.L., Mester, J. and Baulieu, E.E., Nature, 267: 434, 1977.

94. Mester, J., Geynet, C., Binart, N. and Baulieu, E.E., Biochem. Biophys. Res. Commun., 79: 112, 1977.

95. Clark, J.H., Peck, Jr., E.J. and Anderson, J.N., Nature, 251: 446, 1974.

96. Wade, G.N. and Blanstein, S.D., Endocrinology, 102: 245, 1978.

97. Huggins, C. and Jensen, E.V., J. Exper. Med., 102: 335, 1955.

98. Smith, L.D. and Ecker, R.E., Curr. Top. Dev. Biol., 5: 1, 1970.

99. Masui, Y. and Markert, C.L., J. Exp. Zool., 177: 120, 1971.

100. Baulieu, E.E., Godeau, F., Schorderet, M. and Schorderet-Slatkine, S., Nature, 1978, in press.

101. Godeau, F., Schorderet-Slatkine, S., Hubert, P. and Baulieu, E.E., Proc. Nat. Acad. Sci. USA, 75: 1978, in press.

102. Pietras, R.J. and Szego, C.M., Nature, 265: 69, 1977.

103. Masley, M.J., Millar, R.A. and Sutton, J.A. (eds.) "Molecular mechanisms in general anaesthesia" Churchill-Livingstone, London, 1974.

104. Milgrom, E., Atger, M. and Baulieu, E.E., Biochim. Biophys. Acta, 320: 267, 1973.

105. Liang, T., Castaneda, E. and Liao, S., J. Biol. Chem., 252: 5692, 1977.

106. Williams, D. and Gorski, J., Proc. Nat. Acad. Sci. USA, 69: 3464, 1972.

107. Yamamoto, K.R. and Alberts, B.M., Ann. Rev. Biochem., 45: 721, 1976.

108. Baulieu, E.E. In "Hormone and antihormone action at the target cell" (Clark, J.H., Klee, W., Levitzki, A. and Wolff, J., eds.), Life Sciences Research Report 3, Dahlem Konferenzen, p. 51, 1976.

109. Bresciani, F., Nola, E., Sica, V. and Puca, G.A., Fed. Proceed., 32: 2126, 1973.

110. Sherman, M.R., Pickering, L.A., Rollwagen, F.M. and Miller, L.K., Fed. Proceed., 37: 167, 1978.

111. Edelman, I., J. Ster. Biochem., 6: 147, 1975.

112. Munck, A. In "Receptors and mechanism of action of steroid hormones" (Pasqualini, J.R., ed.), Marcel Dekker Inc, New York and Basel, Part I, p. 1, 1976.

113. Terenius, L., Acta Endocrinol., $\underline{64}$: 47, 1970.

Acknowledgments

Grants from the Cnrs', the Dgrst, the Ford Foundation, Who and Roussel-Uclaf are gratefully acknowledged. I also wish to thank Heather Mullis, Eileen Muvihill, Françoise Boussac and Anne Atger for their help with this manuscript. The names of my colleagues appear in the appropriate references.

STEROID HORMONE RECEPTORS IN BRAIN AND PITUITARY

Bruce S. McEwen, Paula G. Davis, Lewis C. Krey, Ivan
Lieberburg, Neil MacLusky and Edward Roy

The Rockefeller University
New York, New York   10021   USA

INTRODUCTION

Steroid hormones act upon the brain and pituitary gland to
regulate hormone secretion rates and to influence particular behaviors (see McEwen, 1976, for overview). Many of these actions
appear to be the result of an interaction of steroids with specific
intracellular receptor proteins which translocate hormone into the
cell nuclear compartment to alter genomic activity (see McEwen et
al, 1978, and Baulieu, this volume, for a discussion of this and
alternative cellular mechanisms of steroid action). Intracellular
steroid receptor proteins have been identified for five major
classes of steroid hormones, estrogens, androgens, progestins,
glucocorticoids, and mineralocorticoids, and have been extensively
studied both with respect to their physical and chemical properties
and their neuroanatomical distribution within the brain (see
McEwen, 1976, 1978). In general, each receptor system binds
preferentially the most active members of that particular class of
steroid hormones and displays a distinctive regional distribution
pattern in the CNS.

Of the steroid receptor systems in the brain, the estradiol
receptors have perhaps the most extensive and best studied role
in neuroendocrine function and behavior. Not only do estrogens
regulate behavioral estrus and gonadotropin secretion in the female, but also they have been implicated (as intracellular metabolites of testosterone in the brain) in the control of male sexual
behavior and in brain sexual differentiation in the rat (see Mc-
Ewen, 1978, for review). This paper will summarize recent aspects
of ongoing research in our laboratory on estrogen receptors and

261

their implications for adult rat brain function and for sexual
differentiation.

QUANTITATION OF ESTROGEN RECEPTOR OCCUPATION AND FUNCTIONAL IMPLICATIONS OF SUCH MEASUREMENTS

Two methods exist by which one can estimate the number of
estrogen receptors occupied in vivo by circulating hormones: direct
radioisotopic methods and receptor exchange assays. Direct radioisotopic methods involve the injection of ^{3}H steroid into the animal, isolation of a receptor-containing fraction and determination
of specifically-bound steroid. With this method, the regional distribution in the brain of estrogen receptors was determined (see
McEwen, 1976, 1978). In subsequent studies with this procedure
we have measured time courses of ^{3}H estradiol retention in brain
and pituitary cell nuclei following the injection of doses of
^{3}H estradiol sufficient to promote lordosis behavior in female rats
(McEwen et al, 1975). In that study, the regional distribution of
estrogen receptors agreed well with the pattern shown by autoradiography (Pfaff and Keiner, 1973) and the time course of estrogen retention demonstrated that estradiol does not have to be
present in the animal at the time the lordosis response can be
elicited. In more recent studies we have studied the rapid feedback action of estradiol in suppressing pituitary gonadotropin
release (Krey et al, 1978). These studies indicate that very low,
subsaturation levels of estradiol-receptor complexes in cell nuclei
are associated with the negative feedback action of estradiol.
Furthermore, this action is blocked by pretreatment of rats with
CI628, an anti-estrogen which prevents estradiol-receptor translocation to the nucleus (Clark et al, 1973). It should be noted
that CI628, as well as other anti-estrogens, also blocks the positive feedback effects of estradiol (see Roy, 1978, for references).
Taken together these studies on behavior and gonadotropin secretion
suggest that the actions of estradiol with the brain and pituitary
are mediated by an alteration of genomic activity by the estrogen-
receptor complex.

Another application of direct radioisotopic methods established
that considerable quantities of injected ^{3}H testosterone are converted in brain (but not in pituitary) into ^{3}H estradiol, which is
then bound by intracellular estrogen receptors and translocated
into cell nuclei (Lieberburg and McEwen, 1977). The in vivo
aromatization of testosterone complements data on the in vitro
aromatization of this steroid by brain homogenates (Naftolin et al,
1975) and also complements data which strongly suggest that aromatization plays a key role in the action of testosterone to activate
male sexual behavior in the rat (see McEwen, 1978; Morali et al,
1977; Davis and Barfield, in preparation).

Thus estrogen receptors in the rat brain appear to participate in male as well as female sexual behavior and control of gonado-tropin secretion. Because the radioisotopic methods used to localize steroid receptors are limited by the necessity to work with cas-trated animals and with pulse injection of isotopic steroid, it became necessary to develop a method of assessing receptor levels under physiological conditions with endogenous steroids if neces-sary. The exchange assay for receptor-bound estradiol was develop-ed in our laboratory (Roy and McEwen, 1977) as a more sensitive modification of existing methods for uterine receptors (Clark et al, 1973). The procedure involves the incubation of a salt extract of purified cell nuclei (containing ≈95% of receptor) with ^{3}H estradiol so as to achieve a one-for-one exchange of unlabeled estradiol on receptor for ^{3}H estradiol. Using this method one can achieve the same estimates of receptor-bound estradiol as are obtained by direct radioisotopic methods described in the first paragraph (see Roy and McEwen, 1977).

One application of this method has been to measure occupation of estrogen receptor sites in female and male rat brains under various physiological conditions. Estrogen receptor occupation fluctuates during the estrous cycle, as expected, with peak levels of occupation in proestrus approaching half the capacity; sur-prisingly, estrogen receptor occupation in brains of intact males is comparable to that in proestrus females (Roy, unpublished).

Another important application of the exchange assay is the estimation of nuclear estrogen receptors translocated by anti-estrogens, which are known to bind to cytoplasmic receptors. Measurements of nuclear receptors following an injection of CI628 (nitromiphene citrate: Parke-Davis) revealed that hypothalamic estrogen receptors are translocated to cell nuclei by anti-estrogens just as uterine receptors are (Roy and MacLusky, 1978). In both tissues the prolonged elevation of nuclear receptor levels is cor-related with prolonged elevation of serum CI628 levels (Katzenellen-bogen et al, 1978). There is, however, a difference between the way estrogen-receptor complexes and anti-estrogen-receptor complexes are bound in the cell nuclei, which is reflected in the salt con-centration required to extract the two types of ligand-receptor com-plexes. That is, anti-estrogen-receptor complexes are more easily extracted. We hope to determine whether the salt extractability property is correlated with the inhibitory activity of anti-estrogens.

Anti-estrogens also have some estrogenic activities. For example, CI628 blocks estrogen induction of lordosis and of ovula-tion, but acts as an estrogen in mimicking the suppression of food intake (see Roy, 1978); at the biochemical level CI628 blocks estrogen induction of glucose-6-phosphate dehydrogenase in pituitary (see Luine and McEwen, 1977) and of the progesterone receptor (Roy

and MacLusky, 1978) in hypothalamus but acts as an estrogen in inducing increases in choline acetyl transferase in preoptic area and
in decreasing monoamine oxidase activity in hypothalamus (see Luine
and McEwen, 1977). All of the above-mentioned effects of CI628
and estradiol require at least 24 hours to be manifested. Future
work must establish whether the estrogenic and anti-estrogenic
properties of CI628 occur in the same or different cells of brain
and pituitary.

A third application of the exchange assay has been the valida-
tion of the action of a steroid inhibitor of aromatization, 1,4,6-
androstatriene-3,17-dione (ATD). In castrated male rats, ATD is
capable of inhibiting testosterone action to facilitate male sexual behavior (Morali et al, 1977). In castrated male rats exposed
to testosterone and ATD, cell nuclear occupation of estrogen
receptors is reduced - virtually completely in hypothalamus and
partially in preoptic area and amygdala. This inhibition had no
adverse effect on the ability of testosterone to inhibit LH release,
suggesting that this aspect of androgen action does not involve
aromatization. This conclusion is supported by further observations that flutamide, an anti-androgen, blocks testosterone
negative feedback without affecting aromatization or estrogen receptor occupation and by the observation that androgen-insensitive
pseudohermaphrodite rats are fully capable of converting testosterone to estradiol and yet show no significant negative feedback effects of testosterone (Krey et al, 1978).

LOCALIZATION OF ESTROGEN ACTION

Two strategies exist for the localization of brain structures
which are the focal points for estrogen action on neuroendocrine
function and behavior: selective inhibition by drugs and localized
hormone implants. The former approach is only now receiving attention in our laboratory, centering around the greater effects of
CI628 on pituitary than on brain estrogen receptors (see Luine and
McEwen, 1977, for background), but there is nothing definitive to
report. The implant approach has been utilized numerous times
(see Barfield and Chen, 1977). Most recently the studies of Goodman (1978) have succeeded in localizing an estrogen-sensitive
region for control of ovulation in the anterior hypothalamus-
preoptic area under conditions in which spread of implanted hormone to pituitary is minimal.

We are attempting to localize estrogen-sensitive neurons,
activation of which is sufficient to activate female sexual behavior in the ovariectomized rat. In so doing we are pursuing the
demonstrated effectiveness of estrogen implants in the ventromedial nucleus of the hypothalamus (Barfield and Chen, 1977) using
a refined technique in which tritium labeled estradiol is diluted

with cholesterol in a ratio of approximately one to 300 (Davis et al, 1978). In this way we can correlate behavioral effectiveness with the spread of the hormone from the region of implantation. Thirty gauge cannulae prepared by this method contain approximately 10-14 ng of estradiol and deliver about 30% of the total hormone to the brain in an 8 day period. In ovariectomized, progesterone-treated rats, both unilateral and bilateral implants produce substantial and significant increases in lordotic behavior accompanied by instances of solicitation behaviors. As determined by scintillation counting methods, there is no evidence of significant spread of hormone to the contralateral side of the hypothalamus (from unilateral implants) or to the POA, cortex, amygdala, pituitary or uterus, with either bilateral or unilateral implants. Moreover, similarly prepared estradiol/cholesterol implants applied to the POA are not effective in stimulating feminine sexual behavior (Rubin and Barfield, unpublished observations). These preliminary findings, taken together, suggest that estrogenic stimulation of the VMN is sufficient to activate feminine sexual behavior.

Among the fascinating unanswered questions about these results is how many of the effects of estradiol can be produced by local application of the hormone in the basal hypothalamus.   Results of Wade and Zucker (1970) suggest that regulation of food intake by estrogen may occur in this area, but that locomotor activity effects of estradiol may not. However, those studies, as many others, suffered from the disadvantage that large cannulae and large amounts of hormone were applied with all of the potential problems of spread of hormone from site of implantation. Thus the reported effects must be reexamined.

CELLULAR MECHANISM OF ESTROGEN ACTION

Most, if not all, of the effects of estradiol under discussion in this article are likely to involve the intracellular estrogen receptors (first mentioned in the Introduction), the localization and occupation of which has been the focus of the first part of this article. Pharmacological evidence with inhibitors of RNA and protein synthesis indicates that estrogen effects on lordosis behavior and on positive and negative responses of the gonadotropin system are indeed mediated, at least in part, by altered genomic activity (for summary, see McEwen et al, 1978).   Therefore the purpose of this section is to examine cellular events distal to estradiol activation of the genomic and RNA synthesis and to consider some of the gene products which are under hormonal control.

But first it is necessary to consider some guidelines. 1) Many estrogen effects are doubtless responsible for the final neuroendocrine or behavioral manifestation. Thus estradiol appears to alter LH release by modulating pituitary sensitivity to LHRH as well as

altering synthesis and/or release of LHRH from hypothalamus (see
McEwen and Luine, 1978, for references). Possibly no single
estrogen effect and no single tissue is sufficient for the preovulatory LH surge. 2) Some estrogen effects which appear to be
involved in neuroendocrine regulation may be indirect. This
appears to be the case for increases in DA turnover, which may
result from estrogen-induced increases in prolactin secretion
(Gudelsky et al, 1976). Thus, estrogen effects must be tested in
hypophysectomized rats to rule out such indirect steroid actions.
3) The recognition of direct effects of estradiol on brain and
pituitary depends in large part on whether they occur in tissues
which contain estrogen receptors and whether they show a steroid
specificity and time course compatible with receptor action via
the genome (see McEwen et al, 1978, for discussion). 4) The temporal course of estrogen effects reveals much about their relevance
to behavioral and neuroendocrine events. Thus alterations in LHRH
sensitivity of pituitary occur within 14 hours (Vilchez-Martinez
et al, 1974; Tang and Spies, 1975), a time course compatible with
the time course of induction of an LH surge. Reduction by estrogen
of Type A MAO activity (Luine and McEwen, 1977) and tyrosine hydroxylase activity (Luine et al, 1977a) in basal hypothalamus occur too
slowly to be involved in ovulation or activation of sexual behavior.
But they may represent long-term priming effects of estradiol, the
absence of which in long-term castrates leads to delays in restor-
ing sexual activity and normal neuroendocrine function (Damassa
and Davidson, 1973). 5) Besides time factors, the effect of anti-
estrogens on a biochemical event reveals whether it could serve as
a critical step in a neuroendocrine event. Thus the estrogen induction of choline acetyltransferase in preoptic area occurs within 24 hours but is mimicked by doses of CI628 which are capable
of blocking sexual behavior and ovulation (see Luine and McEwen,
1977b). On the other hand the estrogen induction of a progesterone
receptor, which will be discussed below, is blocked by CI628 (Roy
and MacLusky, 1978).

ESTROGEN EFFECTS ON PROGESTIN RECEPTORS

It is well known that estradiol and progesterone can act synergistically to promote preovulatory LH secretion and to activate
lordosis behavior. Furthermore, these progesterone effects do not
occur in the absence of prior estrogen priming. A possible mechanism for these effects has emerged recently in the course of
studies demonstrating the existence of a progesterone receptor in
uterus, pituitary and brain (see Kato and Onouchi, 1977; Moguilewsky and Raynaud, 1977; Blaustein and Wade, 1978; MacLusky et al,
1978). Specifically, estradiol is capable of inducing increases
of 10-fold or more in progesterone receptor levels in uterus and
pituitary and of 2- to 3-fold of progesterone receptor levels in
hypothalamus and preoptic area (MacLusky et al, 1978). In the rat,
these increases occur against a background level of receptor which

exists in virtually all regions of the brain and which can be demonstrated in adrenalectomized, gonadectomized animals. Both estrogen-inducible and noninducible receptors show the same physico-chemical properties and steroid specificity (MacLusky et al, 1978). In contrast to the rat, the bonnet monkey (m. Radiata) lacks progestin binding outside of the hypothalamus, but, like the rat, shows a strong estrogen induction of progestin receptors in uterus, pituitary, and hypothalamus. This species difference may be related to species differences in the neural control of ovulation: macaques showing an autonomy within the hypothalamus, pituitary and ovary; rats showing a dependency on afferent neural input entering the hypothalamus through the preoptic area (MacLusky et al, 1978).

We are presently investigating the relationship between aromatization and progestin receptor induction by attempting to induce the progestin receptor with testosterone. Preliminary results indicate that induction occurs in hypothalamus where there is aromatization but that it does not occur in pituitary where there is no aromatization (McEwen and MacLusky, unpublished).

The progestin receptors of uterus, pituitary, and hypothalamus appear to be similar to each other in physical properties and specificity toward various steroids and display a markedly higher affinity for a synthetic progestin, 17α,21-dimethyl-17-norpregna-4,9-diene-3,20-dione (R5020). R5020 is also many times more potent than progesterone in promoting lordosis behavior in estrogen-primed rats (Blaustein and Wade, 1978), and this evidence, together with estrogen-inducibility and presence of the receptor in pituitary and hypothalamus, form the basis of the notion that the receptor participates in the well-known neuroendocrine and behavioral effects of progestins. And yet how does the progestin receptor accomplish this? It has been shown that actinomycin D in the arcuate-ventromedial region of the hypothalamus blocks progesterone action in facilitating LH release (Jackson, 1975) and that cycloheximide attenuates progesterone dependent proestrus refractoriness in guinea pigs (Wallen et al, 1972). Despite this evidence of genomic involvement, in vivo cell nuclear retention of ^{3}H R5020 in brain and pituitary is very limited, though detectable (Blaustein and Wade, 1978) and the latency for progesterone to exert at least one of its effects, namely, the activation of lordosis behavior, can be extremely short (30-60 minutes: Kubli-Garfias and Whalen, 1977) so as to suggest the possibility of a more direct action of progestins at the synaptic level. Studies of this possibility are currently underway.

ORGANIZATION vs. ACTIVATION:
ROLE OF ESTROGEN RECEPTORS IN DEVELOPING RODENT BRAIN

Unlike the effects of estradiol, testosterone, and progestins
on the behavior and neuroendocrine function of the adult rodent,
which are reversible, the action of testosterone and of various
estrogens on the developing rodent brain are irreversible when
these hormones are applied during the first few days of postnatal
life. Testosterone levels in the newborn male rat are at least
20 times higher than in the female (Lieberburg, unpublished) and
it is this testicular secretion which is believed to be responsible
for the enhancement of male traits and the suppression of female
traits such as female sexual behavior and the ability of the hypothalamus to support an ovulatory cycle (see Gustafsson, this symposium). Studies of the androgen-insensitive (Tfm) rodent make
it clear that the absence of androgen receptors is not sufficient
to lead to feminization of the brain; rather the neuroendocrine
characteristics of Tfm rodents is such as to indicate that feminine
traits have been suppressed (Shapiro et al, 1975; Beach and Buehler,
1977). Thus it is not so surprising that the aromatization of
testosterone to estradiol and the interaction of estradiol with
estrogen receptors of the type found also in the adult brain appears to be essential if sexual differentiation (i.e., suppression
of female traits) is to occur (see McEwen et al, 1977).

Estrogen receptors arise in the rat brain prenatally a few
days before birth (MacLusky, unpublished) and are seen to be occupied to about 12% of capacity in limbic brain of males, and to a
far lesser extent in limbic brain of females, from postnatal day 1
to 5 (Lieberburg, unpublished). Some occupation of estrogen receptors is seen in male brains before birth as well (MacLusky,
unpublished). That such occupation of estrogen receptors in newborn male brains is the result of the aromatization of testosterone
is indicated by experiments which show that both castration and ATD
treatment, but not the anti-androgen cyproterone acetate, reduce
estrogen receptor occupation to undetectable levels (Lieberburg,
unpublished). At the same time it can be demonstrated that cyproterone acetate, and not ATD, reduce androgen receptor occupation in
neonatal brains (Lieberburg, unpublished). We are currently investigating the reputed actions of cyproterone acetate in altering
brain sexual differentiation, since this action would suggest that
androgen receptors make some contribution to the process of brain
sexual differentiation.

CONCLUSION

Estrogen receptors in the rodent brain are apparently involved
in far more than female reproduction, namely, in masculine sexual
behavior and in the masculine sexual differentiation of the brain.

They appear to exist in cells in the brain in which one or both
processes occur in addition to the estrogen receptor mechanism,
namely, the aromatization of testosterone and the induction of a
progestin receptor. Future research must decide which estrogen-
sensitive cells have each property. It also should be noted that
diverse biochemical effects of estrogens on enzyme activities
(e.g., increasing choline acetyltransferase activity in preoptic
area and glucose-6-phosphate dehydrogenase activity in hypothalamus,
decreasing monoamine oxidase and tyrosine hydroxylase activities in
hypothalamus) suggest a further, regional heterogeneity of estrogen
effects which will be the object of further scrutiny. All of these
effects, however, may be mediated by a single kind of estrogen receptor, although it should be noted that present analytical methods
preclude the detection of micro-heterogeneity among estrogen
receptors. The large array of effects of estrogens on the adult
brain and pituitary appear to be reversible and this reversibility
may be contrasted to the apparently permanent effects which estrogens produce on the sexual differentiation of the rodent brain.
Future work will undoubtedly focus attention at the cellular level
on the similarities and differences between these rather different
types of steroid hormone action and on the role of the estrogen
receptors in each of them.

SUMMARY

    Steroid hormones act upon the brain and pituitary gland to
regulate pituitary hormone secretion and to influence particular
behaviors. Many of these steroid actions are mediated by intracellular receptor sites which translocate hormone into the cell
nuclei of neurons where alterations of genomic activity ensue.
Intracellular receptor sites have been identified in brain and
pituitary for five major classes of steroid hormones and have been
topographically localized and characterized. Of these receptor
systems those for estrogens have perhaps the most diverse involvement, not only in the mature brain but also in the developing
brain. This article has focused on the functional diversity of
estrogen receptors with respect to their involvement in control of
pituitary gonadotropin secretion, in female sexual behavior, in
male sexual behavior, and in brain sexual differentiation in the
rat. We have concentrated primarily on two events, estrogen
induction of a progestin receptor and the aromatization of testosterone, in relation to their neuroanatomical localization and
independence from each other. We have also discussed a recently
developed exchange assay for measuring the occupation of estrogen
receptors by physiological levels of estradiol which has permitted
us to obtain data pertaining to physiological actions of the hormone
and to monitor the action of drugs that modify steroid metabolism
and action and that also influence pituitary hormone secretion and
hormone-dependent behaviors.

REFERENCES

Barfield, R.J. and Chen, J.J. Endocrinology 101:1716-1725, 1977.
Beach, F.A. and Buehler, M.G. Endocrinology 100:197-200, 1977.
Blaustein, J.D. and Wade, G.N. Brain Res. 140:360-367, 1978.
Clark, J.H., Anderson, J.N., and Peck, E.J. Steroids 22:707-718, 1973.
Damassa, D. and Davidson, J.M. Horm. Behav. 4:269-279k 1973.
Davis, P.G., McEwen, B.S. and Pfaff, D.W. Abstract, Soc. Neurosci. 1978.
Goodman, R.L. Endocrinology 102:151-159, 1978.
Gudelsky, G.A., Simpkins, J., Mueller, G.P., Meites, J., and Moore, K.E. Neuroendocrinology 22:206-215, 1976.
Jackson, G.L. Neuroendocrinology 17:236-244, 1975.
Kato, J. and Onouchi, T. Endocrinology 101:920-928, 1977.
Katzenellenbogen, B.S., Katzenellenbogen, J.A., Ferguson, E.R. and Krauthammer, N. J. Biol. Chem. 253:697-707, 1978.
Krey, L., Roy, E., Lieberburg, I. and Robbins, R. Abstract, Vth Int. Cong. Horm. Steroids, New Delhi, India, 1978.
Kubli-Garfias, C. and Whalen, R.E. Horm. Behav. 9:380-386, 1977.
Lieberburg, I. and McEwen, B.S. Endocrinology 100:588-597, 1977.
Luine, V.N. and McEwen, B.S. J. Neurochem. 28:1221-1227, 1977a.
Luine, V.N. and McEwen, B.S. Endocrinology 100:903-910, 1977b.
Luine, V.N., McEwen, B.S., and Black, I.B. Brain Res. 120:188-192, 1977.
McEwen, B.S. In International Symposium on Subcellular Mechanisms in Reproductive Neuroendocrinology, F. Naftolin, K.J. Ryan and J. Davies (eds), Elsevier, Amsterdam, pp. 277-304, 1976.
McEwen, B.S. In Hormone Receptors, Vol. I: Steroid Hormones, O'Malley, B. and Birnbaumer, L. (eds), Academic Press, New York, pp. 353-400, 1978.
McEwen, B.S. and Luine, V.N. CNRS Colloque International sur La Biologie Cellulaire des Processes Neurosecretoires Hypothalamiques, Bordeaux (in press).
McEwen, B.S., Lieberburg, I., Chaptal, C. and Krey, L.C. Horm. Behav. 9:249-263, 1978.
McEwen, B.S., Krey, L.C. and Luine, V.N. In The Hypothalamus, R.J. Baldessarini and J.B. Martini (eds), Raven Press, New York, pp. 255-268, 1978.
MacLusky, N.J., Krey, L., Lieberburg, I., and McEwen, B.S. Endocrine Soc. 60th Ann. Mtg., Miami, Abst. #447, p.298, 1978.
Moguilewsky, M. and Raynaud, J-P. Steroids 30:99-109, 1977.
Morali, G., Larsson, K. and Beyer, C. Horm. Behav. 9:203-213, 1977.
Naftolin, F., Ryan, K.J., Davies, I.J., Reddy, V.V., Flores, F., Petro, Z. and Kuhn, M. Rec. Prog. Horm. Res. 21:295-315, 1975.
Pfaff, D.W. and Keiner, M. J. Comp. Neurol. 151:121-158, 1973.
Roy, E.J. In Current Studies of Hypothalamic Function, K. Lederis (ed), in press.

Roy, E.J. and McEwen, B.S. Steroids 30:657-669, 1977.
Roy, E.J. and MacLusky, N.J. Endocrine Soc. 60th Ann. Mtg., Miami, Abst. #442, p. 296, 1978.
Shapiro, B.H., Goldman, A.S. and Gustafsson, J.A. Endocrinology 97:487-492, 1975.
Tang, L.K.L. and Spies, H.G. Endocrinology 96:349-386, 1975.
Vilchez-Martinez, J.A., Arimura, A., Debeljuk, L. and Schally, A. V. Endocrinology 94:1300-1303, 1974.
Wade, G.N. and Zucker, I. J. Comp. Physiol. Psych. 72:328-336, 1970.
Wallen, K., Goldfoot, D.A., Joslyn, W.D. and Paris, C.A. Physiol. Behav. 8:221-223.

FEEDBACK EFFECTS ON CENTRAL MECHANISMS CONTROLLING NEUROENDOCRINE FUNCTIONS.

L .Martini, F.Celotti, H.Juneja, M.Motta and M.Zanisi

Department of Endocrinology, University of Milano

21, Via A. Del Sarto - 20129 Milano, Italy.

I.INTRODUCTION.

The peptidergic neurons of the hypothalamus which synthesize the Releasing and Inhibiting Hormones receive three types of endocrine signals: a) from the peripheral target glands (long feedback mechanisms); b) from the anterior pituitary (short feedback mechanisms) (see for references 42 and 54); and c) from the Releasing and Inhibiting Hormones themselves (ultrashort feedback mechanisms)(20, 55; see also 42 and 54 for references). For brevity's sake, the discussion in this paper will be limited to the presentation of some data obtained in the authors' laboratory regarding the effects of sex steroids on the peptidergic neurons synthesizing LH-RH, and on the secretion of gonadotropins and of prolactin.

It has been assumed for many years that testosterone, estradiol and progesterone were the only sex hormones capable to act on the hypothalamic-pituitary complex. This traditional concept has been recently modified, on the basis of two groups of new data. First of all, it has been demonstrated that sex steroids in general, and testosterone and progesterone in particular, may undergo considerable metabolism in the neuroendocrine brain and in the anterior pituitary. Secondly, it has been shown that the androgenic and progestational quota present in the blood includes, in addition to testosterone and progesterone, a series of other steroids which may originate directly from the gonads, or

273

derive from the peripheral conversion of the appropriate precursors.

II. METABOLISM OF TESTOSTERONE AND OF PROGESTERONE IN THE BRAIN AND IN THE ANTERIOR PITUITARY.

It is generally accepted that in the anterior pituitary and in several CNS structures (e.g. hypothalamus, midbrain, amygdala,
etc.) of male mammals, testosterone may be either aromatized
to estradiol (43), or converted into 5α-androstan-17β -ol-3-one
(dihydrotestosterone, DHT) and 5α-androstan-3α,17β -diol(3α-
diol) (see 4 and 30 for references). Such conversions occur under
the influence of a 5α-reductase-3α-hydroxysteroid-dehydrogenase
system (Fig. 1). In the brain and in the anterior pituitary, testosterone may also be converted into 5α-androstan-3,17-dione,
after having been metabolized to Δ_4-androsten-3,17-dione (7)
(Fig. 1).

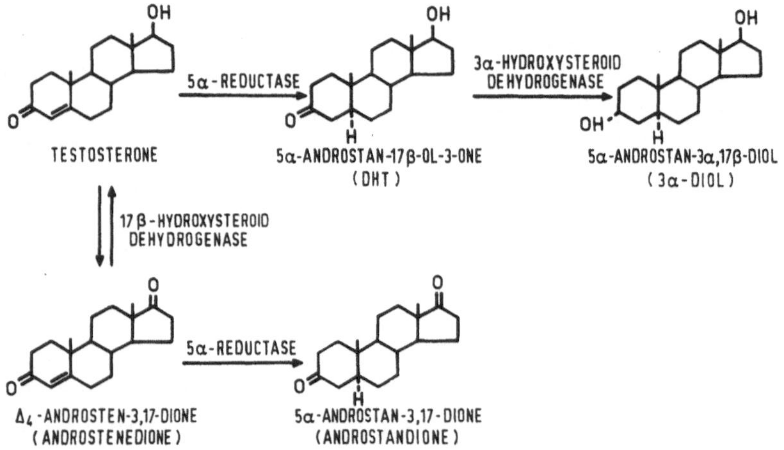

Fig. 1

Similar enzymatic activities have been found in the anterior pituitary and in the brain of female animals, where progesterone
is probably the most important physiological substrate. In vitro
and in vivo studies have demonstrated that, in these structures,
progesterone is converted into 5α-pregnan-3,20-dione (dihydroprogesterone, DHP) and 5α-pregnan-3α-ol-20-one (3α-ol) (Fig.2)
(27, 35, 60). It has also been shown that the neuroendocrine

tissues are able to convert progesterone into pregn-4-en-20α-ol-3-one (20α-P); this steroid may be subsequently converted into 5α-pregnan-20α-ol-3-one (20α-DHP) and 5α-pregnan-3α,20α-diol (3α-20α-DHP) (50, 51, 63).

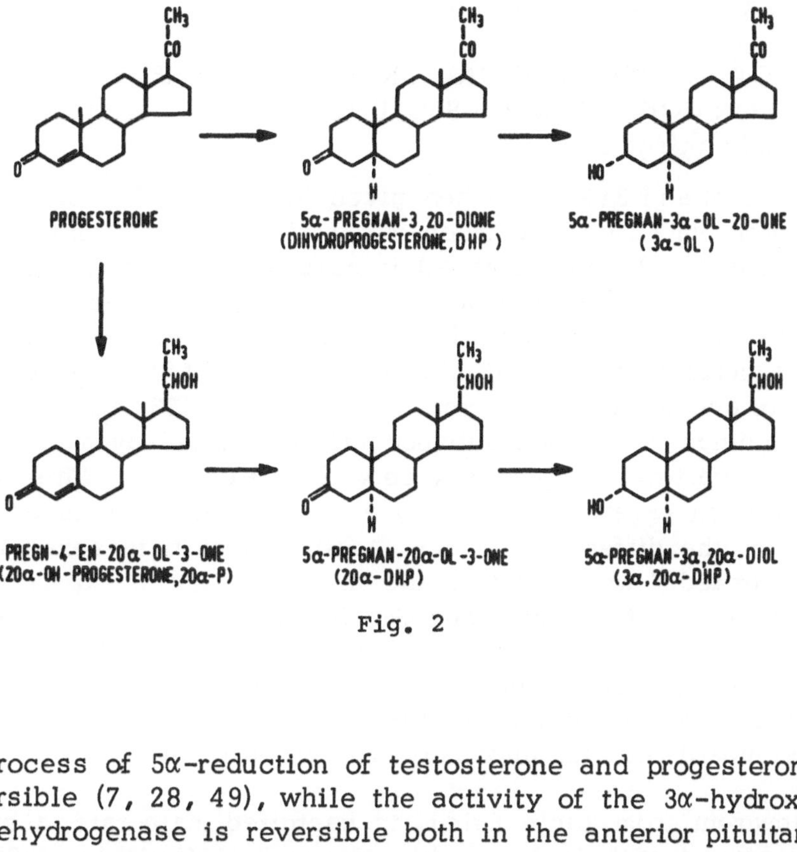

PROGESTERONE

5α-PREGNAN-3,20-DIONE
(DIHYDROPROGESTERONE,DHP)

5α-PREGNAN-3α-OL-20-ONE
(3α-OL)

PREGN-4-EN-20α-OL-3-ONE
(20α-OH-PROGESTERONE,20α-P)

5α-PREGNAN-20α-OL-3-ONE
(20α-DHP)

5α-PREGNAN-3α,20α-DIOL
(3α,20α-DHP)

Fig. 2

The process of 5α-reduction of testosterone and progesterone is irreversible (7, 28, 49), while the activity of the 3α-hydroxyste-roid-dehydrogenase is reversible both in the anterior pituitary and in the hypothalamus (7, 49, 64). For information on the fac-tors which control the 5α-reductase-3α-hydroxysteroid-dehydro-genase system the reader is referred to the reviews of Celotti et al. (4) and of Martini et al. (34).

III. METABOLITES OF TESTOSTERONE AND OF PROGESTERONE PRESENT IN THE GENERAL CIRCULATION.

DHT and 3α-diol have been found in the peripheral blood of the male rat (6, 40), which also contains significant amounts of 5α-androstan-3β,17β-diol (3β-diol) (6), a 5α-reduced metabolite of testosterone, which is not usually formed by the neuroendocrine structures (see 4, for references). Similarly the blood of the systemic circulation of the female rat has been shown to contain, in addition to progesterone, DHP (21, 22), 3α-ol (18, 19, 21, 22), 20α-P (1,11, 16, 21, 22, 53, 56), 20-DHP (18, 19, 21, 22) and 3α-20α-DHP (18, 19, 21, 22). Like in the male, the 3β –isomer of 3α-ol, 5α-pregnan-3β-ol-20-one (3β -ol) is also present in the general circulation of the female rat (18, 19).

IV. EFFECTS OF TESTOSTERONE AND OF ITS 5α-REDUCED META-BOLITES ON GONADOTROPIN SECRETION.
A. Systemic injections.

DHT, 3α-diol and 3β -diol, when given in one single dose 24 hours before sacrifice, are much more effective than testosterone in suppressing LH release in adult castrated male rats. Among the steroids tested, 3α-diol was by far the most effective. Under the same experimental conditions, testosterone and DHT inhibited to a certain extent, but did not suppress, FSH release, while 3α-diol and 3β -diol were totally ineffective (33, 66, 67). These findings have been subsequently confirmed by many authors in a variety of species (see 4, for references). This work has now been extended, utilizing a chronic schedule of administration of the different steroids. Adult rats of both sexes have been castrated 4 weeks before the initiation of the experiment. Testosterone, DHT, 3α-diol and 3β -diol (in the free alcohol form) have been administered subcutaneously in the daily dose of 2 mg/rat for 6 days. The animals have been sacrificed on the seventh day. Serum levels of FSH and LH have been quantitated using standard radioimmunoassay procedures (8, 46).

It is apparent from Table I that, in castrated male rats, testosterone significantly reduced the release of LH. DHT and 3α-diol were both much more effective than testosterone, bringing serum LH close to undetectable levels; 3β -diol was more effective than testosterone, but less potent than the other two 5α-reduced metabolites. In castrated females, testosterone proved to be much more active than in males in inhibiting LH release. DHT and 3α-diol were more effective than testosterone, although less potent than in males; 3β -diol gave a smaller inhibition of this gonadotropin. As shown in Table I, testosterone and DHT inhibited the release of FSH in castrated males, while 3α-diol and 3β -diol were ineffective. No one of the steroids tested proved able to affect the release of FSH in castrated females. The pre-

TABLE I - Effects of chronic treatment with testosterone, DHT, 3α-diol and 3β -diol on the release of LH and of FSH in castrated male and female rats.

Treatment and dose 2 mg/rat/day/6 days	LH(NIAMDD-RAT LH-S]7) ng/ml serum	FSH(NIAMDD-RAT FSH-RP-]) ng/ml serum
Castrated males		
Oil	(16)[a] 27,2+],9[b]	3974,20+63,31
T	(8) 14,2+2,8[x]	2680,50+]80,]0[xx]
DHT	(]0) 0,3+0,2[xx]	]090,00+728,]2[x]
3α-diol	(8) 0,]+0,][xx]	37]2,50+]47,44
3β -diol	(9) 6,7+],4[xx]	4037,00+376,30
Castrated females		
Oil	(34) 27,5+],0	29]9,]9+]74,]7
T	(7) 5,7+2,][xx]	2534,50+]63,6]
DHT	(]0) 4,]+0,8[xx]	3]80,]]+227,]0
3α-diol	(]2) 2,3+0,3[xx]	2600,50+]8],64
3β -diol	(]6)]4,0+2,][xx]	2876,70+23],40

a Number of rats in parentheses
b Mean + SE
x $P < 0,005$
xx $P < 0,00]$

sent results in males are in total agreement with those previously obtained using a single injection of the various steroids (33, 66, 67). The inhibition of LH secretion induced by the various steroids in castrated females followed the pattern observed in males, with some minor quantitative differences. The reason why in castrated female FSH should be totally unresponsive to the administration of testosterone and of its 5α-reduced metabolites remains to be elucidated.

B. Intrahypothalamic and intrapituitary implants.

This portion of the study was aimed at clarifying whether testosterone, DHT, 3α-diol and 3β -diol might influence LH and FSH secretion after having been directly implanted into the median

eminence region of the hypothalamus or into the anterior pituitary. The experiments were performed in adult castrated male rats.
The stereotaxic implantation of the different steroids was performed 3 weeks following castration, using a technique previously described (5). Sham-operated animals received a median eminence or an intrapituitary implant of the vehicle alone (saturated
sucrose solution). Blood was collected from the retroorbital plexus,
immediately before submitting the animals to the implantation procedure and 1, 3, 5, 7 and 9 days after implantation. Figure 3
shows that serum levels of LH were not modified throughout the
experiment in sham-implanted animals. Rats bearing a testosterone implant in the median eminence showed a significant reduction
of serum levels of LH only 1 and 3 days after implantation.

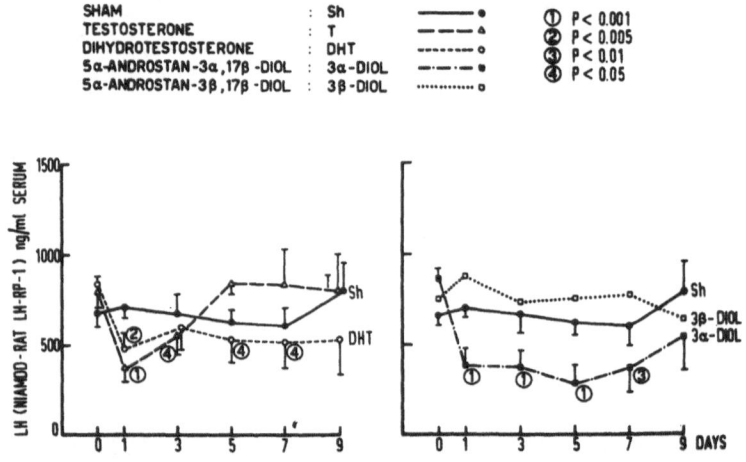

Fig. 3 Effect of median eminence implants of testosterone and of
its 5α-reduced metabolites on LH secretion in castrated ♂ rats.

Median eminence implants of DHT induced a more prolonged decrease of serum LH levels; these were significantly lower than
0 time values, 1, 5 and 7 days after implantation. These results
are in agreement with an observation of Smith and Davidson (59).
Median eminence implants of 3α-diol also exerted a prolonged
suppressive effect on LH release, serum levels of the hormone
being significantly lower than 0 time values 1, 3,5 and 7 days
following implantation. 3β-Diol was totally ineffective. Sham-

implantation did not alter serum FSH titers throughout the experiment (Fig. 4). The intrahypothalamic placement of either testosterone or DHT increased the serum levels of this gonadotropin above control values. However, the pattern of release of FSH

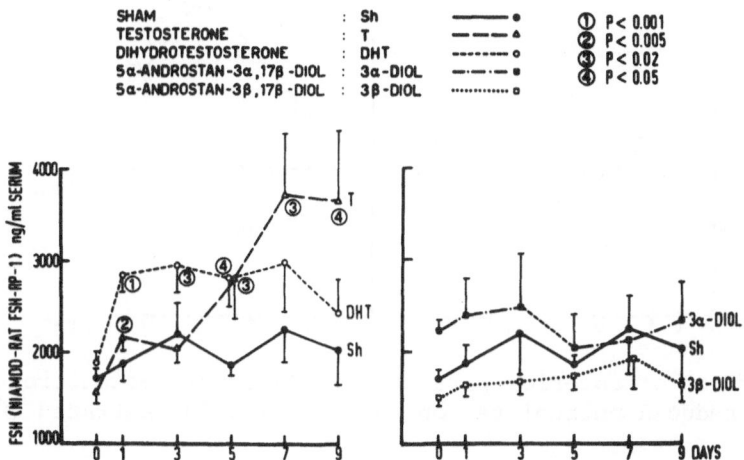

Fig 4   Effect of median eminence implants of testosterone and of its 5α-reduced metabolites on FSH secretion in castrated ♂ rats.

following median eminence implants of the two steroids was different, the increase being rapid but less pronounced in the case of DHT, and retarded but greater in the case of testosterone. The two diols were totally ineffective.

Figure 5 indicates that the intrapituitary implantation of cannulae containing either testosterone or DHT did not change serum LH levels. The intrapituitary placement of 3α-diol induced a significant and long-lasting increase in serum LH levels; on the contrary, 3β -diol significantly depressed serum LH for 7 days. Intrapituitary implants of testosterone resulted in a significant increase of serum FSH, which lasted for 9 days (Fig. 6). An increase of serum FSH was also observed following intrapituitary placement of 3α-diol, but the levels attained were significantly different from controls only at the last time of observation. A minor stimulating effect on FSH secretion was observed after the intrapituitary placement of 3β -diol. DHT implants

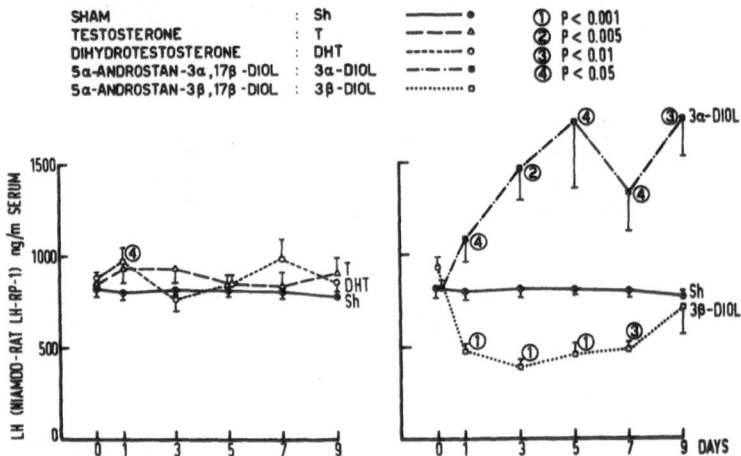

Fig. 5 Effect of anterior pituitary implants of testosterone and of its 5α-reduced metabolites on LH secretion in castrated ♂ rats.

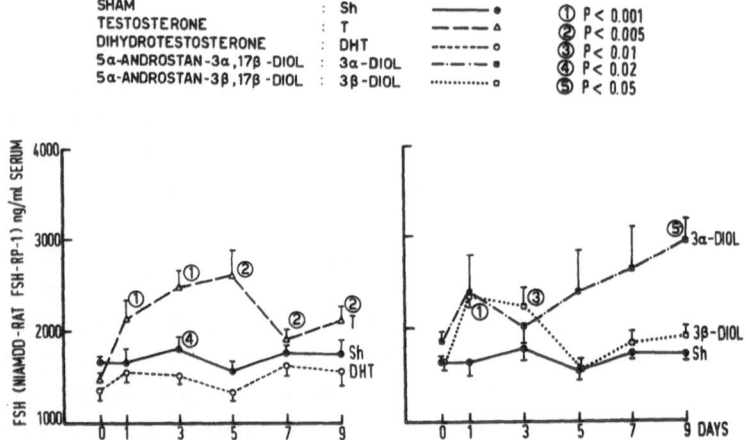

Fig. 6 Effect of anterior pituitary implants of testosterone and of its 5α-reduced metabolites on FSH secretion in castrated ♂ rats.

did not change serum FSH levels throughout the experiment.

The pituitary responsiveness to LH–RH was studied in the groups of animals bearing median eminence implants of the various steroids 1, 3 and 9 days following implantation. Blood was collected for the evaluation of serum LH levels before and 10 minutes after the intrajugular injection of 10 ng of the hypothalamic hormone. It is apparent from Figure 7 that, in general, the sensitivity to LH–RH was reduced in all groups of steroid-implanted animals. However, the decrease of the response was not always significant.

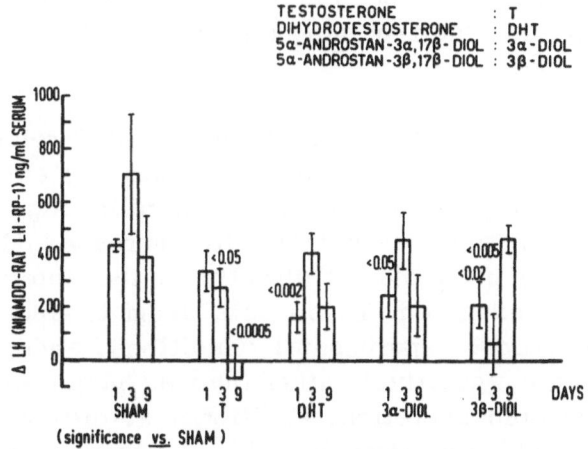

Fig. 7 Effect of intrajugular injections of LH–RH (10 ng/rat) on serum levels of castrated ♂ rats bearing median eminence implants of testosterone and of its 5 -reduced metabolites.

The effects on LH release of the intrahypothalamic implants of testosterone, DHT and 3α-diol are comparable to those obtained after the systemic (acute or chronic) administration of the same steroids. The data seem to suggest that the suppressive effect of these steroids on LH secretion occurs at hypothalamic level. This hypothesis is supported by the observation that no one of these steroids was able to reduce LH secretion following direct placement into the pituitary gland. Actually, 3α-diol proved able to stimulate, rather than to inhibit, LH release when directly implanted in the anterior pituitary. However, the results obtained challenging the anterior pituitary of the intrahypothalamic implan-

ted animals with LH-RH demonstrate that the gland was less
responsive to the stimulating effect of the hypothalamic hormone.
This might suggest that testosterone and its 5α-reduced metabolites might exert, at least in part, their inhibitory effect on LH
release acting directly on the pituitary, possibly after having
been transported to the gland via the portal vessels according
to Bogdanove's hypothesis (2). In the authors' opinion, however,
the decreased sensitivity of the anterior pituitary to LH-RH may
be explained also in a different manner. It is possible that the
presence of the various steroids in the median eminence has altered the secretory function of the hypothalamus, and consequently courtailed the physiological supply of LH-RH to the anterior
pituitary; it is known that the gland is less responsive to exogenous LH-RH when the "self-priming effect" of LH-RH is abolished
(see 31, for references).

The findings: a) that testosterone stimulates FSH release following implantation into the hypothalamus or into the anterior pituitary; b) that DHT enhances FSH secretion following intrahypothalamic placements; and c) that 3α-diol increases LH and FSH
release after having been placed directly in the anterior pituitary were totally unexpected. These data seem to underline the
fact that, in the proper experimental conditions, androgenic molecules may exert a "positive" rather than a "negative" feedback
effect on gonadotropin secretion. Additional studies are obviously
needed to clarify why the same steroid may exert opposite effects
depending on the experimental conditions selected. However, it
is pertinent to recall that some data in the literature support the
possibility that androgens may exert a "positive" feedback effect
on gonadotropin release. Testosterone has been reported to induce cyclic increases of FSH release in castrated female rats (23,
44). The systemic administration of small doses of testosterone
or of DHT is followed by a rise of serum FSH levels in castrated
male and female rats (13, 26, 61). The "in vitro" addition of
either testosterone or DHT to culture media containing rat anterior pituitary tissue results in a stimulation of FSH output
(38, 39, 57). DHT increases the release of FSH induced by LH-RH in
a similar system (10). The same phenomenon occurs also "in
vivo", since pretreatment with either testosterone or DHT enhances the ability of LH-RH to release FSH in normal as well
as in castrated male rats (9, 68). Finally, 3α-diol has been
shown to potentiate LH-RH induced LH release in dogs (24).
The possibility that the "positive" feedback activity of andro-

gens on gonadotropin secretion (and especially on FSH release) might be of physiological importance is shown by the following observations. The administration of an antiserum acting against testosterone blocks the ovulatory response to HCG in intact immature rats; the blockade may be overcome by the simultaneous administration of either testosterone or DHT (41). The same antiserum prevents the increase in serum FSH which normally occurs during the early morning hours of estrus in cycling female rats (14).

V. EFFECTS OF TESTOSTERONE AND OF ITS 5α-REDUCED METABOLITES ON PROLACTIN SECRETION.

It has long been known that testosterone may increase prolactin secretion (36, 37, 58). It is not clear whether this stimulatory effect is due to testosterone as such, or to its aromatization into estrogenic molecules which may occur either in the neuroendocrine structures or in the periphery. It is well established that estrogen can stimulate prolactin secretion (45). In order to clarify the mode of action of testosterone on prolactin secretion, the hormone and its non-aromatizable 5α-reduced metabolites, DHT, 3α-diol and 3β-diol, have been administered for six days in the dose of 2 mg/rat to long-term castrated female rats. Prolactin has been measured using a specific radioimmunoassay (47). It is obvious from Figure 8

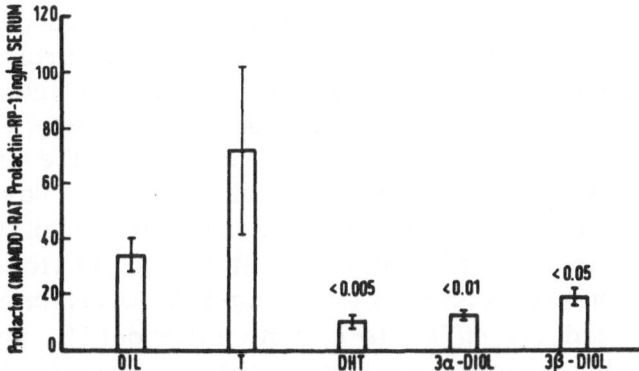

Fig. 8 Effect of 6-day treatment with testosterone (T), dihydrotesterone (DHT), 5α-androstan-3α, 17β-diol (3α-diol) and 5α-androstan-3β, 17β-diol (3β-diol) (2mg/rat/day) on serum prolactin levels of long term castrated (4 weeks) adult ♀ rats.

that, in agreement with previous findings, testosterone induced
an increase of serum levels of prolactin. On the contrary, DHT,
3α-diol and 3β-diol very significantly depressed prolactin release.
The fact that the three non aromatizable androgens inhibit prolactin release indicates that androgens per se block prolactin
secretion and provide indirect support to the hypothesis that the
stimulatory effect of testosterone on prolactin secretion results
from the conversion of the hormone into estrogenic molecules.
The suppressive effect observed with DHT agrees with a similar
finding of Nolin et al. (48) in intact female rats; however, these
authors did not find an inhibitory activity of DHT in castrated
females. The discrepancy is probably explained by differences
in the doses used. Nolin et al. (48) utilized subcutaneous cap-
sules of silastic delivering DHT in amounts lower than those
used in the present study.

VI. EFFECTS OF PROGESTERONE, 20α-P AND THE RESPECTIVE 5α-REDUCED METABOLITES ON GONADOTROPIN SECRETION.

A. Effect of 20α-P and of 3α,20α-DHP.

In a previous paper it has been shown that DHP, 3α-ol and 3β-
ol exert a stimulatory effect on the release of LH and FSH in
ovariectomized estrogen-primed rats (65). These results have
been confirmed by others (3, 12, 52), using different doses and
schedules of administration of estrogens and of the various progestagens.

The present investigation has been performed: a) to gain additional information on the effect of 20α-P in the control of gonadotropin release (15, 17, 62); and   b) to analyze whether its
5α-reduced metabolite   3α,20α-DHP might exert any effect on
gonadotropin secretion. To the authors' knowledge the effects of
this physiological steroid on gonadotropin secretion have not
been studied so far. The schedule of administration of the various steroids has been identical to that of the previous work
(65). It is clear from Figure 9 that long-term ovariectomized
rats have elevated serum levels of both LH and FSH, and that
these can be significantly reduced by the subchronic administration of estrogen (0.4 µg/rat for 5 days). In line with previous
observations (62, 65), progesterone, administered on the morning
of the 5th day of estrogenic treatment, induces a surge of LH
in the afternoon of the same day. A stimulatory effect of comparable magnitude is also observed following the administration of

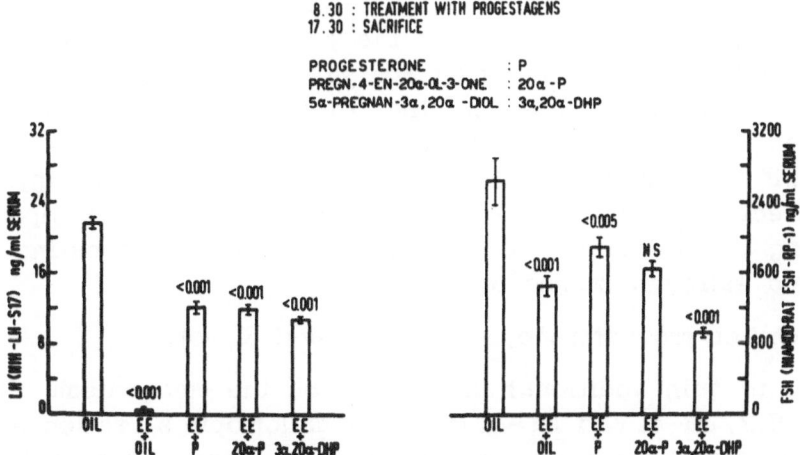

Fig. 9 Effect of treatment with 100 μg/rat of progesterone and of its metabolites on serum levels of LH and FSH of castrated ♀ rats treated for 5 days with 0.4 μg of ethinyl estradiol (EE)

either 20α-P or 3α, 20α-DHP. Progesterone, when given to estrogen-primed animals, significantly increased serum levels of FSH. This is similar to a previous report of this laboratory (65), but conflicts with a finding of Swerdloff et al. (62). FSH levels were not modified by 20α-P in castrated estrogen-primed rats, while 3α, 20α-DHP brought about a significant reduction of the release of this gonadotropin. The present data confirm that, in ovariectomized estrogen-primed rats, 20α-P may exert a strong activatory effect on LH secretion (62). In contrast with the finding of Swerdloff et al. (62), in the present series of experiments 20α-P did not induce a rise of serum FSH. The data have shown in addition that 3α, 20α-DHP may exert a stimulatory action on LH release which is similar to that of either progesterone or 20α-P. These results strongly suggest that 20α-P should be definitely included in the list of the physiological progestagens which participate in the activation of the hypothalamic-pituitary complex at the time of the proestrous surge of LH. Data are available which indicate that serum levels of 20α-P increase at proestrous prior to the LH peak (1, 11, 16, 53, 56). The data also show that 3α, 20α-DHP may play a role in such an activation.

The finding that 3α, 20α-DHP inhibits FSH secretion in castrated estrogen-primed animals, deserves a short comment. First of all, the differential response of the two gonadotropins to this progestin supports the concept that the physiological mechanisms governing LH and FSH secretion are substantially different. Secondly, it must be pointed out that this is the first demonstration that a 5α-reduced metabolite of the progestational series may inhibit FSH release in estrogen-primed animals. A suppressive effect of DHP and 3α-ol on LH secretion has been reported following two injections of large doses of these progestins into castrated estrogen-primed rats (52).

B. Effects of progesterone, DHP, 3α-ol and 3β-ol.

In order to gain additional information on the mechanisms through which DHP, 3α-ol and 3β-ol modify gonadotropin secretion (65), these steroids have been administered subcutaneously to castrated female rats, in which Ethynil Estradiol (EE) had been implanted 5 days earlier in the median eminence. Serum levels of LH and FSH have been measured in the morning of the 6th postimplantation day, 17 hours after the subcutaneous administration of the progestational steroids. The pituitary responsiveness to LH-RH of the various groups of animals was tested, subdividing, at the end of the experiment, each group of treated animals into 2 sub-groups, one receiving intravenous injections of saline, the other 10 ng of the hypothalamic hormone. Animals were killed 20 minutes following injection. To the authors' knowledge, the combination of implantation of estrogen in the median eminence and of the systemic administration of progestational derivatives has never been used in endocrinological studies. Lisk and Barfield (29) have used a similar approach in behavioral experiments. Figure 10 shows that, in castrated female rats, the implantation of EE into the median eminence is followed by a dramatic decrease of serum LH levels. The administration of either progesterone or DHP to EE-implanted animals brings about a very significant increase of serum LH levels. The increase induced by DHP is significantly higher than that induced by progesterone. Neither 3α-ol nor 3β-ol was able to increase serum levels of LH in the EE-implanted animals. The implantation of EE induced also a significant decrease in serum levels of FSH (Fig. 10). The subsequent administration of progesterone did not increase serum FSH. On the contrary, DHP, 3α-ol and 3β-ol were all able to enhance serum levels of FSH.

The dose of LH-RH selected was unable to significantly increa-

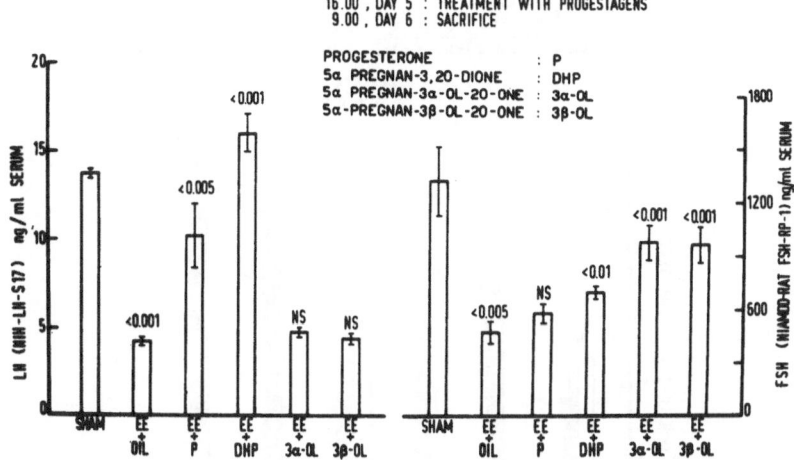

Fig. 10 Effect of treatment with 100 μg/rat of progesterone and of its metabolites on serum levels of LH and FSH of castrated ♀ rats implanted in the median eminence with estradiol (EE) 5 days earlier.

se the levels of LH in sham-implanted animals (Fig. 11). This result is not surprising, since it is known that the anterior pituitary is poorly responsive to LH-RH in the total absence of estrogens (see 31, for references). As expected, LH-RH was highly effective in increasing serum LH in animals bearing median eminence implants of EE. Since, it is generally belived that estrogens facilitate LH-RH responses at pituitary levels (31), the data might indicate that some estrogens have been able to reach the anterior pituitary either via systemic or pituitary portal vessels transport. LH-RH was also effective in the two groups of animals implanted with EE and given progesterone or DHP. However, the increase in serum levels of LH recorded in these two groups of animals was not as large as that observed in animals bearing the implant of EE and receiving oil subcutaneously. Moreover, the response in the animals given DHP was smaller than that observed in the animals treated with progesterone. The two groups of animals bearing median eminence implants of EE and given systemically either 3α-ol or 3β -ol exhibited a response to LH-RH which was much larger than that observed in the animals having only the EE implants in the median eminence. The data confirm once more (with minor differences probably due to methodological reasons) that the 5α-reduced metabolites of progesterone may exert a stimulatory

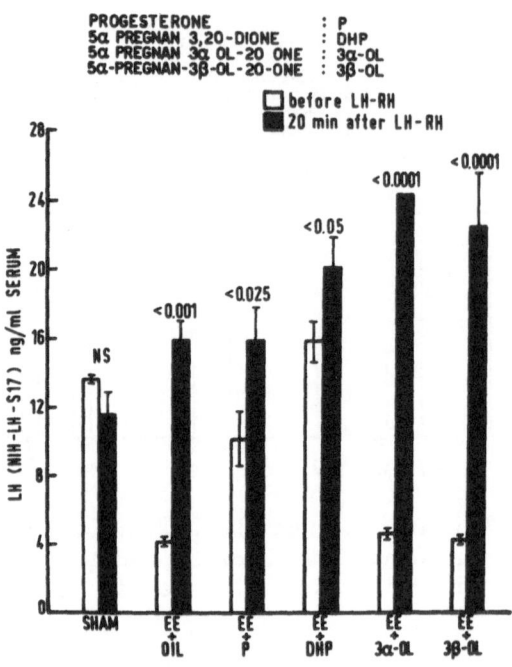

PROGESTERONE   : P
5α PREGNAN 3,20-DIONE   : DHP
5α PREGNAN 3α OL-20 ONE   : 3α-OL
5α-PREGNAN-3β-OL-20-ONE   : 3β-OL

Fig. 11 Effect of intrajugular injections of LH-RH (10 ng/rat)
on serum levels of LH of castrated ♀ rats treated with 100 µg/rat
of progesterone and of its metabolites and implanted in the median
eminence with ethinyl estradiol (EE) 5 days earlier

effect on gonadotropin release. In particular, the data have shown
that there are progestins that increase LH output (progesterone,
DHP) without affecting (progesterone) or having only minor effects
(DHP) on FSH, and others (3α-ol and 3β -ol) which are potent
FSH releasers and which do not exert a stimulatory effect on LH.
This points once more to the existence of separate control me-
chanisms for the release of the two gonadotropins.

The most important finding in the series of experiments with
LH-RH is the observation that the two steroids which induced
a "spontaneous" release of LH (progesterone and DHP) depressed
LH-RH-induced LH release, while the two steroids which did
not facilitate "spontaneous" LH release (3α-ol and 3β -ol) were

able to potentiate the effects of LH-RH. It appears then that the "spontaneous" liberation of the gonadotropins induced by the various progestagens, and the responsiveness of the anterior pituitary to exogenous LH-RH are independent phenomena. Two hypotheses will be put forward to try to explain this dichotomy. Obviously, they are not mutually exclusive. The first one takes into consideration the possibility that the major site of action of all the progestagens studied resides in the anterior pituitary.
It might be suggested that a phase of "hyperresponsiveness" to exogenous LH-RH preceeds the "spontaneous" liberation of gonadotropins, and that after such a phenomenon has occurred, a "refractory" period ensues. Were this interpretation correct, it would seem that animals which had "spontaneous" basal levels of gonadotropins but which overresponded to LH-RH were studied during the early phase of hyperresponsiveness, and that, conversely, the animals which had high "spontaneous" levels of gonadotropins and a low response to LH-RH were tested during the "refractory" period which follows the "spontaneous" release. This hypothesis cannot be either proved or disproved by the pr sent data because, for obvious technical reasons, only one sample of blood was collected in each group of animals. This hypothesis is presently under verification in additional studies. The second hypothesis must consider the possibility that the "spontaneous" release of LH and of FSH induced by the steroids which increase gonadotropin secretion is brought about by an action which occurs in the CNS rather than in the anterior pituitary and that the phenomenon of "hyper"-or "hyporesponsiveness" of the anterior pituitary does not play a major role in their mode of action. A CNS site of action of progesterone is suggested by the finding of Kalra and McCann (25), who have reported that progesterone is able to elevate LH and FSH in castrated estrogen-primed females following implantation into the preoptic-anterior hypothalamic area; according to their data, progesterone may also increase FSH (but not LH) secretion when placed in the amygdaloid nuclei. Moreover, progesterone receptors have been identified in the hypothalamus (see 32, for references). It is not known yet whether 3α-ol and 3β-ol may exert a stimulatory action on gonadotropin release after having been placed in the CNS; experiments of this type will certainly help in clarifying the mode of action of these steroids.

It emerges from the data that the administration of either 3α-ol or 3β-ol may be of clinical usefulness for facilitating the responsiveness of the anterior pituitary in cases of reduced sensitivity of the gland to LH-RH.

SUMMARY AND CONCLUSIONS.

The data presented indicate that several 5α-reduced metabolites of testosterone and of progesterone exert profound influences on the release of LH, FSH and prolactin.
1) Following systemic administration, DHT, 3α-diol and 3β -diol are more effective than testosterone in suppressing LH release in both sexes. Following intrahypothalamic implantation, DHT and 3α-diol block LH release in a more prolonged fashion than testosterone. Testosterone, after systemic administration, is more effective than its 5α-reduced metabolites in inhibiting FSH release in males.
2) In particular experimental conditions, androgens may exert a stimulatory rather than an inhibitory effect on gonadotropin secretion.
3) Prolacin secretion is stimulated by testosterone, but is suppressed by the non-aromatizable 5α-reduced metabolites of the hormone.
4) The 5α-reduced metabolites of progesterone and of 20α-P increase the release of gonadotropins following systemic administration to ovariectomized rats systemically primed with estradiol, or bearing median eminence implants of ethinyl estradiol.
5) The responsiveness of the anterior pituitary to LH-RH is reduced by intrahypothalamic implants of the 5α-reduced metabolites of testosterone. Progestins modify LH-RH responses in opposite directions (facilitation: 3α-ol, 3β -ol; depression: progesterone, DHP), when given systemically to castrated females bearing intrahypothalamic implants of ethinyl estradiol. The efficacy of 3α-ol and 3β -ol in potentiating LH-RH responses may be of clinical value for improving the present clinical LH-RH tests.
6) The reactivity of the anterior pituitary to LH-RH in the presence of the various progestins does not parallel the "spontaneous" effect of each single steroid on gonadotropin release.
7) Testosterone and progesterone metabolites which are not formed in the brain and in the anterior pituitary (3β -diol and 3β -ol), but which are found in the blood respectively of male and female animals, exert important feedback activities.
8) In several instances, the release of the two gonadotropins is modified in divergent directions by the various metabolites of testosterone and of progesterone, suggesting once more that the

feedback mechanisms which control the two gonadotropins are substantially different.

9) The data here reported do not provide a definite answer to the
question whether testosterone and progesterone must be converted
into the corresponding 5α-reduced metabolites in the neuroendocrine tissues in order to exert their feedback effects on gonadotropin secretion. However, they provide some evidence in favor
of such a theory. The validity of this hypothesis may only be
proved by studies in which the process of 5α-reduction and of
3α-hydroxysteroid-dehydrogenation are pharmacologically blocked.
Unfortunately, specific blockers of the 5α-reductase system are
not available so far.

10) Sex steroid feedback mechanisms are much more complicated
than anticipated from previous studies. 5α-Reduced androgens and
5α-reduced progesterone derivatives are not only formed by the
neuroendocrine tissues, but are also present in the general circulation. Consequently, the brain and the anterior pituitary are
constantly exposed to a mixture of androgens and of progestin
derivatives rather than to single hormones. Moreover the neuroendocrine tissues actively participate in the feedback mechanisms by transforming steroids present in the general circulation
into metabolites which possess different endocrine profiles.

ACKNOWLEDGMENTS.

The experiments done in the authors' laboratory and described
in this paper have been supported by grants of the Ford Found-
ation, New York, N.Y., and of the Consiglio Nazionale delle
Ricerche (through the project "Biology of Reproduction"), Roma,
Italy.
All such support is here gratefully acknowledged. Materials for
LH and FSH radioimmunoassay have been kindly provided by Drs.
A.R. Midgley, Jr., L.E. Reichert, Jr., G.D. Niswender and by
the Rat Pituitary Hormone Distribution Program of the National
Institutes of Arthritis, Metabolism and Digestive Diseases of the
National Institutes of Health, Bethesda, Md. Thanks are also
due to Mrs. Paola Assi Brunone and to Mr. Luigi Guadagni for
their skilful technical assistance.

REFERENCES.

1. Barraglough, C.A., R. Collu, R. Massa and L. Martini. Endocrinology 88 : 1437-1447, 1971.
2. Bogdanove, E.M. Vitam. Horm. 22 : 205-260, 1964.
3. Brown-Grant, K. J. Endocr. 62 : 319-332, 1974.
4. Celotti, F., R. Massa and L. Martini. In : Endocrinology : Metabolic Basis of Clinical Practice, edited by L.J. De Groot, New York, Grune and Stratton, pp. 41-53, 1978.
5. Corbin, A., G. Mangili, M. Motta and L. Martini. Endocrinology 76 : 811-818, 1965.
6. Corpéchot, C., B. Eychenne and P. Robel. Steroids 29 : 503-516, 1977.
7. Cresti, L. and R. Massa. Program 5th Intern. Congr. Endocrinology, p. 46, 1976.
8. Daane, T.A. and A.F. Parlow. Endocrinology 88 : 653-663, 1971.
9. Debeljuk, L., J.A. Vilchez-Martininez, A. Arimura and A.V. Schally. Endocrinology 94 : 1519-1524, 1974.
10. Drouin, J. and F. Labrie. Endocrinology 98 : 1528-1534, 1976.
11. Feder, H.H., J.A. Resko and R.W. Goy. J. Endocr. 41 : 563-569, 1968.
12. Fink, G. and S.R. Henderson. J. Endocr. 73 : 157-164, 1977.
13. Gay, V.L. Program 55th Meet. Endocrine Soc., p. 116, 1973.
14. Gay, V.L. and R.L. Tomacari. Science 184 : 75-76, 1974.
15. Goodman, A.L. and J. Neill. Endocrinology 99 : 852-860, 1976.
16. Hashimoto, I., D.M. Henricks, L.L. Anderson and R.M. Melampy. Endocrinology 82 : 333-341, 1968.
17. Hilliard, J., R. Penardi and C.H. Sawyer. Endocrinology 80 : 901-909, 1967.
18. Holzbauer, M. Br. J. Pharmacol. 43 : 560-569, 1971.
19. Holzbauer, M. J. Steroid Biochem. 6 : 1307-1310, 1975.
20. Hyyppa, M., M. Motta and L. Martini. Neuroendocrinology 7 : 227-235, 1971.
21. Ichikawa, S., H. Morioka and T. Sawada. Endocrinology 88 : 372-383, 1971.
22. Ichikawa, S., T. Sawada, Y. Nakamura and H. Morioka. Endocrinology 94 : 1615-1620, 1974.
23. Johnson, D.C. and R.H. Naqvi. Endocrinology 85 : 881-885, 1969.
24. Jones, G.E. and A.R. Boyns. J. Endocr. 61 : 123-131, 1974.
25. Kalra, P.S. and S.M. McCann. Neuroendocrinology 19 : 289-302, 1975.
26. Kalra, P.S., C.P. Fawcett, L. Krulich and S.M. McCann. Endocrinology 92 : 1256-1268, 1973.

27. Karavolas, H.J. and K.M. Nuti. In : Subcellular Mechanisms in Reproductive Neuroendocrinology, edited by F. Naftolin, K. J. Ryan and I.J. Davies, Amsterdam, Elsevier. pp. 305-326, 1976.

28. Karavolas, H.J., D. Hodges and D. O'Brien. Endocrinology 98 : 164-175, 1976.

29. Lisk, R.D. and M.A. Barfield. In : Anatomical Neuroendocrinology, edited by W.E. Stumpf and L.D. Grant, Basel, Karger. pp. 232-244, 1975.

30. Martini, L. In : Subcellular Mechanisms in Reproductive Neuroendocrinology, edited by F. Naftolin, K.J. Ryan and I.J. Davies, Amsterdam, Elsevier. pp. 327-341, 1976.

31. Martini, L. In: Gynecology and Obstetrics, edited by C. Mac Gregor, Amsterdam, Excerpta Medica. pp. 3-25, 1978.

32. Martini, L. In : The Pharmacology of the Hypothalamus, edited by B. Cox, I.D. Morris and A.H. Weston, London, MacMillan. in press, 1978.

33. Martini, L., R. Massa, M. Motta and M. Zanisi. In : Male Fertility and Sterility, edited by R.E. Mancini and L. Martini, London, Academic Press. pp. 359-388, 1974.

34. Martini, L., F. Celotti, R. Massa and M. Motta. J. Steroid Biochem. in press, 1978.

35. Massa, R., E. Stupnicka and L. Martini. Program 4th Intern. Congr. Endocrinology, p. 118, 1972.

36. Meites, J. Recent Progr. Hormone Res. 28 : 471-526, 1972.

37. Meites, J. Vitam. Horm. 30 : 165-221, 1972.

38. Mittler, J.C. Proc. Soc. Exp. Biol. Med. 140 : 1140-1142,1972.

39. Mittler, J.C. Neuroendocrinology 16 : 265-272, 1974.

40. Moger, W.H. Endocrinology 100 : 1027-1032, 1977.

41. Mori, T., A. Suzuki, T. Nishimura and A. Kambegawa. Endocrinology 101 : 623-626, 1977.

42. Motta, M., F. Fraschini and L. Martini. In : Frontiers in Neuroendocrinology, edited by W.F. Ganong and L. Martini, New York, Oxford University Press. Vol. 1, pp. 211-253, 1969.

43. Naftolin, F., K.J. Ryan, I.J. Davies, V.V. Reddy, F. Flores, Z. Petro and M. Kuhn. Recent Progr. Hormone Res. 31 : 295-319, 1975.

44. Naqvi, R.H. and D.C. Johnson. Endocrinology 87 : 418-421, 1970.

45. Neill, J.D. and M.S. Smith. In : Current Topics in Experimental Endocrinology, edited by V.H.T. James and L.

Martini, New York, Academic Press. Vol. 2, pp. 73-106, 1974.

46. Niswender, G.D., A.R. Midgley, Jr., S.E. Monroe and L.E. Reichert, Jr. Proc. Soc. Exp. Biol. Med. 128 : 807-811,1968.

47. Niswender, G.D., C.H.Chen, A.R. Midgley, Jr., J. Meites and S. Ellis. Proc. Soc. Exp. Biol. Med. 103 : 793-797,1969.

48. Nolin, J.M., G.T. Campbell, D.D. Nansel and E.M. Bogda- nove. Endocrine Res. Commun. 4 : 61-70, 1977.

49. Noma, K., B. Sato, S. Yano, Y. Yamamura and T. Seki. J. Steroid Biochem. 6 : 1261-1266, 1975.

50. Nowak, F.V. and H.J. Karavolas. Steroids 24 : 351-357, 1974.

51. Nowak, F.V., K.M. Nuti and H. J. Karavolas. Steroids 28 : 509-520, 1976.

52. Nuti, K.M. and H.J. Karavolas. Endocrinology 100 : 777-781, 1977.

53. Piacsek, B.E., T.C. Schneider and V.L. Gay. Endocrinology 89 : 39-45, 1971.

54. Piva, F., M. Motta and L. Martini. In : Endocrinology : Meta- bolic Basis of Clinical Practice, edited by L.J. De Groot, New York, Grune and Stratton. pp. 21-33, 1978.

55. Renaud, L.P., H.W. Blume and Q.J. Pittman. In : Frontiers in Neuroendocrinology, edited by W. F. Ganong and L. Martini, New York, Raven Press. Vol. 5, pp. 135-162,1978.

56. Rubin, B.L. and H.S. Chase, Jr. Steroids 18 : 59-67, 1971.

57. Schally, A.V., T.W. Redding and A. Arimura. Endocrinology 93 : 893-902, 1973.

58. Shin, S.H., R.B. Aiken, R. Roberts and C. Howitt. J. Endocr. 63 : 257-258, 1974.

59. Smith, E.R. and J.M. Davidson. Endocrinology 95 : 1566-1573, 1974.

60. Stupnicka, E., R. Massa, M. Zanisi and L. Martini. In : Cli- nical Reproductive Neuroendocrinology, edited by P.O. Hu- binont, M. L'Hermite and C. Robyn, Basel, Karger, pp. 88-95, 1977.

61. Swerdloff, R.S., P.C. Walsh and W.D. Odell. Steroids 20 : 13-22, 1972.

62. Swerdloff, R.S., H.S. Jacobs and W.D. Odell. Endocrinolo- gy 90: 1529-1536, 1972.

63. Tabei, T., H. Haga, W.L. Heinrichs and W.L. Herrmann. Steroids 23 : 651-666, 1974.

64. Thieulant, M.L., B. Ducouret, S. Samperez and P. Jouan. C.R. Acad. Sci. 282 : 231-234, 1976.

65. Zanisi, M. and L. Martini. J. Steroid Biochem. 6 : 1021-1023, 1975.
66. Zanisi, M.,M. Motta and L. Martini. J. Endocr. 56 : 315-316, 1973.
67. Zanisi, M. M. Motta and L. Martini. In : The Endocrine Function of the Human Testis, edited by V.H.T. James, M. Serio and L. Martini, New York, Academic Press, pp. 431-438, 1973.
68. Zanisi, M., M. Motta and L. Martini. In : Hypothalamic Hypophysiologic Hormones, edited by C. Gual and E. Rosemberg, Amsterdam, Excerpta Medica. pp. 24-32, 1973.

PITUITARY NEUROPEPTIDES AND BEHAVIOR

D. de Wied

Rudolf Magnus Institute for Pharmacology, Medical Faculty,

University of Utrecht, Vondellaan 6, Utrecht, Netherlands.

INTRODUCTION

The implication of ACTH, α-MSH and the neurohypophyseal hormones on adaptive behavior was first suggested by classical endocrine methods i.e. by the removal of the endocrine gland, in this case, the pituitary, the subsequent demonstration of a deficiency and finally the correction of the deficient function by treatment with hormones produced by the extirpated gland. Removal of the anterior lobe of the pituitary (De Wied, 1964) or the whole gland (Applezweig and Baudry, 1955; De Wied, 1969 ; Weiss et al., 1970) in rats impairs acquisition of shuttle box avoidance behavior. Removal of the posterior pituitary does not materially affect acquisition but interferes with the maintenance of shuttle box avoidance behavior (De Wied, 1965). These behavioral impairments could be corrected with ACTH but also with α-MSH and fragments of ACTH/MSH which are devoid of corticotropic effects, and with [Lys^8]vasopressin (LVP) and [des-Gly^9-Lys^8] vasopressin (DG-LVP) the latter being virtually devoid of the classical endocrine effects of vasopressin (De Wied, 1969; Lande et al., 1971; Bohus et al., 1973). These results suggested that the behavioral effects of pituitary hormones are not mediated through an influence on their endocrine target tissues. Pituitary hormones therefore may act as precursor molecules for neuropeptides involved in acquisition and maintenance of adaptive behavior.

EFFECTS OF ACTH AND RELATED PEPTIDES ON LEARNED BEHAVIOR

In intact rats, ACTH and related peptides delay extinction of shuttle-box avoidance behavior (De Wied, 1969), pole-jumping

297

avoidance behavior (De Wied, 1966), improve maze performance (Flood et al., 1976), facilitate passive avoidance behavior (Levine and Jones, 1965; Lissák and Bohus, 1972; Kastin et al., 1973 ; De Wied, 1974; Flood et al., 1976), delay extinction of food-motivated behavior in hungry rats (Leonard, 1969; Sandman et al., 1969; Gray, 1971; Garrud et al., 1974; Flood et al., 1976), delay extinction of conditioned taste aversion (Rigter and Popping, 1976) and sexually motivated approach behavior (Bohus et al., 1975). On the basis of these observations we postulated that ACTH and related peptides are involved in motivational processes. However, these peptides affect learning and memory processes as well. They alleviate amnesia produced by CO_2 inhalation or electroconvulsive shock (Rigter et al., 1974; Rigter and Van Riezen, 1975) or by intracerebral administration of the protein synthesis inhibitor puromycin (Flexner and Flexner, 1971) or anisomycin (Flood et al., 1976). Rigter et al. (1974) interpreted the effect of $ACTH_{4-10}$ on amnesia as an influence on retrieval processes. Gold and Van Buskirk (1976) found that posttrial injections of ACTH enhance or impair retention of passive avoidance behavior, depending on the dose of ACTH used. These authors suggested that ACTH modulates memory storage processing of recent information. ACTH also improves correct performance rewarded by water through better use of environmental cues without a general effect on learning (Isaacson et al., 1976). These findings do not refute a motivational hypothesis since motivational effects operate in most of the paradigms used. Observations in man suggest that $ACTH_{4-10}$ facilitates selective visual attention (Kastin et al., 1975) but evidence for a motivational influence has been obtained as well (Gaillard and Sanders, 1975).

STRUCTURE ACTIVITY STUDIES WITH ACTH FRAGMENTS

Structure activity studies to determine the essential elements required for the behavioral effect of ACTH revealed that not more than four amino acid residues are needed. Thus, $ACTH_{4-7}$ is as effective as the whole ACTH molecule in delaying extinction of pole-jumping avoidance behavior (Greven and De Wied, 1973; De Wied et al, 1975). The amino acid residue phenylalinine in position 7 plays a key role in this behavioral effect of ACTH. Replacement of this amino acid by the D-enantiomer in $ACTH_{1-10}$ (Bohus and De Wied, 1966) $ACTH_{4-10}$ or in $ACTH_{4-7}$ (Greven and De Wied, 1973) causes an effect on extinction of avoidance behavior opposite to that found with non-substituted ACTH fragments. Such [D-Phe^7]ACTH analogues facilitate extinction of active avoidance behavior or approach behavior motivated by food (Garrud et al., 1974). However, [D-Phe^7]$ACTH_{4-10}$ like $ACTH_{4-10}$ facilitates passive avoidance behavior when given prior to the retention test (De Wied, 1974) but in relatively high doses attenuates passive avoidance behavior when administered immediately following the learning trial (Flood et al., 1976) and delays extinction of conditioned taste aversion (Rigter and Popping, 1976).

Replacement of other amino acid residues in $[Lys^8]$ $ACTH_{4-9}$ does not facilitate extinction of avoidance behavior (Greven and De Wied, 1973). The behavioral activity of ACTH fragments can be completely dissociated from inherent endocrine, metabolic- and opiate-like activities by modification of the molecule. Substitution of Met^4 by methionine sulfoxide, $D-Lys^8$ by Arg and Trp^9 by Phe yields a peptide which is behaviorally a thousand times more active than $ACTH_{4-10}$ (Greven and De Wied, 1973). It possesses however a thousand times less MSH-activity and its steroidogenic action is markedly reduced. It has no fat mobilizing activity nor opiate-like effect as assessed in the guinea pig ileum preparation. The increased potency can be explained in part by protection against metabolic degradation (Witter et al., 1975). This again points to separate and specific receptor sites in the brain for the behavioral effects of ACTH and related peptides.

Although the sequence $ACTH_{4-7}$ is the smallest part of the molecule with full behavioral activity as determined on extinction of pole-jumping avoidance behavior (De Wied et al., 1975), more activity sites are present in ACTH since $ACTH_{7-10}$ and $ACTH_{11-24}$ also contain some activity (table 1). Eberle and Schwyzer (1975) also found a second activity site (region 11-13) in α-MSH. The residual potency of $ACTH_{7-10}$ can be increased to the same level as that of $ACTH_{4-10}$ by extending the carboxyl terminus with $ACTH_{11-16}$ to $ACTH_{7-16}$. Thus, the essential elements for avoidance behavior are not exclusively located in the region $ACTH_{4-7}$ but also in other areas of the molecule. In these areas the information may be present in a latent form which is potentiated by chain elongation.

The tripeptide H-Phe-D-Lys-Phe-OH which is the major break down product of the highly potent hexapeptide (Org 2766) has only minor behavioral effects (De Wied et al., 1975). Again, chain elongation with $ACTH_{10-16}$ restores the potency. Substitution of Lys^{11} by the D-enantiomer further augments the effect on avoidance behavior and extention of the NH_2 terminus with $Met(0)^4-Glu^5-His^6$ further potentiates the action to yield a peptide which is three hundred thousand times more active than $ACTH_{4-10}$ (Greven and De Wied, 1977). The potency of this peptide depends on the residues Gly^{10} and Lys^{16}. Removal of either of these amino acid markedly reduces the effect. This suggests that the requirements for activity in the avoidance test are more related to the structure of ACTH than to that of α-MSH or β-LPH.

Table 1

AMINO ACID SEQUENCES OF A NUMBER OF ACTH ANALOGUES

	1	2	3	4	5	6	7	8	9	10	11	12	13	14	15	16	potency ratio
α-MSH	AC–Ser	Tyr	Ser	Met	Glu	His	Phe	Arg	Trp	Gly	Lys	Pro	Val–OH				1
ACTH$_{4-10}$				H–Met	Glu	His	Phe	Arg	Trp	Gly–OH							1
							H–Phe	Arg	Trp	Gly–OH							0.1
ACTH$_{7-16}$				$\stackrel{0}{\uparrow}$			H–Phe	Arg	Trp	Gly	Lys	Pro	Val	Gly	Lys	Lys–NH$_2$	1
Org 2766				$\stackrel{0}{\uparrow}$ H–Met	Glu	His	^{D}Phe	Lys	Phe–OH								1000
							H–^{D}Phe	Lys	Phe–OH								0.1
							H–^{D}Phe	Lys	Phe	Gly	Lys	Pro	Val	Gly	Lys	Lys–NH$_2$	100
							H–^{D}Phe	Lys	Phe	–––	Lys	Pro	Val	Gly	Lys	Lys–NH$_2$	0.3
							H–^{D}Phe	Lys	Phe	Gly	Lys	Pro	Val	Gly	Lys–NH$_2$		1
							H–^{D}Phe	Lys	Phe	Gly	^{D}Lys	Pro	Val	Gly	Lys	Lys–NH$_2$	10000
				$\stackrel{0}{\uparrow}$ H–Met	Glu	His	^{D}Phe	Lys	Phe	Gly	^{D}Lys	Pro	Val	Gly	Lys	Lys–NH$_2$	100000
				$\stackrel{0}{\uparrow}$ H–Met	Glu	His	^{D}Phe	Lys	Phe	Gly	^{D}Lys	Pro	Val	Gly	Lys	Lys–NH$_2$	300000

EFFECTS OF PEPTIDES RELATED TO β-LPH ON LEARNED BEHAVIOR

Several years ago, a number of peptides which possessed similar behavior effects as ACTH and related peptides as assessed by inhibition of extinction of active, and facilitation of passive avoidance behavior, were isolated from hog pituitary material (Lande et al., 1973). These studies suggested the presence of potent behavioral peptides in the pituitary. One of these peptides which was obtained in pure form, yielded three oligopeptides after tryptic digestion. The amino acid composition of two of these appeared to be similar to those of $\beta\text{-LPH}_{61-69}$ and $\beta\text{-LPH}_{70-79}$ but the amount of material available at the time was insufficient to perform structure analysis studies. Since these peptides became available it was deemed of interest to verify our previous findings on the behavioral effect of peptides related to C-terminal β-LPH fragments.

Extinction of pole-jumping avoidance behavior was used to assay the behavioral effect of Met-enkephalin, $\beta\text{-LPH}_{61-69}$, α-endorphin, β-endorphin and a number of related peptides (table 2). Following subcutaneous injection, α-endorphin ($\beta\text{-LPH}_{61-76}$) appeared to be the most potent peptide. On a molar basis it was more than 30 times more active than ACTH_{4-10}. Oxidation of methionine in Met-enkephalin to the sulfoxide level potentiated the effect on pole-jumping avoidance behavior (De Wied et al., 1978). Intraventricular administration of the respective peptides mimicked the effect of systemic administration, but much less peptide was needed to elicit an equipotent behavioral effect. Interestingly, α-endorphin and ACTH_{4-10} were approximately equiactive following this route. Thus, the difference in potency found following systemic administration seems to be related to biotransformation and/or brain uptake mechanisms rather than to intrinsic behavioral activities. Passive avoidance behavior was affected in a similar fashion by α-endorphin and ACTH_{4-10}. Both peptides facilitated passive avoidance behavior. α-Endorphin was again more active than ACTH_{4-10}.

In view of the morphine-like activities of the endorphins, the effect of morphine and the specific opiate antagonist naltrexone was studied on extinction of pole-jumping avoidance behavior. A tendency to delay extinction of the avoidance response was found following morphine but the effect was slight and not dose dependent. Naltrexone however, facilitated extinction of the avoidance response, in relatively low doses. The effect was clearly demonstrable in rats which were made more resistent to extinction by a 4 days training period. The effect of β-endorphin, α-endorphin, ACTH_{4-10} or the potent ACTH_{4-9} analog (Org 2766) on extinction of pole-jumping avoidance behavior however, could be easily overcome. The same amount of naltrexone was sufficient to block β-endorphin-induced analgesia as measured on the hot plate (De Wied et al., 1978). Thus, the influence of the endorphins and of ACTH and related peptides on avoidance behavior takes place independently of opiate receptor sites in the brain.

Table 2

AMINO ACID SEQUENCE OF VARIOUS β-ENDORPHIN (β-LPH$_{61-91}$) FRAGMENTS

	1	2	3	4	5	6	7	8	9	10	11	12	13	14	15	16	17
γ-endorphin (β-LPH$_{61-77}$)	H-Tyr	Gly	Gly	Phe	Met	Thr	Ser	Glu	Lys	Ser	Gln	Thr	Pro	Leu	Val	Thr	Leu-OH
α-endorphin (β-LPH$_{61-76}$)	H-Tyr	Gly	Gly	Phe	Met	Thr	Ser	Glu	Lys	Ser	Gln	Thr	Pro	Leu	Val	Thr-OH	
β-LPH$_{61-69}$	H-Tyr	Gly	Gly	Phe	Met	Thr	Ser	Glu	Lys-OH								
Met-enkephalin (β-LPH$_{61-65}$)	H-Tyr	Gly	Gly	Phe	Met-OH												

	18	19	20	21	22	23	24	25	26	27	28	29	30	31
β-LPH$_{78-91}$	H-Phe	Lys	Asn	Ala	Ile	Val	Lys	Asn	Ala	His	Lys	Lys	Gly	Glu-OH

NEUROLEPTIC-LIKE ACTIVITY OF β-LIPOTROPIN₆₂₋₇₇ ([des-Tyr]-γ endorphin); IMPLICATION IN THE ETIOLOGY OF PSYCHOPATHOLOGY

Bloom et al. (1976) showed that β-endorphin (β-LPH₆₁₋₉₁) produced a naloxone reversible catatonia following intraventricular administration to rats. Jacquet and Marks (1976) found that β-endorphin injected directly into the periaquaeductal gray of the rat caused profound sedation and catalepsia while fragments of this polypeptide caused attenuated forms of this behavior. These effects could also be blocked by naloxone pre-treatment. The authors suggested that β-endorphin might be an endogenous neuroleptic. This was contradicted by Segal et al (1977) who showed that the spectrum of effects of β-endorphin contrasted with that of the neuroleptic drug haloperidol.

Unlike the analgesic effect of β-endorphin which decreases when the peptide chain is shortened, the influence on avoidance behavior increases. It is possible therefore that β-endorphin possesses opposite behavioral effects and that the eventual activity on extinction of avoidance behavior depends on the fragmentation of the peptide. In view of this the influence of other β-endorphin fragments such as γ-endorphin and β-LPH₇₈₋₉₁ were studied on active and passive avoidance behavior. It appeared that γ-endorphin in doses in which α-endorphin delayed extinction of pole-jumping avoidance behavior had no such effect. In contrast, extinction appeared to be somewhat accelerated. β-LPH₇₈₋₉₁ delayed extinction of the avoidance response but the effect was less potent than that of α-endorphin. To demonstrate an eventual facilitating effect of γ-endorphin, rats were made more resistant to extinction by a training procedure which lasted for 4 instead of for 3 days. Under these conditions, γ-endorphin was shown to facilitate extinction of pole-jumping avoidance behavior. A significant effect was found already following the subcutaneous injection of 30 nanograms. Much smaller amounts (0.3 ng) induced a similar effect when γ-endorphin was injected intracerebroventricularly. Observations on passive avoidance behavior gave essentially the same results. In contrast to α-endorphin which facilitates, γ-endorphin attenuated passive avoidance behavior when injected immediately after the learning trial. When administered 1 h prior to the retention test, γ-endorphin like α-endorphin facilitated passive avoidance behavior. The effect of α-endorphin was still present at the 48 h retention trial but not that of γ-endorphin.

It has been shown that the opiate-like activity of Met-enkephalin is reduced by the removal of tyrosine, the N-terminal amino acid residue (Frederickson, 1977). We found that while [des-Tyr]γ-endorphin (DTγE) had completely lost its opiate-like activity on the guinea pig ileum (preparation of Hughes et al., 1975) it was even more potent in facilitating extinction of pole-jumping avoidance behavior than was γ-endorphin. Similar results with passive avoidance behavior showed that DTγE was more potent than γ-endorphin in attenuating the response.

γ-Endorphin and DTγE in amounts effective on the extinction of
active and on retention of passive avoidance behavior did not affect
gross behavior in an open field. The rate of ambulation, rearing,
grooming and defaecation thus did not differ from that of saline-
treated rats. Further studies designed to explore neuroleptic activities revealed that DTγE was also active in the "grip test". Animals treated with DTγE (50 µg per rat) hung suspended above the floor
of the cage with their front paws grasping a pencil for a signici-
cantly longer period of time than did saline or α-endorphin treated
rats. This effect is characteristic for neuroleptic drugs such as
haloperidol and is not seen in morphine-treated animals. DTγE had
no effect on the corneal reflex, the righting reflex or on rigidity
(bridge test, see Bloom et al., 1976) but slightly decreased the
mobility of the rats in the observation cage.

The most striking result we obtained was that α- and γ-endorphin
had opposite effects on extinction of active and on retention of
passive avoidance behavior. It is surprising that the addition of
one amino acid residue, leucine at the carboxyl end of β-LPH$_{61-76}$
could induce such a dramatic reversal of the behavioral effect.
Moreover, removal of the amino acid residue tyrosine at the N-
terminal end completely dissociated the opiate-like and behavioral
effects.

Austen et al., 1977, showed that membrane bound enzymes from
rat brain and striatal slices degraded β-endorphin in discrete stages. The process yielded successively γ-endorphin, α-endorphin and
Met-enkephalin. The accumulation of these peptides was strongly
favoured by the presence of bacitracin or by an acid pH, conditions
which prevent the loss of the N-terminal amino acid residue tyrosine.
This suggests that DTγE could be generated from β-endorphin in the
brain. The conclusion of Austen et al., 1977, that the shorter peptides were mere degradation products of β-endorphin rather than
physiologically significant natural peptides is untenable in view
of the powerful and specific effects of α- and γ-endorphin and of
DTγE on avoidance behavior.

Our findings provide a clue for the understanding of the mechanism underlying psychopathology. The various effects of DTγE, e.g.
facilitation of extinction of active avoidance behavior and attenuation of passive avoidance behavior after relatively low doses, and
the positive grip test after high doses, are characteristic effects
of neuroleptic drugs. In view of this we postulated that DTγE or a
closely related neuropeptide is an endogenous neuroleptic with a
profile more specific than that of currently used neuroleptic drugs.
It is possible that a reduced availability of DTγE as a result of
an inborn error in its generation or its degradation is an etiological factor in psychopathological states for which neuroleptic drugs
are effective.

To substantiate this hypothesis the influence of DTγE was studied in 6 schizophrenic patients who were practically resistent to treatment with conventional neuroleptics (Verhoeven et al., 1978). Their maintenance therapy was discontinued one week before DTγE treatment. DTγE as Zn-phosphate preparation was given intramuscularly once daily. This preparation was used to prolong the action of the peptide, although it appeared that about 25% of the peptide was present in bound form. Patients were treated for 8 to 14 days. Progressive reduction of all psychotic symptoms was observed on day 4 and these had disappeared from day 6 on in 3 patients. In the other 3 patients improvement was found at day 2 and 3, but thereafter they became progressively psychotic again in spite of the treatment which made it necessary to discontinue the medication on day 8.

All 6 patients showed semipermanent or transient improvement upon treatment with DTγE. This indicates that this neuropeptide is an endogenous neuroleptic and the findings strengthen the hypothesis that derangements in endorphins homeostasis play an essential role in the psychopathology of schizophrenia.

EFFECT OF PEPTIDES RELATED TO ACTH AND TO β-LPH ON EXCESSIVE GROOMING

Peptides related to ACTH, MSH and β-LPH induce a stretching and yawning syndrome following intracerebral administration only (Ferrari et al., 1963; Gessa et al., 1967). In rodents, the onset of the syndrome is preceded by a display of excessive grooming (Ferrari et al., 1963; Izumi et al., 1973; Gispen et al., 1975; Rees et al., 1976). In rats $ACTH_{1-24}$, α- and β-MSH are equipotent in this respect (Gispen et al., 1975). $ACTH_{4-10}$ is less active in rabbits (Baldwin et al., 1974) and inactive in rats (Gispen et al., 1975) and mice (Rees et al., 1976) but [D-Phe^7] $ACTH_{4-10}$ has appreciable activity. $ACTH_{4-10}$ could be activated by C-terminal chain elongation but also by shortening to $ACTH_{4-7}$. This peptide is as active as [D-Phe^7] $ACTH_{4-10}$ in this respect (Wiegant and Gispen, 1977). $ACTH_{4-7}$ is also the shortest sequence which exerts full activity in the pole-jumping avoidance test (Greven and De Wied, 1973). Excessive grooming induced by ACTH analogues can be completely suppressed by pretreatment with specific opiate antagonists. This suggests an interaction with opiate receptors in the brain (Gispen and Wiegant, 1976). In addition, morphine also induces excessive grooming. Terenius (1975) found that $ACTH_{1-28}$ and $ACTH_{4-10}$ had affinity for opiate receptors. Subsequent structure activity studies pointed to an active site within $ACTH_{4-10}$ with some indication for a second site distal to the C-terminus of this heptapeptide (Terenius et al., 1975). Analysis of the binding revealed low selectivity of these peptides for agonist and antagonist binding sites, comparable to that of the partial agonist nalorphine (Terenius, 1976). In view of the relatively low affinity for the opiate receptor (IC50 of the order of $10^{-6}-10^{-5}$ M), $ACTH_{4-10}$ cannot be regarded as a physiological ligand for opiate

receptors in the brain in contrast to C-terminal β-LPH fragments
which exhibit high affinity for brain opiate receptors (Bradbury
et al., 1976b). But the fact that ACTH interacts with opiate receptors in the brain may explain the interaction of these peptides with
the analgesic effect of morphine (Gispen et al., 1976a). Morphine-
induced spinal reflex activity is reduced by ACTH in preparations
in vivo and in vitro (Zimmermann and Krivoy, 1973) and ACTH counter-
acts the analgesic effect of morphine in rodents (Winter and Flataker, 1951). Gispen et al. (1976a) found that ACTH and ACTH fragments reduce the analgesic response of morphine by 50-60 percent
as measured in the hot plate test. The peptides themselves had no
effect on the response of the rats on the hot plate. Interestingly,
this in vivo effect corresponds rather well with the affinity to
brain opiate receptor sites in vitro. If excessive grooming is mediated by opiate receptors in the brain, the endorphins should be
more potent than ACTH. Indeed, intraventricular administration of
doses as low as 10 ng, β-endorphin elicits excessive grooming
(Gispen et al., 1976b). α-Endorphin is much less active while Met-
enkephalin is virtually inactive. β-Endorphin on a weight basis is
ten times as active as $ACTH_{1-24}$. The grooming effect of β-endorphin
is readily blocked by naloxone (Gispen et al., 1976b).

RETROGRADE TRANSPORT OF PITUITARY HORMONES TO THE BRAIN

     Although the effects of ACTH and endorphins on avoidance behav-
ior can be elicited after systemic administration, intracerebroventricular administration is more effective. These observations and
the recently demonstrated presence of ACTH and other pituitary hormones in the cerebrospinal fluid and in the brain (Allen et al., 1974;
Krieger et al., 1977 a, b; Rudman et al., 1974; Oliver et al., 1977)
suggest the production of ACTH or ACTH-like peptides in the brain or
the transport of these hormones from the pituitary to the brain.
In fact, evidence is accumulating that the blood in the portal vessel
system not only flows from the hypothalamus to the pituitary, but
also retrograde from the pituitary to the brain. Several authors
have observed that the blood in some of the vessels of the anterior
pituitary flows to the hypothalamus (Török, 1960; 1964; Szentagothai
et al., 1968). Page et al. (1976) found in the dorsum of the rabbit
pituitary stalk the presence of vessels which connect the posterior
lobe of the pituitary to the hypothalamus. These vessels have also
contact with the long portal vessels. Szentagothai et al. (1968)
suggested that part of the venous outflow of the anterior pituitary
was by way of the vasculature of the posterior pituitary. Such vascular arrangements might offer retrograde transport of pituitary
principles to the hypothalamus and other brain areas. In view of
this Oliver et al. (1977) studied the transport of pituitary hormones in anesthetized rats with intact pituitaries or in which the
anterior, posterior pituitary or the whole gland had been removed.
The authors collected blood from a single long portal vessel and

determined the concentration of various pituitary hormones. They
found that adenohypophysectomy led to a markedly reduced level of
LH, TSH, prolactin and ACTH while the level of α-MSH was slightly
reduced and that of vasopressin was unaffected. Removal of the posterior pituitary induced a marked decrease in prolactin, ACTH, α-MSH
and vasopressin. Total hypophysectomy caused the nearly complete
disappearance of all pituitary hormones in portal blood. These
findings are consistent with the hypothesis that pituitary hormones
are transported retrograde to the brain. Evidence for pituitary-
brain transport of peptides was also obtained by Mezey et al. (1978)
who used a radioactive labelled $ACTH_{4-9}$ analog (Org 2766) which was
injected directly into the pituitary of urethanized rats. Intrasellar
or intrapituitary administration resulted in significantly higher
radioactivity levels in the brain than intravenous injection of an
equimolar dose of the ^3H-$ACTH_{4-9}$ analog 30 min after injection.
Intrapituitary injection induced uptake with clear regional differences which was highest in the hypothalamus. The spinal cord, medulla
oblongata and olfactory bulbs had relatively high uptakes, while uptake in mesencephalon and septum was somewhat less. Labelling in the
hypothalamus was highest in the supraoptic nucleus. Intrasellar administration differed quantitatively and qualitatively from that following intrapituitary injection in that uptake was high in the olfactory bulb and preoptic area but low in the hypothalamus. Twenty
four hours after section of the pituitary stalk the uptake of radioactivity in the hypothalamus but not in other brain regions was
markedly depressed. Hypothalamic uptake however was restored at eight
days after stalk section. These results suggest a significant flow
of radioactivity from the pituitary to the brain in particular to
the hypothalamus. Transport to the hypothalamus presumably is partly
vascular via the stalk while transport to other brain areas may occur
via the cerebrospinal fluid, but a neural route cannot be excluded.

EFFECTS OF VASOPRESSIN AND OXYTOCIN AND
RELATED PEPTIDES ON MEMORY PROCESSES

Vasopressin and related peptides stimulate acquisition of
shuttlebox avoidance behavior (Lande et al., 1970; Bohus et al.,
1973) of hypophysectomized rats, delay extinction of active avoidance behavior of posterior lobectomized and intact rats (De Wied,
1965; De Wied and Bohus, 1966; De Wied, 1971) and facilitate retention of passive avoidance behavior (Ader and De Wied, 1972). The influence of vasopressin is of a long term nature and persists beyond
the actual presence of the hormone in the body. These findings were
interpreted to indicate that vasopressin and related peptides modulate memory processes. Memory is defined as the facilitation and attenuation of consolidation and reproduction (retrieval) of acquired
experience. Vasopressin and related peptides protect against puro-
mycin-induced amnesia in mice (Lande et al., 1972; Walter et al.,
1975) and against amnesia for passive avoidance behavior in rats

induced by CO_2 inhalation or electroconvulsive shock (Rigter et al. 1974; Rigter and Van Riezen, 1975). The maintenance of a food reward-ed approach response in a runway is not affected by vasopressin (Garrud et al., 1974) but sexually motivated behavior is facilitated by vasopressin. Male rats trained in a T-maze to run for a receptive female choose the correct arm of the maze in a higher percentage following treatment with DG-LVP than saline treated controls. Copulation reward is essential for this effect of the peptide (Bohus, 1977).

Oxytocin also increases resistance to extinction of pole-jumping avoidance behavior but it is less active than vasopressin (Walter et al., 1978). However, Schulz et al. (1976) found that oxytocin has an effect opposite to that of vasopressin. It facilitates active and attenuates passive avoidance behavior. Systemically administered oxytocin failed to facilitate extinction of pole-jumping avoidance behavior (Bohus et al., 1978). However, when the peptide was given intraventricularly, facilitation of extinction was found. Intraventricularly administered oxytocin also affects passive avoidance behavior in a way opposite to that of vasopressin. It attenuates passive avoidance behavior (Bohus et al., 1978). The physiological role of the neurohypophyseal hormones was shown in experiments in which intraventricularly injected specific vasopressin or oxytocin antiserum were used to block the action of vasopressin and oxytocin in the brain. Vasopressin antiserum attenuates (Van Wimersma Greidanus et al., 1975) while specific oxytocin antiserum facilitates passive avoidance behavior (Bohus et al., 1978).

STRUCTURE ACTIVITY STUDIES WITH NEUROHYPOPHYSEAL HORMONES AND RELATED PEPTIDES ON MEMORY PROCESSES

Structure activity studies revealed that the covalent ring structure of vasopressin is more important than the linear part of the molecule (table 3) for the long term effect on extinction of pole-jumping avoidance behavior (De Wied, 1976; Walter et al., 1978). However, the linear part of vasopressin and oxytocin is more effective than the covalent ring structures in protecting against retrograde amnesia (Walter et al., 1975). This suggests that extinction of avoidance behavior and protection against retrograde amnesia measures different memory processes. Resistance to extinction is indicative for effects on consolidation processes while protection against retrograde amnesia measures retrieval processes. It may well be therefore that the covalent ring structure of the neurohypophyseal hormones predominantly affect consolidation processes while the linear portions are involved in retrieval mechanisms. This hypothesis was supported in experiments by Bohus and Kovacs (non published observations) in the passive avoidance paradigm in which the effect of intraventricularly administered [Arg[8]]vasopressin (AVP), [Arg[8]] vasotocin (AVT), oxytocin and fragments of these hormones was studied on consolidation and retrieval processes. Peptides were

Table 3

AMINO ACID SEQUENCE OF NEUROHYPOPHYSEAL HORMONES AND RELATED NEUROPEPTIDES

Arginine[8] vasopressin (AVP) H-Cys-Tyr-Phe-Gln-Asn-Cys-Pro-Arg-Gly-NH_2

Desglycinamide[9], arginine[8] vasopressin (LVP) H-Cys-Tyr-Phe-Gln-Asn-Cys-Pro-Arg-OH-Gly-NH_2

Pressinamide (PA) H-Cys-Tyr-Phe-Gln-Asn-Cys-NH_2

Prolyl-argyl-glycinamide (PAG) H-Pro-Arg-Gly-NH_2

Oxytocin (OXT) H-Cys-Tyr-Ile-Gln-Asn-Cys-Pro-Leu-Gly-NH_2

Tocinamide (TA) H-Cys-Tyr-Ile-Gln-Asn-Cys-NH_2

Prolyl-leucyl-glycinamide (PLG) H-Pro-Leu-Gly-NH_2

Arginine[8] vasotocin (AVT) H-Cys-Tyr-Ile-Gln-Asn-Cys-Pro-Arg-Gly-NH_2

considered to influence consolidation processes if avoidance latency
was significantly affected at the 24 and 48 h retention tests following injection immediately after the learning trial (post-trial injection). Peptides were regarded to influence retrieval processes if
avoidance latency was significantly affected at the 24 h retention
test only if given 1 h prior to the retention test (pre-retention
injection). These studies showed that the effect of vasopressin
on memory consolidation is located in the covalent ring structure.
Oxytocin had an effect opposite to that of vasopressin and attenuated
passive avoidance behavior. The same was found for vasotocin. The
covalent ring structure of oxytocin had a similar effect as vasopressin and caused facilitation of consolidation. The linear tri-
peptides of vasopressin and oxytocin had only minor activity in this
respect. These parts of the molecule however, were active on retrieval. In particular retrieval effect of the dipeptide Leucyl-Glycina-
mide of oxytocin corroborates observations with the same peptide
by Walter et al. (1975) which effectively protects against puromycin
induced amnesia in mice. Interestingly, attenuation of retrieval was
found with [des-Gly^9] vasotocin (DGAVT). These findings suggest that
the consolidating influence is located mainly in the covalent ring
structure of both vasopressin and oxytocin, the retrieval mainly in
the linear portion of these molecules while both ring structure and
linear portion of oxytocin or vasotocin are needed for attenuation
of memory consolidation and retrieval.

The different effects of vasopressin, oxytocin, and various
fragments on neurotransmitter activity in different regions in the
brain (Tanaka et al., 1977 a,b) may be related to the multiple CNS
actions of these neuropeptides (Van Ree et al., 1978).

The modulating effects of fragments of neurohypophyseal hormones
on memory processes, suggest that they act as precursor molecules
for neuropeptides which affect respectively consolidation and retrieval processes. However, the two main pathways in the brain which
degrade vasopressin and oxytocin are an aminopeptidaæ pathway which
sequentially releases amino acids from the NH_2-terminus of oxytocin
and vasopressin and a carboxypeptidase pathway which releases amino
acids from the carboxyl end (Pliska et al., 1971a,b,; Walter et al.
1973; Marks et al., 1973; Marks and Stern, 1974). These studies indicate that the covalent ring structures as such are not generated.
More knowledge of the enzymatic degradation of the neurohypophyseal
hormones has to be obtained, and the influence of the respective
degradation products on memory processes to be assessed before the
essential structural requirements for consolidation and retrieval
processes can be determined. This is of extraordinary significance
since clinical studies have shown dramatic effects of vasopressin
on memory recovery in amnesia patients (Oliveros et al., 1978) and
marked effects on several attentional and memory tests in elderly
people (Legros et al., 1978). Vasopressin has powerful effects on
blood pressure and in addition possesses other endocrine and meta-

bolic effects. Fragments of this hormone are devoid of these effects. The development of neuropeptides which stimulate or attenuate consolidation or retrieval processes may yield more specific compounds for the treatment of mental disorders of cognitive origin.

CONCLUDING REMARKS

The study of the influence of pituitary hormones on the central nervous system, basically endocrinological in character, has been of great assistance in gaining more knowledge of various brain functions. Our studies revealed the existence of peptide hormone fragments devoid of classical endocrine effects affecting motivational, learning and memory processes, molecules involved in the development of tolerance to opiates and the development and maintenance of drug seeking behavior. The more recent findings on the behavioral effects of endorphins suggest that psychopathological processes may be the result of disturbances in the hormonal climate of the brain. These processes may thus be regarded as an endocrine or metabolic disease which may be amended in the same way as diabetes mellitus with insuline or myxoedema with thyroid hormones. The studies presented in this paper and those of others have lain a more solid fundament for biological psychiatry and will provide the new leads for the development of highly effective therapeutic agents with more specific actions and without the many side effects of conventional psychoactive drugs.

REFERENCES

Ader, R. & D. de Wied, Psychon. Sci. 29, 46-48, 1972.
Allen, J.P., J.W. Kendall, R. McGilvra & C. Vancura, J. Clin. Endocr. 38,586-593, 1974.
Applezweig, M.H. & F.D. Baudry, Psychol.Rep. 1, 417-420, 1955.
Austen, B.M., D.G. Smyth, C.R.Snell, Nature 269, 619-621, 1977.
Baldwin, D.M., Ch.K. Haun & Ch.H. Sawyer, Brain Res. 80, 291-301, 1974.
Bloom, F., D. Segal, N. Ling & R. Guillemin, Science 194, 630-632, 1976.
Bohus, B., Horm. Behav. 8, 52-61, 1977.
Bohus, B. & D. de Wied, Science 153, 318-320, 1966.
Bohus, B., W.H. Gispen & D. de Wied, Neuroendocrin. 11, 137-143, 1973.
Bohus, B., H.H.L. Hendrickx, A.A. van Kolfschoten & T.G. Krediet, In: Sexual Behavior: Pharmacology and Biochemistry, Eds. M. Sandler and G.L. Gessa, Raven Press, N.Y., 269-275, 1975.
Bohus, B., I. Urban, Tj.B. van Wimersma Greidanus & D. de Wied, Neuropharmacol., 17, 239-247, 1978.
Bradbury, A.F., D.G. Smyth, C.R. Snell, N.J.M. Birdsall & E.C. Hulme, Nature 260, 793-795, 1976.
De Wied, D., Am. J. Physiol. 207, 255-259, 1964.
De Wied, D., Int. J. Neuropharmacol. 4, 157-167, 1965.

De Wied, D., Proc. Soc. exp. Biol. 122, 28-32, 1966
De Wied, D., In: Frontiers in Neuroendocrinology, Eds. W.F. Ganong and L. Martini, Oxford University Press, London/New York, 97-140, 1969.
De Wied, D., Nature 232, 58-60, 1971.
De Wied, D., In: The Neurosciences, Third Study Program, Eds. F.O. Schmitt and F.G. Worden, MIT Press, Cambridge, 653-666, 1974.
De Wied, D., Life Sci. 19, 685-690, 1976,
De Wied, D. & B. Bohus, Nature 212, 1484-1486, 1966.
De Wied, D., A. Witter & H.M. Greven, Biochem. Pharmacol. 24, 1463-1468, 1975.
De Wied, D., B. Bohus, J.M. van Ree & I. Urban, J. Pharmacol. and exp. Ther. 204, 570-580, 1978.
Eberle, A. & R. Schwyzer, Helv. Chim. Acta, 58, 1528-1535, 1975.
Ferrari, W., G.L. Gessa & L. Vargiu, Ann. N.Y. Acad. Sci. 104, 330-345, 1963.
Flexner, J.B. & L.B. Flexner, Proc. Nat. Acad. Sci (Wash.) 68, 2519-2521, 1971.
Flood, J.F., M.E. Jarvik, E.L. Bennett & A.E. Orme, The Neuropeptides; Pharmacol. Biochem. Behav., Vol. 5, suppl. 1, 41-51, 1976.
Gaillard, A.W.K. & A.F. Sanders, Psychopharmacol. (Berl.) 42, 201-208, 1975.
Garrud, P., J.A. Gray & D. de Wied, Physiol. Behav. 12, 109-119, 1974.
Gessa, G.L., M. Pisano, L. Vargiu, F. Crabai & W. Ferrari, Rev. Canad. Biol. 26, 229-236, 1967.
Gispen, W.H. & V.M. Wiegant, Neuroscience Letters 2, 159-164, 1976.
Gispen, W.H., V.M. Wiegant, H.M. Greven & D. de Wied, Life Sci. 17, 645-652, 1975.
Gispen, W.H., J. Buitelaar, V.M. Wiegant, L. Terenius & D. de Wied, Europ. J. Pharmacol. 39, 393-397, 1976a.
Gispen, W.H., V.M. Wiegant, A.F. Bradbury, E.C. Hulme, D.G. Smyth, C.R. Snell & D. de Wied, Nature 264, 794-795, 1976b.
Gold, P.E. & R. van Buskirk, Behav. Biol. 16, 387-400, 1976.
Gray, J.A. Nature, 229, 52-54, 1971.
Greven, H.M. & D. de Wied, In: Drug Effects on Neuroendocrine Regulation, Progress in Brain Research 39, Eds. E. Zimmermann, W.H. Gispen, B.H. Marks & D. de Wied, Elsevier Publ. Comp. Amsterdam 429-422, 1973.
Greven, H.M. & D. de Wied, In: Melanocyte Stimulating Hormone: Control, Chemistry & Effects, Eds. F.J.H. Tilders, D.F. Swaab & Tj.B. van Wimersma Greidanus, Frontiers of Hormone Research Vol. 4, S. Karger, Basel, 140-152, 1977.
Guth, S., S. Levine & J.P. Seward, Physiol. Behav. 7, 195-200, 1971.
Hughes, J., T.W. Smith, H.W. Kosterlitz, L.A. Fothergill, B.A. Morgan & H.R. Morris, Nature 258, 577-579, 1975.
Isaacson, R.L., A.J. Dunn, H.D. Rees & B. Waldock, Physiol. Psychol. 4, 159-162, 1976.
Izumi, K., J. Donaldson & A. Barbeau, Life Sci. 12, 203-210, 1973.
Jacquet, Y.F. & N. Marks, Science 194, 632-634, 1976.
Kastin, A.J., L.H. Miller, R. Nockton, C.A. Sandman, A.V. Schally & L.O. Stratton, In: Drug Effects on Neuroendocrine Regulation,

Eds. E. Zimmermann, W.H. Gispen, B.H. Marks & D. de Wied, Progress
in Brain Research, 39, 461-470, Elsevier Amsterdam, 1973.
Kastin, A.J., C.A. Sandman, L.O. Stratton, A.V. Schally & L.H. Mil-
ler, In: Hormones, Homeostasis and the Brain, Progress in Brain Re-
search 42, Eds. W.H. Gispen, Tj.B. van Wimerma Greidanus & D. de Wied,
Elsevier, Amsterdam, 143-150, 1975.
Krieger, D.T., A. Liotta & M.J. Brownstein, Proc. Natl. Acad. Sci.
74, 648-652, 1977a.
Krieger, D.T., A. Liotta & M.J. Brownstein, Brain Res. 128, 575-
579, 1977b.
Lande, S., A. Witter & D. de Wied, J. Biol. Chem. 246, 2058-2062, 1971
Lande, S., J.B. Flexner & L.B. Flexner, Proc. Nat. Acad. Sci. 69,
558-560, 1972.
Lande, S., D. de Wied & A. Witter, In: Drug Effects on Neuroendocrine
Regulation, Eds. W. Zimmermann, W.H. Gispen, B.H. Marks & D. de Wied,
Elsevier, Amsterdam, Progress in Brain Research 39, 421-427, 1973.
Legros, J.J., P. Gilot, X. Seron, J. Claessens, J. Adam, J.M. Moeglen,
A. Audibert & P. Berchier, The Lancet, January 7, 41, 1978.
Leonard, B.E. Int. J. Neuropharmacol. 8, 427-435, 1969.
Levine, S. & L.E. Jones, J. Comp. Physiol. Psychol. 59, 357-360, 1965
Lissák, K. & B. Bohus (1972), Int. J. Psychobiology 2, 103-115, 1972.
Marks, N. & F. Stern, Biochem. Biophys. Res. Commun. 61, 1458-1463,
1974.
Marks, N., L. Abrash & R. Walter, Proc. Soc. exp. Biol. (N.Y.) 142,
455-460, 1973.
Mezey, E., M. Palkovits, E.R. de Kloet, J. Verhoef & D. de Wied,
Life Sci. 22, 831-838, 1978.
Oliver, C., R.S. Mical & J.C. Porter, Endocrinol. 101, 598-604, 1977.
Oliveros, J.C. M.K. Jandali, M. Timsit-Berthier, R. Remy, A. Benghe-
zal, A. Audibert & J.M. Moeglen, The Lancet, January 7, 42, 1978.
Page, R.B., B.L. Munger & R.M. Bergland, Am. J. Anat. 146, 273-301,
1970.
Pliška, V., R. Barth & N.A. Thorn, Acta Endocr. 67, 1-11, 1971.
Pliška, V., N.A. Thorn & H. Vilhardt, Acta Endocr. 67, 12-22, 1971.
Rees, H.D., A.J. Dunn & P.M. Juvone, Life Sci. 18, 1333-1340, 1976.
Rigter, H. & A. Popping, Psychopharmacol. 46, 255-261, 1976.
Rigter, H. & H. van Riezen, Physiol. Behav. 14, 563-566, 1975.
Rigter, H., H. van Riezen & D. de Wied, Physiol. Behav. 13, 381-388,
1974.
Rudman, D., J.W. Scott, A.E. DelRia, D.H. Houser & S. Sheen, Amer.
J. Physiol. 226, 687-692, 1974.
Sandman, C.A., A.J. Kastin & A.V. Schally, Experientia (Basel) 25,
1001-1002, 1969.
Schulz, H., G.L. Kovács & G. Telegdy, In: Cellular & Molecular Bases
of Neuroendocrine Processes, Ed. E. Endröczi, Akadémiai Kiadók, Bu-
dapest, 555-564, 1976.
Segal, D.S., F. Bloom, N. Ling & R. Guillemin, Science, 198, 411-414,
1977.
Szentagothai, J., B. Flerkó, B. Mess & B. Halász, Akadémiai Kiadó,
Budapest, 1968.

Tanaka, M., D.H.G. Versteeg & D. de Wied, Neuroscience Letters 4, 321-325, 1977a.

Tanaka, M., E.R. de Kloet, D. de Wied & D.H.G. Versteeg, Life Sci. 20, 1799-1808, 1977b.

Terenius, L.,J. Pharm. Pharmacol. 27, 450-452, 1975.

Terenius, L., Europ. J. Pharmacol. 38, 211-213, 1976.

Terenius, L., W.H. Gispen & D. de Wied, Europ. J. Pharmacol. 33, 395-399, 1975.

Török, B., Anat. Anz. 109, 622, 1960.

Török, B., Acta Anat. (Basel) 59, 84-99, 1964.

Van Ree, J.M., B. Bohus, D.H.G. Versteeg & D. de Wied, Biochem. Pharmacol. in press, 1978.

Van Wimersma Greidanus, Tj.B., J. Dogterom & D. de Wied, Life Sci. 16, 637-644, 1975.

Verhoeven, W.M.A., H.M. van Praag, P.A. Botter, A. Sunier, J.M.van Ree & D. de Wied, The Lancet, May 13, 1046-1047, 1978.

Walter, R., E.C. Griffiths & K.C. Hooper, Brain Res. 60, 449-457, 1973

Walter, R., P.L. Hoffmann, J.B. Flexner & L.B. Flexner, Proc. Nat. Acad. Sci (USA) 72, 4180-4184, 1975.

Walter, R., J.M. van Ree & D. de Wied, Proc. Nat. Acad. Sci, in press, 1978.

Weiss, J.M., B.S. McEwen, M. Silva & M. Kalkut, Am. J. Physiol. 218, 864-868, 1970.

Wiegant, V.M. & W.H. Gispen, Behav. Biol. 19, 554-558, 1977.

Winter, C.A. & L. Flataker, J. Pharmacol. exp. Ther. 101, 93, 1951.

Witter, A., H.M. Greven & D. de Wied, J. Pharmacol. exp. Ther. 193, 853-860, 1975.

Zimmermann, E. & W.A. Krivoy, In: Drug Effects on Neuroendocrine Regulation, Progress in Brain Res. 39, Eds. E. Zimmermann, W.H. Gispen, B.H. Marks & D. de Wied, Elsevier, Amsterdam, 383-394, 1973.

SEXUAL DIFFERENTIATING ACTIONS OF STEROIDS ON THE HYPOTHALAMO-PITUITARY-LIVER AXIS

Jan-Åke Gustafsson, Peter Eneroth[x], Barbro Haglund,
Tomas Hökfelt[x], Agneta Mode, Paul Skett and Örjan Wrange

Department of Chemistry and [x]Department of Histology,
Karolinska Institutet, and [x]Hormonlaboratoriet, Depart-
ment of Obstetrics and Gynecology, Karolinska Hospital,
S-104 01 Stockholm 60, Sweden.

In 1953, Hübener and Amelung (1) showed a sex-related difference of hepatic corticosteroid metabolism in the rat. This
finding was substantially confirmed by subsequent reports using
corticosteroids (2-4), androgens (3-5) and estrogens (5) as substrates. Male animals show a higher hydroxylating activity than
females (4,5) whereas females exhibit a higher 5α-reductase activity (3). In 1958, Yates et al. (3) published the first study
indicating that androgens were involved in the control of steroid
metabolism in rat liver. It was found that castration increased
and testosterone treatment decreased the 5α-reductase activity.
This finding was confirmed and extended by DeMoor and Denef (6)
and Gustafsson and Stenberg (7-10) who showed the dependence
of the adult male pattern of metabolism on the presence of neonatal androgen (imprinting). The sexual differences in the hepatic metabolism of 4-androstene-3,17-dione and 5α-androstane-3α,17β-
diol are illustrated in Fig. 1. Although dependent on neonatal
androgen exposure, no sex-related differences in steroid metabolism were observed until 30 days of age (11-15). Sex-related
differences in metabolism and the androgen-dependence of these
differences have also been observed for the metabolism of various
drugs and xenobiotics including ethylmorphine (16-21), aniline
(18,19,23), p-nitroanisole (22) and hexobarbitone (23). The absence of any sex-specific androgen receptor in the liver (24)
(similar to those found in other androgen-responsive tissues (25))
coupled to the fact that male animals are more sensitive to androgen action as regards hepatic steroid metabolism in the adult
period (9,10) seems to indicate an indirect action of androgen
on hepatic steroid metabolism.

315

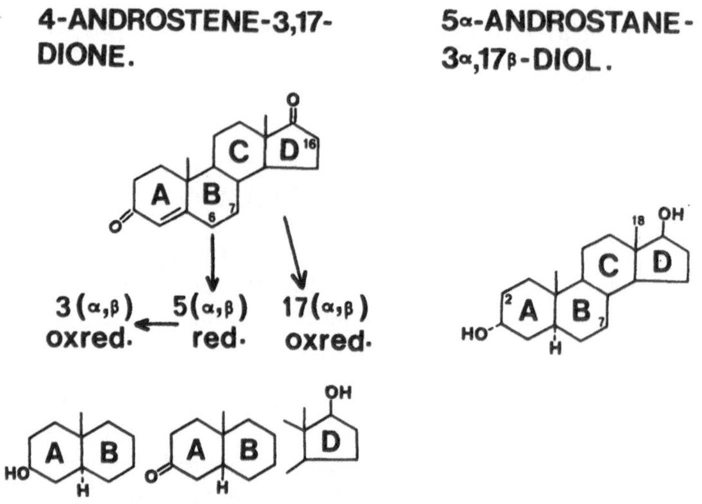

5α-REDUCTASE	♀ > ♂
5β- –"–	♀ < ♂
3α-OXIDORED.	♀ = ♂
3β- –"–	♀ < ♂
17α- –"–	♀ < ♂
17β- –"–	♀ = ♂
6β-OHASE	♀ < ♂
7α- –"–	♀ = ♂
16α- –"–	♀ < ♂

2α-OHASE	♀ < ♂
2β- –"–	♀ < ♂
7α- –"–	♀ ≥ ♂
7β- –"–	♀ < ♂
18- –"–	♀ < ♂

Fig. 1. Sexual differences in hepatic metabolism of 4-androstene-3,17-dione and 5α-androstane-3α,17β-diol.

A possible explanation of the indirect action of androgens on hepatic steroid metabolism was put forward in 1974 when Denef (26) and Gustafsson and Stenberg (27) published the effects of hypophysectomy on hepatic steroid metabolism. Both groups showed that hypophysectomy of female animals resulted in a male pattern of metabolism, indicating that the female pattern of metabolism is dependent on the presence of an intact pituitary. It is well known that the hypothalamus in the rat is sex-differentiated with regard to control of pituitary secretion (28,29) and thus it is likely that the sex-related difference in hepatic steroid metabolism observed is secondary to the sexual differentiation of the brain at birth and that androgens have their action via the hypothalamo-pituitary axis rather than directly on the liver. This is further supported by the fact that electrothermal destruction of the hypothalamus causes a feminization of steroid metabolism in male rats (30). The pituitary dependence of hepatic metabolism is not confined to steroid metabolism although both androgen (26,27) and corticosteroid (31) metabolism are involved. Kim et al. (32) have shown that protein demethylase II activity in the rat liver is pituitary-dependent and there is evidence to support the pituitary dependence of drug and xenobiotic metabolism, including that of ethylmorphine (33,34). The basal hepatic lipolysis (35) and synthesis of plasma proteins (36) have also been shown to be controlled, at least partially, by the pituitary gland. Kramer et al. (33) have shown that the effect of testosterone on ethylmorphine metabolism is not manifested in the absence of the pituitary gland. A similar observation was seen in the study of Gustafsson and Stenberg (27) where neither testosterone nor 17β-estradiol had any major effects on hepatic steroid metabolism in hypophysectomized rats. It thus seems likely that the effects of the sex steroids on hepatic metabolism is indirect and is effectuated via the hypothalamo-pituitary axis.

Implantation of a pituitary gland under the kidney capsule of a hypophysectomized male or female rat resulted in feminization of the host's hepatic steroid metabolism (26,37). We also showed that different time periods were required for the changes in different enzyme activities (38), emphasizing the importance of the sensitivity of the respective enzyme system to the controling factor(s). This would seem to indicate that the ectopic pituitary gland secretes a factor into the general circulation that is normally only secreted by the female pituitary gland in situ. The ectopic pituitary tissue is known to secrete prolactin and growth hormone and small amounts of the other anterior pituitary hormones (39-42). In our experiments only prolactin and growth hormone could be measured in the serum of implanted animals (38). The hypophyseal factor responsible for the maintenance of the female-type of hepatic metabolism is, as yet, an open question.

A number of reports have been published showing effects of
various purified pituitary hormones or combinations of hormones
on drug and steroid metabolism in the rat. Gustafsson and Stenberg
(43,44) have shown an effect of FSH but no effect of LH or prolactin on hepatic steroid metabolism using 4-androstene-3,17-dione
as substrate. These results are in agreement with those of Colby
et al. (31) where corticosterone was used as substrate and gain
further confirmation from the study of Lax et al. (45) where the
effect of prolactin on reductive metabolism was examined. In the
report of Colby et al. (31), a combination of GH and ACTH decreased reductive metabolism (an effect similar to hypophysectomy),
while Lax et al. (45) found that prolactin had this effect. A
later report by Kramer and Colby (46) indicated that GH alone increased reductive metabolism of steroids (an effect dramatically
opposite to that of GH combined with ACTH). In the field of
drug metabolism, ACTH and GH seem to be the most active. GH alone
was thought to reduce the hepatic metabolism of ethylmorphine
(46,47) and aminopyrine (47), whereas combinations of ACTH and GH
have been shown to increase the carcinogenicity of N-hydroxy-N-2-
fluorenylacetamide (48,49) (thereby indicating a decrease in de-
hydroxylation and deacetylation of the carcinogen by the liver).
Kramer et al. (50) have reported that both LH and FSH, when given
to castrated male and female rats, increased ethylmorphine demethylase activity in the liver.

None of the above mentioned effects are, however, as marked
as that seen after implantation of a pituitary gland under the
kidney capsule. These data would indicate, and it is our opinion,
that the control of hepatic steroid and xenobiotic metabolism is
mediated via an, as yet, unidentified pituitary factor (or
possibly an unidentified combination of factors) - we have labelled
this factor, "feminizing factor", or "feminotropin". The purification of this factor is underway in our laboratory and it is hoped
by the purification and characterization of this polypeptide to
prove the existence of a new hormone.

Studies by our group (51,52) on the ontogenesis of the pituitary control of hepatic steroid metabolism have indicated that
the liver and pituitary are not per se sex differentiated with
regard to their response to and production of "feminizing factor",
respectively. Thus the masculinized ("imprinted") control centre
must lie elsewhere.

The possibility that hepatic steroid metabolism is under indirect hypothalamic control was first indicated by our above-
mentioned finding (53) that complete electrothermal destruction
of the hypothalamus in male rats resulted in "feminization" of
the steroid metabolism in the liver of the operated animals.
This indicated for the first time that, in the male rat, release

of the factor(s) maintaining a female-type steroid metabolism in female animals was inhibited by the hypothalamus. This inhibitory control is similar to that of prolactin, the release of which is inhibited by prolactin release inhibiting factor (PIF) (54).

The site of the sexual differentiation in the hypothalamus appears to lie predominantly in the preoptic area where, in fact, anatomical differences in nerve structure can be found. Greenough et al. (55), on studying the hamster preoptic area, found different patterns of dendrite density corresponding to different afferent inputs to the dorsomedial preoptic area. In addition, many reports have been published on the sex differences in hypothalamic structure and their relation to neonatal androgen exposure (56-59).

Even the brain transmitter systems have been shown to be dependent on neonatal exposure to androgen. Libertun et al. (60) presented evidence that the cholinergic transmitter system in the suprachiasmatic area of the hypothalamus was affected by exposure to androgen in the neonatal period.

Recently, a great deal of interest has been focused on the specificity of imprinting. Is it solely androgens that cause imprinting or can other types of compounds have this effect? In earlier work, testosterone (normally as an ester) alone was used to test for imprinting (for review see 29). The discovery by Naftolin et al. (61) that androgens could be converted to estrogens in the brain, led to an investigation of the active compounds in imprinting. In respect to gonadotropin secretion and sexual behaviour it was found that (in the rat and hamster at least) estrogens were as effective, if not more effective, than androgens (for review see 62). It was also found that persistent estrus (characteristic of "androgen" imprinting in the female animal) could be caused by clomiphene (63) and o,p'-DDT (64). It, thus, seems likely that estrogens are the active hormones in imprinting, the brain having enzymes to convert neonatal androgen to estrogen and receptors to mediate the action of the formed estrogen (63,66). To test this hypothesis, 5α-dihydrotestosterone (DHT) (an androgen that cannot be aromatized in the brain) was tested for its imprinting ability. It was found that DHT could not induce persistent estrus or suppression of sexual behaviour in female rats (67-69). The specificity of imprinting has also been tested in relation to hepatic steroid metabolism. It was shown by our group (70) that, in contrast to imprinting of sexual behaviour, dihydrotestosterone, as well as testosterone and estradiol, was effective in inducing a male pattern of hepatic steroid metabolism while o,p'-DDT was without effect on the imprinting of microsomal metabolism of steroids in rat liver. These facts would seem to indicate that the mechanisms involved in imprinting of hepatic steroid metabo-

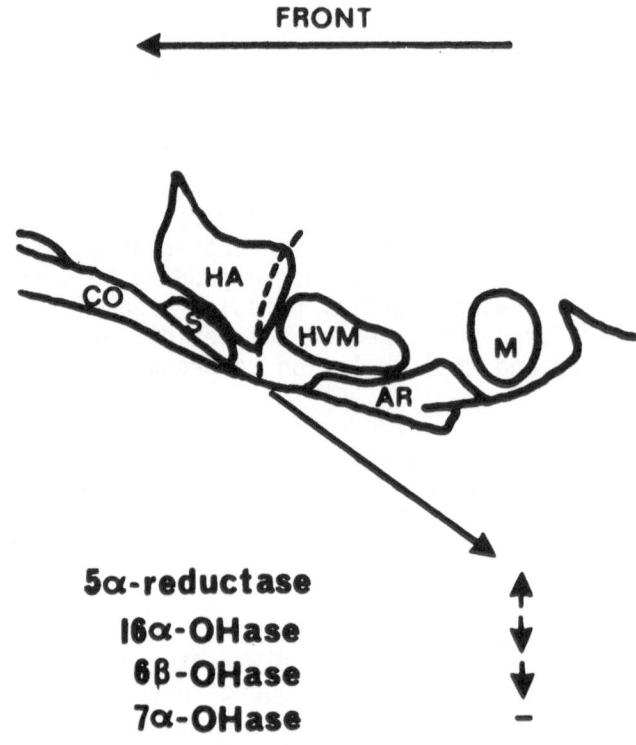

FRONT

5α-reductase
16α-OHase
6β-OHase
7α-OHase

Fig. 2. Diagrammatic representation of the level of deafferentation used. Deafferentation at the retrochiasmatic level caused
complete "feminization" of hepatic steroid metabolism. Abbreviations used: CO - optic chiasma, HA - anterior hypothalamic
nucleus, S - suprachiasmatic nucleus, HVM - ventromedial nucleus,
AR - arcuate nucleus, M - mamillary body.

lism are different than those involved in gonatotropin secretion
and sexual behaviour and that these have different sensitivity
and selectivity with regard to the inducing agent.

        In order to ascertain more exactly the area of the hypothala-
mus that controls the secretion of "feminizing factor", a series
of experiments was performed by our group (71) whereby the effects
of various brain lesions on hepatic steroid metabolism were investigated. Two distinct types of lesions were performed - one,
anterior hypothalamic deafferentation and two, discrete lesions
in the rostral parts of the hypothalamus. One week after the
lesioning, the animals were killed and their hepatic steroid metabolism investigated. The effects of the lesions on steroid meta-

bolism were correlated to the changes in secretion of pituitary hormones.

Hypothalamic deafferentation about the level of A5660μ (according to the atlas of Jacobovitz and Palkovits, 1974) (posterior to the preoptic area and thus separating the rostral parts of the hypothalamus from the medial basal hypothalamus including the median eminence) (Fig. 2) caused complete "feminization" of hepatic steroid metabolism in the operated male animals while having little effect on the female animals (71). These results indicate that the centre of control lies anterior to the level of A5660μ, possibly in the suprachiasmatic nucleus or in the surrounding periventricular area. This was the area described by Libertun et al. (60) to have sex-differentiated cholinergic transmitter systems. This finding is in contrast to the proposed regulatory site for prolactin secretion which lies within the boundaries of the deafferentation (72). In order to further investigate the possible involvement of the hypothalamic areas, a number of discrete lesions were placed in various positions in this area and their effects on hepatic steroid metabolism studied. A comparatively large midline lesion placed in the periventricular region and stretching between levels A7190μ and A5660μ (i.e. roughly located anterior to the level of the deafferentation) caused complete "feminization" of steroid metabolism in the livers of operated male animals while having only small effects on female animals. However, lesions in the midline but occupying only the more rostral aspects of the lesion mentioned above (stretching from levels A7890μ to A6670μ) were without effect indicating that the essential areas involved in the control of the secretion of "feminizing factor" lies caudally to the level A6670μ; i.e. approximately extending from the beginning of the suprachiasmatic nucleus in a caudal direction.

Bilateral lesions essentially involving the nuclei interstitialis striae terminalis but also extending medially including the anterior commissure caused a moderate degree of "feminization" indicating that higher centres may be involved in the regulation of the secretion of "feminizing factor". Small bilateral lesions in the lateral preoptic area were without effect on steroid metabolism in the livers of the lesioned rats.

Overall, it can be said that the control of secretion of "feminizing factor" is located or mediated via the caudal suprachiasmatic area and/or overlying periventricular area of the hypothalamus (levels A6670μ to A5780μ). It is possible from the above data that the suprachiasmatic nucleus is involved (and is, indeed, sex-differentiated) but this is not yet certain.

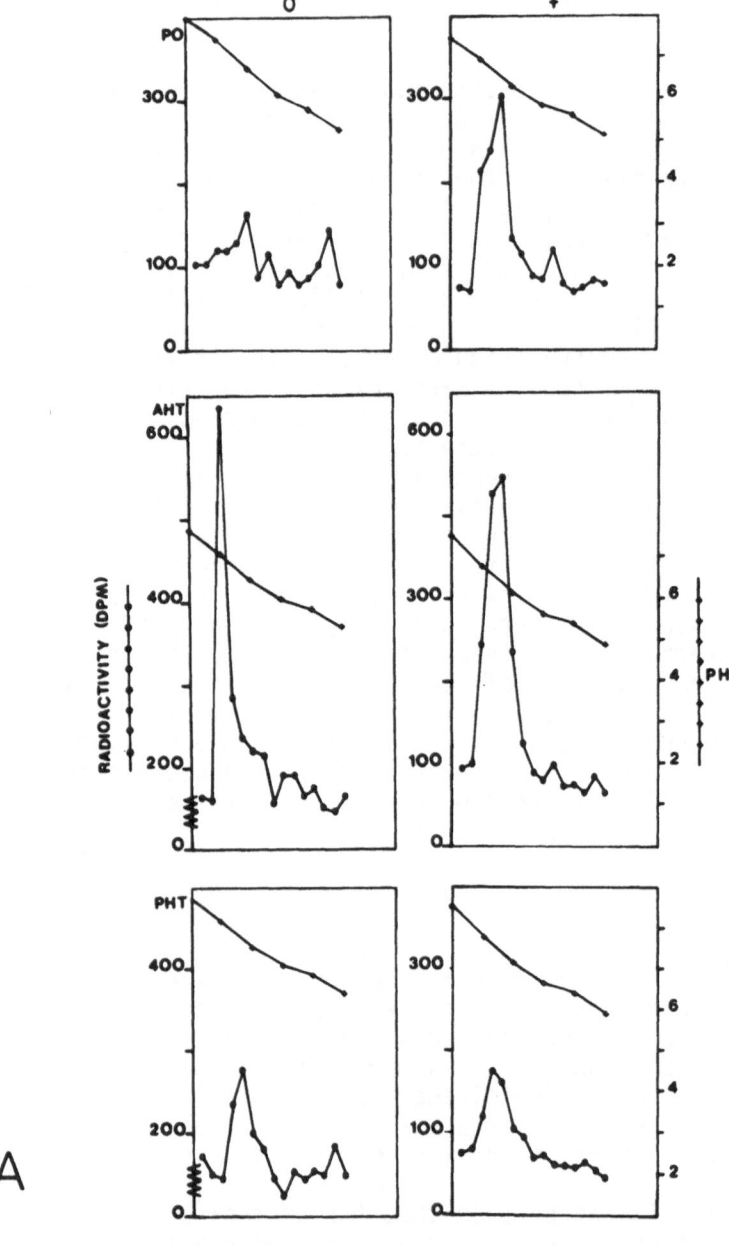

Fig. 3A and 3B. Isoelectric focusing of estradiol receptor in the preoptic area (PO), anterior hypothalamus (AHT), posterior hypothalamus (PHT), amygdala (AMYG), septum (SEP) and hippocampus (HIPP). The following amounts of protein (obtained from two animals) were applied in the respective analyses: 0.50 and 0.45 mg (males and females, respectively; PO), 0.58 and 0.60 mg (AHT), 0.42 and 0.52 mg (PHT), 1.01 and 0.72 mg (AMYG), 0.35 and 0.44 mg (SEP), and

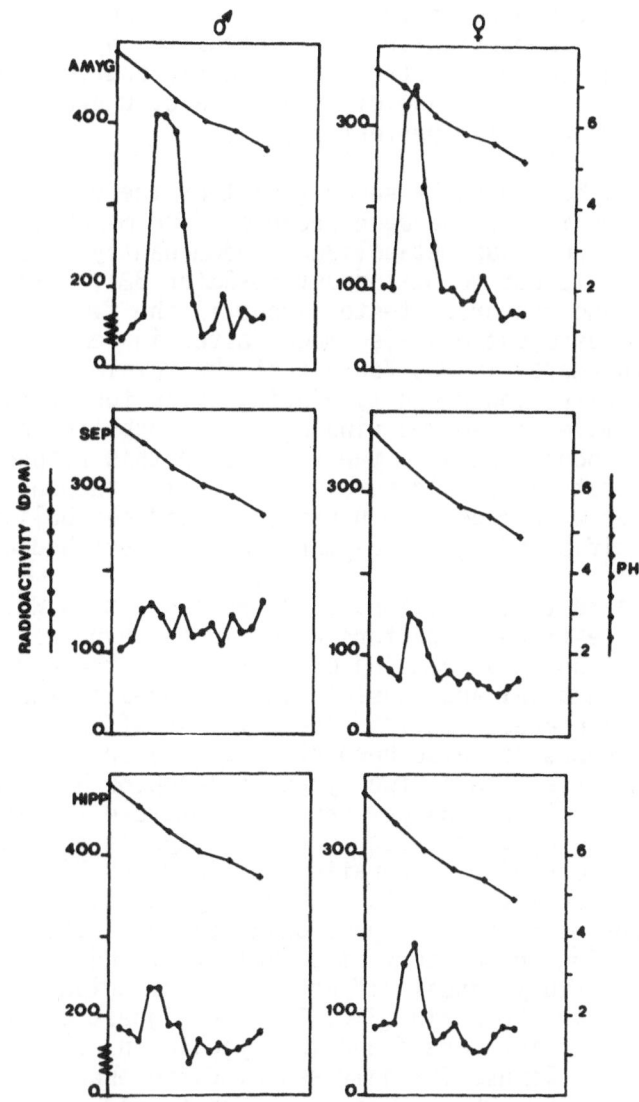

Fig. 3A and 3B (legend continued)

0.71 and 0.73 mg (HIPP). By quantitation of the area under the peak focused at about pH 6.6 the following receptor concentrations were calculated (expressed in fmol per mg protein): 1.4 and 7.5 (PO), 6.5 and 8.8 (AHT), 4.1 and 2.9 (PHT), 5.0 and 4.9 (AMYG), 2.5 and 2.3 (SEP), and 1.4 and 1.8 (HIPP).

The question may now be asked whether the effects of androgens and estrogens on the liver may be mediated via the hypothalamus. Much work has been done on the presence of receptors for sex steroids in the brain particularly as regards the hypothalamus. Naess et al. (73) showed the existence of specific androgen receptors in rat hypothalamus, cortex and pituitary gland. Characterization of these receptors was also performed (74).

Our group has recently shown (75) that there appears to be a sexual dimorphism of androgen receptors and metabolism in rat brain. Testosterone was metabolized predominantly to androstenedione in the male but to 5α-androstane-3α(or 3β),17β-diol, epitestosterone and dihydroepitestosterone in the female. The metabolism of testosterone was also much faster in female than male brains. Furthermore, it was found that the female brain lacked the high-affinity, low-capacity binding sites for testosterone found in the male. These two findings could explain the relative androgen unresponsiveness in the female. In this respect it is of interest that 28-day-old female rats (which respond to androgen as adult males with regard to hepatic steroid metabolism) have the male-specific androgen receptor protein described above.

With regard to the actions of estrogens, it is well accepted that specific estrogen receptors exist in the hypothalamus (see 65 for review) and that binding of estrogen to these receptors can cause biochemical and physiological changes in the receptor-containing brain areas (76,77). The ontogeny of the hypothalamic estrogen receptors has also been studied (66) and the findings indicated the existence of two types of receptor - a prepubertal and an adult type. The change-over from immature to mature receptor type appeared to occur between 20 and 30 days of age (the age at which sexual dimorphism of hepatic steroid metabolism is first seen). We have recently developed a highly sensitive and specific method for estradiol receptor quantitation, based on isoelectric focusing in polyacrylamide gel (78), and we have used this method to study sexual differences in distribution of estradiol receptors in the brain (Fig. 3). The highest concentrations of estradiol receptor were found in amygdala and anterior and posterior hypothalamus. The only sexual difference observed was in the preoptic area where female rats contained considerably higher receptor levels than male rats. The regulation of the estradiol receptor in the preoptic area is now being further investigated in our laboratory.

All of these data would seem to indicate that the hypothalamus (and pituitary gland) is capable of reacting to estrogens (in both sexes) and androgen (only in the male). It is thus likely that most, if not all, effects of sex steroids on hepatic steroid metabolism are indirect and mediated via the hypothalamo-pituitary

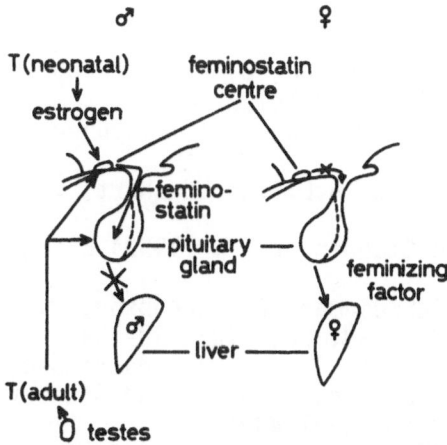

Fig. 4. Proposed model for the control and imprinting of hepatic steroid metabolism in the rat. T = testosterone.

system. Furthermore, it is the hypothalamus that is imprinted at birth by androgens exactly as has been demonstrated for the adult patterns of sexual behaviour and release of gonadotropins.

Our proposed theory for the imprinting of hepatic steroid metabolism in the rat is shown in Fig. 4. Imprinting occurs entirely in the hypothalamus where androgen in the neonatal period alters the anatomical, physiological and biochemical workings - possibly - of the suprachiasmatic area. This area, which we have labelled the feminostatin centre since we think it produces the feminizing factor release-inhibiting factor, is only active in the male. Androgen in the neonatal period is essential for activation of the centre. In the normal adult male the activated feminostatin centre produces and releases feminostatin which in turn inhibits the release of feminizing factor from the pituitary gland. This leads to a masculine pattern of hepatic metabolism. In the female, the feminostatin centre is inactive and thus feminizing factor is released from the pituitary gland giving a feminine pattern of hepatic metabolism. Androgens are active in the male in the adult period by interaction with the sex-specific androgen receptors in the brain while estrogens are active in both sexes via the estrogen receptors in the hypothalamus and pituitary gland. It is probable that the effects of the sex steroids are mediated via the feminostatin centre but a direct effect on the pituitary gland cannot be ruled out.

ACKNOWLEDGEMENTS

This work was supported by a grant from the Swedish Medical Research Council (No. 13X-2819).

REFERENCES

1. Hübener, H.J. and Amelung, D. (1953) Hoppe-Seyler's Z. Physiol. Chem. 293, 137-141.
2. Forchielli, E. and Dorfman, R.I. (1956) J. Biol. Chem. 223, 443-448.
3. Yates, F.E., Herbst, A.L. and Urquhart, J. (1958) Endocrinology 63, 887-902.
4. Leybold, K. and Staudinger, H. (1959) Biochem. Z. 331, 389-398.
5. Conney, A.H., Schneidman, K., Jacobson, M. and Kuntzman, R. (1965) Ann. N.Y. Acad. Sci. 123, 98-109.
6. DeMoor, P. and Denef, C. (1968) Endocrinology 82, 480-492.
7. Einarsson, K., Gustafsson, J.-Å. and Stenberg, Å. (1973) J. Biol. Chem. 245, 4987-4997.
8. Begue, R.-J., Gustafsson, J.-Å. and Gustafsson, S. (1973) Eur. J. Biochem. 40, 361-366.
9. Gustafsson, J.-Å. and Stenberg, A. (1974) J. Biol. Chem. 249, 711-718.
10. Gustafsson, J.-Å. and Stenberg, A. (1974) J. Biol. Chem. 249, 719-723.
11. Björkhem, I., Eriksson, H., Gustafsson, J.-Å., Karlmar, K.-E. and Stenberg, Å. (1972) Eur. J. Biochem. 27, 318-326.
12. Berg, A. and Gustafsson, J.-Å. (1973) J. Biol. Chem. 248, 6559-6567.
13. Gustafsson, J.-Å. and Gustafsson, S.A. (1974) Eur. J. Biochem. 44, 225-233.
14. Gustafsson, J.-Å. and Ingelman-Sundberg, M. (1974) J. Biol. Chem. 249, 1940-1945.
15. Stenberg, A. (1976) J. Endocrinol. 68, 265-272.
16. Castro, J.A. and Gillette, J.R. (1967) Biochem. Biophys. Res. Comm. 28, 426-430.
17. Davies, D.S., Gigon, P.L. and Gillette, J.R. (1968) Biochem. Pharmacol. 17, 1865-1872.
18. Gram, T.E., Guarino, A.M., Schroeder, D.H. and Gillette, J.R. (1969) Biochem. J. 113, 681-685.
19. El Defrawy El Masry, S., Cohen, G.M. and Mannering, G.J. (1974) Drug Metab. Dispos. 2, 267-278.
20. El Defrawy El Masry, S. and Mannering, G.J. (1974) Drug Metab. Dispos. 2, 279-284.
21. Erickson, R.R. and Holtzman, J.L. (1976) Biochem. Pharmacol. 25, 1501-1506.
22. Bell, J.U. and Ecobichon, D.J. (1974) Can. J. Biochem. 53, 433-437.
23. Quinn, G.P., Axelrod, J. and Brodie, B.B. (1958) Biochem. Pharmacol. 1, 152-159.
24. Gustafsson, J.-Å., Pousette, Å., Stenberg, A. and Wrange, Ö. (1975) Biochemistry 14, 3942-3948.
25. Feigelson, M. (1973) Enzyme 15, 169-176.

26. Denef, C. (1974) Endocrinology 94, 1577-1582.
27. Gustafsson, J.-Å. and Stenberg, Å. (1974) Endocrinology 95, 891-896.
28. Flerkö, B. In "The Hypothalamus" Eds. Martini, L., Motta, M. and Fraschini, F. (1970) Academic Press, p. 351-363.
29. Harris, G.W. (1964) Endocrinology 75, 627-648.
30. Gustafsson, J.-Å., Ingelman-Sundberg, M., Stenberg, Å. and Hökfelt, T. (1976) Endocrinology 98, 922-926.
31. Colby, H.D., Gaskin, J.H. and Kitay, J.I. (1974) Steroids 24, 679-686.
32. Kim, S., Wasserman, L., Lew, B. and Ki Paik, W. (1975) FEBS Lett. 51, 164-167.
33. Kramer, R.E., Greiner, J.W., Canady, W.J. and Colby, H.D. (1975) Biochem. Pharmacol. 24, 2097-2099.
34. Wilson, J.T. and Spelsberg, T.C. (1976) Biochem. J. 154, 433-438.
35. Schillinger, E. and Gerhards, E. (1974) Acta Endocrinol. (Kbh) 77, 502-508.
36. Griffin, E.E. and Miller, L.L. (1974) J. Biol. Chem. 249, 5062-5069.
37. Gustafsson, J.-Å. and Stenberg, Å. (1976) Proc. Natl. Acad. Sci. USA 73, 1462-1465.
38. Eneroth, P., Gustafsson, J.-Å., Skett, P. and Stenberg, Å. (1977) Mol. Cell Endocrinol. 7, 167-175.
39. Chen, C.L., Amenomori, Y., Lu, K.H., Voogt, J.L. and Meites, J. (1970) Neuroendocrinology 6, 220-227.
40. Lam, P.C.O., Morishige, W.K. and Rothchild, I. (1976) Proc. Soc. Exptl. Biol. Med. 152, 615-617.
41. Arimura, A., Debeljuk, L., Shino, M., Rennels, E.G. and Schally, A.V. (1973) Endocrinology 92, 1507-1514.
42. Evans, J.S. (1972) Endocrinology 90, 123-130.
43. Gustafsson, J.-Å. and Stenberg, Å. (1975) Acta Endocrinol. (Kbh) 78, 545-553.
44. Gustafsson, J.-Å. and Stenberg, Å. (1975) Endocrinology 96, 501-504.
45. Lax, E.R., Ghraf, R., Schriefers, H., Herrmann, M. and Petutschnigk, D. (1976) Acta Endocrinol. (Kbh) 82, 774-784.
46. Kramer, R.E. and Colby, H.D. (1976) J. Endocrinol. 71, 449-454.
47. Wilson, J.T. (1970) Nature 225, 861-863.
48. Shirasu, Y., Grantham, P.H., Yamamoto, R.S. and Weisburger, J.H. (1966) Cancer Res. 26, 600-606.
49. Shirasu, Y., Grantham, P.H., Weisburger, E.K. and Weisburger, J.H. (1967) Cancer Res. 27, 81-87.
50. Kramer, R.E., Greiner, J.W. and Colby, H.D. (1977) Biochem. Pharmacol. 26, 66-69.
51. Gustafsson, J.-Å. and Skett, P. (1978) J. Endocrinol. 76 187-191.
52. Skett, P., Eneroth, P. and Gustafsson, J.-Å. (1978) Mol. Cell.

Endocrinol. 10, 21-27.

53. Gustafsson, J.-Å., Ingelman-Sundberg, M., Stenberg, A. and Hökfelt, T. (1976) Endocrinology 98, 922-926.

54. Kamberi, I.A., Mical, R.S. and Porter, J.C. (1971) Endocrinology 88, 1003-1011.

55. Greenough, W.T., Carter, C.S., Steerman, C. and de Voogd, F.J. (1977) Brain Res. 126, 63-72.

56. Pfaff, D.W. (1966) J. Endocrinol. 36, 415-416.

57. Dörner, G. and Staudt, J. (1968) Neuroendocrinology 3, 136-140.

58. Dörner, G. and Staudt, J. (1969) Neuroendocrinology 4, 278-281.

59. Raisman, G. and Field, P.M. (1973) Brain Res. 54, 1-29.

60. Libertun, C., Timiras, P.S. and Kragt, C.L. (1973) Neuroendocrinology 12, 73-85.

61. Naftolin, F., Ryan, K.J. and Petro, Z. (1971) J. Clin. Endocrinol. Metab. 33, 368-370.

62. Plapinger, L. and McEwen, B.S. (1973) Endocrinology 93, 1119-1128.

63. Gellert, R.J., Bakke, J.L. and Lawrence, N. (1971) Fert. Steril. 22, 244-249.

64. Gellert, R.J., Heinrichs, W.L. and Swerdloff, R. (1974) Neuroendocrinology 16, 84-94.

65. McEwen, B.S., Gerloch, J.L., Luine, V.N. and Lieberburg, I. (1977) Psychoneuroendocrinology 2, 249-255.

66. Spona, J., Bieglmayer, C., Adamiker, D. and Gettmor, W. (1977) FEBS Lett. 76, 306-310.

67. Luttge, W.G. and Whalen, R.E. (1970) Horm. Behav. 1, 265-281.

68. McDonald, P.G. and Doughty, C. (1974) J. Endocrinol. 61, 95-103.

69. Whalen, R.E. and Rezek, D.L. (1974) Horm. Behav. 5, 125-128.

70. Gustafsson, J.-Å. and Stenberg, A. (1976) Science 191, 203-304.

71. Gustafsson, J.-Å., Eneroth, P., Hökfelt, T. and Skett, P. (1978) Endocrinology in press.

72. Krulich, L., Hefco, E. and Aschenbrenner, J.E. (1975) Endocrinology 96, 107-118.

73. Naess, O., Attramdal, A. and Aakvaag, A. (1975) Endocrinology 96, 1-9.

74. Naess, O. (1976) Steroids 27, 167-185.

75. Gustafsson, J.-Å., Pousette, A. and Svensson, E. (1976) J. Biol. Chem. 251, 4047-4054.

76. Pfaff, D. In "Primer of Neuroendocrine Function and Behaviour" Ed. Adler, N. (1978) in press.

77. Kubo, K., Gorski, R.A. and Kawakami, M. (1975) Neuroendocrinology 18, 176-181.

78. Wrange, Ö., Nordenskjöld, B., Gustafsson, J.-Å. (1978) Anal. Biochem. 85, 461-475.

NEUROTRANSMITTERS IN THE CONTROL OF ANTERIOR PITUITARY FUNCTION

S.M. Mc Cann, L. Krulich, S.R. Ojeda, A. Negro-Vilar
and E. Vijayan
Univ Tx Hlth Sci Ctr at Dallas, Southwestern Med Sch

Dallas, Texas U.S.A.

INTRODUCTION

Earlier papers in this volume have amply documented the fact that the anterior pituitary is controlled by a family of hypothalamic releasing and inhibiting hormones which are secreted into the hypophyseal portal vessels in the median eminence (ME) to stimulate or inhibit the release of individual pituitary hormones. The distribution of the peptidergic neurons which synthesize and release the various releasing hormones within the hypothalamus brings these neurons in close contact with a variety of putative synaptic trans- mitters. The purpose of this chapter will be to review the evidence that some of these transmitters may be involved in controlling release of the releasing and inhibiting hormones. We will stress results from our laboratory but will include crucial experiments from other laboratories so as to present a balanced picture of the present status of knowledge in this field.

We will concentrate on the evidence implicating various transmitters in the release of gonadotropins and PRL because more work has been done in these areas but will also briefly discuss putative transmitters involved in control of TSH, GH and ACTH secretion.

Putative synaptic transmitters controlling gonadotropin release. The first evidence for neurotransmitter control of gonadotropin release was that of Sawyer, Markee and Everett (1) who found that the adrenergic blocking drug, dibenamine, or large doses of atropine could block ovulation. They postulated both adrenergic and cholin- ergic involvement in the control of gonadotropin secretion.

a. <u>Dopamine (DA)</u>. The juxtaposition of dopaminergic and
LHRH containing terminals in the same regions in the ME clearly
suggests the possibility of a dopaminergic involvement in the
control of gonadotropin secretion (2) but in spite of much investigation, the role of DA in control of gonadotropin secretion
remains controversial . We will first present the evidence that DA
stimulates the release of LHRH and then the evidence that it may
inhibit release of the neuropeptide. Finally, we will attempt to
reconcile these opposing positions.

It was first reported in 1969 that addition of DA to pituitaries conincubated with hypothalamic fragments increased release
of LH (3).   Intraventricular injection of DA also released LH and
increased LRH activity in portal and in peripheral blood of hypophysectomized rats (4).

Recent <u>in vitro</u> experiments in which LHRH was measured directly
by RIA support the stimulatory role for DA in the control of LH
release since DA was found to elicit LHRH release by hypothalamic
fragments in a dose-related fashion (5) and also from hypothalamic
synaptosomes (6).   Recently, we also found that incubation of
median eminence tissue with DA resulted in a dose-related release
of LHRH into the medium (7).

We have also re-evaluated the effects of intraventricular
injection of DA both in ether-anesthetized and conscious rats.
Dopamine or its agonist, apomorphine, was capable of releasing LH
in the ovariectomized, steroid-primed rat in agreement with earlier
experiments; however, norepinephrine (NE) and epinephrine (E) were
slightly more effective (8). Furthermore, low doses of DA or
apomorphine infused intravenously also elicited LH release in
steroid-primed but not in ovariectomized animals. Pulse i.v.
injection of DA elevated LH in the steroid-primed and also in
ovariectomized animals (9).

Certain other experiments are consistent with the stimulatory
role for DA. For example, blockade of DA receptors with pimozide
although failing to modify pulsatile LH release in ovariectomized
rats, nonetheless, decreased the elevated LH levels in these
animals after a delay of several hours (10,11). Pimozide also
blocked the LH induced by gonadal steroids in ewes (12) and
suppressed the preovulatory surge of gonadotropins on proestrus in
the rat (13).   Injection of lergotrile mesylate, a dopaminergic
receptor stimulating drug, to proestrous rats in which the preovulatory surge had been blocked by reserpine reinitiated the discharge
(14). Lastly, animals treated neonatally with monosodium glutamate
showed impairment in ovulatory LH release as adults which correlated
with decreased hypothalamic DA and not NE levels (15).

At the same time that evidence for a stimulatory role of DA on LH release was being presented, Fuxe and Hokfelt obtained evidence of decreased turnover of tuberoinfundibular dopaminergic neurons in situations associated with enhanced LH release and vice versa and postulated that dopaminergic neurons inhibited rather than elicited release of LHRH (2). In a series of papers, using quantitative histofluorometric measurement of tuberoinfundibular DA turnover, Fuxe and associates have obtained additional evidence that DA may be an inhibitory transmitter to suppress LHRH release (16). Rubinstein and Sawyer (17) reported that E released LH in the rat but that DA was ineffective. More recently Sawyer et al. reported that DA injected into the 3rd ventricle of the rabbit not only failed to elicit LH release but blocked the increase in LH induced subsequently by NE. In a ME pituitary superfusion system, Miyaki et al. reported that DA inhibited LH release. Krieg and Sawyer even found that intraventricular injection of DA in doses reported earlier to evoke LH release in ovariectomized estrogen progesterone-treated rats had no effect; however, it must be pointed out that only three rats were used (for references and review see 18).

The first direct evidence that DA could inhibit LH release was provided by Gnodde and Schuiling (19) and Drouva and Gallo (10) who reported that systemic injection of large doses of apomorphine, a DA receptor stimulant, inhibited pulsatile LH release in castrates. Intravenous infusion of DA was also found to lower circulating LH levels in man (20).

Several DA receptor agonists also inhibited the LH surge induced in immmature females by pregnant mare's serum gonadotropin, but some DA receptor blockers (chlorpromazine) had the same effect (21). Spurred by these reports, we evaluated the effects of very large doses of DA and apomorphine in ovariectomized conscious rats and found that i.p. injection of 5 or 50 mg/kg of either agent could suppress LH release under these circumstances (9). It should be pointed out that these doses are much larger than those which had been found either to have no effect or to stimulate LH release in our experiments carried out concurrently in ovariectomized or OEP rats. In our experiments the inhibitory action appeared to be exerted on the hypothalamus since the response to LHRH was not blocked; however, in the experiments in patients, there was a diminished release of LH in response to LHRH raising the possibility that these doses of DA may inhibit the gonadotrophs directly.

Thus, we conclude that DA has a stimulatory action at relatively low doses particularly in animals under the influence of gonadal steroids and that high doses can suppress gonadotropin release particularly in the castrate. The physiological significance of these actions of DA is still not clear since results with receptor blockers are not completely unambiguous. There is little evidence that the inhibitory effect has physiological significance since administration of receptor blockers seldom leads to augmented LH release; however, implantation of pituitaries under the kidney capsule which leads to a lasting elevation of serum prolactin levels, which in turn stimulate the activity of the tuberoinfundi-bular DA system (22) prevented the postcastration rise of LH levels (23) and temporarily decreased LH levels in castrated rats (22,24).

The mechanism by which dopamine and its agonists either stimulate or inhibit LH release is also not clear. It is possible that the inhibitory action might be mediated via DA auto-receptors (25) which would lead to reduced synthesis of DA followed by decreased DA release and decreased dopaminergic stimulation of LH release. Alternatively, relatively low doses of DA may act on post-synaptic-excitation mediating hormone receptors (DA_e) and higher doses on the inhibition-mediating post-synaptic dopamine receptors (DA_i) (26). The ability of steroid priming to enhance the LH releasing action of DA could be related to an action of estrogen to alter the population of DA receptors.

b. Norepinephrine (NE). In contrast to the controversy regarding the role of DA in gonadotropin control, a stimulatory role for NE in the control of gonadotropin release appears to be reasonably well established. Turnover of hypothalamic NE is in-creased during proestrus at the time of preovulatory release of gonadotropins (27) and this increased turnover has been localized to the suprachiasmatic nucleus (28). Norepinephrine synthesis is elevated in the anterior hypothalamus in ovariectomized rats (29) and an increase in NE content of the median eminence precedes the rise in gonadotropins following orchidectomy (30). In very recent studies the elevated NE turnover rate in the suprachiasmatic-preoptic region of ovariectomized rats was acutely suppressed by estradiol indicating that the acute negative feedback of estradiol may be mediated at least in part through decreased noradrenergic transmission (31).

Injection of NE into the 3rd ventricle can elevate LH release (8) by an effect apparently mediated via alpha receptors since alpha receptor blockers inhibit not only ovulation but also the post-castration rise in gonadotropin release in male rats (11) and the pulsatile release of LH in the ovariectomized monkey (32).

Studies with inhibitors of catecholamine synthesis also support the role of NE in control of LH release. Administration of α-methyltyrosine (α-MT) to inhibit tyrosine hydroxylase and produce a reduced synthesis of catecholamines blocked the post-castration rise in gonadotropins, progesterone-induced LH release in estrogen-primed rats and the preovulatory discharge of LH. In these instances, the inhibition could be at least partially reversed by administration of either L-dopa to reinitiate the synthesis of both DA and NE or of dihydroxyphenylserine (DOPS) to reinitiate synthesis of only NE. Selective blockade of NE synthesis has also prevented the LH surge induced by gonadal steroids, the preovulatory LH discharge and reduced elevated LH titers in castrates.

Since blockade of NE synthesis prevented the elevation in LH release following stimulation of the preoptic area but not stimulation of the median eminence and this blockade could be reversed by the administration of L-dopa or DOPS to replenish NE stores it was suggested that noradrenergic terminals in the preoptic or anterior hypothalamic area synapse with LHRH neurons to mediate the release of LHRH which occurs at the time of the preovulatory release of gonadotropins. Further experiments have shown that acute lesions in the ventral noradrenergic tract which carries most of the axons of noradrenergic neurons which project to the hypothalamus abolished the LH surge on proestrus or that induced by progesterone in estrogen-primed rats (33). Therefore, it is probable that the noradrenergic neurons involved in the LH surge have their cell bodies in the brain stem and not in the hypothalamus.

Since E is also capable of releasing LH following its intra-ventricular injection (8) and axons of epinephrinergic neurons terminate in the hypothalamus, it is possible that the effects attributed to NE may be mediated by E instead (for further references see 34).

c. Serotonin (5HT) and Melatonin. It appears quite clear that 5HT can inhibit gonadotropin release since intraventricular injection of this amine has decreased circulating LH levels, the effect being particularly apparent in castrates (see 34). Stimulation of the raphé nuclei to activate 5HT release in the hypothalamus has been found to suppress pulsatile release of LH by Gallo and Moberg (35) which indicates that endogenous 5HT can suppress LH release. Whether 5HT plays any physiological role in the control of LH release remains to be determined since administration of p-chlorophenylalanine to block 5HT biosynthesis has failed to modify gonadotropin release in the castrate. It is even possible that serotonergic tone may be needed for the expression of the preovulatory discharge of LH since administration of p-chlorophenyl-alanine or production of raphé lesions to deplete 5HT stores resulted in an inhibition of estrogen-induced LH release (36).

The pineal indole, melatonin, has been found capable of inhibiting gonadotropin and augmenting PRL release following its injection into the 3rd ventricle (see 34).

d. Acetylcholine (Ach). It was reported recently that Ach could increase FSH release from hypothalamic-pituitary co-incubates and that intraventricular injection of acetylcholine induced LH

release (37). We have similarly observed that Ach bromide induced
LH release in ovariectomized, estrogen progesterone-treated (OEP)
rats. Furthermore, intraventricular injection of atropine can
block the response to intraventricular acetylcholine as well as the
preovulatory surge of gonadotropins. Intrahypothalamic implants
of atropine blocked the post-castration rise in LH and FSH (38).
That the effects of Ach may be mediated through the dopaminergic
system is suggested by the fact that the elevation in LH induced
by intraventricular Ach is blocked in the pimozide-treated animal.
Conversely, activation of nicotinic receptors with nicotine inhibited
the proestrous surge (18).

        e. Histamine. Histamine is present in the hypothalamus in
high concentration and appears to be localized in synaptosomes.
Although it appears possible that histamine is a stimulatory transmitter to release prolactin (PRL) a role for histamine in control
of LH release appears rather doubtful since only very high doses
of histamine injected intraventricularly induced LH release (34).

        f. γ-Aminobutyric Acid (GABA). GABA is found in high quanti-
ties in the diencephalon and its involvement in LH control was
suggested by the observation that injection of GABA into the 3rd
ventricle induced LH but not FSH release in the Nembutal-anesthetized
male rat (39). Since the amino acid was inactive when injected
directly into the pituitary it was thought to act on the hypothalamus
to stimulate LHRH release. No effect of GABA was observed in unanesthetized males but we have subsequently found elevation of plasma
LH following 3rd ventricle injection of GABA in unanesthetized
ovariectomied and OEP rats which were blocked by the GABA blocker
bicuculline suggesting that GABA was acting specifically. Since
bicuculline alone did not alter LH it is still not clear if GABA
plays a physiological role in control of LH release (40). Glutamic
acid, lysine and β-alanine have recently also been reported to
stimulate LH release following 3rd ventricular injection in male
rats (41). The specificity of the effects is suggested by the
failure of glycine and α-alanine to alter LH in similar doses.

        g. Peptides and the Release of LH. There is evidence to sug-
gest that certain peptides found in the brain may be involved in
the control of LHRH release. For example, the pineal peptide,
arginine vasotocin, given systemically or infused into the ventricle
inhibited the preovulatory discharge of LH while having no effect
on plasma LH in the castrate (42). Vasoactive intestinal peptide
(VIP) has recently been demonstrated to be present in the hypothalamus and intraventricular injections of VIP have been found to elevate
plasma LH whereas its systemic injection or incubation with pituitaries in vitro was without effect (43). These results suggest a
possible role of VIP in stimulating LHRH release. Substance P,
neurotensin and gastrin have also been extracted from hypothalamus
and we have found that intraventricular injection of substance P

can elevate plasma LH whereas similar doses of neurotensin and
gastrin lowered LH titers (44). Since there was no effect of systemic injection of the peptides, and in vitro LH release was un-
modified, these effects are probably mediated via alterations in
LHRH release.

Putative synaptic transmitters controlling prolactin (PRL)
release. Similar approaches have been used to evaluate the participation of transmitters in the control of PRL release.

a. Dopamine. Early studies utilizing systemic administration
of drugs which alter catecholamine synthesis pointed to an important
inhibitory role of DA in the control of PRL release (for reference
see 34). Many subsequent studies with a variety of drugs have
always pointed to an inhibitory role of DA in the control of PRL
release.

In later studies in unanesthetized animals to eliminate the
effects of stress, infusion or injection of either apomorphine,
the dopamine receptor stimulator, or DA itself lowered PRL (9).
Furthermore, pimozide, the DA receptor blocker elevated plasma PRL
providing further evidence for dopaminergic inhibitory control.

That DA acts centrally to inhibit prolactin release was shown
by injection of DA or apomorphine into the 3rd ventricle which
resulted in suppression of PRL release (8). The effect of DA was
not blocked by DDC to block its conversion into NE and ET495,
another DA receptor stimulator, was also effective to lower PRL.
Implantation of pimozide into the ME-arcuate region evoked a marked
release of PRL and there was only a small elevation of PRL following
implantation of the drug into the adenohypophysis. These in vivo
results then suggested that DA may exert a dual inhibitory control
over PRL release by stimulating the release of a PRL-inhibiting
factor (PIF) from the ME and by acting directly on the gland to
suppress PRL release.

There can be little doubt that DA can act directly on the
pituitary to suppress PRL release. First, in patients with stalk
sections or in rats with lesions in the ME, L-dopa still produces
a dramatic decline in serum PRL presumably by its conversion to DA
which than acts directly on the pituitary. Second, it has been
clearly shown that DA can inhibit PRL release by the pituitary in
vitro as demonstrated by MacLeod and Lehmeyer (45). Dopamine is
effective in a dose-related fashion and NE also suppresses release.
Since DA has now clearly been demonstrated in portal vessels (46),
it may play a physiological role as a PIF. Recent studies show
that the quantity found in portal blood is sufficient to suppress
PRL release when administered systemically (47). This has led
some to postulate that DA is PIF; however, it is possible that DA

may also release a peptide PIF in the ME which can act to inhibit
PRL. For example, co-incubation of hypothalami with pituitaries
gave an inhibition of PRL release even in the presence of haloperidol to block the effect of DA and the inhibition increased when DA
was added to the medium. This inhibition of release of PRL is
therefore presumably independent of DA and due to another PIF (48).
Furthermore, Enjalbert et al. (49), have obtained a catecholamine
free PIF activity in hypothalamic synaptasomes. Consequently, it
appears that DA acts to inhibit PRL release either by releasing a
PIF or by a direct action on the gland or by both actions.

     The physiological role of the DA system appears to be to hold
PRL secretion in check under resting conditions. Lesions in the
ME lead to a rise in PRL as expected but complete deafferentation
of the medial basal hypothalamus is not followed by an elevation
of serum PRL (50).   The low PRL levels in these animals probably
result from continued release of DA from the tuberoinfundibular
dopaminergic neurons resident within the island. Consistent with
this view are the findings that systemic injection of L-dopa led
to a further decrease of PRL levels in the deafferented animals,
whereas reserpine, to deplete catecholamines, elevated PRL. Norepinephrine is absent from these islands because of severance by
the deafferentation of the noradrenergic axons entering the MBH.
The activation of PRL secretion by stress or suckling is probably
not via inhibition of the dopaminergic systems since both stimuli
still activated PRL release after the influence of this system had
been eliminated by reserpine, pimozide or a α-methyl-p-tyrosine
(51).

     b. Norepinephrine and Epinephrine. Early evidence suggested
that NE can stimulate PRL release since injection of DOPS in α-
methyl-tyrosine treated animals to normalize only NE synthesis
leaving a deficit in the synthesis of DA resulted in a further
elevation in PRL levels. Treatment with DOPS alone to elevate NE
levels selectively also raised PRL. Furthermore, intraventricular
injection of NE at a relatively high dose elevated PRL (for reference
see 34). Clonidine injection also was effective to elevate PRL (52)
but the results are clouded since intraventricular injection of
α-receptor blockers also stimulated PRL release (52). Systemic
injection of alpha and beta receptor blockers inhibited periodic
PRL surges in ovariectomized estrogen-treated rats which would be
consistent with a role for NE in stimulating PRL release (53).
Similarly, lesions in the ventral noradrenergic tract blocked the
preovulatory rise in PRL suggesting that it may involve a noradrenergic link (33). The preponderance of evidence favors a stimulatory
role of NE in control of PRL release.

     c. Serotonin. Considerable evidence supports a stimulatory
role for 5HT in the control of PRL release. Systemic injection of
5-hydroxytryptophane (5HTP), the immediate precursor of 5HT, leads

to a marked elevation of PRL levels concomitant with increased concentrations of brain 5HT. The specificity of this effect is evidenced by the finding that the PRL releasing effect of small doses of 5HTP is greatly augmented by treatment with fluoxetine which specifically blocks the re-uptake of 5HT by serotoninergic terminals. Quipazine, a 5HT receptor stimulator, also gives a marked stimulation of PRL release following its systemic administration. Although 5HT itself is a potent stimulus for PRL release following its systemic injection, its effects may be a secondary consequence of the powerful stress induced by 5HT, particularly since 5HT does not readily cross the blood brain barrier. It appears that the PRL releasing effect of 5HT is independent of the DA system since stimulation of PRL release by 5HTP or quipazine still occurs in animals in which DA synthesis has been blocked with α-MT (for reference see 48).

Serotonin has been thought to mediate the PRL-releasing effect of stress and suckling since the 5HT receptor blocker, methysergide, has been found capable of blocking these responses; however, methysergide inhibits PRL secretion in animals with lesions in the ME indicating that it has a direct inhibitory effect on the pituitary lactotrophs which may not be related to its anti-serotoninergic properties. Furthermore, depletion of central 5HT stores by p-chlorophenylalanine or p-chloroamphetamine does not prevent the PRL-releasing effect of stress. The suckling-induced release of PRL has been blocked by p-chlorophenylalanine which certainly supports an important role for 5HT in this phenomenon (for references see 48).

d. Acetylcholine (Ach). An inhibitory role for Ach in PRL release is supported by the ability of systemic injection of pilocarpine or eserine to decrease plasma PRL in OEP rats. There was an unexplained delayed increase in PRL at 6 hours following injection. The lowering of PRL was blocked by treatment of the animals with atropine sulphate which crosses the blood brain barrier but not with atropine methyl nitrate which cannot penetrate it suggesting that the drugs were acting on the CNS muscarinic receptors to produce their effect on PRL release. Similarly, injection of Ach into the 3rd ventricle can inhibit PRL release. The inhibition of PRL secretion following activation of muscarinic receptors probably involves stimulation of the dopaminergic system since systemically administered pilocarpine and intraventricularly administered Ach failed to affect PRL release in animals pretreated with pimozide or α-methyltyrosine whereas conversely pimozide restored the PRL response to stress which had been inhibited by pilocarpine. In contrast to the effects following systemic administration of cholinergic drugs, intraventricular injection of atropine blocked the proestrous surge of PRL. This result does not fit with those previously discussed and could be related to other actions of atropine at these relatively high doses (for references see 48).

e. Histamine. Intraventricular injection of histamine
produced an elevation of PRL in OEP rats. In other studies carried
out concurrently it was shown that diphenhydramine, an H1 histamine
receptor blocker, could block the restraint stress-induced elevation
of PRL suggesting that histamine may play a role in mediating the
elevation of PRL which follows a variety of stressful stimuli.
Later studies have shown that blockade of the H1 but not H2 histamine
receptor can partially block suckling induced PRL release. Surprisingly, intraventricular injection of a H2 receptor blocker actually
elevated PRL (for references see 48).

f. γ-Aminobutyric Acid. Intraventicular injection of GABA
has been reported to elevate plasma PRL in male rats (54) but in
ovariectomized, steroid-primed females low doses of GABA injected
into the ventricle lowered plasma PRL whereas high doses elevated
PRL (40). These elevations were blocked by the GABA blocker, bicuculline, which by itself had no effect on plasma PRL. Thus, it is
clear that GABA can influence PRL but its physiological role remains
to be clarified. The effects obtained by intraventricular injection
were thought to be mediated by the hypothalamus via alterations in
PIF or PRL-releasing factor release since injection into the pituitary had no effect; however, it must be noted that several workers
have reported that high doses of GABA can lower PRL release by the
pituitary incubated in vitro (55). The effect of GABA on plasma
PRL in ovariectomized rats was reversed by pretreatment of the
animals with pimozide, the elevation which supervened in intact
animals was converted into a lowering. Thus, the elevation in PRL
induced by GABA may involve, in part, removal of dopaminergic
inhibitory tone.

g. Polypeptides. A variety of peptides found in the brain
have been shown to be capable of modifying PRL release. For example,
vasoactive intestinal peptide (VIP) elevated PRL following its
injection into the 3rd ventricle at certain doses in ovariectomized
rats (43). Higher doses were effective to elevate PRL after intravenous administration. Substance P injected into the 3rd ventricle
also elevated PRL whereas similar doses of neurotensin and gastrin
induced significant declines in plasma PRL (44). Intravenous injection of similar doses of each of these three peptides produced
elevations of PRL which contrasted with the lowering induced by
neurotensin and gastrin when injected intraventricularly. The effects
with intravenous injection could be related to stress whereas the
lowering induced by neurotensin and gastrin injected into the
ventricle is presumably a central action to suppress PRL release.
In the case of substance P, since systemic injection also produced
elevations, it is not clear if substance P acts centrally or peripherally to stimulate PRL release. Recently, in male steroid-
primed rats, bombesin and alatesin were found to be extremely
active in stimulating PRL release (56); however, their minimal
effective dose was lower after intravenous than after intracisternal

administration. Therefore it is still possible that these are
peripheral rather than central actions of these peptides. The
stimulatory effects of these peptides on PRL release were not modified by diphenhydramine, the H1 histamine receptor blocker or the
opiate blocker, naloxone. Intraventricular injection of endorphins
and encephalins has also been found to evoke PRL release and the
responses can be blocked by the opiate receptor blocker, naloxone
(57). This was expected since it had earlier been reported that
morphine injected into the ventricle could stimulate PRL release.
That the opiate system may play a role in mediating stress-induced
PRL release is suggested by the finding that naloxone can block
stress-induced PRL release in the rat (58).

Synaptic transmitters involved in the control of growth hormone (GH) secretion. Extensive studies indicate that DA can
stimulate GH release in primates. Contrary to early studies (59),
recent findings fit with a dopaminergic stimulation of GH release
in the rat as well. Recently, we observed that intraventricular
or i.p. injection of DA or its agonists, ET495 or apomorphine, was
capable of elevating plasma GH in ovariectomized or OEP rats and
that DA was effective whether the animals were anesthetized or
conscious (60). Our conclusion that activation of the central
dopaminergic system can stimulate GH secretion is supported by
earlier results from systemic administration of DA receptor agonists
in rats (61) or in human subjects (62). A DA receptor blocker
decreased the amplitude of GH pulses in the rat which suggests a
physiological role for DA in GH release (63). The earlier findings of
suppression of GH release after intraventricular injection of DA
(59) may be related to the use of urethane anesthesia.

The evidence that activation of the central noradrenergic
systems leads to GH secretion is quite good. In our own experiments,
intraventricular injection of NE induced a significant rise in plasma
GH levels in ovariectomized as well as OEP rats (60). Similarly in
the baboon injection of NE into the ventromedial nucleus elevated
plasma GH (64) and it appears on the basis of experiments with
receptor blockers that the noradrenergic control is mediated by an
alpha receptor mechanism in humans and other primates (65,66). In
agreement with our results Muller et al. (67) reported enhancement
of GH secretion following injection of NE into the lateral ventricle
and systemic injection of the alpha receptor stimulant, chlonidine,
was found effective by Durand et al. (68). Blockade of α-receptors
or inhibition of catecholamine synthesis abolished pulsatile GH
release in the rat (63). The surges could be reinstated with
clonidine and not apomorphine, an indication that the central NE
system represents a major drive for GH release. In man, α-receptors
also appear to stimulate GH release whereas β-receptors inhibit it
(69). Since E was also effective to release GH following its
intraventricular injection in our experiments, it is possible
that E may play a physiological role in stimulating GH secretion.

Serotonin may be involved in stimulating GH secretion since elevation of plasma GH has been observed following administration of 5HTP, its immediate precursor, in monkey and man (70) and since blockade of serotonin receptors prevented activation of GH secretion by hypoglycemia in man (71). The serotonin system appears to stimulate GH secretion in the rat since intraventricular injection of serotonin or quipazine elevated plasma GH (72). It is probable that the serotoninergic system participates in physiological regulation of GH secretion, because in humans blockade of serotonin receptors suppressed activation of GH secretion by insulin hypoglycemia (73) and the sleep-related activation of GH secretion (74).

Recently we have observed that intraventricular injection of Ach bromide will elevate GH release and the effect is blocked by intraventricular administration of atropine (75). This effect may be mediated via the dopaminergic system since the administration of pimozide to block DA receptors blocked the stimulatory action of Ach. Injection of GABA into the 3rd ventricle in ovariectomized or OEP animals also stimulated GH secretion and the effect was blocked by the GABA blocker, bicuculline, which by itself had no effect on GH secretion (76). Consequently, GABA appears capable of stimulating GH secretion but may not have a physiological role. Pimozide failed to modify the action of GABA which appears to be exerted independently of the dopaminergic system.

Several peptides found in the hypothalamus may also play a role in controlling GH release but much more work is needed to establish this concept. For example, significant elevations in plasma GH levels were observed within five minutes of intraventricular injection of various doses of substance P and neurotensin whereas only a relatively high dose of gastrin elevated GH titers (44). Systemic injection of similar doses of each of these peptides had no effect on plasma GH establishing a central action. Similarly, VIP elevated GH following its intraventricular but not its systemic administration (43). In this case, large doses given systemically lowered GH probably by lowering blood pressure and inducing stress which is known to lower GH in the rat. As in the case of PRL, encephalins and endorphins elevate GH by a central action (58) and the central opiode peptide system may be involved in the control of the release of this trophic hormone. Very recently bombesin and alatesin, peptides isolated from the skin from amphibia, were also shown to elevate GH following systemic or intracisternal injection; it cannot be stated with certainty that this was a central action of the peptides. The action may be mediated via the central opiate system since naloxone reversed the GH releasing activity of bombesin (57). There was no activity on GH release from the pituitary in vitro of any of these peptides suggesting that their action is on the hypothalamus.

Synaptic transmitters controlling TSH release. It is now clear that DA has an inhibitory effect on TSH release. For example, activation of the DA system either by DA itself or by the receptor stimulants, ET495, or apomorphine, whether they are given intraventricularly or systemically markedly inhibits release of TSH (60). The action is probably exerted on the hypothalamus rather than the pituitary since DA receptor stimulants failed to inhibit TSH secretion by the pituitary thyrotrophs directly (77). It is not clear whether DA plays a physiological role in the regulation of TRH release since blockade of DA receptors has no effect on basal or cold stimulated TSH release (77).

On the other hand, the evidence is mounting that TRH release is stimulated by the central noradrenergic system. Increased release of TSH follows intraventricular injection of NE or the α-receptor stimulant, clonidine (60,78), whereas selective inhibition of NE synthesis or blockade of alpha receptors inhibits cold-induced release of TRH (77). Intraventricular injection of E is effective in releasing TSH (60), but the role of E remains to be established.

Limited studies have not clarified the role of serotonin in control of TSH release in the rat. Although large doses of tryptophane inhibited TSH secretion, 5HTP, a more direct serotonin precursor had no effect (79) and an inhibition of TSH release followed blockade of serotonin biosynthesis with p-CPA (80). Similar treatments had no effect on TSH secretion in man.

Acetylcholine bromide injected into the 3rd ventricle was capable of inhibiting TSH release (81). This effect may be mediated via activation of the tuberoinfundibular dopaminergic system since treatment with the DA receptor blocker, pimozide, which did not affect TSH levels itself, blocked the action of Ach.

TSH release was suppressed by injection of GABA into the 3rd ventricle, an effect blocked by the GABA receptor blocker, bicuculline (76). GABA may act by stimulating the release of DA since its action was also blocked by the administration of pimozide to block DA receptors (82).

The same doses of a variety of peptides which altered release of other pituitary hormones following their injection into the 3rd ventricle had little or no effect on the release of TSH. This would include VIP, substance P, neurotensin, and gastrin (43,44).

Synaptic transmitters involved in control ACTH release. Considerable evidence has been amassed particularly by Ganong and his colleagues for an inhibitory role of NE in the control of corticotropin-releasing hormone (CRH) release (for references see 83). Administration of drugs which altered catecholamine synthesis led to an inverse correlation between levels of central NE and plasma

corticosterone titers in the rat (84). At the same time, there
was no correlation between hypothalamic DA concentrations and
corticosterone titers. The proposed noradrenergic inhibition of
CRH release appears to be mediated by alpha adrenergic receptors.
Unfortunately, there is some controversy since other workers showed
that hypothalamic implants or injections of NE stimulated adreno-
cortical secretion (85) and that there was no alteration in stress-
induced secretion of corticoids by depletion or elevation of central
NE stores (86). In monkeys neither L-dopa nor clonidine altered
basal corticoid output or the stimulating action of 5HTP (66).
There appears to be little or no effect of DA on CRH release.

Serotonin appears to be an important stimulant of CRH release.
Hypothalamic implants of 5HT in the rat (85) as well as systemic
administration of 5HTP in man and monkey stimulated ACTH release
(83,70).

Further evidence for a role of serotonin was provided by
results of Scampanini et al. (87) that 5HTP content in the limbic
system fluctuated with the same circadian rhythm as plasma corticosterone. Furthermore, the blocker of serotonin biosynthesis, p-
chlorophenylalanine, blocked the diurnal rhythm of corticosterone
(88). Earlier studies of Krieger and Rizzo had shown that drugs
interfering with serotonin transmission blocked the circadian rise
in plasma corticosteroids in the cat but did not block the response
to stress (89). Later studies by Vernikos-Danellis et al. (88)
showed that inhibition of serotonin synthesis in the rat was
associated with an enhanced response to stress even though the
diurnal rhythm of corticosterone was blocked.

It appears that Ach can stimulate CRH release and this has
been demonstrated recently in an in vitro system (90). The action
of Ach appears to be mediated via nicotinic receptors. These
workers also obtained evidence that serotonin acts indirectly via
a cholinergic interneuron. Norepinephrine inhibited both basal
and Ach stimulated release in agreement with the earlier in vivo
results.

Implantation of atropine into the anterior hypothalamus blocked
stress-induced ACTH discharge in the rat; however, in the cat
systemic administration of atropine blocked the circadian variation
in corticoids but not stress-induced elevations (91). Thus, the
physiological significance of Ach in ACTH release is not yet fully
established.

ABSTRACT

The discharge of the releasing and inhibiting hormones appears
to be controlled by a number of transmitters which are presumably

released at synapses which impinge on the releasing hormone producing
neurons. The evidence for a dopaminergic inhibitory control of
prolactin release is overwhelming. This control may be mediated
both by a direct action of DA on the lactotrophs after its release
into portal vessels and also by stimulation of release of a peptidic
PIF which in turn inhibits the lactotrophs. The role of DA in controlling gonadotropin release is controversial but the evidence is
strong that depending on the condition it may either stimulate or
inhibit the release of LHRH. The dose required for the inhibitory
action appears to be higher and the inhibitory effect is most
clearly seen in the castrate animal. It would appear that gonadal
steroids modify the response to DA. Dopamine also can inhibit TSH
release but whether this has physiological significance has yet to
be determined since pimozide to block DA receptors does not alter
TSH release. The preponderance of evidence indicates that DA can
also stimulate growth hormone release but it appears to have little
effect on ACTH secretion.

In general, norepinephrine (NE) appears to have a stimulating
action on release of the hypothalamic hormones. Gonadotropin release
is under stimulatory noradrenergic control which probably also
extends to GH, TSH, and prolactin. The preponderance of evidence
indicates an inhibitory role of NE in control of ACTH release.
Although epinephrine can stimulate the release of LH, GH, and TSH,
and inhibit release of prolactin, its physiological significance
remains to be determined. Acetylcholine is capable of stimulating
the release of LH, ACTH, and GH and inhibiting the release of TSH and
prolactin. The action appears to be on muscarinic receptors and
may involve the dopaminergic system since the effects are blocked
by the DA receptor blocker, pimozide. Serotonin appears to have a
stimulatory role in the secretion of prolactin, GH, and ACTH and to
inhibit LH release. GABA can stimulate LH, ACTH, and GH release.
It inhibits TSH release and has a dose-related effect on prolactin,
low doses inhibiting and high doses stimulating its release. These
actions are antagonized by the GABA blocker bicuculline but the
physiological significance of GABA in the control system remains
to be elucidated. A variety of peptides found in the CNS alter
pituitary hormone release by a central action, but the physiological
significance of these various actions remains to be elucidated.

REFERENCES

1. Sawyer, C.H., J.E. Markee and J.W. Everett. J. Exp. Zool. 113:
   659-682, 1950.
2. Fuxe, K. and T. Hokfelt. In: Frontiers in Neuroendocrinology,
   Vol. I, edited by W.F. Ganong and L. Martini. London: Oxford
   University Press, 1969, p. 47-96.

3. Schneider, H.P.G. and S.M. McCann. Endocrinology 85: 121-132, 1969.

4. Schneider, H.P.G. and S.M. McCann. Endocrinology 87: 249-253, 1970.

5. Rotsztejn, J.L., E. Charli, E. Pattou, J. Epelbaum and C. Kordon. Endocrinology 99: 1663-1666, 1976.

6. Bennett, G.W., J.A. Edwardson, D. Holland, S.L. Jeffcoate and N. White. Nature 257: 323-325, 1975.

7. Negro-Vilar, A., S.R. Ojeda and S.M. McCann. Fed. Proc. 37: 296, 1978

8. Vijayan, E. and S.M. McCann. Neuroendocrinology 125: 150-165, 1978.

9. Vijayan, E. and S.M. McCann. Neuroendocrinology 125: 221-236, 1978.

10. Drouva, S.V. and R. Gallo. Endocrinology 99: 651-658, 1978.

11. McCann, S.M. and S.R. Ojeda. In: Reviews of Neuroscience, Vol. 2, edited by S. Ehrenpreis and I.J. Kopin. New York: Raven Press, 1976, p. 91-110.

12. Jackson, G.L. Biol. Reprod. 16: 543-548, 1977.

13. Beattie, C.W., M.I. Gluckman and A. Corbin. Proc. Soc. Exp. Biol. Med. 153: 147-150, 1976.

14. Clemens, J.A., F.C. Tinsley and R.W. Fuller. Acta Endocrinol. 85: 18-24, 1977.

15. Nemeroff, C.B., R. Konkol, G. Bissette, W. Youngblood, J.B. Martin, P. Brazeau, M.S. Rone, A.J. Prange, Jr., G.R. Breese and J.S. Kiser. Endocrinology 101; 613-622, 1977.

16. Fuxe, K., T. Hokfelt, A. Lofstrom, O. Johansson, L. Agnati, B. Everitt, M. Goldstein, S. Jeffcoate, N. White, P. Eneroth, J-A. Gustafsson and P. Skett. In: Subcellular Mechanisms in Reproductive Neuroendocrinology, edited by F. Naftolin, K.J. Ryan and J.J. Davies. Amsterdam: Elsevier Scientific Publishing Co., 1976, p. 193-246.

17. Rubinstein, L. and C.H. Sawyer. Endocrinology 86: 988-995, 1970.

18. Sawyer, C.H. Neuroendocrinology 17: 97-124, 1975.

19. Gnodde, H.P. and G.A. Schuiling. Neuroendocrinology 20: 212-223, 1976.

20. Yen, S.S.C. In: Clinical Reproductive Neuroendocrinology, edited by P.O. Hubinot, M. L'Hermite and C. Robyn. Basel: Karger, 1977, p. 150-157.

21. Agnati, L., K. Fuxe, A. Lofstrom, and T. Hokfelt. Adv. Biochem. Psychopharm. 16: 159-168, 1977.

22. Gudelsky, G.A., J. Simpkins, G.P. Mueller, J. Meites and K.E. Moore. Neuroendocrinology 22: 206-215, 1976.

23. Grandison, L., C. Hodson, H.T. Cheu, J. Advis, J. Simpkins, J. Meites. Neuroendocrinology 23: 312-322, 1977.

24. Beck, W., S. Engelbart, M. Gelato and W. Wuttke. Acta Endocrinol. 84: 62-71, 1977.

25. Carlsson, A. In: Advances in Biochemical Psychopharmacology, edited by G.L. Gessa and E. Costa. New York: Raven Press, 1977, p. 439-442.

26. Cools, A.R. In: Advances in Biochemical Psychopharmacology, edited by G.L. Gessa and E. Costa. New York: Raven Press, 1977, p. 215-226.
27. Donoso, A.O. and M.B. DeGutierrez-Moyano. Proc. Soc. Exp. Biol. Med. 135: 633-641, 1970.
28. Selmanoff, M.K., L.D. Brodkin, R.I. Weiner and P.K. Siiteri. Endocrinology 101: 841-848, 1977.
29. Bapna, J., N.H. Neff and E. Costa. Endocrinology 89: 1345-1349, 1971.
30. Chiocchio, S.R., A. Negro-Vilar and J.H. Tramezzani. Endocrinology 99: 629-635, 1976.
31. Advis, J.P. and S.M. McCann. Prog. 60th Mtg. Endocrine Soc. 1978, p. 339.
32. Bhattacharya, A.N., D.J. Dierschke, T. Yamaji and E. Knobil. Endocrinology 90: 778-786, 1972.
33. Martinovic, J.V. and S.M. McCann. Endocrinology 100: 1206-1213, 1977.
34. McCann, S.M. and S.R. Ojeda. In: Reviews of Neuroscience, Vol. 2, edited by S. Ehrenpreis and I.J. Kopin. New York: Raven Press, 1976, p. 91-110.
35. Gallo, R.V. and G.P. Moberg. Endocrinology 100: 945-954, 1977.
36. Hery, M., E. Laplante and C. Kordon. Endocrinology 99: 496-503, 1976.
37. Fiorindo, R., G. Justo, M. Motta, I. Simonovic and L. Martini. In: Hypothalamic Hormones, edited by M. Motta, P.G. Crosignani and L. Martini. London: Academic Press, 1975, p. 195-204.
38. Libertun, C. and S.M. McCann. Proc. Soc. Exp. Biol. Med. 152: 143-146, 1976.
39. Ondo, G.J. Science 186: 738-739, 1976.
40. Vijayan, E. and S.M. McCann, Brain Res. 1978, in press.
41. Ondo, J.G., K.A. Pass and R. Baldwin. Neuroendocrinology 21: 79-87, 1976.
42. Cheesman, D.W., R.B. Osland and P.H. Forsham. Endocrinology 101: 1194-1202, 1977.
43. Vijayan, E., et al., unpublished data
44. Vijayan, E. and S.M. McCann. Prog. 60th Mtg. Endocrine Soc. 1978, p. 271.
45. MacLeod, R.M. and J.E. Lehmeyer. Endocrinology 94: 1077-1085 1974.
46. Ben-Jonathan, N., C. Oliver, H.J. Weiner, R.S. Mical and J.C. Porter. Endocrinology 100: 452-458, 1977.
47. Gibbs, D.M. and J.D. Neil. Fed. Proc. 37: 555, 1978.
48. McCann, S.M., L. Krulich, S.R. Ojeda and E. Vijayan. In: Progress in Prolactin Physiology and Pathology, edited by C. Robyn and M. Harter. Holland: Elsevier, 1978, p. 137-148.
49. Enjalbert, A., M. Priam and C. Kordon. Europ. J. Pharmacol. 41: 243-244, 1977.
50. Krulich, L., E. Hefco and J.E. Aschenbrenner. Endocrinology 96: 107-118, 1975.

51. Marchlewska-Koj, A. and L. Krulich. Fed. Proc. 39: 252, 1975.
52. Stevens, R.W. and D.M. Lawson. Life Sci. 20: 261-265, 1977.
53. Subramanian, M.G. and R.R. Gala. Endocrinology 98: 842-848,
       1976.
54. Pass, K.A. and J. Ondo. Endocrinology 100: 1473-1442, 1977.
55. Schally, A.V., T.W. Redding, A. Arimura, A. Dupont and G.L.
       Linthicum. Endocrinology 100: 681-691, 1977.
56. Rivier, C. and W. Vale. Endocrinology 101: 506-511, 1977.
57. Bruni, J.F., D. Van Vugt, S. Marshall and J. Meites. Life Sci.
       21: 461-466, 1977.
58. Van Vugt, D.A., J.F. Bruni and J.F. Meites. Life Sci. 22: 85-90,
       1978.
59. Collu, R., F. Fraschini and L. Martini. Progr. in Brain Res.
       39: 289-300, 1973.
60. Vijayan, E., L. Krulich and S.M. McCann. Neuroendocrinology,
       in press.
61. Muller, E.E., D. Cocchi, H. Jalando and G. Udeschini. Endo-
       crinology 92: A248, 1973.
62. Lal, S., C.E. De La Vega, T.L. Sourkes and H.G. Friesen. J.
       Clin. Endocrinol. Metab. 37: 719-724, 1973.
63. Martin, J.B., D. Durand, W. Gurd, G. Faille, J. Audet and P.
       Brazeau. Endocrinology 102: 106-113, 1978.
64. Toivola, P.T.K. and C.C. Gale. Fed. Proc. 32: 265, 1973.
65. Martin, J.B., In: Frontiers in Neuroendocrinology, Vol. 4,
       edited by L. Martini and W.F. Ganong. New York: Oxford
       University Press, 1976, p. 129-168.
66. Chambers, J.W. and G.M. Brown. Endocrinology 98; 420-428, 1976.
67. Muller, E.E., P. Dal Pra and A. Pecile. Endocrinology 83:
       893-896, 1968.
68. Durand, D., J.B. Martin and P. Brazeau. Endocrinology 100:
       722-728, 1977.
69. Blackard, W.G. and S.A. Heidingsfelder. J. Clin. Invest. 47:
       1407-1414, 1968.
70. Imura, H., I. Nakai and T. Hoshimi. J. Clin. Endocrinol.
       Metab. 36: 204-206, 1973.
71. Smythe, G.A. and L. Lazarus. J. Clin. Invest. 54: 116-121,
       1974.
72. Vijayan, E., L. Krulich and S.M. McCann, in preparation.
73. Smythe, G.A. and L. Lazarus. J. Clin. Invest. 54: 116-121, 1974.
74. Chihara, K., Y. Kato, K. Maeda, S. Matsuhura and H. Lumura.
       J. Clin. Invest. 57; 1393-1402, 1976.
75. Vijayan, E. and S.M. McCann, unpublished data.
76. Vijayan, E. and S.M. McCann, Endocrinology, submitted.
77. Krulich, L., A. Giachetti, A. Marchlewska-Koj, E. Hefco and
       H.E. Jameson. Endocrinology 100: 496-505, 1977.
78. Annunziato, L., G. Di Renzo, G. Lombardi, E. Scopocasa, G.
       Preziosi and U. Scapagnini. Endocrinology 100: 738-744, 1977.
79. Mueller, G.P. C.P. Twohy, H.T. Chen, J.P. Advis and J. Meites.
       Life Sci. 18: 715-724, 1976.

80. Shopsin, B., L. Shenkemean, I. Saughoi and C.S. Hollander. Adv. Biochem. Psychopharmacol. 10: 279-286, 1974.

81. Vijayan, E. and S.M. McCann, unpublished data.

82. Vijayan, E. and S.M. McCann, unpublished data.

83. VanLoon, G.R. In: Frontiers in Neuroendocrinology, Vol. 3, edited by W.F. Ganong and L. Martini. New York: Oxford University Press, 1973, p. 209-248.

84. VanLoon, G.R., U. Scapagnini, G.P. Moberg and W.F. Ganong. Endocrinology 89: 1464-1469, 1971.

85. Krieger, H.P. and D.T. Krieger. Am. J. Physiol. 218: 1632-1641, 1970.

86. Abe, K. and T. Hiroshige. Neuroendocrinology 10: 195-211, 1974.

87. Scapagnini, U., G.P. Moberg, G.R. VanLoon, J. DeGroot and W.F. Ganong. Neuroendocrinology 7: 90-96, 1971.

88. Vernikos-Danellis, J., P. Berger and J.D. Barchas. Prog. in Brain Res. 39: 310-310, 1973.

89. Krieger, D.T. and F. Rizzo. Am. J. Physiol. 217: 1703-1707, 1969.

90. Jones, M.T., E.W. Hillhouse and J. Burden. J. Endocrinol. 69: 1-10, 1976.

91. Krieger, D.T., A.I. Silverberg, F. Rizzo and H.P. Krieger. Am. J. Physiol. 215: 959-967, 1968.

NEUROTRANSMITTER MECHANISMS IN THE CONTROL OF THE SECRETION OF HORMONES FROM THE ANTERIOR PITUITARY

K. Fuxe, K. Andersson, A. Löfström, T. Hökfelt,
L. Ferland, L.F. Agnati, M. Pérez de la Mora, R. Schwarcz,
Dept. of Histology, Karolinska Institutet, Stockholm,
Sweden

P. Eneroth, Hormone Laboratory, Dept. of Obstetrics and
Gynecology, Karolinska Hospital, Stockholm, Sweden

J.-Å. Gustafsson and P. Skett, Dept. of Medical
Chemistry, Karolinska Institutet, Stockholm, Sweden

Over the last decades evidence has accumulated that hypo-thalamic catecholamines (CA) play an important role in the regula-tion of the secretion of hormones from the anterior pituitary gland (1-4). The mapping out of CA and peptidergic pathways such as LHRH pathways, TRH pathways and somatostatin containing pathways (5-7) has indicated that the noradrenaline (NA) and adrenaline (A) systems control the activity in the hypothalamic hormone containing systems via actions in the regions containing the LHRH-, TRH- and somato-statin-positive nerve cell bodies. The dopamine (DA) neurons, on the other hand, mainly seem to control the activity in the hypo-thalamic hormone containing pathways via an axo-axonic influence at the level of the external level of the median eminence (8,9). In the present paper the evidence will be summarized indicating that the various CA systems within the hypothalamus and the preoptic area can integrate hormonal and neuronal information, a property which makes them very suitable as controllers of the secretion of hypothalamic hormones from the anterior pituitary. The paper will mainly deal with the evidence that the hypothalamic DA pathways can inhibit the secre-tion of prolactin and LH and that noradrenergic pathways can facili-tate the secretion of LH. Adrenaline pathways within the hypothalamus may instead both inhibit the secretion of prolactin and facilitate the secretion of LH.

349

MORPHOLOGICAL STUDIES

CA and Hypothalamic Hormone Containing Systems

Immunohistochemical studies on the localization of LHRH perikarya and nerve terminals have demonstrated that the LHRH containing
nerve terminals are concentrated within the lateral palisade zone
of the median eminence which also contains the highest density of
DA nerve terminals (9). Two populations of LHRH containing perikarya
have been demonstrated using antisera generated against different
LHRH conjugates. One population is present within the preoptic and
septal regions (6), and the other population is present within the
region of the median basal hypothalamus including the arcuate nucleus
and the subependymal layer of the median eminence (7).

DA cell bodies are found all over the hypothalamus, the preoptic
area and the zona incerta (10). So far, however, mainly the functional role of the tubero-infundibular DA systems have been evaluated.
These systems have their cell bodies within the arcuate nucleus and
within periventricular hypothalamic regions. They project to the
external layer of the median eminence, both the medial and lateral
palisade zone, and to the external layer of the infundibular stem
(2). It has been suggested that at least part of the tubero-infundibular pathways to the medial palisade zone release DA as a prolactin
inhibitory factor into the primary capillary plexus (11).

The NA nerve terminals within the median eminence are mainly
found in the subependymal layer and a few also within the medial
palisade zone (10,12). Thus, there is little morphological evidence
for any axo-axonic interaction between NA and LHRH containing nerve
terminals at the level of the median eminence. On the other hand,
A and NA nerve terminal plexa exist within the LHRH cell body rich
nuclei (preoptic regions and medial basal hypothalamus), within TRH
cell body rich nuclei (e.g. the parvo-cellular part of the nuc. paraventricularis hypothalami) and within the anterior periventricular
hypothalamic region, which contains the somatostatin-positive nerve
cell bodies projecting to the median eminence. Thus, the morphological evidence support the possibility that NA and A terminals can
control activity in the hypothalamic hormone containing pathways
via an action at the cell body level. In view of the pioneering work
of Bloom and collaborators (13), it is likely that NA and A cause
a hyperpolarization of the postsynaptic membrane, producing an inhibitory effect on the neuron that it innervates. Since the pharmacological evidence suggests that NA facilitates the secretion of
LHRH (1), it seems likely that NA controls the activity in the LHRH
neurons via an inhibitory interneuron located in the same nucleus as
indicated in Fig. 1. So far the hypothalamus and preoptic area has
been found to be rich inter alia in GABA interneurons and enkephalin

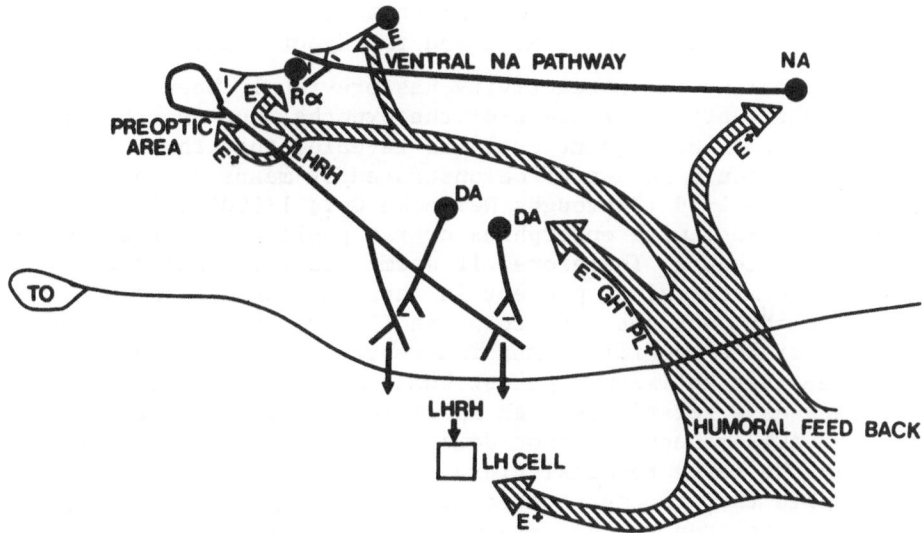

Fig. 1. Schematic illustration of the involvement of hormones and catecholamines and their interactions in control of LHRH secretion. The ventral NA pathways facilitate phasic secretion of LH, and the DA system of the lateral palisade zone of the median eminence may produce inhibition of LHRH release via presynaptic inhibition (Table I). A positive estradiol feedback may involve an activation of the NA pathway while the inhibitory estrogen feedback may involve an activation of the DA system in the lateral palisade zone. It is also indicated that hypophyseal hormones such as growth hormone and prolactin can affect the tubero-infundibular DA neurons. By means of various arrows it is indicated that estrogen (E) can influence the circuitary controlling LHRH secretion at many levels. Thus, estrogen could influence interneurons controlling release of NA or interneurons innervated by the NA pathways. Estrogen may also directly influence the LHRH secreting neurons. Finally, estrogen has an important effect on the LH gland cell, enhancing its reactivity to the secretion of LHRH. The true nature of the actions of various types of hormones on nervous circuitaries within the brain remains to be established. R_α = α-adrenergic receptor.

interneurons. The nature of the transmitters of the inhibitory interneurons particularly involved in the control of the hypothalamic
hormone containing pathways remains to be established.

A Peptide Containing Neuron System Projecting from the Arcuate Nucleus and the Median Eminence to Periventricular Areas of the Preoptic Area and the Hypothalamus

Prolactin-like immunoreactivity has previously been described
within the periventricular areas of the hypothalamus and the preoptic region (14). Subsequent analysis revealed that these nerve
terminal plexa could be better demonstrated by means of an antiserum against ACTH 1-39 (Burroughs Wellcome Co.; 1/500) (11), which
does not crossreact with endorphins or rat prolactin in the radioimmunoassay procedure. Therefore, it seems possible that the ACTH
immunoreactive material is present in these neurons. Also ACTH
1-39 (2×10^{-10}M) but not rat prolactin (10^{-7}M) prevents the demonstration of the specific immunofluorescence. It is important to
note that another antiserum against human ACTH 1-39 (Sorin, France)
which possesses a relatively high specificity for the N-terminal
part of the ACTH molecule cannot demonstrate the immunoreactive
material in the nerve terminal plexa described above. It therefore
seems possible that immunoreactive material demonstrated is a pep-
tide-sequence related to the C-terminal part of the ACTH molecule.
It was suggested by Prof. Besser in the discussion that the material might in fact be a CLIP-like material (ACTH 18-39). In agreement ACTH 1-39 (2×10^{-10}M) but not ACTH 1-24 (10^{-6}M) prevented
the development of the specific immunofluorescence.

After colchicine injections into the lateral ventricle, this
ACTH-like immunoreactive material can be demonstrated also within

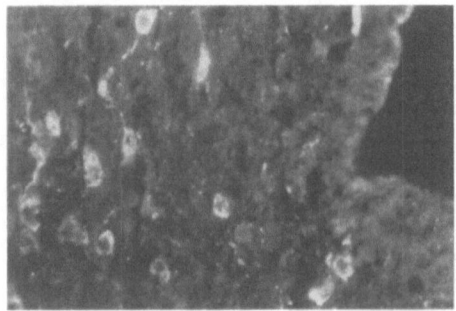 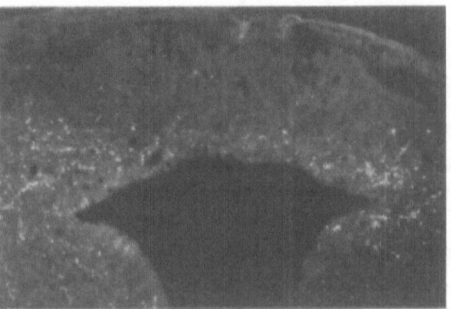

A   B

Fig. 2. ACTH-like immunoreactivity demonstrated within nerve cell
bodies (A) in the arcuate nucleus and the lateral part of the subependymal layer (colchicine pretreatment, 24 h, 75 µg intraventricularly) and within nerve terminals (B) of the subependymal layer
by means of antibodies against human ACTH 1-39 (Burroughs Wellcome Co.; 1/500). X100.

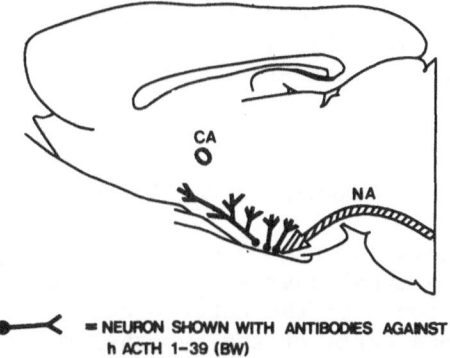

= NEURON SHOWN WITH ANTIBODIES AGAINST
h ACTH 1-39 (BW)

Fig. 3. Schematic illustration of a neuron system showing ACTH-
like immunoreactivity within the medial basal hypothalamus and
which projects towards various regions of the hypothalamus and the
preoptic area. It is demonstrated by means of antibodies against
rat prolactin and especially by means of antibodies against ACTH
1-39 (Burroughs Wellcome Co.). The exact nature of the peptides involved remains to be established. The ACTH peptide may be related
to the C-terminal part of the ACTH molecule, since an antibody
against human ACTH 1-39, which preferentially detected the N-terminal part (Sorin, France), did not demonstrate the above neuron
system, and since ACTH 1-39 had at least a thousand times higher
potency in preventing the development of the ACTH-like immunoreactivity than ACTH 1-24.

nerve cell bodies of the arcuate nucleus and the adjacent ventro-
lateral region close to the ventral surface of the brain and within the lateral parts of the subependymal layer (SEL) of the median
eminence (see Fig. 2). Thus, we can describe a system within the
medial basal hypothalamus which projects not to the external layer
of the median eminence but to the SEL and in the opposite direction towards the hypothalamus particularly the periventricular regions of the hypothalamus and the preoptic area. A minor projection is also found to the amygdaloid cortex and to certain nuclei
of the thalamus. This system is schematically illustrated in Fig.3.
In the schematic figure it is also indicated the possibility that
NA nerve terminal plexa within the subependymal layer and the arcuate nucleus and surrounding areas can influence this neuron
system containing ACTH-like immunoreactivity. The topography of
this system is new within the hypothalamus, and it implies that
it can act as feedback system from the medial basal hypothalamus
up towards mainly periventricular regions where the cell bodies
of many peptide containing neurons are found. It may be speculated
that this system showing ACTH-like immunoreactivity can receive
information from collaterals belonging to the hypothalamic hormone containing systems projecting to the external layer of the

median eminence. The arcuate-periventricular system described
above could then transfer this information back towards the regions containing the cell bodies of the hypothalamic hormone containing system.

AMINE TURNOVER CHANGES IN VARIOUS ENDOCRINE STATES

In order to understand the functional role of the various types
of DA and NA nerve terminal systems in the hypothalamus, it is help-
ful to study the turnover of the catecholamines within discrete terminal systems in various parts of the hypothalamus and the preoptic
area. The CA turnover estimation will give an indication if there is
a high or a low release of catecholamines and thus of the activity
in the CA terminal system under evaluation. To evaluate CA turnover
we have studied the disappearance of catecholamines following inhibition of tyrosine hydroxylase by means of the tyrosine hydroxylase
inhibitor α-methyl-tyrosine methylester (H 44/68; see 15). The faster
the disappearance rate, the higher the turnover of the CA. The CA
stores in the discrete terminal systems were measured by means of
quantitative microfluorimetry performed on sections taken from brains
treated according to the Falck-Hillarp procedure to demonstrate CA
stores at the cellular level (see 12, 15). To illustrate the results
obtained on CA turnover with this specific methodology, the rate
constants obtained for CA turnover in the various parts of the median
eminence of the normal male rat are shown in Fig. 4. The half-lives
are also indicated in the figure. The highest turnover is found
within the medial palisade zone which contains a mixture of DA
and NA terminals, where DA terminals predominate, and the lowest
turnover is found within the subependymal layer, where mainly
NA nerve terminal plexa are present. The lateral palisade zone,
on the other hand, contains almost exclusively DA nerve terminals.

Using this approach it has been possible to show that under
conditions of phasic secretion of LH such as in proestrus or in the
critical period induced by pregnant mare serum gonadotrophin (PMS)
in the immature female rat, there occurs a marked increase of NA
turnover within the medial preoptic area and within the subependymal
layer, while a reduction of DA turnover takes place in the medial
and lateral palisade zone of the median eminence. These results
support the view that there exists an excitatory noradrenergic
influence on the secretion of LHRH facilitating the peak secretion
of LHRH during proestrus and that there exists an inhibitory DA
influence on the secretion of LHRH. The NA nerve terminals in the
medial preoptic region and in the subependymal layer can thus facilitate activity in the LHRH cell bodies found in these two types of
regions respectively, while the DA terminals can inhibit the release
of LHRH via an axo-axonic interaction in the lateral palisade zone
of the median eminence. Agnati et al. (16) have in fact established

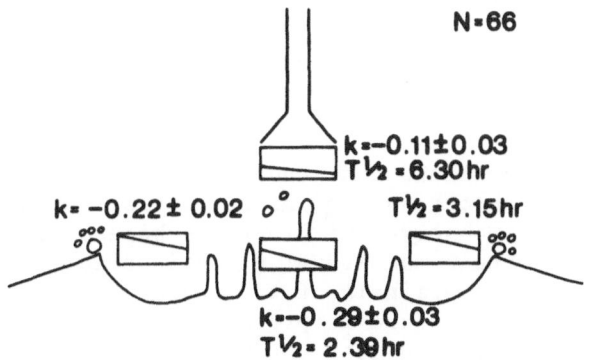

k values ± S.D.

Fig. 4. Schematic presentation of results from CA turnover measurements in nerve terminals in the central part of the median eminence in normal male rat, based on microfluorimetric quantitation of changes in the formaldehyde induced CA fluorescence following tyrosine hydroxylase inhibition produced by α-methyl-p-tyrosine methylester (250 mg/kg, i.p.). The disappearance of the fluorescence was exponential and the calculated T1/2 for the external layer was 2.39 h (medial part) and 3.15 h (lateral part) while T1/2 for the subependymal layer was 6.30 h. The k-values ± S.D. for the slope are also given.

that there exists a significant correlation in the pattern of distribution of DA and LHRH-positive nerve terminals in this area. Thus, one important factor in the control of the secretion of LHRH could be the ratio of activity between NA and DA nerve terminals in the areas described above. In relation to phasic LHRH secretion there occurs a switch from a preferential activation of DA terminals to a preferential activation of NA terminals. In view of the fact that the NA turnover increase somewhat proceeds the reduction of DA turnover, it seems possible that the activation of the NA systems in part at least contributes to the inactivation of the DA neurons within the external layer of the median eminence. Thus, it seems possible that at least under certain conditions NA terminal plexa can exert an inhibitory influence of the tubero-infundibular DA systems. In line with this view, it has been shown (17) that immobilization stress increases NA turnover in the hypothalamus and reduces DA turnover in the external layer of the median eminence.

In view of the fact that no LHRH terminals or few LHRH terminals are present within the medial palisade zone, we have speculated that the DA terminals in this region in part at least releases DA as an prolactin inhibitory factor to act on DA receptors of the prolactin containing gland cells. The reduction of DA turnover found in the medial palisade zone in proestrus and during immobilization stress could therefore explain the increases of prolactin secretion found under these conditions.

It should be underlined that under conditions of tonic hyper-secretion of LH, which occurs e.g. after castration, no increases of NA turnover have been observed within the subependymal layer of the median eminence or within the medial preoptic area (15). The amine turnover within the DA terminals of the median eminence, on the other hand, is similar to that found in proestrus. Thus, it seems as if tonic hypersecretion of LHRH does not require an activation of the NA terminal systems.

EFFECTS OF STEROID HORMONES ON DA AND NA TURNOVER IN DISCRETE CA TERMINAL SYSTEMS OF THE HYPOTHALAMUS AND THE PREOPTIC AREA

The amine turnover changes described above in DA and NA terminal systems in relation to proestrus may be brought about by changes in hormone levels of e.g. estrogen, progesteron, prolactin, LH and FSH. Therefore, the effects of various hormones have been evaluated on CA turnover in the median eminence and in the hypothalamus and preoptic area.

Effects of Estrogen

Estradiol benzoate has been repeatedly shown to increase DA turnover in the medial and lateral palisade zone of the ovariecto-mized female rat (18-21). At the same time there was a reduction of NA turnover within the subependymal layer of the median eminence (19). Under these conditions estrogen exerts a central inhibitory feedback action on the secretion of LHRH. It therefore seems likely that the increase of DA turnover observed in the median eminence in part mediates the inhibitory feedback action of estrogen. The reduction of NA turnover observed in the subependymal layer could also contri-bute to such inhibitory effects of estrogen on LH secretion. Again we see that a preferential activation of the DA terminal systems within the median eminence leads to a reduction of LH secretion. Under similar conditions corticosterone and progesteron do not increase DA turnover in the median eminence (2).

It is well known that estrogen treatment also increases pro-lactin secretion from the anterior pituitary probably via an action beyond the DA receptors in the prolactin containing gland cells leading to a failure of DA receptor activity to exert its inhibitory control of prolactin secretion (see Labrie, this symposium). It is, therefore, possible that the hypersecretion of prolactin produced by estrogen can contribute to the increases of DA turnover observed in the median eminence following estrogen treatment, since prolactin has been found to increase DA turnover in the median eminence (21-23). However, it must also be emphasized that the degree of prolactin secretion is not correlated to the activation of the DA systems in the median eminence after estrogen treatment. Furthermore, increases of DA turnover in the median eminence following estrogen treatment precedes the hypersecretion of prolactin produced by estrogen. It is obvious from the above that the DA systems in the median eminence can respond directly to treatment with estrogen. In view of the work of Stumpf, Grant and collaborators (24,25) this action of estrogen may be mediated via estrogen receptors in the DA cell bodies, since these workers have demonstrated that estradiol can accumulate in nuclei of arcuate DA cells. Using autoradiography they were able to show that about 50 per cent of the DA cell bodies in this area were labelled with estrogen. It seems possible that the activation of the estrogen receptor via a change in messenger RNA and protein synthesis can lead to changes in the number and configurations of various types of receptor proteins controlling the sensitivity of the DA neurons to surrounding excitatory and inhibitory nerve terminals.

Under the estrous cycle, however, the NA terminal systems do not appear to be inhibited by estrogen. Instead the low concentrations of estrogen present in the estrous cycle appear to exert a stimulatory effect on NA turnover in the preoptic area (26). Thus, the positive estrogen feedback producing the LH surge in proestrus may involve an activation of the NA systems. Thus, the positive estradiol feedback produces a completely opposite pattern of changes in CA turnover to that observed after estrogen treatment which results in an inhibitory feedback action on LH secretion. These findings may be explained on the basis that under the estrous cycle with low circulating amounts of estrogen, estrogen preferentially activates a high affinity estrogen receptor e.g. in the medial preoptic area. The estrogen receptor located in the DA cell bodies, which can mediate part of the inhibitory feedback action of estrogen may have a lower affinity for estrogen and therefore operates only at higher concentrations of estrogen (see 18).

Pharmacological Analysis

If the above interpretations are correct, it should be possible to inhibit ovulation by means of DA receptor agonists and NA receptor antagonists and to inhibit the hypersecretion of LH in castrated animals by means of DA receptor agonists. This has in fact been demonstrated in recent papers (20,27). As seen in Fig. 5, a DA receptor agonist, ET 495, reduces the secretion of LH in ovariectomized female rats and this effect is counteracted by a DA receptor blocking agent, pimozide. Pimozide by itself produces a certain reduction of LH secretion, probably via its ability to block also NA receptor sites (see also Table II). Recent papers also show that DA receptor agonists can block the cyclic discharge of LH in the

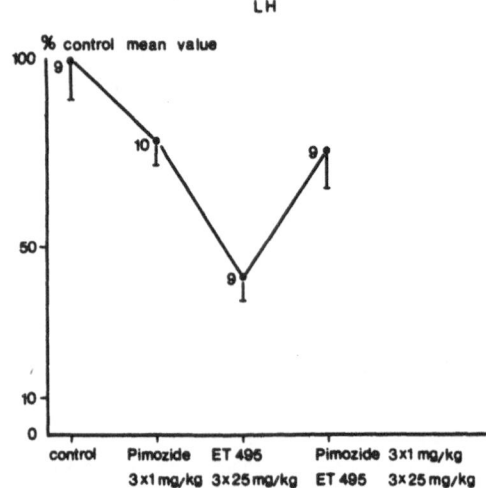

	CONTROL	ET 495	PIMOZIDE	PIMOZIDE+ET 495
CONTROL	–	S	NS	NS
ET 495	–	–	S	S
Pimozide	–	–	–	NS
Pimozide+ET 495	–	–	–	–

Dunn test for multiple comparisons
(α = .10)

LH

Fig. 5. Effect of ET 495 and pimozide on LH secretion in castrated female rats. Female Sprague-Dawley rats were castrated 2 months earlier. Pimozide was given i.p. 2 h before every injection of ET 495. ET 495 was given i.p. 24, 18 and 2 h before decapitation. Values are expressed as means ± S.E.M. and taken in per cent of control mean value. Multiple comparisons have been made according to Dunn's test (non-parametrical procedures) with experimental error α = 0.05 (Agnati, Wuttke, Fuxe et al. to be published).

ovariectomized female rat (see 27), effects which are antagonized by DA receptor blocking agents. Furthermore, we have performed an extensive pharmacological analysis of the effects of DA receptor agonists on ovulation in immature female rats treated with PMS (see 2,8,20). A number of DA receptor agonists were found to inhibit ovulation when administered during the critical period. These effects were antagonized by pretreatment with pimozide. The pharmacological analysis therefore gives support to the interpretation made above of the findings obtained in the CA turnover experiments. Thus, the axonic DA receptors probably located on the LHRH containing nerve terminals may have an inhibitory influence on the release of LHRH into the primary, capillary plexus. It must be emphasized, how-

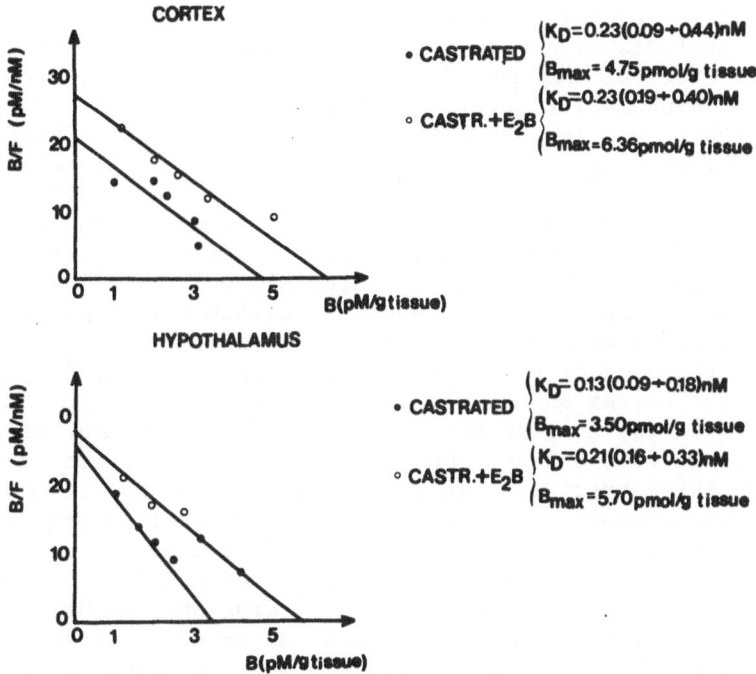

Fig. 6. Effects of estrogen treatment of ovariectomized female rats on ^{3}H-WB-4101 binding in membranes from hypothalamus plus preoptic area and from cerebral cortex. Estrogen was administered in a dose of 5 μg/rat twice. The injections were made subcutaneously 24 and 2 h before killing. The animals had been ovariectomized two months before the experiment. The assay was performed as described by U'Prichard et al. (30). The binding of ^{3}H-WB-4101, an α-adrenergic blocking agent, was evaluated by means of saturation analysis. Five to seven concentrations of ^{3}H-WB-4101 were used in the absence or presence of 100 μmol of (-)-NA. Five values were obtained for each concentration. The 95% confidence inervals are indicated around the dissociation constants.

ever, that the number of DA receptors and their dissociation constants may vary depending upon the endocrine state. Thus, DA receptor agonists are much more potent in inhibiting ovulation in the immature female rat than in the normally cycling female rat. Wuttke and collaborators (28,29) have also postulated that desensitization of DA receptors on the LHRH containing nerve terminals must occur.

Effects of Estrogen on the Specific Binding of the
α-Adreno Receptor Blocking Agent ^{3}H-WB-4101

In view of the above discussion it becomes of particular interest to evaluate whether estrogen can influence not only presynaptic mechanisms in the CA systems but also postsynaptic mechanisms. In a preliminary analysis it has therefore been evaluated if estrogen given in two doses of 5 µg/rat (24 and 2 h before killing, s.c., 17β-estradiol benzoate) to ovariectomized female rats (two months before killing) could influence the specific binding of ^{3}H-WB-4101 (30) in the hypothalamus plus preoptic area and in the cerebral cortex. The results are summarized in Fig. 6. The Scatchard analysis reveals that there may be an increase in the dissociation constant (K_D) of the ^{3}H-WB-4101 hypothalamic binding site after estrogen treatment. Thus, the 95% confidence intervals only slightly overlap. (This approach is conservative). However, within the cerebral cortex, which does not contain any estradiol concentrating neurons, no change in the dissociation constant was observed. However, both in the cerebral cortex and within the hypothalmus including the pre-optic area there was a trend for an increase in the number of binding sites for ^{3}H-WB-4101, a trend, however, which could be statistically evaluated in this experiment. In view of these findings it may be speculated that estrogen under conditions of which it exerts an inhi-bitory feedback action on LH secretion and increases prolactin secre-tion may influence selectively the dissociation constant of certain α-adreno receptor sites within the hypothalamus and the preoptic area. It may also be speculated that the number of α-adreno receptor sites can be influenced by estrogen treatment. Thus, it seems possible that hormones such as estrogen can have important regulatory effects on CA transmission, not only via presynaptic influences but also via post-synaptic influences on CA receptors. It is possible that such effects of estrogen can be produced via an estrogen cytosol receptor leading to changes in protein kinase activity and in phosphorylation pro-cesses (see Greengard, this symposium) or via an estrogen receptor in the membrane (see Baulieu, this symposium).

In view of the above it seems likely that steroidal hormones such as estrogen can control the number and the dissociation con-stants of axonic DA receptors located e.g. on the LHRH containing nerve terminals. Thus, the hormonal environment can via such actions control the sensitivity of e.g. DA receptor sites to its transmitter. This could of course represent one general mechanism by which ste-roidal hormones can influence neurotransmission within the peri-pheral and central nervous systems.

When discussing the regulation by estrogen of LH secretion, it must be emphasized, however, that estrogen exerts its regulatory effects at several levels. It should be considered (see Fig. 1) that estrogen can influence all links in the neural circuitary controlling the release of LHRH. In addition it controls the sensitivity of the LH containing gland cells to LHRH. Thus, estrogen is taken up and accumulated within NA and DA cell bodies as demonstrated by Grant, Stumpf and coworkers (25). It may also directly influence the LHRH containing nerve cells as well as interneurons controlling their activity. Thus, estrogen seems to tune the circuitary at many synaptic links and not only at one synapse. In this way a proper coordination of the various neuronal systems involved can take place.

Mechanisms by which DA Can Inhibit the LHRH Secretion

As indicated in Table I, the existence of presynaptic inhibition of LHRH release by DA can explain the available data on the influence of DA on LHRH secretion. Thus, in states of high activity in LHRH pathways DA will by way of depolarization reduce the release of LHRH produced by the action potentials reaching the nerve terminals. However, when the LHRH pathways are silent, the depolarization produced by DA will produce a certain release of LHRH (Table I). This hypothesis could explain why McCann, Kordon and collaborators in their in vitro analysis and in their studies on the effects of intraventricular injections of DA on LHRH and LH secretion have observed increases in the secretion of these hormones (31-33). Of special interest is the fact that this LHRH release is observed mainly when the LHRH systems are silent such as after priming of the animals with large doses of estrogen and progesteron (see McCann, this symposium).

In line with the present hypothesis are the elegant experiments of Yen and collaborators (see Yen, this symposium) demonstrating in humans that inhibition of LH secretion by DA mainly occurs in states with high secretion of LH such as in the preovulatory period. The LHRH containing pathways must be activated before the presynaptic inhibition produced by DA can be demonstrated.

The axonic DA receptors in the neostriatum do not seem to be linked to the DA sensitive adenylate cyclase. Therefore, also the axonic DA receptors in the external layer of the median eminence may not be linked to the DA sensitive adenylate cyclase but to some other biological effector mechanisms.

Effects of Androgens

As seen in Fig. 7, various types of androgens like estrogen can increase the DA turnover in the medial and lateral palisade zone of the median eminence (see 34). In these experiments, however, the NA turnover within the subependymal layer did not seem to be effected. Experiments were performed in castrated male rats and repeated doses of androgens as seen in Fig. 7 markedly reduced the LH secretion. Testosterone propionate and 5-α-dehydrotestosteron propionate did not influence prolactin secretion while the androgenic steroid R 1881 markedly reduced the serum levels of prolactin. Thus, the increases of DA turnover in the external layer of the median eminence could not be related to changes in secretion of prolactin but rather to inhibi-

TABLE I

Schematic outline of DA action on LHRH nerve terminals

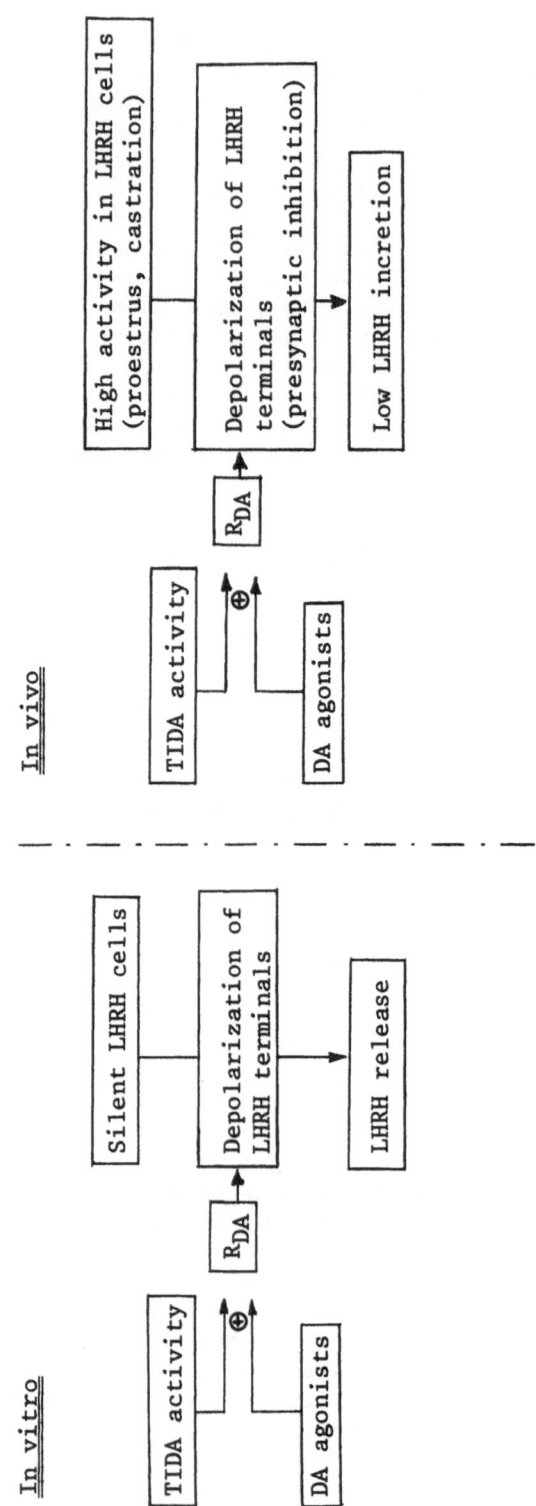

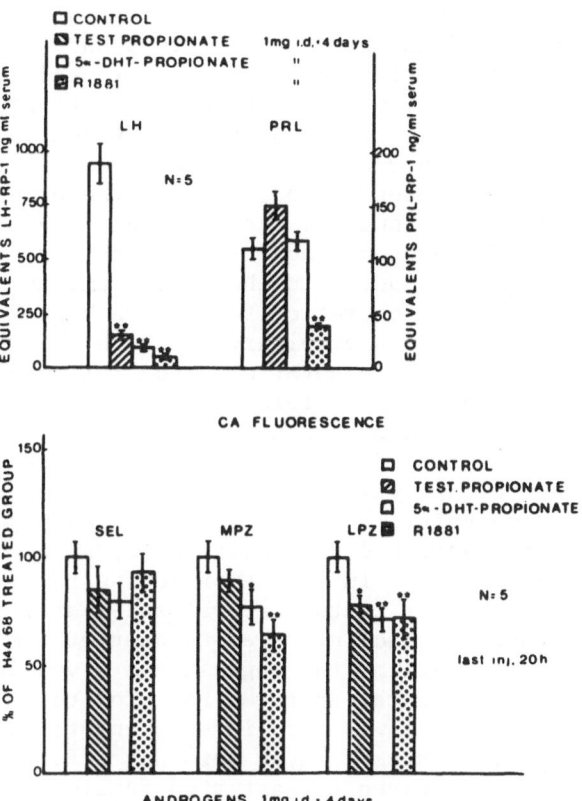

Fig. 7. The effects of testosterone propionate, 5-α-dehydrotestoste-
rone propionate and R 1881 (androgenic steroid) on the serum levels
of LH and prolactin and on the H 44/68 induced CA fluorescence disappearance in the median eminence of castrated male rats. The rats
were castrated two to four weeks prior to the experiment. Steroids
were dissolved in propylene glycol containing 3-5% absolute ethanol.
Injections were made subcutaneously in the mornings and the rats were
killed 20 h after the last injection. H 44/68 was given in a dose of
250 mg/kg i.p. 2 h before killing. The fluorescence values are given
in per cent of H 44/68 alone group mean value. The LH and prolactin
serum levels are indicated as equivalents of the NIAMDD Rat-LH-RP-1
and Rat-prolactin-RP-1 preparation, respectively. Means ± S.E.M.
All statistical comparisons are made with the H 44/68 alone group.
Student's t-test: after one way analysis of variance. ×: $p < 0.05$;
××: $p < 0.01$. (Ferland, Fuxe, Andersson, Eneroth, Gustafsson and Skett,
to be published).

tion of LH secretion. It is to be noted, however, that unlike estrogen the androgens tested only produced a small but significant
increase of the DA turnover in this region. This may be related to
the fact that the androgens only affect a minor population of the

tubero-infundibular DA systems. This possibility is supported by the
observations of Stumpf, Grant and collaborators showing accumulation
of androgens over only a small number of DA nerve cells in the
arcuate nucleus (25, and Stumpf, private communication). The androgen receptor in this small population of DA nerve cells seems to
have a high affinity for androgen, since similar increases were
observed even with small doses of androgens in the order 50–100 μg/
rat. The findings therefore suggest that the reduction of LH secretion with low doses of androgens could involve an activation of an
inhibitory dopaminergic mechanism in the median eminence, while the
increasing inhibition of LH secretion found with higher doses of
androgens probably is due to a pituitary action by the androgens
(see 35). The increase of DA turnover observed following treatment
with the androgens cannot involve formation of estrogens, since
5-α-dehydrotestosteron propionate cannot be converted into estrogen.
The same is true for R 1881. Studies of the time course indicated
an effect by the androgens 6–12 h following a single dose of the
androgen. At this time point, the levels of LH were significantly
reduced. It should be mentioned that corticosteron in doses of
1–10 mg/rat and dexamethasone in a dose of 100 μg/rat for 4 days
did not increase DA turnover in the lateral and medial palisade zone
of the median eminence when administered to castrated male rats.
These results further underline the suggestions given above that
the androgens act via a central androgen receptor to produce their
effects on the DA systems. It is also interesting to note that the
NA terminals in the subependymal layer were not affected by the
androgens, since estrogen given to castrated female rats can produce
a reduction of NA turnover within the subependymal layer. It may be
speculated that the NA neurons are more efficiently controlled by
estrogens than by androgens.

EFFECTS OF HYPOPHYSEAL HORMONES ON THE NA AND DA TERMINAL SYSTEMS OF THE HYPOTHALAMUS

Effects of Prolactin

The first hormone shown to influence the DA turnover within the
median eminence was prolactin when given in repeated doses to hypophysectomized rats (21,22). It produces a marked increase of DA turnover in the medial and lateral palisade zone of the median eminence
(11). In immature female rats prolactin has been shown to increase
DA turnover after two doses daily of each 0.5 μg/kg of ovine prolactin (11), indicating that this hypophyseal hormone in physiological doses can influence DA turnover in the median eminence. Wuttke
and collaborators (28,29) have shown that the preovulatory type of LH
peaks in the immature female rat can be blocked by treatment with
ovine prolactin. Therefore, it seems likely that prolactin secretion
at least in part may control the LH secretion in the immature female

rat via increasing DA turnover in the lateral palisade zone, in this
way inhibiting the secretion of LH. It is also well-known that hyperprolactinemia in animals and man will lead to a reduction of LH
secretion and inhibition of ovulation. It seems likely that also in
the latter case the mechanism for the lowering of LH secretion
involves at least in part an activation of the DA system in the
lateral palisade zone of the median eminence. Low doses of DA
receptor agonists, such as Bromocriptine, which has a high affinity
for the pituitary DA receptor, inhibiting prolactin secretion, will
also restore serum LH levels and ovulation. It seems likely that
prolactin acts via prolactin receptors within the central nervous
system after retrograde transport in the infundibular stem (36).

These prolactin receptors may control not only the tuberoinfundibular DA system but also the DA terminal systems in the nuc.
accumbens and nuc. caudatus, since treatment with prolactin or endogenous hypersecretion via pituitary transplants will increase or
decrease DA turnover within the nuc. accumbens and nuc. caudatus
depending upon the endocrine state of the animal (11,37). The fact
that both increases and decreases of DA turnover have been observed
in these regions of the forebrain may indicate the existence of two
types of prolactin-like receptors in the central nervous system and
that their sensitivity is controlled by the endocrine and the developmental state of the animal. The existence of two types of prolactin receptors could in part explain the complex behavioural actions
of prolactin (38). It should be mentioned that also NA systems may
be influenced by prolactin secretion, since pituitary transplants in
castrated female rats can preferentially increase NA turnover in the
median eminence. These results are obtained four days following the
transplantation of the pituitary and the serum prolactin levels were
only slightly elevated at this time interval. It is well known that
pituitary transplants in the anterior chamber of the eye of hypophysectomized rats can markedly increase DA turnover in the median
eminence, probably via endogenous secretion of prolactin.

From the above it is clear that the actions of prolactin on the
CA systems in the hypothalamus and in the forebrain are complex and
varies with the endocrine state of the animal. It remains to be
established if prolactin receptors do exist within the central nervous system and if so, if at least two types of prolactin receptors
can be demonstrated.

Effects of Growth Hormone and ACTH 1-24

Subsequent work has revealed that not only prolactin can
influence the CA systems but also growth hormone and ACTH. Thus,
growth hormone given to hypophysectomized animals produces a reduction of DA turnover within the medial and lateral palisade zone of
the median eminence (39). Within the posterior periventricular hypo-

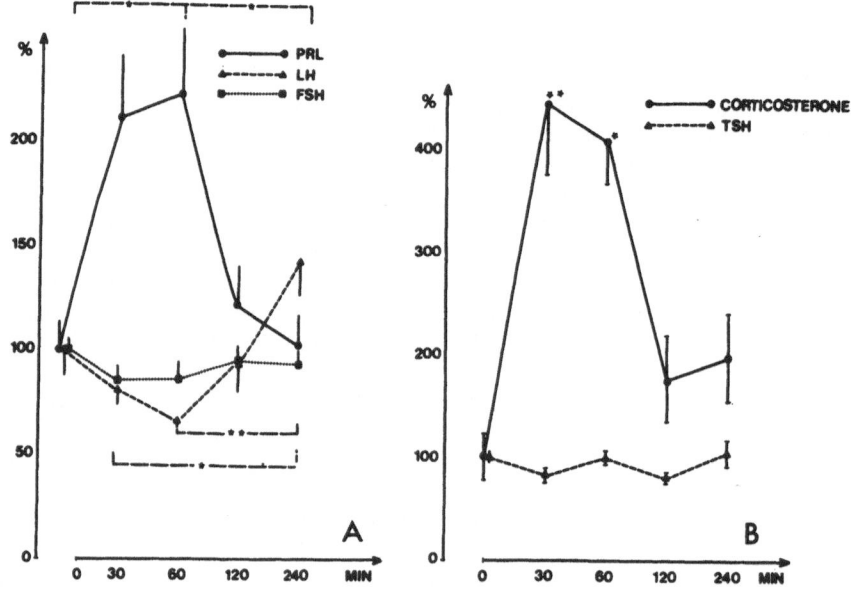

<u>Fig. 8</u>. The effects of the PNMT inhibitor SK&F 64139 on the secre-
tion of LH, FSH, prolactin (A), TSH and corticosteron (B). The PNMT
inhibitor was given in a dose of 40 mg/kg i.p., and the animals were
killed at various time intervals following the injection. The hormones were determined in serum according to standard radioimmunoassay
procedures. The levels of hormones are given in per cent of the
levels at the time of the injection of the PNMT inhibitor. The absolute levels of prolactin at time zero was 22 ng/ml, of LH 30 ng/ml,
of FSH 1022 ng/ml, of TSH 2250 ng/ml and of corticosteron 212 nmol/1.
Means ± S.E.M. are shown. Statistical analysis was performed using
a multiple comparison procedure according to Wilcoxon one way
classification. ×: p<0.05; ××: p<0.01.

thalamic region, there occurs a reduction of NA turnover following
the growth hormone administration. ACTH 1-24, on the other hand,
caused a selective increase of NA turnover in the subependymal layer
and the medial palisade zone of the median eminence. Thus, each
hypophyseal hormone produces its own pattern of changes of turnover
within the various DA and NA terminal systems. We speculate that
these changes are produced due to the existence of receptors for
growth hormone and ACTH within the hypothalamus and that these
receptors can directly or indirectly control the release of NA and
DA in discrete terminal systems of the hypothalamus. In view of the
morphological features of the NA systems with one neuron system
innervating large numbers of nuclei, it seems unlikely that the
discrete changes observed in CA turnover can be due to changes in
nervous impulse flow. Instead the possibility is favoured that the
receptors for in this case hypophyseal hormones can control activity

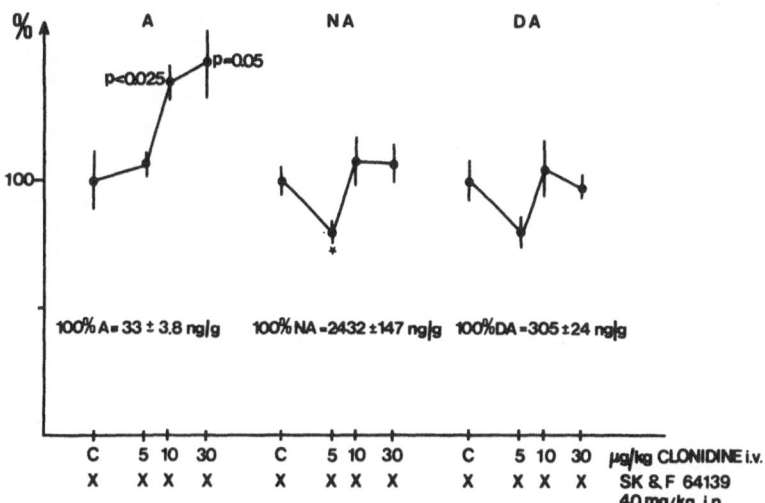

Fig. 9. The effects of clonidine on the depletion of the A stores
produced by the PNMT inhibitor SK&F 64139 in the nuc. tr. solitarius
area. The PNMT inhibitor was given in a dose of 40 mg/kg i.p., and
approximately at the same time clonidine was administered in several
doses intravenously into the jugular vein. The rats were killed by
rapid decapitation 2 h later. It must be pointed out that the piece
taken out also contains several other brain nuclei. The A, NA and
DA levels determined are expressed in per cent of the mean value of
the group treated with the PNMT inhibitor alone. Means ± S.E.M. are
shown out of 3-5 rats. Statistical analysis was made by means of
Student's t-test after one-way analysis of variance. All comparisons
have been made with the group treated with the PNMT inhibitor alone.
×: $p < 0.05$.

within interneurons of the hypothalamus which via axo-axonic contacts
can control the release of NA and DA from the nerve terminals. In
this way, each NA terminal system within a certain nucleus can
participate in a local circuitary independent of the NA nerve terminal systems of other regions. An alternative explanation is that
various NA nerve terminal systems contain different types of receptors on their presynaptic membrane.

One interneuron to be considered in this respect could be the
GABA interneuron. Such GABA interneurons may at least operate within
the parvocellular part of the paraventricular hypothalamic nucleus,
since the GABA receptor agonist muscimol preferentially increases NA
turnover within this nucleus without influencing the turnover in
adjacent NA terminal rich nuclei (40).

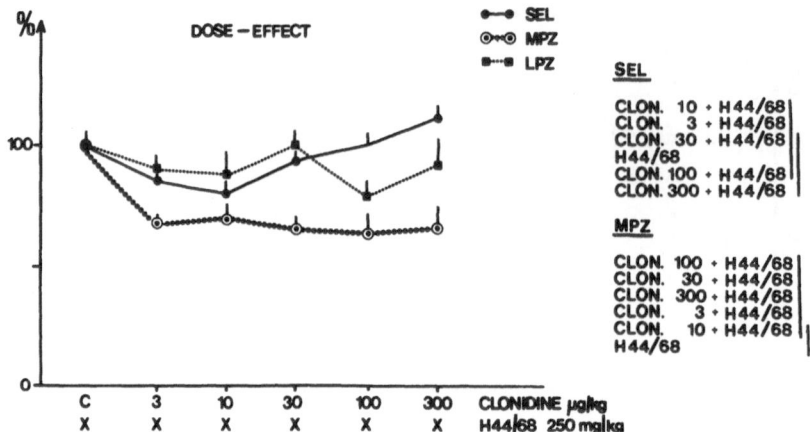

<u>Fig. 10</u>. The effect of clonidine on the H 44/68 induced CA fluo-
rescence disappearance in the various regions of the median eminence.
SEL = subependymal layer, LPZ = lateral palisade zone, MPZ = medial
palisade zone. Clonidine was given i.p. at the same time as the
H 44/68 injection (250 mg/kg, i.p., 2 h before killing). The DA
fluorescence values are given in per cent of the values found at
the time of the H 44/68 injection. Means ± S.E.M. are shown. Sample
size is 4-5. Statistical analysis was made according to one way
analysis of variance and subsequently multiple comparisons were
made: Sum of squares simultaneously test procedure (experimental
error rate, α = 0.05). The means not significantly different from
each other are covered by the same line in the tables of the figure.

ON THE ROLE OF ADRENALINE PATHWAYS IN THE CONTROL OF
HORMONE SECRETION FROM THE ANTERIOR PITUITARY

The role of the A systems have <u>inter alia</u> been evaluated by
studying the effects of a PNMT inhibitor (SK&F 64139) on the secretion of hormones from the anterior pituitary. A dose of 40 mg/kg
given i.p. produces a selective depletion of A stores by about
50 per cent (see Goldstein, this symposium, and 41). As seen in
Fig. 8, this PNMT inhibitor produces a marked increase in the secretion of prolactin and corticosteron while at the same time reducing
the secretion of LH. These results indicate that A systems in the
hypothalamus may exert an inhibitory control of secretion of prolactin and of ACTH, while having a facilitory influence on the secretion
of LH. The latter suggestion is an agreement with previous findings
that A given intraventricularly can release LH from the anterior
pituitary (see 42 and McCann, this symposium).

In order to further understand the neuroendocrine role of the
A systems, the α-adreno receptor agonist clonidine, which in low

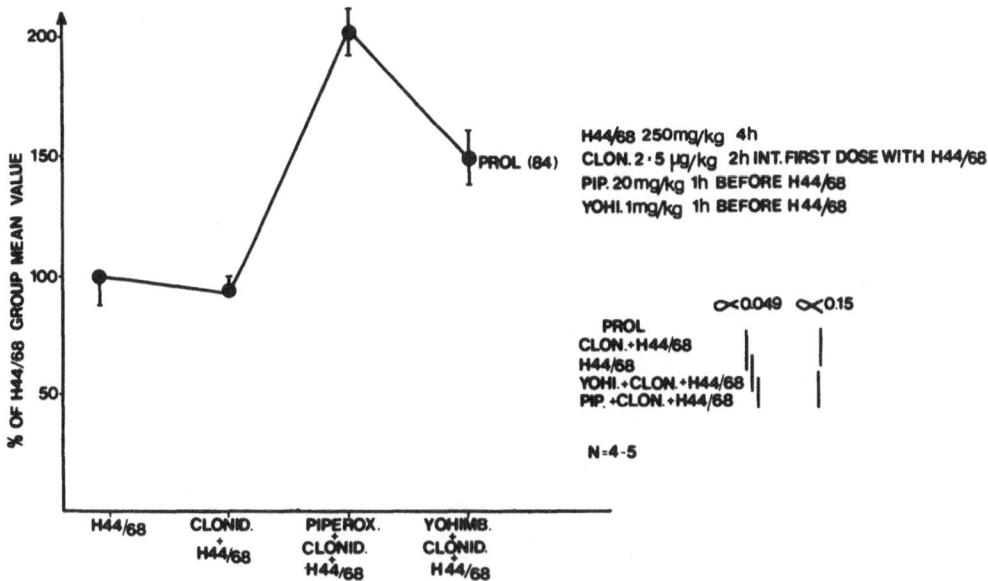

Fig. 11. The effects of piperoxane and yohimbine on the secretion of prolactin in normal male rats treated also with a tyrosine hydroxylase inhibitor (H 44/68) and clonidine. Piperoxane and yohimbine were given i.p. 1 h before the H 44/68 injection (250 mg/kg, i.p., 4 h before killing). Clonidine was administered in two doses intraperitoneally of 5 μg/kg, the first dose given together with H 44/68 and the second dose given 2 h later. The prolactin levels are given in per cent of the group treated with H 44/68 alone. Means ± S.E.M. are shown. Sample size is 4-5. The absolute levels of serum prolactin in the H 44/68 treated group is shown in parenthesis. Statistical analysis was performed according to one way analysis of variance and subsequently multiple comparisons were made. Sum of squares simultaneous test procedure (experimental error rate, α = 0.05 or α = 0.15). The means not significantly different are covered by the same line in the tables of the figure.

doses may preferentially activate A receptors (41) has been studied on CA turnover in the median eminence and other regions of the hypothalamus and also on the secretion of prolactin and LH. As seen in Fig. 9, clonidine seems in the low dose range to be capable of reducing A turnover. On the other hand, NA turnover within the hypothalamus was only reduced in doses around and above 30 μg/kg. Thus, in doses below 30 μg/kg clonidine may preferentially activate A receptors leading to a compensatory reduction of A turnover. Clonidine was found to reduce prolactin secretion. This reduction could at least in part be mediated via activation of an A receptor, increasing activity within the tubero-infundibular DA system innervating the medial palisade zone (see Fig. 10). Thus, clonidine in

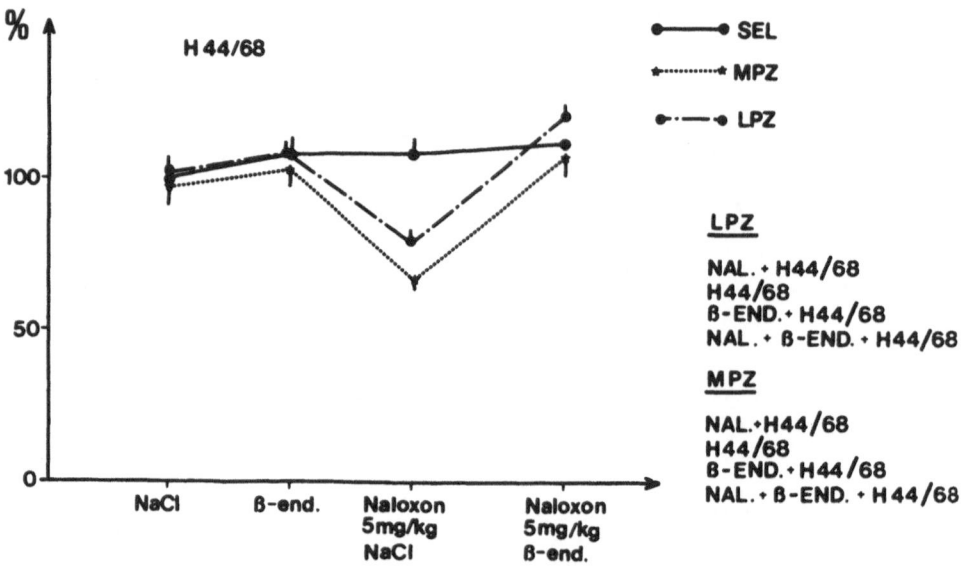

Fig. 12. The effects of naloxone on the H 44/68 induced DA fluorescence disappearance in the lateral and medial palisade zone and
in the subependymal layer of the median eminence of normal male rats.
Naloxone was given in a dose of 5 mg/kg, i.p., 30 min before the
H 44/68 injection (250 mg/kg, i.p., 2 h before killing). β-Endorphin
was given intraventricularly to unanesthetized rats at the same time
as the H 44/68 and naloxone injection. The fluorescence values are
given in per cent of the values found in the H 44/68 alone group.
Means ± S.E.M. are shown. The statistical analysis was made according
to one way analysis of variance and subsequently multiple comparisons
were made: sum of squares simultaneous test procedure (experimental
error rate, α = 0.05). The means not significantly different are
covered by the same line in the tables of the figure.

doses from 3 μg/kg selectively enhanced CA turnover within the medial
palisade zone without influencing the DA turnover in the lateral
palisade zone and the NA turnover in the subependymal layer in doses
up to 300 μg/kg. Consequently, within the medial basal hypothalamus
there may occur an important facilitory adrenergic control of the
tubero-infundibular system to the medial palisade zone. In agreement
clonidine does not reduce prolactin secretion after tyrosine hydroxylase inhibition. α-Adreno receptor blocking agents, which we have
postulated to be preferential blockers of A receptors, are also
capable of increasing the secretion of prolactin in agreement with
the present hypothesis. Yohimbine (1 mg/kg) and piperoxane (20 mg/kg)
produced an hypersecretion of prolactin, even in the presence of a
tyrosine hydroxylase inhibitor and clonidine (2 x 5 μg/kg) (see
Fig. 11).

TABLE II. DRUGS INFLUENCING NEURONAL CONTROL OF ANTERIOR PITUITARY
          FUNCTION.

Legend to the Table

The Table shows only the results obtained in our laboratory in
normal male rats or in ovariectomized female rats.

↑ ≡ increase in serum hormone levels
↓ ≡ decrease in serum hormone levels
~ ≡ not significant changes in serum hormone levels
NT ≡ not tested

Neuronal systems		Changes in hormone secretion			
MA systems		LH	FSH	PL	GH
A	Clonidine (low doses)	↑	~	↓	NT
	Piperoxane	~	~	↑	NT
	Yohimbine	~	~	↑	NT
	PNMT inhibitor (SK&F 64139)	↓	~	↑	NT
NA	Clonidine (high doses)	~	NT	↓	NT
	Phenoxybenzamine	↓	~	~	NT
DA	Apomorphine	↓	~	↓	NT
	Ergot drugs	↓	~	↓	NT
	Pimozide	↓	↑	↑	NT
5-HT	GEA 654	↑	~	↑	↓
	Zimelidine	↑		↑	↓
	Parachloramphetamine	NT	NT	↑	↓
ACh system					
ACh	Scopolamine (low doses)	↑	↑	~	NT
GABA system					
GABA	Muscimol	↓	~	↓	NT
	γ-Acetylenic GABA	↓	~	↓	NT
Peptidergic systems					
Endorphins	Morphine	NT	NT	↑	NT
	β-Endorphine	~	~	↑	↑
	Naloxone	~	~	~	~

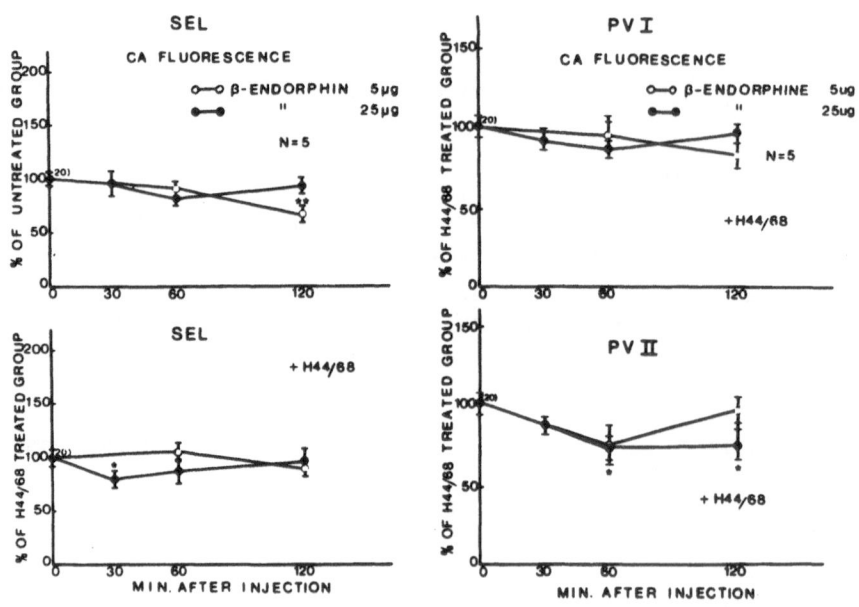

Fig. 13. The effect of β-endorphin on the H 44/68 induced NA fluorescence disappearance in the subependymal layer (SEL), in the anterior periventricular hypothalamic region (PV I) and in the posterior
periventricular hypothalamic region (PV II) of normal male rats.
β-Endorphin was given intraventricularly to unanesthetized male rats
in a dose of 5 or 25 μg/rat. The time after injection of β-endorphin
is shown on the x-axis. On the y-axis the fluorescence values are
given in per cent of the values found in the respective H 44/68
alone group or in per cent of untreated rats if H 44/68 had not been
injected (in SEL). Means ± S.E.M. are shown. All statistical comparisons have been made with the H 44/68 alone group. Student's t-test;
×: p<0.05; ××: p<0.01. (Ferland, Fuxe, Andersson, Eneroth, Gustafsson and Skett, to be published).

EFFECTS OF ENDORPHINS ON THE AMINE TURNOVER IN VARIOUS NA
AND DA NERVE TERMINAL SYSTEMS OF THE HYPOTHALAMUS

Met-enkephalin, β-endorphin and morphine have all been found
to markedly reduce the turnover of DA within the lateral and medial
palisade zone of the median eminence (34,43,44). These experiments
were performed using intraventricular injections of endorphins and
morphine. The reduction of DA turnover within the external layer of
the median eminence may in part be responsible for endorphin induced
increases of prolactin and growth hormone secretion (see 45-49).
These hormonal effects by endorphins and morphine may thus in part
involve an inhibition of activity within the DA terminal systems of
the median eminence. The increase of prolactin secretion seems to be
related to a reduction of turnover of DA within the medial palisade

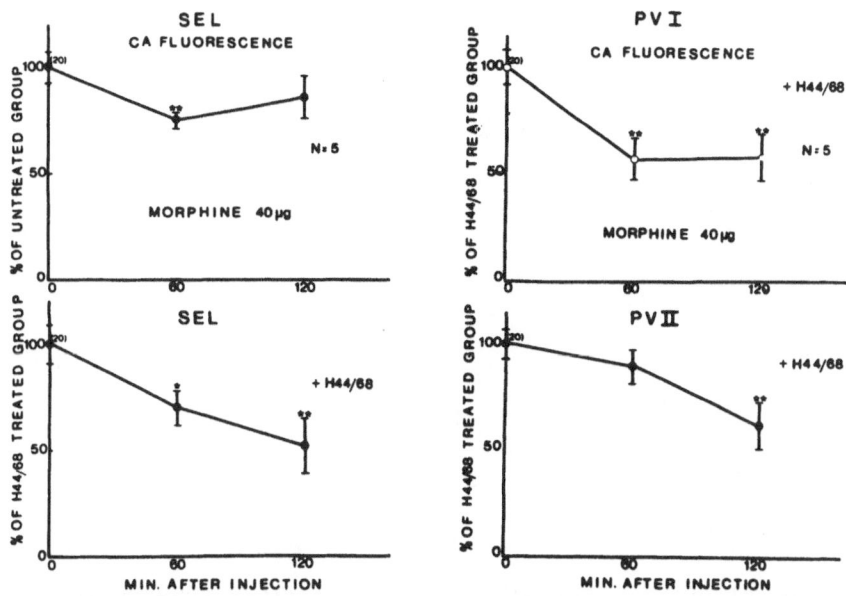

Fig. 14. The effect of morphine on the H 44/68 induced NA fluorescence disappearance in the subependymal layer (SEL), in the
anterior periventricular hypothalamic region (PV I) and within the
posterior periventricular hypothalamic region (PV II) of normal
male rats. Morphine was given intraventricularly to unanesthetized
male rats in a dose of 40 µg/rat. For further details, see text to
previous figure. Means ± S.E.M. are shown. All statistical comparisons have been made with the H 44/68 alone group or with the
untreated group mean value. Student's t-test. ×: $p<0.05$; ××: $p<0.01$.
(Ferland, Fuxe, Andersson, Eneroth, Gustafsson and Skett, to be
published).

zone, since β-endorphin produced a selective increase of the CA
stores in this layer without affecting the CA stores in the lateral
palisade zone. The increase of growth hormone secretion may instead
in part be due to a loss of release of DA onto axonic DA receptors
located on growth hormone releasing hormone containing nerve terminals within the external layer of the median eminence. It is thus
speculated that there may exist inhibitory DA receptors on growth
hormone releasing hormone containing nerve terminals. In agreement
with this view Labrie's group have observed increases of growth
hormone secretion by endorphins also in the presence of antibodies
against somatostatin. In agreement with the above findings, we have
recently observed that naloxone can produce an increase of DA turnover within the external layer of the median eminence (see Fig. 12).
However, it seems as if the neuroendocrine effects produced by
endorphins also can involve a 5-HT mechanism, since 5-HT receptor
blocking agents can reduce the release of prolactin by endorphins

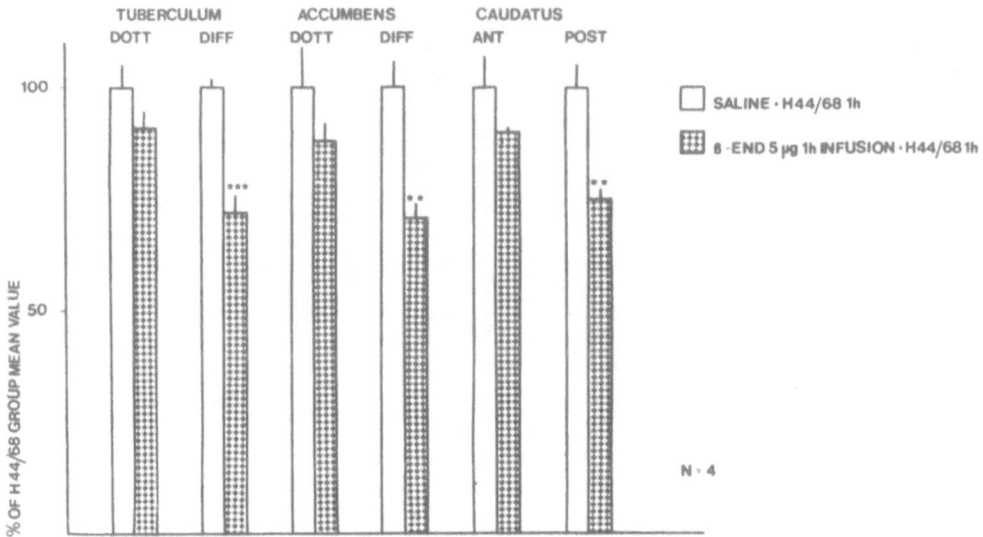

Fig. 15. The effect of β-endorphin on the H 44/68 induced DA fluorescence disappearance in various regions of the forebrain. β-Endorphin was infused during a period of 1 h into the lateral ventricle,
the total dose being 5 µg. At the onset of infusion the H 44/68
injection was made (250 mg/kg, i.p., 1 h before killing). During the
infusion period the saline or β-endorphin injected animals were in
fluothane-air anesthesia. The fluorescence values are given in per
cent of H 44/68 alone treated group. Within the tuberculum olfacto-
rium and the nuc. accumbens the effects of β-endorphin were evaluated
on both the diffuse and dotted types of DA nerve terminals. In the
caudate nucleus effects were evaluated in the anterior and posterior
dorsal part of the nucleus. Means ± S.E.M. are shown. Sample size
is 4. Statistical analysis was made according to Student's t-test.
××: p<0.01.

(see Müller, this symposium). Thus, a reduction of DA activity and
an increase of 5-HT activity may produce the dramatic increase of
prolactin secretion observed following treatment with intraventricular injections of endorphins.

     Also the NA systems appear to be controlled by opiate receptors.
However, this opiate receptor seems to be of another type than the
one controlling the DA systems, since only morphine produced marked
increases of NA turnover within the subependymal layer and within the
anterior and posterior periventricular hypothalamic regions (see
Figs. 13-14). β-Endorphin, on the other hand, produced much weaker
effects and no effects at all within the anterior periventricular
hypothalamic region. Also naloxone did not influence the NA turnover
within these NA systems in a dose of 5 mg/kg which enhanced turnover

ENK.–DA INTERACTION IN CONTROL OF LHRH
SECRETION

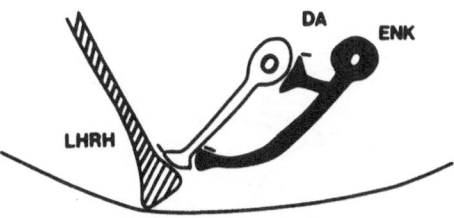

ENK.–DA INTERACTION IN CONTROL OF PROLACTIN
SECRETION

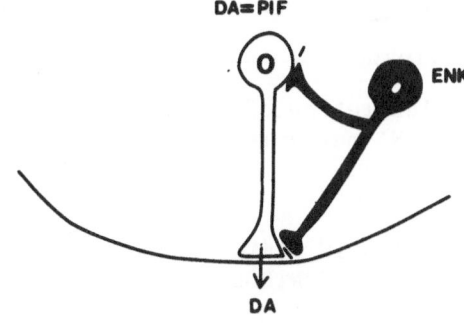

Fig. 16. Schematic illustration of possible enkephalin-DA interactions in the median eminence. Enkephalin terminals may control and
reduce the activity within the tubero-infundibular DA systems either
via actions at the nerve terminal or the cell body level. An action
at both levels is also possible. Enk = enkephalin.

of DA in the median eminence. It is therefore speculated that at
least two types of opiate receptors can be involved in the control
of the CA systems of the hypothalamus, and thus also in the control
of neuroendocrine mechanisms. Finally, it should be mentioned that
the endorphins increase DA turnover within certain DA terminal
systems of the forebrain, in contrast to the findings in the median
eminence (44; Fig. 15).

     These results suggest that opiate peptides can exert a neuro-
transmitter or a modulator role within the hypothalamus. A schematic
diagram is shown in Fig. 16, indicating the possibility that the DA
terminals in the median eminence are controlled by enkephalin
positive nerve terminals demonstrated in this region (50). Thus, the
possibility exists that DA terminals controlling release of LHRH in
turn can be controlled by other nerve terminals in the same layer,
having at least in part the function to control the controllers of
the secretion of the hypothalamic hormones.

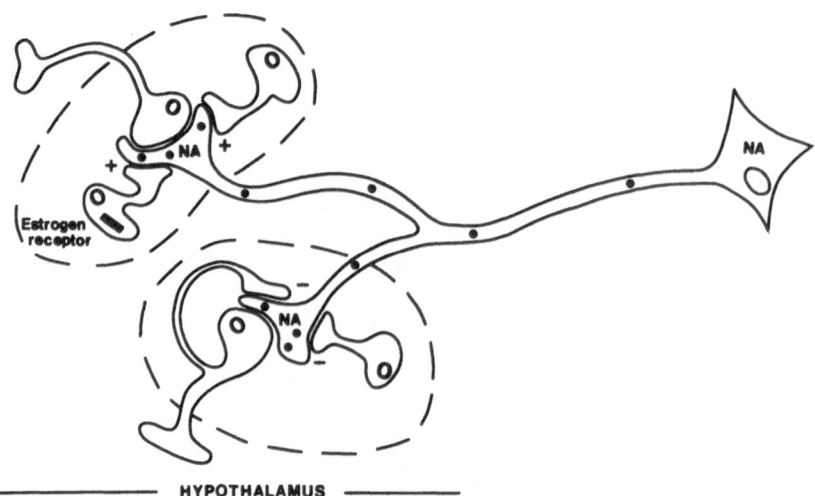

Fig. 17. Schematic illustration of possible local circuits controlling
activity in individual NA nerve terminal plexa of the hypothalamus.

NA NERVE TERMINALS IN VARIOUS HYPOTHALAMIC NUCLEI: INDEPENDENT UNITS CONTROLLED BY LOCAL CIRCUITS

Recent studies in this laboratory on the effect of hormones and
drugs on the NA turnover within the various hypothalamic and preoptic
NA nerve terminal systems have clearly indicated that the NA nerve
terminal plexus present in one hypothalamic nucleus is independently
controlled from NA terminal systems in adjacent regions of the hypothalamus. As examples it can be mentioned that the GABA transaminase
inhibitor γ-acetylenic GABA (200 mg/kg) increases NA turnover in the
parvocellular but not the magnocellular part of the paraventricular
hypothalamic nucleus (40), and intraventricular injections of β-endorphin (25 μg/rat) only increases NA turnover within the posterior but
not the anterior periventricular hypothalamic region. Furthermore,
four weeks of thyroidectomy causes a profound increase of NA turnover
within the parvocellular part of the paraventricular hypothalamic
nucleus, while the increase in the magnocellular part is considerably
less pronounced.

These results can best be explained on the assumption that each
individual NA nerve terminal plexus in a nucleus is to a large extent
controlled by local circuits which via axo-axonic or dendro-axonic
influences regulate the release of NA. Some of these local circuit
neurons may contain receptors for hormones, e.g. for estrogen, which
can explain why hormone treatment preferentially influences only
certain types of NA nerve terminal systems. These local circuitries
are schematically indicated in Fig. 17. In this figure it is also

pointed out that the NA nerve terminal plexa may be independently
controlled, even if they belong to one and the same NA nerve cell
body, and thus should receive a similar number of action potentials.
It may be that only in certain situations the nerve impulses have a
decisive role in controlling NA release, making possible a coordination of activities in adjacent NA nerve terminal regions.

ACKNOWLEDGEMENTS

This work was supported by a grant (04X-715) from the Swedish
Medical Research Council, by a grant (MH25504-04) from the National
Institute of Health, and by a grant from Magn. Bergvalls Stiftelse.

REFERENCES

1. Sawyer, C.H. Neuroendocrinology 17, 97-124, 1975.
2. Fuxe, K., Hökfelt, T., Löfström, A., Johansson, O., Agnati, L.,
    Everitt, B., Goldstein, M., Jeffcoate, S., White, N., Eneroth,
    P., Gustafsson, J.-A. and Skett, P. In Subcellular Mechanisms
    in Reproductive Neuroendocrinology, pp. 193-246, Ed. Naftolin,
    F., Ryan, K.J. and Davies, J., Elsevier, Amsterdam, 1976.
3. Kordon, C., Epelbaum, J., Enjalbert, A. and McKelvy, J. In
    Cellular and Molecular Processes in Neuroendocrine Tissues,
    pp. 167-184, Ed. Naftolin, F., North Holland Publishing, Amster-
    dam, 1976.
4. McCann, S.M., Ojeda, S.R., Martinovic, J. and Vijayan, E. In
    Advances in Biochemical Psychopharmacology Vol. 16, pp. 109-114,
    Ed. Costa, E. and Gessa, G.L., Raven Press, New York, 1977.
5. Hökfelt, T., Elde, R., Johansson, O., Ljungdahl, Å., Schultz-
    berg, M., Fuxe, K., Goldstein, M., Nilsson, G., Pernow, B.,
    Terenius, L., Ganten, D., Jeffcoate, S.L., Rehfield, J. and
    Said, S. In "Psychopharmacology", pp. 39-66, Ed.
    Lipton, M.A., DiMascio, A. and Killam, K.F., Raven Press,
    New York, 1978.
6. Barry, J., Dubois, M.P. and Carette, B. Endocrinology 95,
    1416-1423, 1974.
7. Hoffman, G.E., Melnyk, V., Hayes, T., Bennet-Clarke, C. and
    Fowler, E. In Brain-Endocrine Interaction III. Neural Hormones
    and Reproduction. 3rd Int. Symp., Würzburg, pp. 67-82, Ed.
    Scott, D.E., Rochester, N.Y., Kozlowski, G.P., Fort Collins,
    Colo. and Weindl, A., Karger, Basel, 1978.
8. Hökfelt, T. and Fuxe, K. In Brain Endocrine Interaction.
    Median Eminence: structures and function, pp. 181-223, Ed.
    Knigge, Scott and Weindl, Karger, Basel, 1972.
9. Hökfelt, T., Fuxe, K., Goldstein, M., Johansson, O., Park, D.,
    Fraser, H. and Jeffcoate, S.L. In Anatomical Neuroendocrinology
    pp. 381-392, Ed. Stumpf and Grant, Karger, Basel, 1975.

10. Hökfelt, T., Elde, T., Fuxe, K., Johansson, O., Ljungdahl, Å.,
    Goldstein, M., Luft, R., Efendic, S., Nilsson, G., Terenius, L.,
    Ganten, D., Jeffcoate, S.L., Rehfield, J., Said, S., Perez de
    la Mora, M., Possani, L., Tapia, R., Teran, L. and Palacios, R.
    In The Hypothalamus, pp. 69-135, Ed. Reichlin, S., Baldessarini,
    R.J. and Martin, J.B., Raven Press, New York, 1978.
11. Fuxe, K., Andersson, K., Hökfelt, T., Agnati, L., Ögren, S.-O.,
    Eneroth, P., Gustafsson, J.-Å. and Skett, P. In Progress in
    Prolactin Physiology and Pathology, pp. 95-109, Ed. Robyn, C.
    and Harter, M., Elsevier/North-Holland Biomedical Press, Amster-
    dam-New York, 1978.
12. Löfström, A., Jonsson, G. and Fuxe, K. Histochem. Cytochem.
    24, 415-429, 1976.
13. Bloom, F.E. In Psychopharmacology: A Generation of Progress,
    pp. 131-141, Ed. Lipton, M.A., DiMascio, A. and Killam, K.F.,
    Raven Press, New York, 1978.
14. Fuxe, K., Hökfelt, R., Eneroth, P., Gustafsson, J.-Å. and
    Skett, P. Science 196, 899, 1977.
15. Löfström, A., Jonsson, G., Wiesel, F.-A. and Fuxe, K. J. Histo-
    chem. Cytochem. 24, 430-442, 1976.
16. Agnati, L., Fuxe, K., Hökfelt, T., Goldstein, M., Jeffcoate,
    S.L. and Elde, P. J. Histochem. Cytochem. 25, 1222-1236, 1978.
17. Lidbrink. P., Corrodi, H., Fuxe, K. and Olson, L. Brain
    Research 45, 507-524, 1972.
18. Fuxe, K., Löfström, A., Eneroth, P., Gustafsson, J.-Å., Skett,
    P., Hökfelt, T., Wiesel, F.-A. and Agnati, L. Psychoneuro-
    endocrinology 2, 203-225, 1977.
19. Löfström, A., Eneroth, P., Gustafsson, J.-Å. and Skett, P.
    Endocrinology 101 (5), 1559-1569, 1977.
20. Fuxe, K., Löfström, A., Eneroth, P., Gustafsson, J.-Å., Hökfelt,
    T., Skett, P., Wuttke, W., Fraser, H. and Jeffcoate, S. In
    Proc. 1st Int. Symp. on Basic Applications and Clinical Uses of
    Hypothalamic Hormones, pp. 165-177, Ed. Salgado, A.L.Ch.,
    Durango, R.F. and Lopez del Campo, J.G. Excerpta Medica Inter-
    national Congress Series No. 37Y, 1976.
21. Fuxe, K. and Hökfelt, T. In Aspects of Neuroendocrinology,
    pp. 192-205, Ed. Scharrer, B., Springer, Berlin-Heidelberg-
    New York, 1970.
22. Hökfelt, T. and Fuxe, K. Neuroendocrinology 9, 100, 1972.
23. Olsson, L., Fuxe, K. and Hökfelt, T. Acta Endocrinol. 75,
    233, 1972.
24. Grant, L.D. and Stumpf, W.E. J. Histochem. Cytochem. 21, 404,
    1973.
25. Stumpf, W.E., Sar, M., Grant, L.D. and Heritage, A.S. In
    Brain-Endocrine Interaction III. Neural Hormones and Reproduc-
    tion, pp. 212-227, Ed. Scott, D.E., Kozlowski, G.P. and Weindl,
    A., Karger, Basel, 1978.
26. Löfström, A. and Bäckström, T. Psychoneuroendocrinology, 1978
    (in press).
27. Drouva, S.V. and Gallo, R.V. Endocrinology 99, 651-658, 1976.

28. Wuttke, W. In Endocrinology Vol. 2, pp. 287-292, Ed. James,
     V.H.T. Excerpta Medica, Amsterdam-Oxford, 1977.
29. Wuttke, W. Intern. Rev. Physiol. Pharmacol. Biochem. 76,
     59-101, 1976.
30. U'Prichard, D.C., Greenberg, D.A. and Snyder, S.H. Mol. Pharma-
     col. 13, 454-473, 1977.
31. Kordon, C. and Rotsztejn, W.H. In Advances in Biochemical
     Psychopharmacology Vol. 16, pp. 89-97, Ed. Costa, E. and Gessa,
     G.L., Raven Press, New York, 1977.
32. Rotsztejn, W.H., Charli, J.L., Patton, E., Epelbaum, J. and
     Kordon, C. Endocrinology 99, 1663, 1976.
33. McCann, S.M., Moss, R.L., Ojeda, S.R., Martinovic, J. and
     Vijayan, E. Prog. Reprod. Biol. 2, 105-114, 1977.
34. Fuxe, K., Ferland, L., Andersson, K., Eneroth, P., Gustafsson,
     J.-Å. and Skett, P. In Brain-Endocrine Interaction III. Neural
     Hormones and Reproduction, 3rd Int. Symp., Würzburg, pp. 172-182
     Ed. Scott, D.E., Rochester, N.Y., Kozlowski, G.P., Fort Collins,
     Colo. and Weindl, A., Karger, Basel, 1978.
35. Ferland, L., Drouin, J. and Labrie, F. In Hypothalamus and
     Endocrine Functions, pp. 192-209, Ed. Labrie, F., Meites, J.
     and Pelletier, G., Plenum Press, New York, 1976.
36. Porter, J.C., Eskay, R.L., Oliver, C., Ben-Jonathan, N., War-
     berg, J., Parker, Jr., R. and Barnea, A. In Advances in
     Experimental Medicine and Biology Vol. 87, pp. 181-201, Ed.
     Porter, J.C., Plenum Press, New York, 1977.
37. Fuxe, K., Eneroth, P., Gustafsson, J.-Å., Löfström, A. and
     Skett, P. Brain Research 122, 177, 1977.
38. Hutchison, R.E. In Progress in Prolactin Physiology and
     Pathology, pp. 243-251, Ed. Robyn, C. and Harter, M., Elsevier/
     North-Holland Biomedical Press, Amsterdam-New York, 1978.
39. Andersson, K., Fuxe, K., Eneroth, P., Gustafsson, J.-Å. and
     Skett, P. Neurosci. Lett. 5, 83-89, 1977.
40. Fuxe, K., Andersson, K., Ögren, S.-O., Perez de la Mora, M.,
     Schwarcz, R., Hökfelt, T., Eneroth, P., Gustafsson, J.-Å. and
     Skett, P. In GABA-Neurotransmitters Alfred Benzon Symposium 12,
     Munksgaard, Copenhagen, 1978 (in press).
41. Fuxe, K., Bolme, P., Jonsson, G., Agnati, L., Goldstein, M.,
     Hökfelt, T., Schwarcz, R. and Engel, J. In Perspectives in
     Nephrology & Hypertension, Ed. Schmitt, H. and Meyer, P., 1978
     (in press).
42. Rubinstein, L. and Sawyer, C.H. Endocrinology 86, 988-995,
     1970.
43. Ferland, L., Fuxe, K., Eneroth, P., Gustafsson, J.-Å. and Skett,
     P. Eur. J. Pharmacol. 43, 89-90, 1977.
44. Fuxe, K., Ferland, L., Agnati, L.F., Eneroth, P., Gustafsson,
     J.-Å., Labrie, F. and Skett, P. Acta pharmacol. et toxicol.
     41, Suppl. 4, 48, 1977.
45. Lien, E.L., Fenichel, R.L., Garsky, V., Sarantakis, D. and
     Grant, N.H. Life Sci. 19, 837-840, 1976.

46. Dupont, A., Cusan, L., Garon, M., Labrie, F. and Li, C.H.   Proc.
    natn. Acad. Sci. USA 74, 358-359, 1977.
47. Dupont, A., Cusan, L., Labrie, F., Coy, D.H. and Li, C.H.
    Biochem. Biophys. Res. Commun. 75, 76-82, 1977.
48. Rivier, C., Vale, W., Ling, N., Brown, M. and Guillemin, R.
    Endocrinology 100, 238-241, 1977.
49. Rivier, C., Brown, M. and Vale, W.   Endocrinology 100, 751-754,
    1977.
50. Elde, R., Hökfelt, T., Johansson, O. and Terenius, L.   Neuro-
    science 1, 349-355, 1976.

CONTROL OF PEPTIDERGIC NEURONS IN HUMANS:

AN INTRODUCTION

Rolf Luft

Department of Endocrinology
Karolinska Hospital
104 01 Stockholm, Sweden

During the preceding sessions we have been brought up to the
front-line of research in the area covered by this Symposium. We
already know that the development of this field has had a considerable
impact on clinical medicine and, especially, on clinical endocrinology.

This last session of the Symposium mainly concerns the clinical
implications of the basic discoveries already discussed. It is grati-
fying to notice that clinical research, to some extent, has also
reached the forefront of this field, and that findings in clinical-
experimental medicine have sometimes been the incitement for basic
research. Therefore, to some extent, we are able today in clinical
work to utilize available information regarding biogenic amine pathways - dopaminergic, serotonergic as well as noradrenergic ones -
for the function of the specialized peptidergic pathways and, thereby,
for peripheral endocrine functions in health and disease.

This last session of the Symposium is devoted to the interplay
of neurotransmitter and peptidergic neurons, i.e., to the critical
link between brain and hypophysiotropic functions, in man. Emphasis
will be put on neuropharmacological agents - alone or in combination
with hypothalamic hormones - as diagnostic aids in endocrine disorders
of probable hypothalamic origin in which altered neurotransmitter
function may be envisaged. Furthermore, we shall consider neuropharmacologic approaches to the treatment of neuro-endocrine
disorders.

However, from the very onset we have to limit the scope of our
discussion. The anterior pituitary comprises seven discrete functions, each of which probably may be controlled by inhibiting and

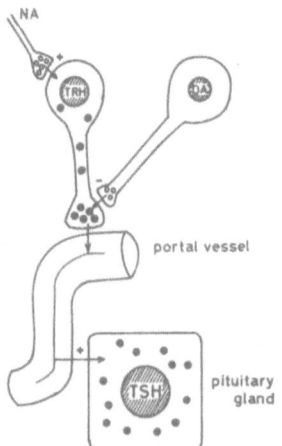

Fig. 1. Regulation of TSH secretion; NA - noradrenaline,
DA - dopamine, + denotes stimulation, - inhibition

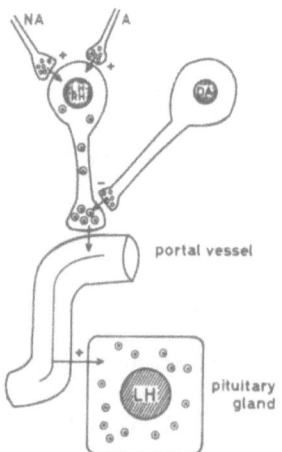

Fig. 2. Regulation of LH secretion; NA - noradrenaline,
Da - dopamine, + denotes stimulation, - inhibition

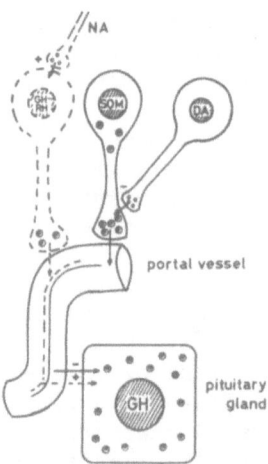

Fig. 3. Regulation of GH secretion; NA - noradrenaline,
DA - dopamine, SOM - somatostatin, + denotes stimulation,
- inhibition

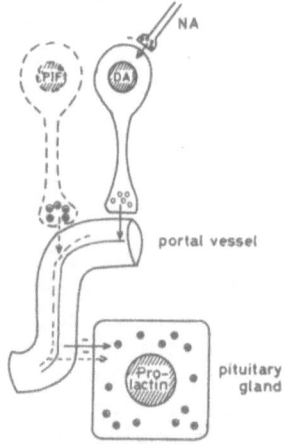

Fig. 4. Regulation of prolactin
secretion; NA - noradrenaline,
DA - dopamine, PIF - prolactin
inhibiting factor, + denotes
stimulation, - inhibition

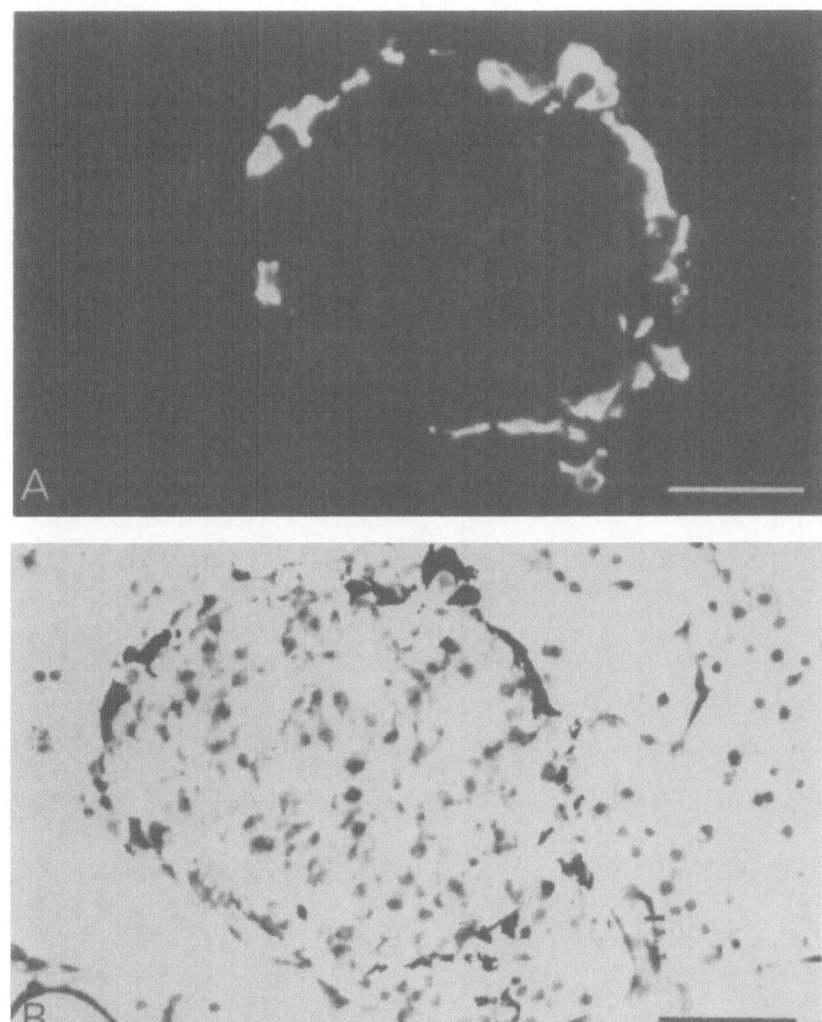

Fig. 5A,B. (A) Immunofluorescence micrograph of an islet of Langerhans after incubation with somatostatin antiserum. Numerous peripherally located cells are immunoreactive. (B) Silver staining of the same section. Note complete correlation between somatostin positive and argyrophilic cells. Bars indicate 50 μm.

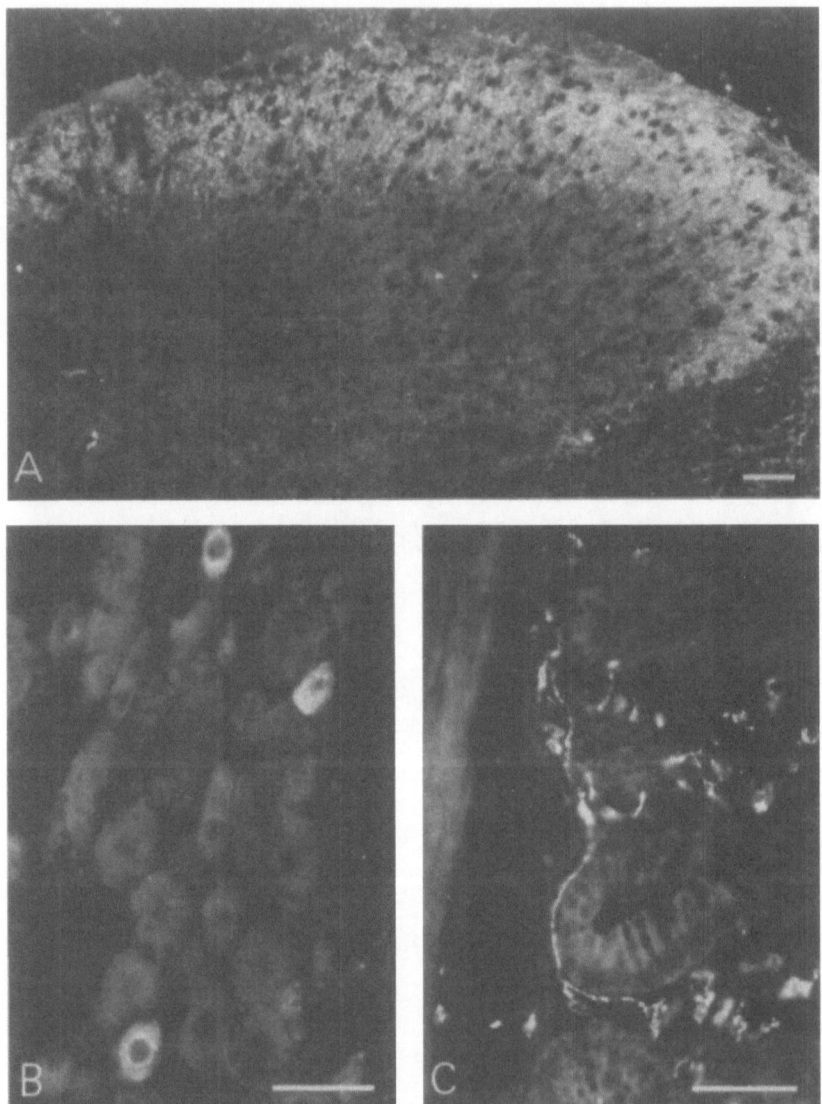

Fig. 6A, B & C. Immunofluorescence micrographs of the spinal cord
(A), spinal ganglion (B), and colon (C), after incubation with somatostatin antiserum. Somatostatin positive fibers are seen in the
dorsal horn of the spinal cord (A), and in the lamina mucosa of
colon (C). In the spinal ganglion immunoreactive cell bodies are
found (B). Bars indicate 50 μm.

stimulating hypothalamic regulators. We shall today concertrate on
the regulators for the release of prolactin, growth hormone, TSH and
the gonadotrophins, mainly because most information is available
regarding these regulating mechanisms.

I shall introduce the clinical session with a few slides
summarizing some basic, mostly established facts regarding the regulations of the secretions of TSH, LH, PRL, and GH. (Figs. 1,2,3,4).
In these figures I have omitted regulators such as serotonin, GABA
and enkephalins, since much more work is needed until their roles in
this connection are firmly established. I am sure that the pictures
will be more complicated next time we meet.

Another relevant area in which clinical research has been
involved is what we may call the "non-specificity" of the hypothalamic
peptides, i.e., that they do not exert one single action but several.
In addition, or parallel, to this is the identification of production
sites for these peptides to more than one area within and even outside
the CNS. Somatostatin is an example of this. It was first isolated
from the hypothalamus but it has also been demonstrated in the D-cells
of the pancreas (Fig. 5), in the gastrointestinal canal, in the vagus
nerve and in prevertebral autonomic ganglia in the mesenteric-coeliac
region. In other words, somatostatin is produced or is present in
areas that probably are involved in the secretion of hormones regulating blood glucose homeostasis or in the nerve supply to such areas.
We may visualize the hypothalamus, stomach and pancreas and their
nerve connections as a "golden triangle" in this respect.

But what would then be the meaning of the same peptide being
produced in a system of sensory neurons (Figs. 6 A-C)? Is the latter
somatostatin really the same as the former compound?

Some but certainly not all of such problems will be dealt with
during this session, and many more exciting questions will present
themselves during the next few years. We can predict that in the
future we shall find in our textbooks a series of specific "named"
clinical syndromes with specific functional and/or anatomical substrates in the limbic system and hypothalamus.

STUDIES OF THE ROLE OF DOPAMINE IN THE CONTROL OF PROLACTIN AND GONADOTROPIN SECRETION IN HUMANS

Samuel S.C. Yen

Department of Reproductive Medicine, University of

California, San Diego, School of Medicine, La Jolla, CA

INTRODUCTION

The participation of the catecholaminergic system of the brain in the central regulation of gonadotropin (FSH and LH) and prolactin (PRL) release is now established in all mammalian species studied except in humans (39). That the tuberoinfundibular dopamine (DA) system functions as a hypothalamic PRL inhibitor by a direct action of the DA-receptor of the lactotrope, delivered by way of the hypophyseal portal system is unequivocally demonstrated (8,13,20,37,38, 53,54). By contrast, catecholamines which influence gonadotropin secretion appear to operate by way of a central site(s), affecting the peptidergic neuron that produces LRF (19,21,22,25,35,41,48,51). Through the use of immunohistochemistry, Fuxe and associate and Löfstrom et al. have demonstrated a morphological relationship which serves as a basis for the interaction between DA, norepinephrine (NE), and LRF nerve terminals in the median eminence of the rat (18,19,35). Recently, axo-axonic contact of LRF and DA terminals was demonstrated in close proximity to the portal vessels of the rat median eminence suggesting the synaptic influence of DA on the release of LRF into the portal system (1a,40).

While a reproducible α-adrenergic stimulatory role of NE on LRF release has been convincingly demonstrated (19,22,39,49,50), there exist contradictory data concerning the inhibitory (6,7,18,19, 21,22,35,41) vs. stimulatory effect (39,48,51) of DA. Experimental results derived from pharmacological manipulations and direct intraventricular administration of DA support the hypothesis of a stimulatory role of DA on LRF release (39). These findings were extended by the observation of Rotsztejn et al. (1977) that the addition of DA to the hypothalamic fragments in vitro induces a dose-related LRF release which can be blocked by a DA-antagonist, pimozide (48). On the other hand, in vitro studies using pituitary-ME unit (41) and pituitary hypothalamus co-incubation system (22), and DA-agonist and antagonist studies in vivo (6,7), DA was found to exhibit a clear-cut inhibitory effect on LH release. These latter observations are consonant with those of immunohistochemical studies indicating that NE stimulates and DA inhibits LH release by a direct action on hypothalamic LRF producing neurons (19,35). Perhaps the most cogent argument against a stimulatory action of DA on LRF release is the absence of any evidence that DA administration in a variety of in vivo experiments to induce LH surge.

We were not discouraged by these apparently conflicting data, and during the past few years, we have conducted studies on the role of DA in the regulation of gonadotropin and PRL secretion in humans. The preliminary results of these clinical investigations suggest that DA inhibits the secretion of both LH and PRL by the adenohypophysis. Ovarian steroids, PRL and possibly LRF appear to modify this inhibitory action of DA.

Effects of Dopamine Infusion

Our first human investigation to determine the effect of DA on pituitary release of PRL and gonadotropin was initiated in 1974 and reported in 1976 (32). We found that in normal and hyperprolactinemic subjects, infusion of DA at 4 μg/kg/min, devoid of cardiovascular effect, induces a progressive decline in circulating PRL concentration (Fig. 1). By the end of 3 hrs, a greater suppression of PRL release is seen in women (33%, from 13.2 ± 1.0 ng/ml to 8.8 ± 0.8 ng/ml) than in men (20%, from 13.7 ± 1.0 ng/ml to 10.8 ± 1.0 ng/ml). This suppression is statistically significant both in the men (P < 0.005) and the women (P < 0.00001). The mean percent deviation from average baseline values was significantly different (P = 0.026) for normal men vs. normal women. In women with hyperprolactinemia, a mean decrease of 45% from their pre-infusion levels was reached at the end of the infusion period (Fig. 1). This represents a 10-fold greater fall in average circulating PRL levels (45 ng/ml) than is observed in normal women (4.4 ng/ml). The percentage decrease in PRL during DA infusion was also significantly greater for the hyperprolactinemic women than for the normal women (P < 0.0005). After discontinuation of the DA infusion, circulating PRL levels rebound significantly (P < 0.05) to levels higher than basal in normal subjects and return promptly to the high base level in the hyperprolactinemic patients.

The mean (± SE) basal (5 determinations) serum LH and FSH concentrations in the 8 normal subjects were 9.4 ± 1.6 and 9.8 ± 1.3 mIU/ml, respectively. LH levels show a progressive and

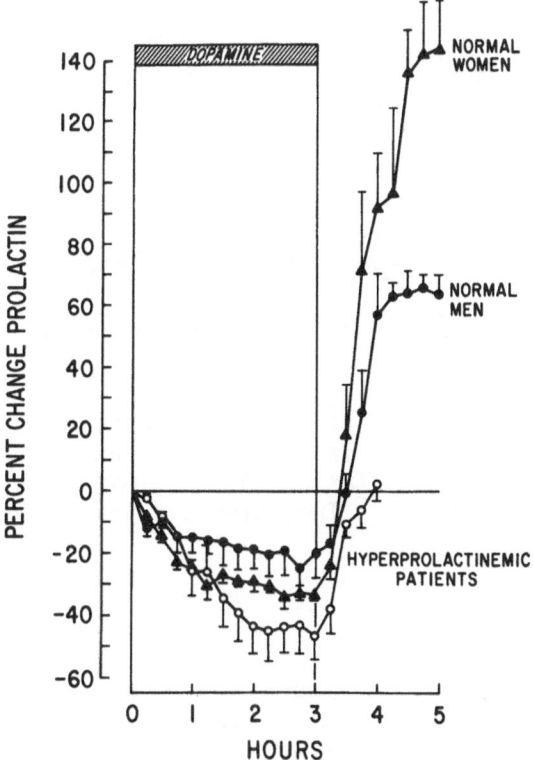

Figure 1. Percent change in serum PRL concentrations during and after dopamine infusion in normal women and men in hyperprolactinemic patients. (Analyses of variance revealed the percent deviation from the baseline resulting from DA infusion was significantly greater for the hyperprolactinemic women than in normal women (P = 0.00035) and the normal women are significantly greater than in the normal men (P = 0.026).

significant (P < 0.00001) decline during DA infusion and an immediate rebound (within 15 min) following termination of the infusion (Fig. 2). The LH values during the first hour of the rebound are significantly greater than the values found during the last hour of DA infusion (P < 0.00001) as well as the pre-infusion values (P < 0.0005). FSH levels do not change significantly during DA infusion. No significant changes in either serum LH or FSH levels were found in 8 control adult subjects during a saline infusion.

These data indicate that a dopaminergic mechanism exists in humans for regulation of PRL and LH release and that for LH, the effect of DA is to suppress its secretion. We were encouraged by

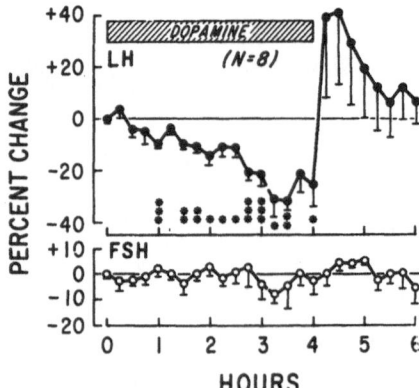

Figure 2.   Mean (± SE) percent change in basal (mean of 5 values for each subject) LH and FSH in 4 normal men and 4 normal women during and after dopamine infusion (4 µg/kg/min).   * P < 0.05, ** P < 0.005, *** P < 0.001.   Indicates significantly different from baseline value.

these preliminary findings, and our studies were then extended to determine the relationships between gonadal steroids and DA in the regulation of PRL and gonadotropin, and to examine the effect of DA agonists and antagonists.

Effect of DA-Agonists

While the acute inhibitory effects of orally administered DA agonists, L-dopa and 2-bromo-α-ergocryptine (CB-154), on PRL release are well established (11,15,17,33,36,55,56,61) the effects of L-dopa or CB-154 on gonadotropin secretion have not been unequivocally demonstrated (23,45,46,55,60).   In view of our observation that DA infusion inhibits LH release, the effect of L-dopa and CB-154 on gonadotropin secretion was studied in experiments designed to parallel our DA infusion studies.

An inhibitory effect of both L-dopa and CB-154 on LH release was convincingly demonstrated in these studies (28).   Following the administration of L-dopa (Fig. 3), the expected suppression of mean PRL levels (P < 0.00001) is found and this is followed by a significant rebound (P < 0.00001) to levels above basal value after the 6th hour.   L-dopa also causes significant depression in mean LH levels, followed by a significant rebound above basal levels.   The greatest suppression in mean LH levels is seen between 1 and 4 hours

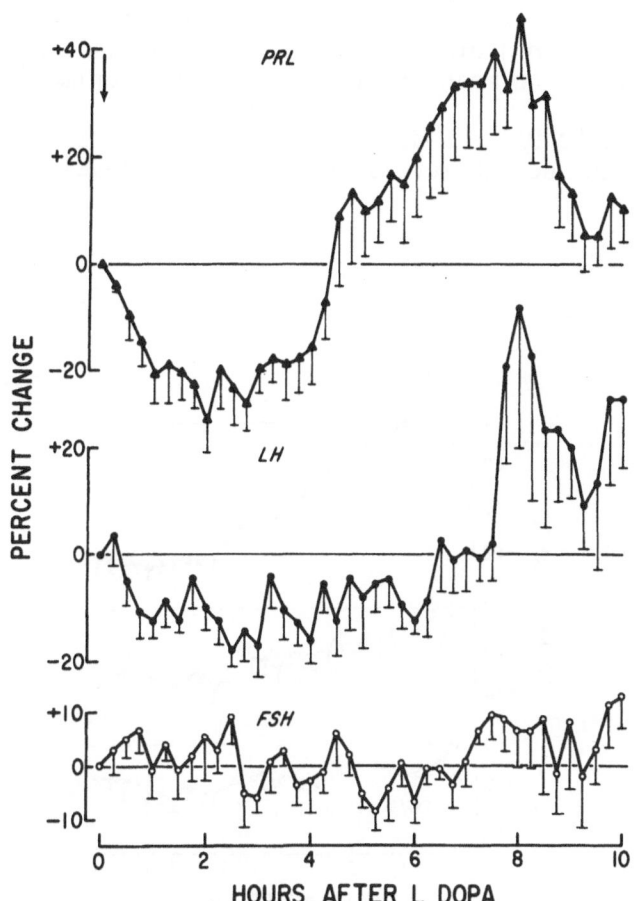

Figure 3.   Mean (± SE) percent changes from basal levels (mean of
5 basal samples) in serum PRL, LH and FSH following the oral administration of L-dopa (0.5 g) in 6 normal women in the early follicular
phase.

(P < 0.00005).   Response to L-dopa administration appears to vary
from one subject to another both in terms of the time course of the
LH decline and the rebound.   This variability among subjects was
assessed by the presence or absence of significant interaction in
an analysis of variance.   Significant interaction was not found for
the degree of LH decline in response to L-dopa.   For degree of LH
rebound, however, significant interaction (P < 0.0002) was found,
indicating subject variability of post suppression rebound (a clear
rebound was found in 4 of 6 subjects).   L-dopa, like DA, had no
significant effect on mean FSH levels in our experiments.

CB-154 unlike DA or L-dopa induces a significant fall in FSH
as well as in LH and PRL levels (Fig. 4).   The levels of both

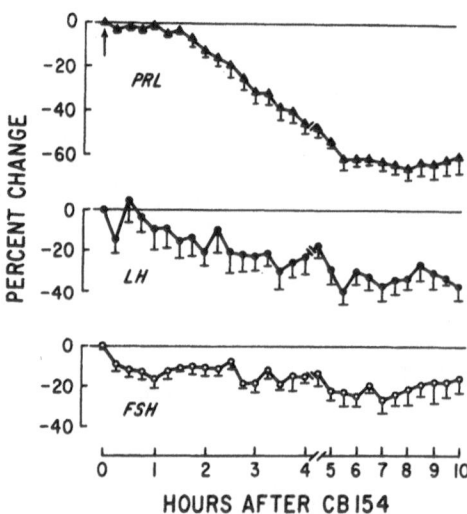

Figure 4.   Mean (± SE) percent changes from basal levels (mean of
5 basal samples) in serum PRL, LH and FSH in 6 patients with hyperprolactinemic amenorrhea following the oral administration of
CB-154 (2.5 mg).

gonadotropins are significantly lower between 5 and 10 hours after
the administration of CB-154 than the mean basal levels
(P < 0.00001). As with L-dopa, responses of different subjects
vary in time course and degree, and significant interaction was found
for both LH (P < 0.00002) and FSH (P < 0.0002). Further, the longer
duration of action for CB-154 as compared to L-dopa and DA itself
is evident. Although the effect of L-dopa on LH secretion has been
investigated in several laboratories in the past (23,45,46,55,60),
an unequivocal response has not been found. This may be due to the
fact that most of these earlier investigators anticipated a rise
rather than for a fall in LH levels basing their reasoning on the
experimental data which indicated that dopamine enhances the release
of hypothalamic LRF in rats (39,51). Further, these earlier studies
were limited by the relatively short duration of the experiments
(6 hrs or less), infrequent blood sampling and limited methods of
statistical analysis. In our studies frequent blood sampling (at
15 min intervals) and an extended period of observation (10 hrs)
clearly disclosed that L-dopa suppresses secretion of LH and PRL.

The Inhibitory Action of Dopamine - a Physiological Consideration

        Although it is well established that in the mammal, including
humans, PRL release is under the tonic inhibition of an hypothalamic
PIF, the identity of which remains uncertain. Evidence has recently
been presented which implicates hypothalamic DA as a physiological
PIF. Incubation of rat pituitary glands with DA in minute doses
causes an 85% decrease in PRL release in vitro (37,53) and infusion
of DA into rat hypophyseal portal vessel significantly reduces the
PRL levels in vivo (54). Further, the degree of PIF activity, in
a purified porcine hypothalamic preparation, is correlated with the
amount of catecholamine contained in that preparation (54), and when
rat hypothalamic extracts are subjected to preincubation with monoamine oxidase or to aluminum oxide catecholamine adsorption, the
extracts lose their ability to inhibit PRL release in vitro (53).
Moreover, in the rat the presence of DA (but not NE or E) in the
hypophyseal-portal blood has been demonstrated (8,20) and the concentration in portal blood ($\sim$ 6 ng/ml) is sufficient to inhibit PRL
secretion in vivo (20). These findings link hypothalamic DA and the
pituitary lactotrope as a functional system that controls PRL secretion. In this context, hypothalamic DA, in addition to its neurotransmitter function, may be viewed as a neurohormone delivered by
the portal blood to its target cells, the lactotropes, located at a
distance from the site of DA secretion.

        DA receptors have been found in both the median eminence and
the adenohypophysis (14) of the rat. Although DA presumably does
not cross the blood-brain barrier, both the median eminence and
adenohypophysis lie outside the blood-brain barrier (2,9). Thus,
the effect of DA and DA agonists on PRL release, as demonstrated in

our studies, could represent either an action directly on the hypothalamus with a resulting increase in PIF release or, more likely,
a direct action on the pituitary lactotrope DA receptors (13).
Although the findings in our study do not distinguish between these
two possibilities, the presence of a dopaminergic mechanism in the
regulation of PRL secretion is clearly established for humans.

The mechanism by which DA and DA-agonists exert their inhibitory action on LH release has not been entirely clear. Although
substantial evidence has suggested a direct inhibitory action of DA
on hypothalamic LRF neurons as discussed earlier, a pituitary site
of action could not be excluded. In order to determine whether DA
acts directly on the pituitary, LH responses to pulses of LRF
(10 μg) were measured before and during DA-infusion. DA was found
to blunt the LRF-mediated LH release (Fig. 5). Although this
finding is consistent with a pituitary site of action for DA, it is
also compatible with a reduction of endogenous LRF release during
the period of DA infusion resulting in a decreased accumulation of

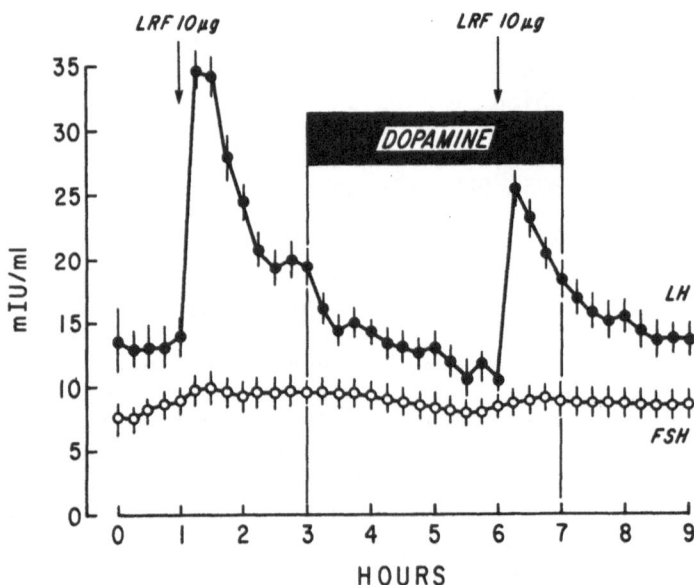

Figure 5.   The effect of dopamine-infusion on the pituitary
responses to LRF in 6 normal women during the early follicular
phase of the cycle.

acutely releasable LH (24,59).   In vitro studies, using hemipituitary
preparations have shown that DA lacks a direct inhibitory effect on
the pituitary response to LRF (Fig. 6); this confirms the earlier
work by Schneider and McCann and Miyachi et al. (41,51).   The earlier
proposal by Fuxe et al. (19) of an interaction between the LRF and
DA neuronal systems has recently been extended.   McNeill and Sladek
(40) and Ajika and Okinaga (1a) using fluorescence-immunohistochem-
istry for the simultaneous visualization of DA and LRF have afforded
a functional support of a dual distribution of DA in the endocrine
hypothalamus; one discrete DA terminal was seen in the outermost
region of the median eminence at the zone in contact with the
portal capillaries, and it probably represents the DA pathway that
controls PRL release via portal delivery.   A second diffuse inner
band of DA varicosities was seen in association with LRF terminals.
Thus, the evidence currently available strongly suggests that the

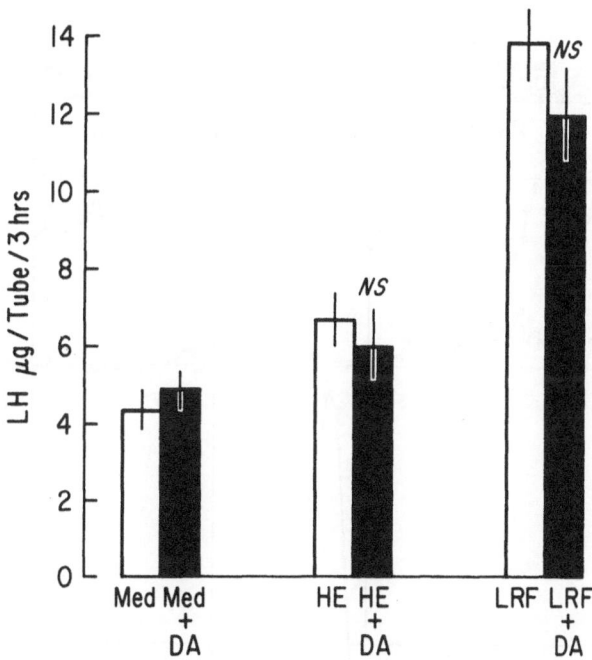

Figure 6.   The effect of DA on the in vitro release of LH from the
rat hemipituitary incubated in medium (Kreh Ringer bicarbonate
buffer).   DA (5.3 x 10^{-6}M) was added to the medium with the presence
of hypothalamic extract (HE) or synthetic LRF (10 ng/tube).

primary site of DA influence on LH release is mediated through
communications by way of Axo-axonal or terminals of DA and LRF
neurons within the arcuate-ME region of the hypothalamus which is
distinct from the neurohormonal mechanism of DA receptors on the
lactotrope for the inhibition of PRL release.

The mechanism for the rebound release of LH observed after the
7th hour of L-dopa administration as well as immediately following
DA infusion (Figs. 1-3) is not clear but may reflect changes in the
functional activities of endogenous DA and/or pituitary gonadotropin.
The possibility that L-dopa and DA, in time, are converted to
norepinephrine in the hypothalamus which may then cause LRF release
appears to be excluded since consistent suppressive effect of DA can
be maintained by extending the infusion beyond 4 hours (Fig. 7)
To account for the observed rebound, one may deduce that exogenous
DA in sufficient concentrations in the median eminence may inhibit
DA neuronal activity through the re-uptake negative feedback-loop

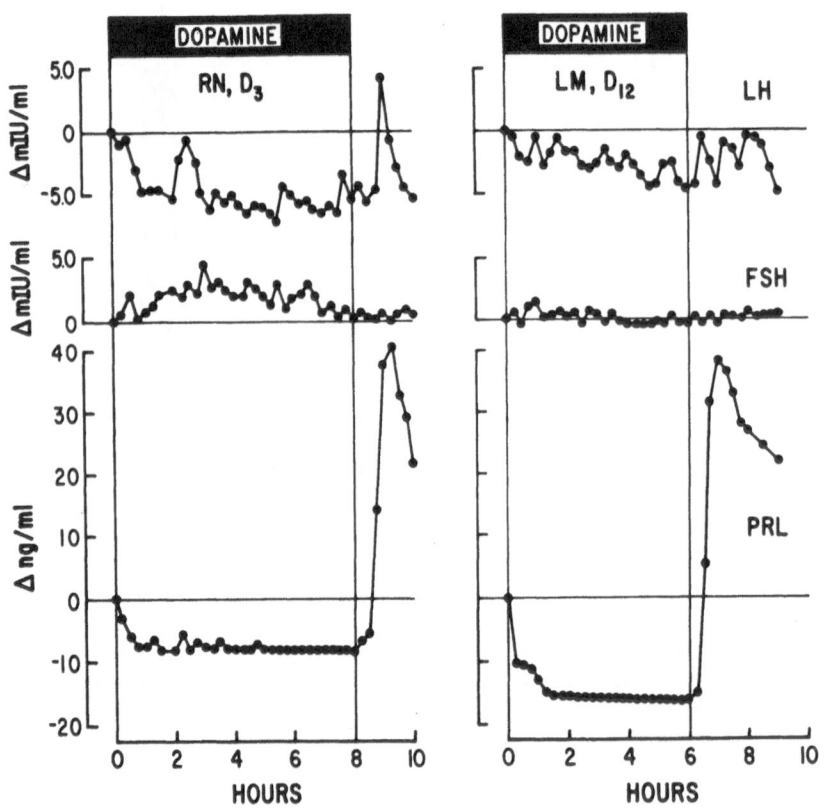

Figure 7.   The effect of 6 or 8 hours DA infusion on serum LH, FSH
and PRL levels.

(53a). As a consequence, a transient and relative reduction of
endogenous DA inhibition occurs following the rapid disappearance
of exogenous DA. Alternatively, the inhibitory effect of DA may
result in an accumulation of·acutely releasable pituitary LH, and
this may be followed by an autonomous discharge from the enlarged
releasable pool of LH as the level of DA within the hypothalamic-
pituitary tissue falls.

Effect of Endogenous Estradiol Levels on the Inhibitory Action of DA

Endogenous estradiol (E_2) was not included as one of the
studied variables in our earlier investigations. In subsequent
experiments, however, the relationship between endogenous E_2 levels
and gonadotropin and PRL sensitivity to DA infusion was assessed by
using subjects at different stages of the follicular phase of their
menstrual cycles (27). Basal LH and FSH levels were found to be
comparable at day 2 and day 12, and despite a 4-fold difference in
E_2 concentration, the inhibition of LH by DA is small and similar
for the two stages of the cycle. No effect on FSH release was seen on
either occasion. In marked contrast, day 14 subjects, with an elevated basal LH level, exhibit a dramatic increase in the sensitivity of
both LH and FSH to DA inhibition (Fig. 8). At this time a significant correlation ($r = 0.979$) between the basal LH and the degree of
inhibition by DA was found. Further, a remarkable rebound release
of LH but not FSH occurs on the termination of DA infusion. The
increased sensitivity to DA at a time when hypothalamic LRF secretion is assumed to be elevated (42,52) adds credence to the concept
that the inhibitory effect of DA on gonadotropin release is through
a direct action on LRF release. This is consistent with the
hypothesis that a reduction of endogenous DA on the day of proestrus
in the rat constitutes one component of the complex mechanism for
triggering the midcycle surge (19).

The inhibition of PRL release by DA is correlated with endogenous E_2 levels ($r = 0.685$) as well as basal PRL levels ($r = 0.878$).
Rebound release of PRL, on termination of the DA infusion, occurs
on all three stages of the cycle studied, but the magnitude is
greatest on Day 14 when the endogenous E_2 levels are highest (Fig.9).
These observations are at variance with the in vitro finding that
E_2 exhibits a potent antidopaminergic activity on PRL release by
pituitary cells in culture (47).

Effects of Ovariectomy and Estrogen Treatment
on the Inhibitory Action of DA

Although estrogen is known to have a profound tropic effect on
both lactotropes (58) and gonadotropes (31), estrogen withdrawal in
women, as a result of ovariectomy causes an elevated release of

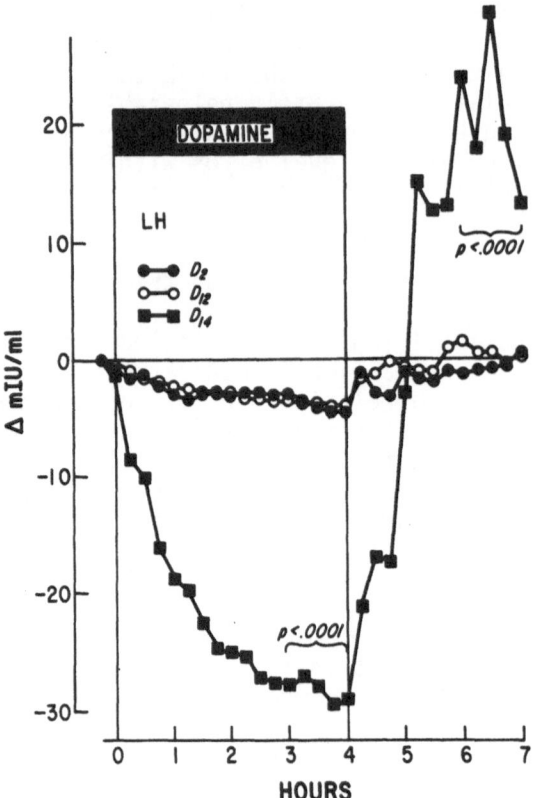

<u>Figure 8.</u>   The effect of DA infusion on circulating LH levels
during different stages of the follicular phase of the cycle. Data
expressed as net change (Δ) from mean basal concentrations of LH
in mIU/ml.

gonadotropin (59), but a reduced pituitary synthesis and storage of
PRL (3). Estrogens have been found also to increase hypothalamic
DA turnover in the rat (16,19,35), and this in turn is correlated
with lowering of the plasma LH and elevation of PRL (16,35). Thus,
a relationship between estrogen and the hypothalamic DA system in
the regulation of pituitary release of PRL and gonadotropin is
strongly suggested. Thus, our further studies in the human were
designed to explore the effect of estrogen on the response of
gonadotropin and PRL release to DA infusion in agonadal women.

In agonadal women, the mean basal levels of LH and FSH are 9
and 10-fold higher respectively and the E_2 is significantly
($P < 0.001$) lower than in normal women on day 2 of the cycle.

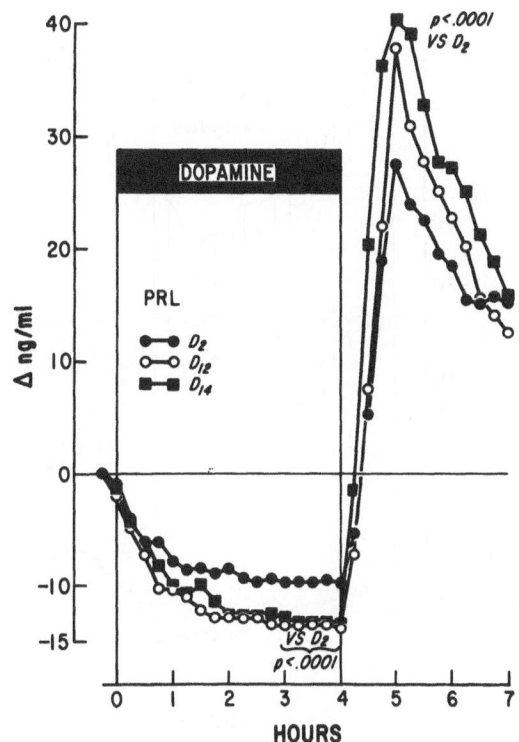

<u>Figure 9.</u>   The effect of DA infusion on the net change (Δ) of
serum PRL levels during different stages of the follicular phase of
the cycle.   The P values indicate significance between responses in
day 14 subjects and day 2 subjects.

Differences for E_1 and PRL are not statistically significant.

        The mean maximum decrement of LH in response to DA infusion is
9-fold greater in agonadal subjects than in normal women (Fig. 10).
While FSH levels remain unchanged in normal women, a significant
decline in response to DA infusion is observed in agonadal women.

        The elevated rate of gonadotropin secretion in agonadal women,
in all probability, reflects an increased rate of endogenous LRF
release (42,52).   In the present study, the greater suppression of
LH and FSH, the positive correlation between the LH and FSH response
to DA and their respective basal levels (r = 0.985 for LH; r = .975

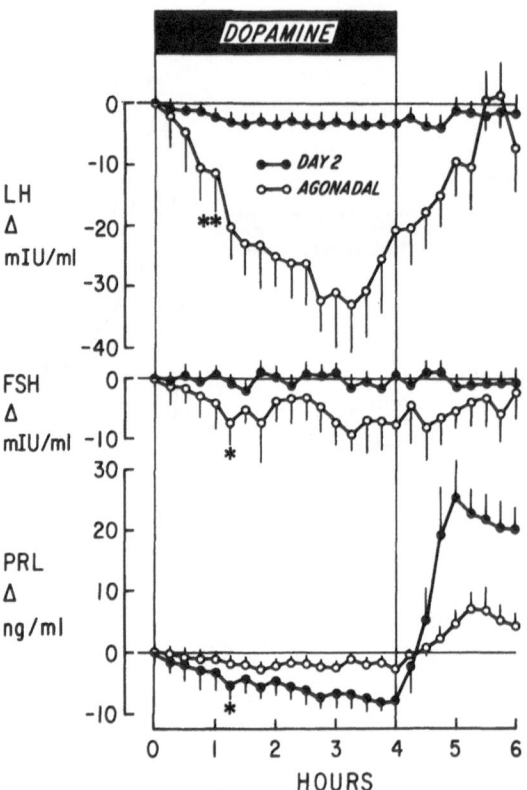

Figure 10.   Mean (± SE) changes in serum LH, FSH and PRL from mean
basal levels during and following a 4 hr DA infusion in agonadal
women as contrasted to normal cycling women (Day 2).   The differences for all three hormones become significant by 75 min after
initiation of DA infusion (* P < 0.05; ** P < 0.01).   The release
of PRL following DA withdrawal is significantly greater (P < 0.01)
in Day 2 subjects than in agonadal women.

for FSH), are in keeping with the proposal that DA suppresses
gonadotropin release at midcycle by inhibiting LRF secretion
(Fig. 8), a mode of action is consonant with the experimental
findings in animal models (22,41).   Thus, when endogenous LRF activity is increased, the sensitivity to DA inhibition appears to be
enhanced.   By inference, this observation may reflect a reduced DA
activity is accompanied by an increased LRF release.

     In terms of responsiveness to DA inhibition, the well-known
inverse relationship between the secretion of gonadotropin and PRL
is maintained.   The DA induced decline in PRL levels is significantly less in agonadal women than in normal women on day 2 of their

cycles (Fig. 10). Since estrogen has a direct tropic effect on the lactotrope, the very low circulating estrogen, resulting in a reduced number of lactotropes and of lactotropic DA receptors in agonadal women may explain the relative decrease in the DA response observed.

Administration of ethinyl estradiol (EE) to agonadal women was found to induce the expected decrease in basal gonadotropin secretion and to augment basal PRL release (Fig. 11). Concomitant with

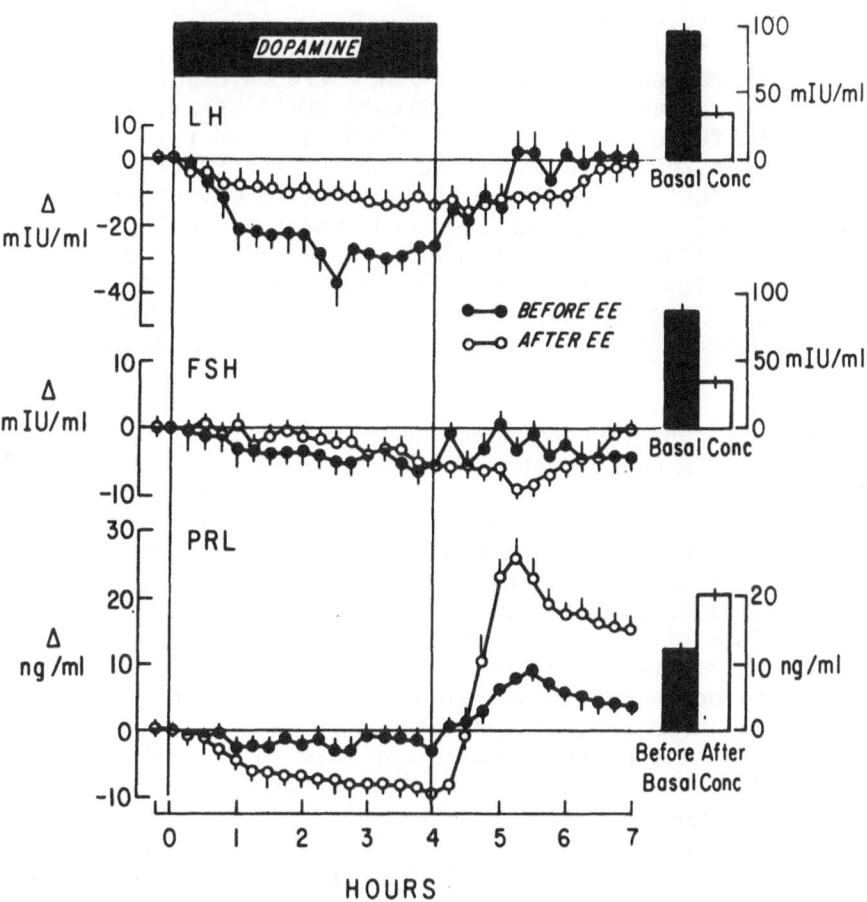

Figure 11. Mean (± SE) changes in basal concentrations of LH, FSH and PRL (bars) and their decrements during a 4 hr DA infusion in subjects before and after treatment with EE. Differences become significant at 60 mins for LH (P < 0.01) and 75 mins for PRL (P < 0.01). Upon withdrawal of DA, PRL (but not LH and FSH) rise is significantly (P < 0.001) greater in agonadal subjects after EE treatment.

the EE induced changes in basal secretion, a reversal in the responsiveness of all three pituitary hormones to DA inhibition is evident and the pattern of responses approaches those seen on day 2 of normally cycling women.

These findings demonstrate for the first time that estrogen modifies the DA inhibition of gonadotropin and PRL secretion in humans. The mechanism(s) by which estrogen modifies dopaminergic control of pituitary release of these reproductive hormones require further study. The attenuation of DA inhibition of LH and FSH release by estrogen in ovariectomized women (Fig. 11), may be the result of increased endogenous hypothalamic DA activity. Consequently, LRF neuronal activity is inhibited, which, in turn, would reduce gonadotropin secretion. This concept of estrogenic action is consistent with the evidence that estrogen administration increases hypothalamic DA turnover in ovariectomized rats and that hypothalamic DA plays an inhibitory role on LRF neuronal activity (19,42,52). The possibility that the response to DA infusion may provide an index for the indirect assessment of endogenous LRF activity deserves further investigation. On the other hand, the elevated plasma PRL level seen in estrogen-treated ovariectomized women (Fig. 11) probably reflects the antidopaminergic effect of estrogen on PRL release (47) balanced by the increased DA syntheses in the hypothalamus.

Progesterone-Induced Acute Release of Prolactin - A Dopaminergic-Mediated Event?

As indicated earlier, estrogen promotes tuberoinfundibular dopamine (DA) turnover. Progesterone, on the other hand, may reduce DA activity by virtue of its ability, within 4 to 6 hrs, to inhibit tyrosine hydroxylase (a rate-limiting enzyme for catecholamine synthesis). This effect of estrogen has been demonstrated in the median eminence of ovariectomized rats (4,5). These data, together with growing evidence that hypothalamic DA inhibits the release of both LH and PRL in humans, has prompted us to determine whether progesterone (P) also plays a regulatory role in the release of PRL. Accordingly, serum LH, FSH, and PRL were measured in estrogen-primed ovariectomized women receiving P (46a). As can be seen in Fig. 12a, the mean circulating concentrations or PRL, as well as FSH and LH, exhibit substantial and significant rises (P < 0.001) after a latent period of about 4 hrs and last for at least 5 hrs following P administration. When P is given prior to EE priming, no effect on the circulating levels of PRL, LH, and FSH is observed. The pattern of incremental changes of these hormones in a single individual is shown in Fig. 12b. It is evident that P administration induces a remarkable episodic increase in circulating levels of PRL, as well as LH, with an accompanied modest increase in FSH.

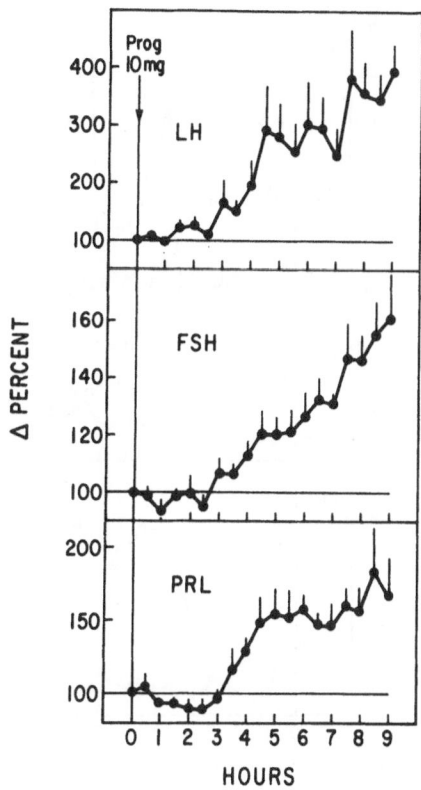

Figure 12a.   Mean (± SEM) percent change from mean basal circulating levels (mean of 3 determinations) for LH, FSH and PRL in 5 hypogonadal women in response to progesterone (10 mg, im) following EE priming.

These findings provide evidence for the first time that P induces a rapid release of PRL, as well as the expected gonadotropin rise in estrogen-primed women (34,43,44,57) and rats (12). To our knowledge, simultaneous determinations of PRL and gonadotropin following P administration have not previously been reported in either the rat or human.

The possibility that acute incremental changes in these pituitary hormones may be related to estrogen treatment itself rather than to P appears to be excluded. Comparable doses of EE treatment in hypogonadal subjects do not cause acute release of LH or FSH (30).

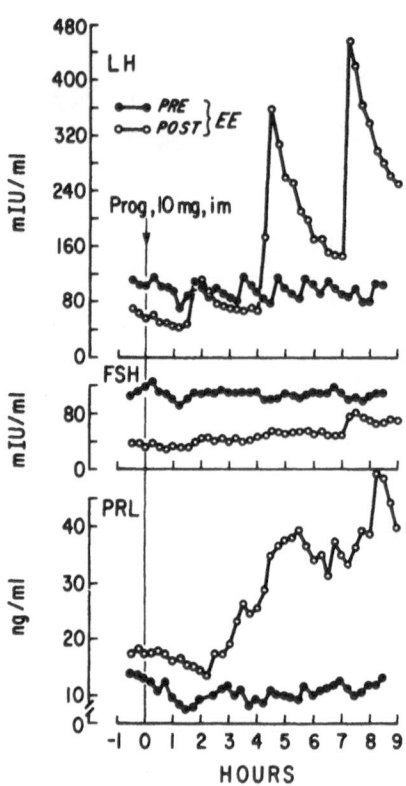

<u>Figure 12b</u>.   Changes in serum LH, FSH and PRL concentrations in response to progesterone (10 mg, im) before and following estrogen priming (EE, 300 µg/d for 5 days) in an ovariectomized woman.

        The mechanism(s) to account for the concurrent effect of P on PRL, FSH, and LH release are unknown. It is likely that this effect of P is exerted at the level of both the pituitary and the hypothalamus. Lasley et al. (31) have shown that in estrogen-primed women, progesterone administration at the same dose (10 mg) is capable of augmenting pituitary sensitivity to LRF. In estrogen-primed ovariectomzied rats, P administration induced a reduction of hypothalamic tyrosine hydroxylase activity (4,5) and the release of PRL (12). In all these animal experiments, the P effects are manifested after a lag time of 4 hrs.; this is in good agreement with our finding in humans. It may be postulated that the effect of P on the acute release of all three reproductive hormones by the pituitary is mediated through a reduction of the inhibitory

influence of hypothalamic DA combined with an augmented pituitary
sensitivity to LRF. Further investigation involving the direct
assessment of changes in neurotransmitter and LRF in appropriate
regions of the brain in animal models could yield data to validate
this postulated mechanism for the action of progesterone.

Assessments of Endogenous DA Activity

The inhibitory role of DA on LRF secretion was further
assessed in hyperprolactinemic patients with documented pituitary
microadenoma. The PRL and gonadotropin responses to DA and to
DA-antagonist, metoclopramide (MCP), were studied (Quigley et al.,
unpublished). We have previously postulated that hyperprolactinemia
in humans, as in the rat, induces an elevation of endogenous DA in
the tuberoinfundibular system through a short-loop feedback. The
increased dopaminergic inhibition of LRF neuroanl activity would
then account for the low and acyclic gonadotropin secretion (29).
This postulated interrelationship is diagrammed in Fig. 13.

The infusion of DA (Fig. 14) is associated with a significant
decrease in serum LH, but not FSH, in both the hyperprolactinemic
group and normal controls. However, the decrease in LH is significantly greater in normal controls than in the hyperprolactinemic
patients, and the difference becomes significant after the 1st hour
of infusion ($P < 0.001$). While a continued decline in LH during
the 4-hour infusion was seen in normal subjects, the LH response
in the hyperprolactinemic group exhibited an initial decline
followed by a significant increase during the last hour of the
infusion. A significant correlation between the basal level of LH
and maximum decrement in LH was found ($r = 0.881$, $P < 0.001$).
Following discontinuation of the DA infusion, LH returns to basal
levels by the 3rd hour in normal controls. Serum FSH levels showed
no significant fluctuation in either group during the DA infusion.

PRL responses to DA infusion are much greater in the hyperprolactinemic patients than in controls (Fig. 14). The differences
become significant after 30 minutes ($P < 0.001$); and by the end of
the infusion the mean decrement in PRL is 8-fold greater in the
hyperprolactinemic group (83 ± 7.4 ng/ml) than in controls
(11.5 ± 2.8 ng/ml). There is a positive correlation between the
maximum decline of PRL and basal PRL levels ($r = 9.685$, $P < 0.05$).
On withdrawal of DA, there is a rapid return in serum PRL levels
during the first hour. The relative degree of rebound release
(above basal value) is disproportionately greater in normal women
(38.6 ± 8.6 ng/ml) than in hyperprolactinemic patients
(18 ± 2.0 ng/ml, $P < 0.05$). These findings suggest that hyperprolactinemic patients with microadenoma have endogenous hypothalamic DA excess and paradoxically relative DA deficiencies on the
lactotrope as compared to normal subjects.

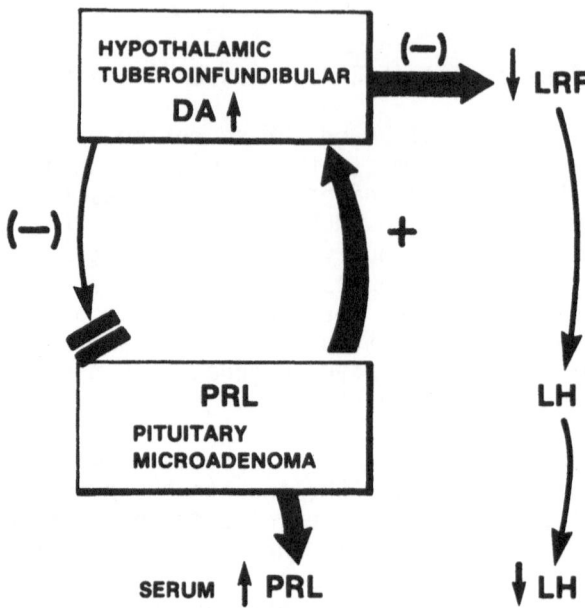

Figure 13.   Schematic representations for the proposed mechanism of DA as a link between hyperprolactinemic patients with pituitary microadenomas and the occurrence of acyclic and reduced LH secretion (see text for details).

Assuming that MCP can be delivered to the target sites of the arcuate-median eminence unit, as well as to the adenohypophysis, the responses to DA-antagonist alone or during DA infusion may disclose the functional status of tuberoinfundibular DA and LRF neurons. MCP is an excellent DA-antagonist (44a,45a) with further advantage of the lack of a psychotropic effect (45a), a rapid action on prolactin release (26) and it can be administered intravenously.

In hyperprolactinemic patients, the administration of MCP (10 mg) induces a rapid release of LH with a time course similar to that elicited by exogenous LRF (Fig. 15). Since discernible LH

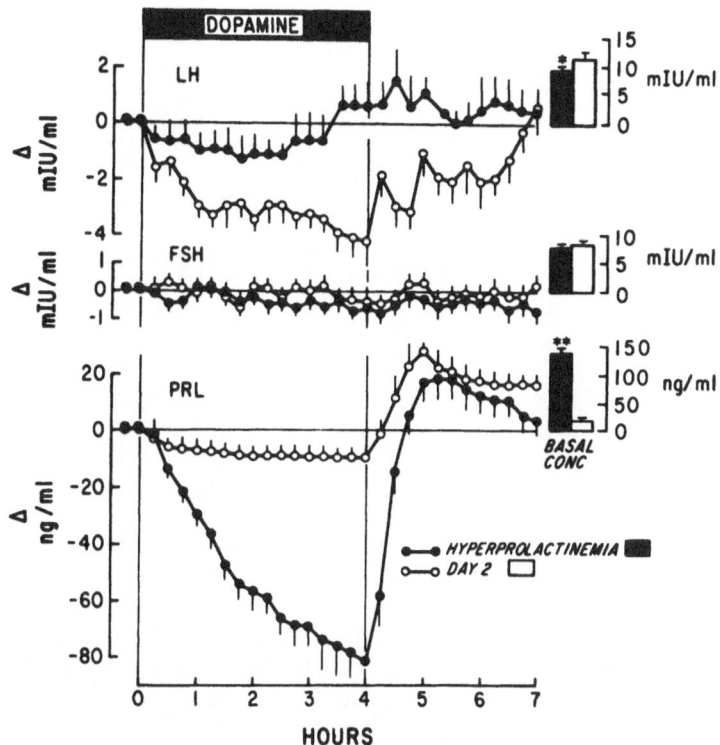

Figure 14.   The effect of DA infusion on the decrement of LH and PRL in hyperprolactinemic patients with microadenoma as compared to normal women on day 2 of the cycle. Data expressed as net changes (Δ) from the mean basal levels.

release in response to MCP was not found in normal women during the early follicular phase (unpublished observations), the present finding strongly suggests the existence of DA excess in hyperpro- lactinemic-microadenoma patients. In contrast, the elevated PRL levels in these patients show no response to MCP, suggesting either a lack of DA or the presence of a DA receptor abnormality in the lactotrope. However, both LH and PRL release are responsive to DA-inhibition which is much greater for PRL than for LH (Fig. 15). Administration of the same dose of MCP during the 3rd hour of DA infusion overcomes the DA suppression of both LH and PRL. These findings suggest that the inhibitory action of DA on the release of both LH and PRL is mediated through a DA-receptor mechanism and that

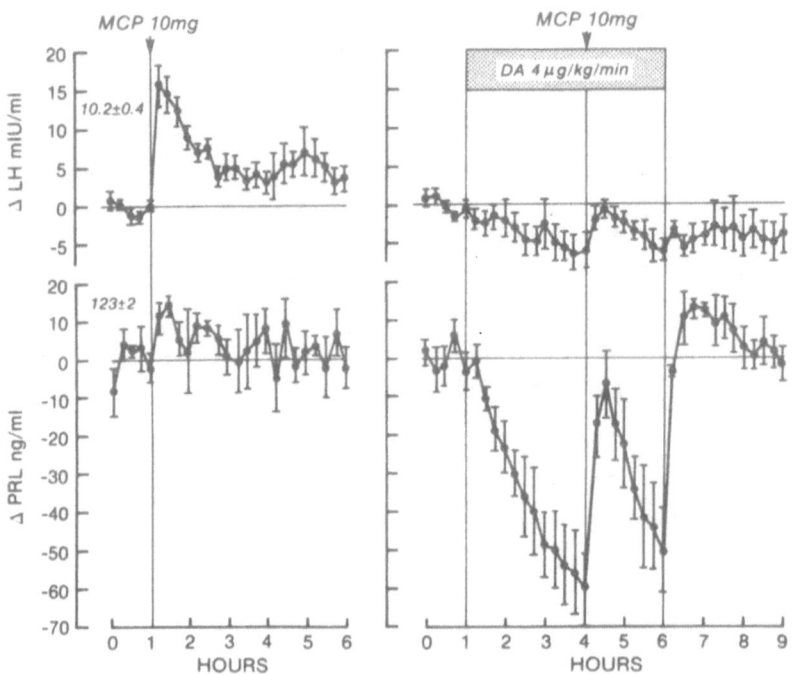

<u>Figure 15.</u>   The effect of DA-antagonist (MCP, 10 mg, iv) admini-
stration on the release of LH and PRL with (right) or without DA
infusion (left) in 5 hyperprolactinemic patients with microadenoma.
The basal LH and PRL levels are also shown on the left.   Data
expressed as net changes (Δ).

the DA receptors controlling both LH and PRL are intact in these
patients. Similar studies in hyperprolactinemic rats (6,7) have
revealed comparable results. The apparent paradox of a DA excess
in the LH system and a DA deficiency in the PRL system may be re-
solved by either an inadequate vascular delivery system or an
increased local enzymatic (MAO) DA degradation by the adenoma (38),
or both. In any case, our findings implicate DA in the regulation
of both LH and PRL and support our hypothesis that DA excess plays
a role in the development of the acyclic gonadotropin secretion in
patients with hyperprolactinemia.

Prolactin Release During LRF Infusion

This aspect of our investigation was designed to test the
possibility that LRF neurons may also communicate with neighboring
DA neurons within the arcuate median eminence unit. During prolonged
LRF infusion (0.2 μg/min) for 18 hours in hypogonadal women, as well

as in cycling women, LH level rises rapidly during the first hour, levels off to reach a maximum at 8 hours, and is followed by a progressive decline throughout the remainder of the infusion. The refractoriness to LRF stimulation during the last 10 hours of infusion is an example of the phenomenon of "down regulation" (Hoff, Lasley, and Yen, unpublished observation). However, the most interesting observation during LRF infusion is the occurrence of a significant rise in PRL level at the time when LH release is maximal in both hypogonadal (Fig. 16a) and cycling women (Fig. 16b) (Lasley, Hoff and Yen, unpublished observations). The mechanism for the rise of PRL during LRF infusion is unknown, but one is tempted to postulate that the acute elevation of either LRF itself or LH within the arcuate-median eminence region may inhibit dopaminergic neuronal activity and thus account for the rise of

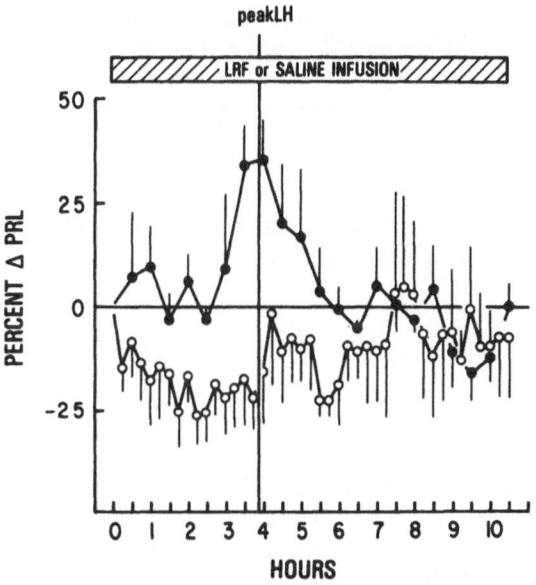

Figure 16a.   The effect of long-term LRF infusion on circulating PRL levels in normal cycling women.

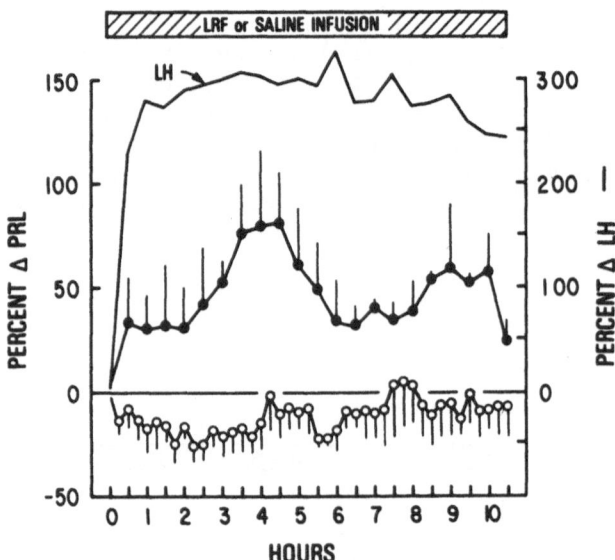

<u>Figure 16b.</u>   The effect of long-term LRF infusion on circulating
PRL levels in hypogonadal subjects.

PRL. We favor a direct action of LRF since administration of
either hCG or purified human pituitary LH producing high circulating
concentrations fails to induce an elevation of PRL levels. The
potential for interaction between the secretory activities of LRF
and DA neurons requires further <u>in</u> <u>vivo</u> and <u>in</u> <u>vitro</u> study.

Summary and Conclusion

The results of the foregoing experiments provide compelling
evidence of an inhibitory role of the hypothalamic dopaminergic
system in the regulation of pituitary secretion of LH, as well as
PRL in humans. The site of DA action appears to be the LRF neurons
for LH release and the pituitary lactotrope for PRL secretion. The

inhibitory role of DA on gonadotropin release becomes more evident
at those times when endogenous LRF is assumed to be elevated. The
activity of the DA system appears to be influenced by estrogen,
progesterone, and possibly by LRF as well. It is postulated that
PRL may play a key role in modifying the estrogen and progesterone
feedback action on gonadotropin secretion by influencing the hypothalamic DA via a PRL-DA short-loop feedback system. Further
support of this postulate is found in a series of experiments in
the rat; administration of PRL was found to prevent the post
castration rise in LH (21) and the estrogen-induced increased
hypothalamic DA turnover is abolished by hypophysectomy (25). On
the other hand, PRL induced elevation of DA turnover is unaffected
by hypophysectomy (16). Thus, the regulation of the release of
gonadotropin and PRL involves a system the elements of which
include estrogen, DA, LRF, and PRL itself. The interactive
functions of these components of the system are responsible for
normal operation of the system, including the well-known inverse
relationship between the secretion of PRL and gonadotropin. Additional elements, such as norepinephrine, serotonin, and endogenous
opioid peptides are undoubtedly involved in this complex system and
are subjects of current investigation. Nonetheless, the malfunction
of one or more of these components leads to system disturbances in
the form of recognized clinical syndromes. The hyperprolactinemic-
pituitary adenoma patient appears to serve as an example.

ACKNOWLEDGEMENT

The data presented in the paper are contributed by my associates, Drs. H. Leblanc, G. Lachelin, S. Judd, E. Quigley, J. Hoff
and B. Lasley and supported by Rockefeller Foundation Grant RF-75029
and the National Institutes of Health Grant HD-09728-03. A portion
of the patient studies was performed in the Clinical Research
Center, supported by NIH Research Grant RR-00827. I wish to thank
Professor A. Lein for the review and suggestions.

REFERENCES

1. Advis, J.P., T.R. Hall, C.A. Hodson, G.P. Mueller, and J. Meites,
    Temporal relationship and role of dopamine in "short-loop"
    feedback of prolactin. Proc. Exp. Biol. Med. 155:567, 1977.
1a. Ajika, K., and S. Okinaga, Simultaneous localization of LHRH
    and catecholamine neuronal system in the rat hypothalamus:
    Immunohistochemical identification by light and electron
    microscope: Program of Endocrine Society Annual Meeting
    Ab #183, 1978.
2. Axelrod, J., H. Weil-Malherbe, and R. Tomchick, The physiologi-
    cal disposition of H^3-epinephrine and its metabolite metane-
    phrine. J. Pharmacol. Exp. Ther. 127:251, 1959.

3. Baker, B.L., A.R. Midgley, Jr., B.E. Gersten, and Y.Y. Yu,
    Differentiation of growth hormone and prolactin containing
    acidophils with peroxidase labelled antibody. Anat. Res.
    164:163, 1969.
4. Beattie, C.W., C.H. Rodgers, and L.F. Soyka, Influence of
    ovariectomy and ovarian steroids on hypothalamic hydroxylase
    activity in the rat. Endocrinology 91:276, 1972.
5. Beattie, C.W., and L.F. Soyka, Influence of progesteronal
    steroids on hypothalamic tyrosine hydroxylase activity in
    vitro. Endocrinology 93:1453, 1973.
6. Beck, W., and W. Wuttke, Desensitization of the dopaminergic
    inhibition of pituitary luteinizing hormone release by
    prolactin in ovariectomized rats. J. Endocrinol. 74:67-74,
    1977.
7. Beck, W., J.L. Hancke, and W. Wuttke, Increased sensitivity of
    dopaminergic inhibition of luteinizing hormone release in
    immature and castrated female rats. Endocrinology 102:
    837-843, 1978.
8. Ben-Jonathan, N., C. Oliver, H.J. Winer, R.S. Mical, and J.C.
    Porter, Dopamine in hypophysial portal plasma of the rat
    during the estrus cycle and throughout pregnancy. Endocrinology
    100:452, 1977.
9. Bertler, A., B. Falck, and E. Rosengren, The direct demonstra-
    tion of a barrier mechanism in the brain capillaries. Acta
    Pharmacol. Toxicol. 20:317, 1963.
9a. Bjorklund, A., B. Falck, F. Hromek, C. Owman, and K.A. West,
    Identification and terminal distribution of the tubero-
    hypophyseal monoamine system in the rat by means of stereotaxic
    and microspectro-fluorometric techniques. Brain Res. 17:1, 1970.
10. Boden, G., L.E. Lundy, and O.E. Owen, Influence of levodopa on
    serum levels of anterior pituitary hormones in man,
    Neuroendocrinology 10:309, 1973.
11. Buckman, M.T., N. Kaminsky, M. Conway and G.T. Peake, Utility
    of L-dopa and water loading in evaluation of hyperprolactinemia,
    J. Clin. Endocrinol. Metab. 36:911, 1973.
12. Caligaris, L., J.J. Astrada, and S. Taleisnik, Oestrogen and
    progesterone influence on the release of prolactin in ovari-
    ectomized rats. J. Endocrinol. 60:205-215, 1974.
13. Caron, M.G., M. Beaulieu, V. Raymond, B. Gagne, J. Drouin,
    R.J. Lefkowitz, and F. Labrie, Dopaminergic receptors in the
    anterior pituitary gland. Correlation of [^{3}H] dihydroergo-
    cryptine binding with the dopaminergic control of prolactin
    release. J. Biochem. Chem. 253:2244-2253, 1978.
14. Cronin, M.J., J.M. Roberts, and R.I. Weiner, Dopamine and
    dihydroergocryptine binding to the anterior pituitary and
    other brain areas of the rat and sheep. Endocrinology
    103:302-309, 1978.
15. del Pozo, E., R. Brun del Re, L. Varga, and H. Friesen, The
    inhibition of prolactin secretion in man by CB-154 (2-Br-α-
    ergocryptine). J. Clin. Endocrinol. Metab. 35:768, 1972.

16. Eikenburg, D.C., A.J. Ravitz, G.A. Gudelsky, and K.E. Moore, Effects of estrogen on prolactin and tuberoinfundibular dopaminergic neurons. J. Neural. Transmission 40:235, 1977.
17. Friesen, H., H. Guyda, P. Hwang, J.E. Tyson, and A. Barbeau, Functional evaluation of prolactin secretion: A guide to therapy. J. Clin. Invest. 51:706, 1972.
18. Fuxe, K., T. Hökfelt, and O. Nilsson, Effect of constant light and androgen-sterilization on the amine turnover of the tubero-infundibular dopamine neurons: Blockade of cyclic activity and induction of a persistent high dopamine turnover in the median eminence. Acta Endocrinol. 69:625-639, 1972.
19. Fuxe, K., T. Hökfelt, L. Agnati, A. Löfström, B.J. Everitt, O. Johansson, G. Jonsson, W. Wuttke, and M. Goldstein, Role of monoamines in the control of gonadotropin sedretion, In: Neuroendocrine Regulation of Fertility, edited by T.C. Anand Jumar, Basel, Karger, 1976, pp. 124-140.
20. Gibbs, D.M., and J.D. Neill, Dopamine levels in hypophysial stalk blood in the rat are sufficient to inhibit prolactin secretion. Endocrinology 102:1895-1900, 1978.
21. Grandison, L., C. Hodson, H.T. Chen, J. Advis, J. Simpkins, and J. Meites, Inhibition by prolactin of post-castration rise in LH. Neuroendocrinology 23:312-322, 1977.
22. Hall, T.R., J.P. Advis, J.F. Bruni, C.A. Hodson, and J. Meites, Neurotransmitter regulation of LH release in pituitary-hypo-thalamus co-incubation system. Fed. Proc. 36:321, 1977.
23. Hayek, A., and J.D. Crawford, L-dopa and pituitary hormone secretion. J. Clin. Endocrinol. Metab. 34:764, 1972.
24. Hoff, J.D., B.L. Lasley, C.F. Wang, and S.S.C. Yen, The two pools of pituitary gonadotropin. Regulation during the menstrual cycle. J. Clin. Endocrinol. Metab. 44:302-312, 1977.
25. Jimenez, A.E., J.L. Voogt, and L.A. Carr, Plasma luteinizing hormone and prolactin levels and hypothalamic catecholamine synthesis in steroid-treated ovariectomized rats. Neuroendocrinology 23:341-351, 1977.
26. Judd, S.J., L. Lazarus, and G. Smythe, Prolactin secretion by metoclopramide in man. J. Clin. Endocrinol. Metab. 43:313, 1976.
27. Judd, S.J., J.S. Rakoff, and S.S.C. Yen, Inhibition of gonado-tropin and prolactin release by dopamine: Effect of endogenous estradiol levels. J. Clin. Endocrinol. Metab. 47:000, 1978.
28. Lachelin, G.C.L., H. Leblanc, and S.S.C. Yen, The inhibitory effect of dopamine agonists on LH release in women. J. Clin. Endocrinol. Metab. 44:728, 1977.
29. Lachelin, G.C.L., S. Abu-Fadil, and S.S.C. Yen, Functional delineation of hyperprolactinemic amenorrhea. J. Clin. Endocrinol. Metab. 44:1163, 1977.
30. Lachelin, G.C.L., and S.S.C. Yen, The biphasic change in pitui-tary capacity induced by estrogen in hypogonadal women. J. Clin. Endocrinol. Metab. 46:369, 1978.
31. Lasley, B.L., C.F. Wang, and S.S.C. Yen, The effects of estrogen and progesterone on the functional capacity of the gonadotrophs.

J. Clin. Endocrinol. Metab. 41:820, 1975.

32. Leblanc, H., G.C.L. Lachelin, S. Abu-Fadil, and S.S.C. Yen, Effects of dopamine infusion on pituitary hormone secretion in humans. J. Clin. Endocrinol. Metab. 43:668, 1976.

33. Leblanc, H., and S.S.C. Yen, The effect of L-dopa and chlorpromazine on prolactin and growth hormone secretion in normal women. Am. J. Obstet. Gynecol. 128:162, 1976.

34. Leyendecker, G., S. Wardlow, and W. Nocke, Experimental studies on the endocrine regulation during the periovulatory phase of the human menstrual cycle. Acta Endocrinol. 71:169-175, 1972.

35. Löfström, A., P. Eneroth, J.A. Gustafsson, and P. Skett, Effects of estradiol benzoate on catecholamine levels and turnover in discrete areas of the median eminence and the limbic forebrain and on serum luteinizing hormone, follicle stimulating hormone and prolactin concentrations in the ovariectomized female rat. Endocrinology 101:1559, 1977.

36. Malarkey, W.B., L.S. Jacobs, and W.H. Daughaday, Levodopa suppression of prolactin in nonpuerperal galactorrhea. N. Engl. J. Med. 285:1160, 1971.

37. MacLeod, R.M., and J.E. Lehmeyer, Studies on the mechanism of the dopamine-mediated inhibition of prolactin secretion. Endocrinology 94:1077, 1974.

38. MacLeod, R.M., H. Kimura, and I. Login, Inhibition of prolactin secretion by dopamine and piribedit (ET-495), In: Growth Hormone and Related Peptides, edited by A. Pecile, and E.E. Uuller, New York, American Elsevier, 1976, pp. 443-453.

39. McCann, S.M., and R.L. Moss, Putative neurotransmitters involved in discharging gonadotropin-releasing neurohormones and the action of LH-releasing hormone on the CNS. Life Sci. 16:833, 1975.

40. McNeill, T.H., and J.R. Sladek, Jr., Fluorescence-Immunocytochemistry: Simultaneous localization of catecholamines and gonadotropin-releasing hormone. Science 200:72-74, 1978.

41. Miyachi, Y., R.S. Mecklenburg, and M.B. Lipsett, In vitro studies of pituitary-median eminence unit. Endocrinology 93:492-496, 1973.

42. Mortimer, C.H., A.S. McNeilly, L.H. Rees, P.J. Lowry, D. Gilmore, and H.G. Dobbie, Radioimmunoassay and chromatographic similarity of circulating endogenous and hypothalamic extracts in man. J. Clin. Endocrinol. Metab. 43:882, 1976.

43. Nillius, S.J., and L. Wide, Effects of progesterone on the serum levels of FSH and LH in postmenopausal women treated with oestrogen. Acta Endocrinol. 67:362-370, 1971.

44. Odell, W.D., and R.S. Swerdloff, Progestogen induced luteinizing and follicle stimulating hormone surge in postmenopausal women: A simulated ovulatory peak. Proc. Nat. Sci. 61:529-536, 1968.

44a. Peringer, E., P. Jenner, I.M. Donaldson and C.D. Marsden, Metoclopramide and dopamine receptor blockade. Neuropharmacol. 15:463, 1976.

45. Pinter, E.J., G. Tolis, and H.G. Friesen, L-dopa, growth hormone and adipokinesis in the lean and the obese. Int. J. Clin. Pharmacol. 12:277, 1975.

45a.Pinder, R.M., R.N. Brogden, P.S. Sawyer, T.M. Speight and G.S. Avery, Metoclopramide: A review of its pharmacological proper- ties and clinical use. Drugs 12:81, 1976.

46. Polansky, S., E. Muechler, and S. Sorrentino, Jr., The effect of L-dopa and clomiphene citrate on peripheral plasma levels of luteinizing hormone-releasing factor. Obstet. Gynecol. 48: 79, 1976.

46a.Rakoff, J.S., and S.S.C. Yen, Progesterone induced acute release of prolactin in estrogen-primed ovariectomized women, J. Clin. Endocrinol. Metab., in press.

47. Raymond, V., M. Beaulieu, F. Labrie, and J. Boissier, Potent antidopaminergic activity of estradiol at the pituitary level of prolactin release. Science 200:1173-1176, 1978.

48. Rotsztejn, W.H., J.L. Charli, E. Pattou, and C. Korden, Stimula- tion by dopamine of luteinizing hormone releasing hormone (LHRH) release from the mediobasal hypothalamus in male rats. Endocrinology 101:1475, 1977.

49. Sawyer, C.H., J.E. Markee, and W.H. Hollinshead, Inhibition of ovulation in the rabbit by the adrenergic-blocking agent dibenamine. Endocrinology 41:395, 1947.

50. Sawyer, C.H., J. Hilliard, S. Kanematsu, R. Scaramuzzi, and C. Blake, Effects of intraventricular infusions of norepineph- rine and dopamine on LH release and ovulation in the rabbit. Neuroendocrinology 15:328, 1974.

51. Schneider, H.P.G., and S.M. McCann, Possible role of dopamine as transmitter to promote discharge of LH releasing factor. Endocrinology 85:121, 1969.

52. Seyler, L.E., Jr., and S. Reichlin, Luteinizing hormone-releasing factor (LRF) in plasma of postmenopausal women. J. Clin. Endocrinol. Metab. 37:197, 1973.

53. Shaar, C.J., and J.A. Clems, The role of catecholamines in the release of anterior pituitary prolactin in vitro. Endocrinology 95:1202, 1974.

53a.Starke, K., H.A. Taube, and E. Borowski, Presynaptic receptor system in catecholamine transmission. Biochem. Pharmacology 26:289, 1977.

54. Takahara, J., A. Arimura, and A.V. Schally, Suppression of prolactin release by a purified porcine PIF preparation and catecholamines infused into a rat hypophysial portal vessel. Endocrinology 95:462, 1974.

55. Tolis, G., E.J. Pinter, and H.G. Friesen, The acute effect of 2-bromo-α-ergocryptine (CB-154) on anterior pituitary hormones and free fatty acids in man. Int. J. Clin. Pharmacol. 12:281, 1973.

56. Turkington, R.W., Inhibition of prolactin secretion and success- ful therapy of the Forbes-Albright syndrome with L-dopa. J. Clin. Endocrinol. Metab. 34:306, 1972.

57. Yen, S.S.C., C.C. Tsai, G. VandenBerg, and R. Rebar, Gonadotropin dynamics in patients with gonadal dysgenesis: A model for the study of gonadotropin regulation. J. Clin. Endocrinol. Metab. 35:897-904, 1972.

58. Yen, S.S.C., Y. Ehara, and T.M. Siler, Augmentation of prolactin secretion by estrogen in hypogonadal women. J. Clin. Invest. 53:652, 1974.

59. Yen, S.S.C., B.L. Lasley, C.F. Wang, H. Leblanc, and T.M. Siler, The operating characteristics of the hypothalamic-pituitary system during the menstrual cycle and observations of biological action of somatostatin. Recent Progr. Hormone Res. 31:321, 1975.

60. Zarate, A., L.S. Jacobs, E.S. Canales, A.V. Schally, A. de la Cruz, J. Soria, and W.H. Daughaday, Functional evaluation of pituitary reserve in patients with the amenorrhea-galactorrhea syndrome utilizing luteinizing hormone-releasing hormone (LH-RH), L-dopa and chlorpromazine. J. Clin. Endocrinol. Metab. 37:855, 1973.

61. Zarate, A., E.S. Canales, J. Soria, P.J. Maneiro, and C. MacGregor, Effect of acute administration of L-dopa on serum concentrations of follicle-stimulating hormone (FSH) and luteinizing hormone (LH) in patients with amenorrhea-galactorrhea syndrome. Neuroendocrinology 12:362, 1973.

NEUROTRANSMITTER CONTROL OF GROWTH HORMONE AND PROLACTIN SECRETION

E.E. Müller,[°]D. Cocchi,[°]V. Locatelli, [°°]E.A. Parati and [°]P. Mantegazza.

Institute of Pharmacology and Pharmacognosy University of Cagliari, [°]Department of Pharmacognosy University of Milan and [°°] Istituto Neurologico C. Besta, Milan Italy.

It is less than 20 years that the problem of the separate identity of human and monkey prolactins was posed with the observations of Lyons et al. (1961), extended by Chadwich et al. (1961) and Rivera et al. (1967) that purified human growth hormone (hGH) possessed intrinsic prolactin (PRL) bioactivity. Since then as a result of an impressive series of biochemical, biological and clinical studies and the development of heterologous and homologous radioimmunoassays of human prolactin in plasma, the concept has emerged that human and monkey prolactins exist as separate peptides in the anterior pituitary (AP) gland and that are released by stimuli different from those causing GH release, although both are released in stress (see Pasteels and Robyn, 1973; Bohnet and Friesen, 1976; Clemens, 1976; Pecile and Müller, 1972, 1976). Determination of the entire amino acid sequence of hPRL has revealed that it bears very little identity with hGH (Shome and Parlow, 1977).

The knowledge of the separate identity of hGH and hPRL has greatly stimulated the search for the identification of specific neurohormonal influences responsible for the control of their secretion. For GH there is convincing evidence that this control is operated by at least two hypothalamic peptidergic hormones: a GH-releasing hormone (GH-RH), which is still unidentified, and a GH-release inhibiting hormone (GH-IH), which has been isolated, identified and synthesized (Brazeau et al., 1973). For PRL, it is a widespread notion that its secretion in mammals is mainly under a tonic inhibitory influence of the CNS, although the true identity of the PRL-inhibiting hormone or factor PR-IH or PIF so far has eluded investigators, and there is the possibility that dopamine (DA) may be a physiologic regulator of PRL secretion (see below). In addition, evidence has also been presented supporting the concept of a hypothalamic mechanism

417

capable of stimulating the secretion of PRL and mediated by a PRL-
releasing hormone (PRH).

Parallel to the studies on the identification of specific neurohormone influences for the secretion of AP hormones, physiologic,
histofluorescence and immunohistochemical studies have led to the
identification and topographical localization of complex networks of
neurotransmitter neurons, which provide the critical intermediate
link between brain and hypophysiotrophic function (see Hökfelt et
al., This Symposium).

This contribution besides overviewing experimental evidence previously reported will emphasize some of the latest observations on
the role of brain neurotransmitters in the control of GH and PRL
secretion in animals and humans in health and disease states and,
in addition, will consider some interactions between neurotransmitters and peptide hormones at AP level.

Details on materials and methods have been incorporated in the
Results section. Double antibody radioimmunoassays for rat GH (rGH),
hGH, rat PRL (rPRL), hPRL and rat FSH and LH were used to measure plasma GH in rats (Schalch and Reichlin, 1966), and man (Molinatti et al.,
1969), plasma PRL in rats (Niswender et al., 1969) and man (Kit Bio-
data, Rome, Italy) and plasma gonadotropins in rats (Niswender et al.,
1968).

I. MONOAMINERGIC CONTROL OF GH SECRETION
A. Norepinephrine and Dopamine

A vast literature has appeared in the last 10 years on GH which
suggests that brain monoamines, notably norepinephrine (NE), DA and
serotonin (5-HT), may regulate GH secretion by functioning as neurotransmitters in the CNS of subprimate and primate species. Experimental evidence for an adrenergic component in the hypothalamic control
of GH has been provided by studies performed in conscious baboons.
They showed that pharmacological blockade of α-adrenergic receptors
consistently lowered baseline GH concentration and suppressed the
stimulated release; conversely, β-adrenergic blockade with propranolol was associated with a prompt rise in GH (Werrbach et al., 1970).
Mapping of the hypothalamus by microinjection of NE into discrete regions localized α-adrenergic receptors to the mediobasal hypothalamus (MBH), primarily in the vicinity of the ventromedial nucleus
(VMN) (Toivola and Gale, 1972). In keeping with these findings infusion of the α-adrenergic blocker phentolamine into the third ventricle or anterior hypothalamus induced, at systemically ineffective doses, a fall in GH levels (Toivola et al., 1972). In baboons stimulation
of GH release was induced by systemic infusion of NE (Illner et al.,
1976), DA (Steiner et al., 1977) and epinephrine (E) (Steiner, unpublished observations). Interestingly, the DA-induced GH release
was suppressed by concomitant infusion of FLA 63, an inhibitor of

DA- β -hydroxylase (D-β -H) (Corrodi et al., 1970) and hence blocker
of NE synthesis, and by phentolamine, thus indicating α -adrenergic
mediation of the DA-induced GH release (Steiner et al., 1977). In
all, these findings agree with α -adrenergic stimulation of GH secretion shown recently in rats (Ruch et al., 1976; Durand et al., 1977),
cats (Ruch et al., 1977), dogs (Lovinger et al., 1976) and man (Kansal
et al., 1972; Lancranjan and Marbach, 1977),the only exception being
goats (Ruch et al., 1977),and challenge the view that GH release following L-DOPA in monkeys (Ruch et al., 1973) is mediated by a DA-receptor mechanism in the hypothalamus. In fact, subemetic doses of apomorphine (Apo),a stimulant of DA receptors,failed to release GH in
monkeys (Jacoby et al., 1974; Chambers and Brown, 1976) and direct
microinjection of DA into the MBH (Steiner et al., 1977) or its intraventricular (IVT) infusion (Toivola and Gale, 1970) suppressed plasma
GH concentrations. That DA receptors in monkeys may actually mediate
GH suppression is further suggested by the finding that infusion of
pimozide, a selective blocker of DA receptors (Andén et al., 1970),
was followed by a sudden GH rise; the drug failed to block the GH
rise induce by infusion of DA (Illner et al., 1977).

The dual role for NE and DA in the control of GH secretion evidenced in monkeys by no means can be expanded to include man; in the
latter both NE and DA seem to exert a stimulatory influence on GH release. In addition to L-DOPA, which could act to increase also NE levels in the hypothalamus or limbic system and/or affect the brain
5-HT system (see indoleaminergic control) and whose effect is blunted by concomitant phentolamine administration (Kansal et al., 1972),
Apo in submetic doses, 2-Br- α -ergocriptine (bromocriptine, CB 154,
Sandoz), piribedil and nomifensine, all DA agonist drugs (see Müller
et al., 1977a) induced a rise in hGH levels (Lal et al., 1972; Brown
et al., 1973; Camanni et al., 1975; Thorner et al., 1976; E.E. Müller,
unpublished results).

Fig. 1 shows the results of an experiment in which the GH-releasing effect of CB 154 (2.5 mg po), an ergot derivative capable of inducing a long-lasting activation of DA receptor sites in the rat
striatum (Corrodi et al., 1973), was compared with that induced in
the same individuals by oral administration of L-DOPA (Larodopa,
Hoffmann La Roche, 500 mg). CB 154 increased hGH after 210 minutes,
an effect still present 300 minutes after drug ingestion; this pattern contrasted with that of L-DOPA, which induced a more prompt
but short-lived increase of hGH levels.

Also 5-butylpicolinic acid, a natural enzyme inhibitor of D-
β -H (Hidaka, 1973) and hence potential activator of DA neurotransmission, induced by itself a slight elevation in hGH levels (Hidaka
et al., 1973) and haloperidol or chlorpromazine, two neuroleptics
antagonist to DA (and NE) inhibited insulin-or Apo-induced hGH rise
(Kim et al., 1971; Sherman et al., 1971; Lal et al., 1973).

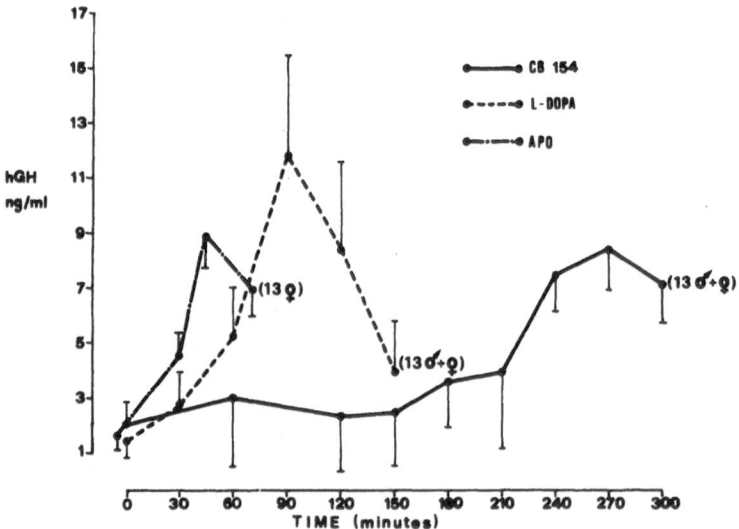

Fig. 1 Plasma hGH levels (mean ± SE) in 13 male and female healthy
subjects after administration of CB 154 (2.5 mg po) or L-
DOPA (500 mg po). (From Liuzzi et al. "Growth Hormone and
Related Peptides", 1976, by courtesy of Excerpta Medica).
Depicted is also the rise in plasma hGH evoked by apomorphine
(Apo 0.75 mg sc) in 13 female subjects (From Ettigi et al.,
J. Clin. Endocrinol. Metab., 1975, by courtesy of J.B. Lip-
pincott Co.,)

As in primates also in subprimate species considerable data support the concept of central α-adrenergic stimulation of GH. In conscious (Takahashi et al., 1973) or anesthetized (Ganong, 1975) dogs,
GH release induced by systemic L-DOPA could be abolished by α-re-
ceptor blockade induced by iv phentolamine or by phenoxybenzamine
infused into the brain ventricle. Blockade of DA receptors by pimozide did not prevent GH release (Lovinger et al., 1976). In dogs,
like dopamine, the DA-receptor stimulant Apo, administered intraventricularly, did not modify plasma GH (Lovinger et al., 1976).
Also in cats, hypothalamic DA receptors do not appear to stimulate
GH release; systemic Apo infusion failed to alter basal GH levels
and, phenoxybenzamine but not pimozide completely abolished L-DOPA-
induced GH release (Ruch et al., 1977).

The concept of α-adrenergic stimulation of GH release in mammals includes also rodents. Ruch et al. (1976) showed in rats that
the α-adrenergic receptor stimulant drug clonidine given systemically or into the third ventricle stimulates GH release and Durand
et al. (1977) induced a significant suppression of GH bursts by
blocking catecholamine biosynthesis with α-methyl-p-tyrosine(α -
MpT).

The influence of dopaminergic neurons on GH release in the rat is unclear. Direct instillation of DA into the lateral brain ventricle reportedly reduced plasma GH (Collu et al., 1972) and pimozide elevated plasma GH (Müller et al., 1973). Intraventricular injection of DA was also reported to induce a depletion of pituitary GH content as determined by bioassay, indicating a stimulation of GH release (Müller et al., 1968). L-DOPA has been reported to either stimulate (Chen et al., 1974), inhibit (Müller et al., 1973) or have no effect (Kato et al., 1973) on GH secretion in the rat.

The fact that in the adult unanesthetized rat GH is secreted in episodic bursts (Martin et al., 1975) suggests caution in interpreting results obtained in this species from neuropharmacologic studies (Müller et al., 1973; Kato et al., 1973; Chen et al., 1974) in which the frequency of blood sampling was lower than the frequency of the pulsatile GH secretion (Martin et al., 1975; Tannenbaum and Martin, 1976). In this context, longitudinal studies should be more rewarding for the assessment of the effect of a given experimental procedure on GH regulation in the rat.

With this in mind it was decided to evaluate the effect of dopaminergic stimulation induced by Apo in regulation of physiological episodic GH secretion in the unanesthetized male rat. Fig. 2 shows the results of an experiment in which Apo was injected iv into rats prepared with chronic indwelling jugular cannulae and sampled for 2hr beginning at 1515h. Individual rats were used as their own controls receiving control injection on one experimental day and Apo on a subsequent day. Apo (.1mg/kg) induced a striking and prompt rise in plasma rGH when compared to the baseline secretory pattern of the same rats ($p < .005$ at 15 and 30 min). Administration of Apo at a higher dose (1 mg/kg) induced a delayed rise in plasma rGH ($p < .05$ vs baseline values at 75 min) (data not shown). An opposite pattern was evident when the drug was injected at a dose of 5.0 mg/kg (Fig. 3). In this instance Apo induced a striking and sustained decline in rGH values, evident from 15 to 105 min ($p < .05$ at 15 min; $p < .01$ at 30 min; $p < .02$ at 45 min; $p < .005$ at 105 min vs controls). In keeping with these findings, Mueller et al. (1976a) have recently reported that Apo and Piribedil, another DA agonist (Corrodi et al., 1971), release GH at low doses but are ineffective to release GH at high doses in rats killed by decapitation.

Even though interpretation of the biphasic neuroendocrine effect of apomorphine is uneasy, it is tempting to speculate that in a "low" dose range the drug may predominantly stimulate "presynaptic" or "regulatory" DA receptors, thereby turning off the firing of the dopaminergic neurons (Bunney and Aghajanian, 1975). At high doses the postsynaptic effect of Apo would predominate with ensuing inhibition of plasma GH levels. If the concept of "inhibitory" DA receptors, which has been mutuated by recent biochemical and neurophysiologic studies on striatal DA receptors (Bunney and Aghajanian, 1975; Carlsson,

1975; Di Chiara et al., 1976) may be applied to the limbic and/or hypothalamic DA receptors involved in GH regulation, a possible interpretation is envisaged of the discrepancy existing on the role of DA on GH release in man or the other mammals. It may be that the DA stimulant drugs at the low doses which are being currently used in man for stimulating GH release might have the primary effect of turning off the DA function acting predominantly on "regulatory" DA receptors. This would imply that, as in the other subprimates and primates, the role of DA in the control of GH secretion in man is actually inhibitory in nature. Consistent with this hypothesis is the finding that administration of gamma-hydroxybutyric acid, a drug capable of blocking the release of DA from nerve endings and of increasing the DA concentration of the brain (Menon et al., 1974), induced in normal subjects a "paradoxical" GH rise (Takahara et al., 1977). Further studies in both laboratory animals and man are needed to clarify the possible role of pre-and postsynaptic DA receptors in the control of GH secretion.

Recent findings obtained in our laboratory in subjects affected by Huntington's chorea (HC), an autosomal dominantly-inherited progressive neurological disorder, suggest that (hypothalamic) DA receptors involved in the control of GH secretion may be hyperresponsive to dopaminergic stimuli. In these studies, the GH-releasing effect of CB 154 (2.5 mg po) was investigated in 7 rendomly selected hospitalized male and female patients, who had a progressive dementia of varying degree, choreiform movements and a positive family history for the disease. Six nonobese hospitalized female and male patients without endocrine or metabolic diseases, age-and sex-matched, were selected as controls. Baseline plasma GH values were not significantly different in choreic and control subjects; however after CB 154 administration the mean hGH peaks were 14.2 $\pm$ 5.5 and 3.9 $\pm$ 1.8 ng/ml respectively (F = 12.41, p $<$ .01) and the highest concentrations occurred at 120-150 and 180-240 min in choreic and control subjects, respectively (data not shown). In a further study two different doses of Apo (.75 and 1.0 mg) were administered to choreic and control subjects. In 5 patients with HC, Apo (.75 mg) did not induce a rise in plasma GH significantly higher than that present in control subject although a trend was present towards higher GH levels (Fig.4). In 6 patients with HC, Apo (1.0 mg) induced a significant higher increase in GH levels than in controls (F = 5.61, p $<$ .05) (Fig. 4).

In all these data point to the existence in the CNS of choreic subjects of supersensitive DA receptors involved in GH regulation, and are in keeping with the exaggerated GH response of these subjects to insulin-induced hypoglycemia (Keogh et al., 1976), a stimulus acting presumably through dopaminergic mediation (Kim et al., 1971; Ettigi et al., 1975). The possible "presynaptic" nature of these DA receptors awaits clarification.

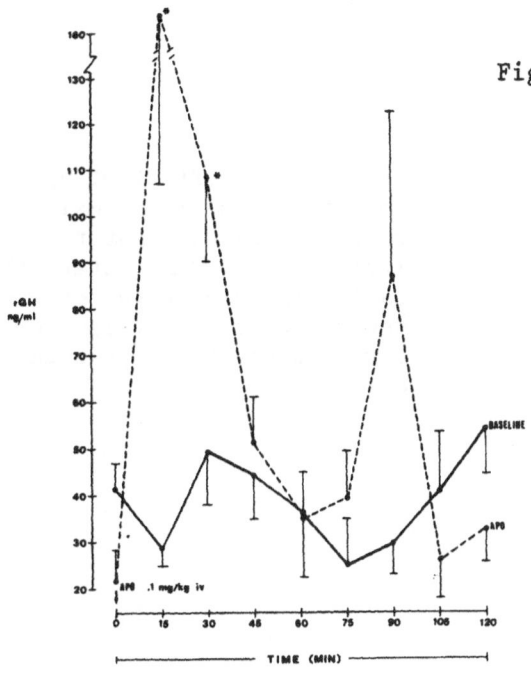

Fig. 2 Averaged rGH levels in 8 rats either given diluent injection (solid line) or .1 mg/kg of apomorphine (Apo) at time 0,24h later (dotted line), and sampled at 15 min intervals from 1515 to 1715h. Asterisks indicate a difference statistically significant vs corresponding baseline values. The same description applies also to Fig. 3

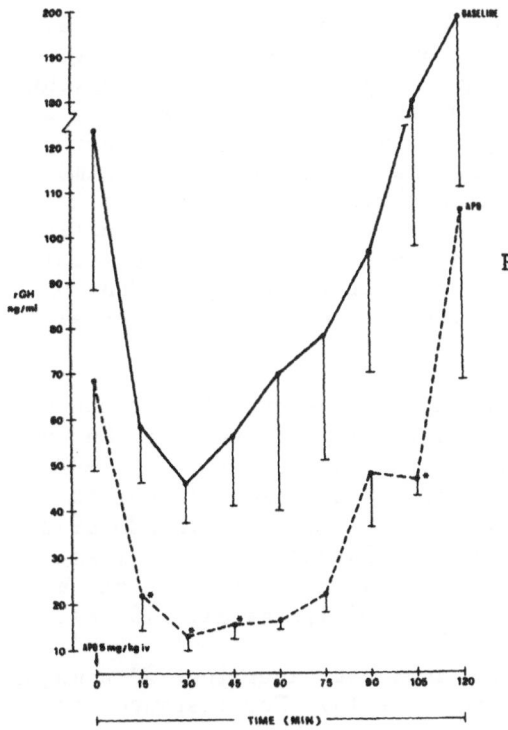

Fig. 3 Averaged rGH levels in 7 rats either given a diluent injection (solid line) or 5.0 mg/kg iv of Apo at time 0,24h later (dotted line).

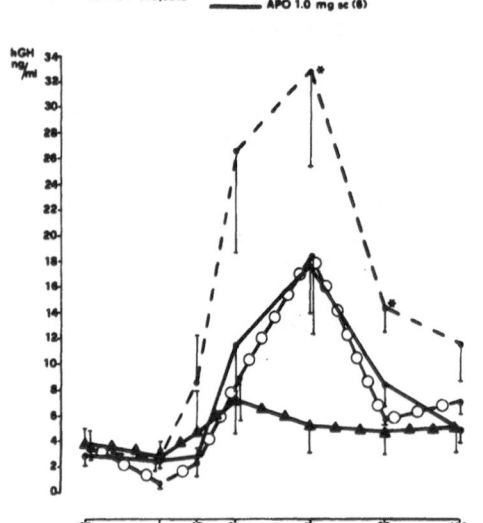

choreic patients ○—○—○ APO 0.75 mg sc (5)
                  — — APO 1.0 mg sc(6)

normal subjects ▲—▲—▲ APO 0.75 mg sc(5)
                 ——— APO 1.0 mg sc (6)

Fig. 4 GH-releasing effect of apomorphine (Apo) in normal or
choreic subjects. Asterisks
indicate differences statistically significant vs corresponding control levels. Number of subjects in parentheses.

B. Indoleamines

In spite of many studies on the involvement of brain indoleamines
on GH secretion, the exact role played by the brain serotoninergic
system is still an extremely controversial problem (see also Smythe
et al., 1976; Müller et al., 1977a). On this subject two extreme at-
titudes have been expressed with regard expecially to the effect of
5-HT on GH secretion in man. The former maintains that DA is a monoamine releasing per se GH (but see also above), while 5-HT has no
stimulatory role (Müller et al., 1974) or even an inhibitory effect
(Liuzzi et al., 1976). According to the latter 5-HT is the physiological stimulus for GH release and all stimuli including L-DOPA, DA
and CB 154 release GH by stimulating 5-HT receptors (Smythe et al.,
1976).

Most of the experiments in which a stimulatory role for the serotoninergic system on GH secretion has been claimed are based on the
use of 5-hydroxytryptophan (5-HTP), as specific stimulant of 5-HT receptors after its conversion to 5-HT, and of alleged antagonists of
5-HT as methysergide (Meth) or cyproheptadine (Cy). 5-HTP administered systemically is capable of inducing consistent rises in plasma GH
in both subprimate and primate species; Meth and Cy inhibit the stimulated GH release in rodents and man (see Müller et al., 1977a).

A numerous series of experiments does not favor instead a major
role for 5-HT in the GH releasing mechanism(s). For instance, in

dogs, infusion of 5-HTP induced a prompt rise in plasma GH (Müller et al., 1976a) but this was associated to an impressive assembly of side effects which argued against the specificity of the applied stimulus. Moreover, in other studies in man, oral administration of 5-HTP given either alone (Müller et al., 1974) or in combination with a peripheral decarboxylase inhibitor (Benkert et al., 1973; Handwerger et al., 1975) failed to induce plasma hGH changes. Oral administration of tryptophan (Trp), the amino acid precursor of 5-HT, induced in man (Müller et al., 1974) and dogs (Müller et al., 1976a) only a slight rise in plasma GH titers.

Since systemically administered 5-HTP increases brain 5-HT in part owing a selective accumulation by 5-HT neurons (Fuxe, 1965; Corrodi et al., 1967), the GH-releasing effect of 5-HTP should be conceivably attributed to an enhancement of 5-HT-mediated neuronal transmission. On the other hand, loading doses of 5-HTP can enter catecholamine (CA) neurons, be decarboxylated to 5-HT and displace endogenous catecholamines (Ng et al., 1972), a phenomenon which is even magnified when a peripheral inhibitor of aromatic amino acid decarboxylase is associated to 5-HTP (Lancranjan et al., 1977), since in this instance higher amounts of unmetabolized 5-HTP are shunted to the CNS. Thus, 5-HTP-induced changes in CNS function may largely reflect actions on CA or other central neurotransmitter systems.

To verify the specificity of action of 5-HTP on 5-HT system with regard to GH release, it was thought of interest to study GH release in the rat after administration of 5-HTP alone or in combination with drugs (1) either selectively diminishing or potentiating its neurochemical effects on brain 5-HT system or (2) capable of blocking the effect of its potential activity on brain CA systems. To avoid the variability in plasma GH titers present in the adult rat due to the pulsatile secretion (see page 5), these investigations were performed in the infant rat, in which the episodic release of GH is absent (Cocchi et al., 1976a). The effect of treatments on plasma PRL was concomitantly evaluated.

Administration of 5-HTP induced a marked elevation of plasma GH and PRL levels. Pretreatment with Meth or metergoline (MCE), another 5-HT receptor blocker (Beretta et al., 1965), markedly reduced the 5-HTP-induced rPRL rise but did not affect the rGH response to 5-HTP alone (data not shown).

The effect of pretreatment with fluoxetine and chlorimipramine (CIM), two rather selective inhibitors of 5-HT reuptake (Wong et al., 1975; Iversen, 1967), on 5-HTP-induced GH and PRL rises was then studied. Administration of fluoxetine markedly potentiated the 5-HTP-induced PRL rise but significantly reduced the GH response to 5-HTP. Similarly to fluoxetine, pretreatment with CIM greatly reduced the 5-HTP induced GH rise (Table 1).

Table 1   Effect of Fluoxetine or CIM on 5-HTP-induced GH and PRL release in 10-day-old pups[a]

Group	Pretreatment[1]	Treatment[2]	rGH ng/ml	Significance of differences	rPRL ng/ml	Significance of differences
1	Saline	Saline	21.7 ± 3.4	-	4.9 ± 0.5	-
2	Saline	5-HTP	85.6 ± 18.7	p < 0.005 vs 1	48.1 ± 8.2	p < 0.001 vs 1
3	Fluoxetine (10 mg/kg ip)	Saline	13.5 ± 3.7	-	5.6 ± 0.9	-
4	Fluoxetine (10 mg/kg ip)	5-HTP	16.9 ± 5.2	p < 0.05 vs 2	119.0 ± 8.5	p < 0.001 vs 2
5	CIM (10 mg/kg sc)	Saline	39.0 ± 13.4	-	-	-
6	CIM (10 mg/kg sc)	5-HTP	29.1 ± 9.7	p < 0.025 vs 2	-	-

[1] Fluoxetine and CIM were given 4h and 30 min, respectively, before 5-HTP or saline administration.

[2] 5-HTP was administered at a dose of 50 mg/kg ip. Rats were killed 30 min after 5-HTP or saline (10 ml/kg) administration.

[3] Mean ± SEM of 8-10 determinations.

[a] From Cocchi et al., Neuroendocrinology (1977), by courtesy of S. Karger.

Table 2 shows that pretreatment with either pimozide (PIM) or phentolamine (Phent) blocked or reduced the 5-HTP-induced GH rise without modifying the 5-HTP-induced PRL release. Similarly to Pim and Phent, central sympathectomy by the neurotoxic drug 6-hydroxy-dopamine (6-OHDA) counteracted the 5-HTP-induced GH release; 6-OHDA, instead, potentiated although not significantly the PRL response to 5-HTP. In all these data indicate that in the infant rat a nonspecific activation of the catecholaminergic system is responsible for the GH response to 5-HTP, whereas 5-HTP-induced PRL release appears to be mediated via the brain 5-HT system. Therefore, results of experiments in which 5HTP is used as "selective" activator of 5-HT neurotransmission and the GH response is studied (see above) have to be interpreted cautiously unless careful neuropharmacologic analysis of the neuroendocrine response investigated is performed.

Table 2   Effect of Pim, Phent, or 6-OHDA on 5-HTP-induced GH and PRL release in 10-day-old pups[a]

Group	Pretreatment[1]	Treatment[2]	rGH ng/ml	Significance of differences	rPRL ng/ml	Significance of differences
1	Saline	Saline	28.6 ± 4.1[3]	-	3.4 ± 0.7	-
2	Saline	5-HTP	102.8 ± 25.7	p < 0.01 vs 1	12.8 ± 4.7	p < 0.01 vs 1
3	Pim (500 μg/kg ip)	Saline	56.4 ± 23.1	-	7.5 ± 2.3	-
4	Pim (500 μg/kg ip)	5-HTP	35.2 ± 9.3	p < 0.02 vs 2	17.7 ± 3.6	p < 0.05 vs 3
5	Phent (25 μg/10 μl ivt)	Saline	7.8 ± 2.3	p < 0.001 vs 1	1.5 ± 0.2	p < 0.05 vs 1
6	Phent (60 μg/10 μl ivt)	5-HTP	54.6 ± 26.7	-	17.9 ± 5.0	-
7	6-OHDA (60 μg/10 μl ivt)	Saline	20.3 ± 3.2	-	16.3 ± 3.9	p < 0.05 vs 1
8	6-OHDA (60 μg10 μl ivt)	5-HTP	35.5 ± 11.2	p < 0.05 vs 2	24.9 ± 13.8	-

[1] Pim and Phent, were given 3h and 10 min, respectively, before 5-HTP or saline administration. 6-OHDA was injected when rats were 5 day-old. As vehicle, 10 μl of 0.1% ascorbic acid was used.

[2] 5-HTP was administered at the dose of 50 mg/kg ip. Rats were killed 30 min after 5-HTP or saline (10 ml/kg) administration.

[3] Data represent pooled results (mean ±SEM) of 2 experiments in which similar results were obtained; 8-10 animals per group were used in each experiment.

[a] From Cocchi et al., Neuroendocrinology (1977), by courtesy of S. Karger.

II. NEUROTRANSMITTER CONTROL OF PRL SECRETION

A. Catecholamines

Prolactin is unique among AP hormones in that its secretion in mammals is tonically inhibited by the CNS. The existence of a PIF has been proved (see Schally et al., This Symposium); however, the nature of PIF still remains elusive and, as will be discussed, the possibility that under physiological conditions CA's, namely DA, may be PIF has not been excluded.

An enormous literature on PRL has appeared in the last 7 years which indicates that brain CA's are inhibitory on PRL release (see Clemens, 1976; Meites, 1977; Mac Leod and Login, 1977). First experimental data linking in a reciprocal way the catecholaminergic tonus in the CNS with PRL secretion go back to the studies of Barraclough and Sawyer (1959) and Coppola et al. (1965), which showed that treatments of rats with chlorpromazine or CA synthesis inhibitors resulted in the induction of pseudopregnancy. Van Maanen and Smelik (1968) first proposed that DA was the PIF on the basis of the ability of reserpine inplants in the MBH to deplete DA and induce pseudopregnancy. Since then an exhaustive series of studies involving more specifically pharmacologic agents and direct measurements of circulating PRL levels have strengthened the theory that brain CA's exert an inhibitory action on PRL secretion. Intraventricular administration of relatively high doses of DA (1.25 μg) or of high doses of E and NE (100 μg) produced a striking reduction in the circulating levels of PRL and increased PIF in the portal circulation. DA, NE and E, perfused into the AP via a hypophyseal portal vein, had no effect on PRL release (Kamberi et al., 1970, 1971a). A single ip injection of L-DOPA elevated hypothalamic PIF activity and also evoked the presence of PIF (in vitro assay) in the systemic blood of intact and hypophysectomized (hypox) rats bearing a pituitary graft under the kidney capsule (Lu and Meites, 1971, 1972). In contrast blockade of NE and DA synthesis by α-MpT induced a rapid elevation in serum PRL (Donoso et al., 1971; Meites et al., 1972) which could be reversed by administration of L-DOPA (Donoso et al., 1971). Drugs such as Apo, ergocornine, CB 154, lisuride and piribedil, all possessing DA-stimulant activity, suppressed baseline or stimulated PRL secretion in rats (Wuttke et al., 1971; Lawson and Gala, 1975; Gräf et al., 1976; Mac Leod et al., 1976). The same effect was produced in sheep by direct infusion of DA (Davis et al., 1973).

These observations and others (see Meites, 1973; Mac Leod and Login, 1977; Clemens and Meites, 1977), plus the failure to demonstrate unequivocally that free DA from the brain can reach in significant amounts the AP (Ben-Jonathan et al., 1975) led some authors to postulate that DA acts primarily via PIF which then inhibits PRL release. However, the possibility that CA's may also act directly on the lactotrops is supported by abundant experimental evidence. It includes the fact that: 1) either DA (Birge et al., 1970), DA pre-

cursor (Mac Leod and Lehemeyer, 1972) or mimetic drugs (Floss et al.,
1973; Smalstig et al., 1974) were shown capable of inhibiting PRL
release by AP incubated in vitro, an effect which could be prevented
by specific DA blockers (Mac Leod and Lehemeyer, 1974); 2) suppression of elevated baseline PRL levels was also induced by DA-mimetic
drugs administered to hypophysectomized rats bearing an ectopic pituitary (Gräf et al., 1977; see also Fig. 5) or electrolytic
lesions of the median eminence (ME) (Shaar and Clemens, 1976) or
when administered to stalk-lesioned rhesus monkeys (Diefenbach et
al., 1976) or humans (Woolf et al., 1974); 3) specific receptor binding sites for DA, DA-like agents, and DA receptor blockers have
been detected at AP level (Kimura and Mac Leod, 1975; Brown et al.
1976); 4) DA injected directly into hypophyseal portal vessels inhibited PRL secretion (Takahara et al., 1974a), in contrast to the earlier,
negative findings of Kamberi et al. (1971a) (see above); 5) DA, but
not E and NE, has been finally detected in hypophyseal portal blood
of female rats, although meaningful corrections with PRL release as
yet have not demonstrated (Ben-Jonathan et al., 1977).

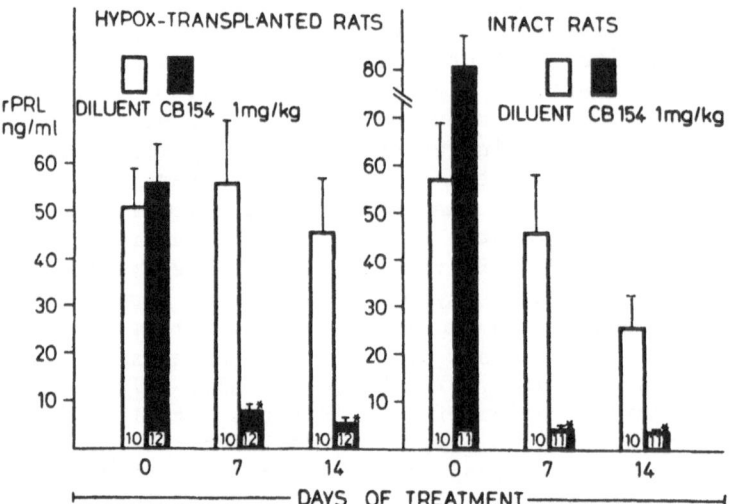

Fig. 5 Effect of chronic CB 154 administration (50 µg/100g BW sc
       t.i.d.) on plasma PRL levels of hypophysectomized (hypox)
       Sprague-Dawley rats bearing an entire anterior pituitary
       under the kidney capsule or of intact female rats. Asterisk
       indicates a difference statistically significant (p< .05)
       vs corresponding control values.Numbers at the bottom of
       each column indicate the number of rats for each group.

In conclusion these series of studies support the view that DA can act directly on the AP to inhibit tonic PRL release. Undoubtedly on considering the amount of DA contained in the ME of adult rats, with a half-life in the hypothalamus of approximately 1.5hr (see Weiner, 1975), the ME possesses the potential capability of producing DA in amounts sufficient to inhibit PRL secretion in the event of a consistent release into the portal blood. Presently, however the existence of a hypothalamic PIF distinct from DA cannot be negated; PIF activity has recently been reported in DA-free purified hypothalamic extracts (Greibrokk et al., 1975; Dupont and Redding, 1975; see also pag.20) and quantitative studies have suggested that only half of the total MBH PIF activity is accounted for by DA (Enjalbert et al., 1977).

1. Hypothalamic-Prolactin-Gonadotropin Relationship - The anatomical arrangement of the tuberoinfundibular DA (TIDA) neurons with the capillaries in the external layer of the ME provides a suitable neuroanatomical substrate for a direct control of pituitary lactotrophs, by DA released into the portal circulation, not so for a neurotransmitter (DA)-neurohormone (PIF) interaction. Althoug, similar to other dopaminergically innervated CNS areas, a DA-sensitive adenylate cyclase (AC) has been detected at ME level (Clement-Cormier and Robison, 1977), its true significance has yet to be clarified. Whereas high affinity receptors for haloperidol and DA have been found in the rat pituitary, they are apparently lacking at the level of the MBH (Brown et al., 1976), which suggests that the TIDA neurons do not produce postsynaptic effects in this region. Certainly, the regulatory mechanism(s) governing the activity of TIDA neurons is not modulated by a neuronal feedback mechanism subsequent to dopaminergic receptor blockade or activation (Gudelski and Moore, 1976). However, ME-DA nerve terminals although unreactive to specific neuropharmacologic manipulations are highly endocrine-responsive. Changes in the turnover rate of DA have been measured during the estrus cycle, in pregnant, pseudopregnant and lactating rats (Fuxe et al., 1974). Single or repeated injections of high doses of PRL into normal (Fuxe and Hökfelt, 1969) or ovariectomized (Gudelski et al., 1977) rats markedly increased DA turnover in the ME, thus suggesting that DA terminals at this level may mediate the inhibitory feedback action of PRL on its own secretion.

It is a widespread notion that spontaneous or experimentally-induced hyperprolactinemia exerts antigonadotropic effects in experimental animals (Fang et al., 1974; Lu et al., 1976) and man (see Besser and Mortimer, 1976). Hyperprolactinemic amenorrhea is a common disorder, occurring in about 20% of women with amenorrhea (Franks and Jacobs, 1977). In view of the reported inhibitory action of dopaminergic stimulation over LH release (Fuxe and Hökfelt, 1969; Sawyer et al., 1974), the anti-gonadotropic effect of hyperprolactinemia may result ultimately from a sustained activation of dopaminergic neuro-

transmission causing a blockade of LH-RH discharge from the hypothalamus.

    With this in mind studies were performed in intact or castrated
(CX) Wistar/Furth (W/Fu) rats made hyperprolactinemic by a transplan-
table mammosomatotropic tumor MtTW$_5$ to ascertain: 1) the effect of
high circulating PRL levels on the LH-RH system; 2) the effect in
these rats of a chronic treatment with bromocriptine, which is known
to reduce PRL levels and repair the reproductive defects (see Crosignani and Robyn, 1977).

    Hypothalamic LH-RH content in intact W/Fu rats treated or not
with CB 154 was significantly higher than that of intact W/Fu controls, except for female rats treated with CB 154 in which hypothalamic LH-RH content was not significantly different from that of
W/Fu controls. Castration completely antagonized the increased LH-RH
content present in the W/Fu male rats but was completely unable to
modify the increased LH-RH content in the W/Fu female rats (Table 3).

    Chronic treatment with CB 154 markedly decreased hypothalamic
LH-RH content in both W/Fu male controls and W/Fu male rats and in
W/Fu intact and CX female rats. The drug was instead unable to modify
the already low LH-RH content of CX W/Fu male rats and cycling W/Fu
female controls (Table 3). In W/Fu male controls a single injection
of CB 154 (2.5 mg/kg sc) did not significantly modify 2hr later LH-
RH content in MBH (9.22 ± 1.23 vs 6.9 ± 1.15 ng/ME in diluent-and
CB 154-treated rats, respectively) (data not show).

Table 3   LH-RH levels in the Median Eminence (ME) of male and female
          non-bearing or bearing tumor rats treated with Diluent or
          CB 154[1]

Group and Treatment		No. of rats	LH-RH ng/ME	Group and Treatment		No. of rats	LH-RH ng/ME	Significance of difference vs respective Diluent treated group.
Males								
W/Fu Controls	Diluent	3	14.63 ± 0.56	W/Fu Controls	CB 154	4	4.05 ± 0.70	$p < 0.001$
W/Fu	Diluent	6	21.40 ± 1.90●	W/Fu	CB 154	6	12.40 ± 1.90●●	$p < 0.02$
CX W/Fu	Diluent	3	5.30 ± 0.88***	CX W/Fu	CB 154	4	5.28 ± 1.25*	NS
Females								
W/Fu Controls	Diluent	4	8.32 ± 1.28	W/Fu Controls	CB 154	4	6.05 ± 1.38	NS
W/Fu	Diluent	4	18.80 ± 3.30●●	W/Fu	CB 154	4	8.80 ± 1.20	$p < 0.02$
CX W/Fu	Diluent	3	18.20 ± 1.20	CX W/Fu	CB 154	4	3.65 ± 0.80●●	$p < 0.001$

*$p < 0.05$ vs respective W/Fu group   ●$p < 0.05$ vs W/Fu Controls Diluent
**$p < 0.02$ vs respective W/Fu group   ●●$p < 0.02$ vs W/Fu Controls Diluent
***$p < 0.005$ vs respective W/Fu group

[1]CB 154: 2.5 mg/kg sc, b.i.d., 6 days a week for 90 days.

These data indicate that the state of hyperprolactinemia was associated in the intact W/Fu rats with increased LH-RH stores; there was following castration a sex-related difference in both the decrease of ME LH-RH and in the increase in plasma gonadotropins (data not shown), since these events occurred only in the male rats, were pratically absent in the females. On recalling the more exhuberant tumoral growth and, hence, more elevated plasma PRL levels of the CX female than male rats of this study (data not shown), it is apparent that the alteration of the CNS-controlled gonadotropin secretion induced by high PRL levels is a graded phenomenon depending on degree and duration of hyperprolactinemia. Bromocriptine restored to normal the increased ME LH-RH content of W/Fu intact and CX rats, an effect which might result from the drug's ability to lower plasma PRL levels and, hence, diminishing the feedback action of the latter on the CNS. However, from our data the possibility cannot be dismissed that chronic CB 154 administration had per se stimulated LH-RH discharge from the hypothalamus since its effect was also evident in W/Fu male controls, in which baseline PRL levels were normal. That CB 154 through dopaminergic stimulation may induce hypothalamic LH-RH discharge is in keeping both with experimental findings in the rat (Bennet et al., 1975; Rotszejn et al., 1976) and with the observation that in hypothalamic amenorrhea it restored ovarian function also when plasma PRL was normal (Nencioni et al., 1978).

B. Indoleamines

Several studies indicate that brain 5-HT is involved in stimulation of PRL secretion. A threefold increase in plasma PRL levels was reported after IVT injection of 5-HT or melatonin (Kamberi et al., 1971b; Caligaris and Taleisnik, 1974). Subsequent, it was shown that systemic injection of the 5-HT precursors, Trp and 5-HTP, but not 5-HT itself, could induce a rise in serum PRL in rats (Lu and Meites, 1973; Marchlewska-Koj and Krulich, 1975; Mueller et al., 1976b). Although the neuropharmacologic specifity of 5-HTP as "selective" activator of 5-HT neurotransmission has to be questioned, the extra-serotoninergic action of administered 5-HTP apparently does not involve the neuroendocrine system which controls PRL secretion (see pag. 10). Pretreatment of rats with fluoxetine or CIM, two 5-HT' re-uptake inhibitors, strongly potentiated the PRL-releasing effect of 5-HTP (this study, pag. 9) as well as the PRL release induced by either stress or blood withdrawal (Krulich, 1975). Similarly, iv infusion of L-Trp or oral administration of 5-HTP stimulated PRL secretion in man (Mc Indoe and Turkington, 1973; Kato et al., 1974), an effect whose specificity rests on the finding that the alleged 5-HT blocker Cy inhibited 5-HTP-induced PRL secretion (Kato et al., 1974). PRL secretion can also be affected by decreasing in several ways central 5-HT activity. Administration of parachlorophenylalanine (PCPA), a drug which blocks 5-HT biosynthesis was found ineffective in changing plasma PRL levels in post-partum female (Kordon et al., 1973, 1974) ovariectomized female (Gallo et al., 1975) and CX male

rats (Donoso et al., 1971). However, in the study of Gil-Ad et al.
(1976a) acute impairment of central 5-HT activity by PCPA or persistent depletion of brain 5-HT and 5-hydroxyindoleacetic (5-HIAA) levels
by IVT administration of the neurotoxic drug 5,7-dihydroxytryptamine
(5,7-DHT) were associated to a marked reduction in baseline plasma
PRL levels. Similarly, ovariectomized, estrogen-primed rats respond-
ed to p-chloroamphetamine, a depletor of brain 5-HT, with a decrease
in serum PRL (Chen and Meites, 1975). Baseline plasma PRL levels were
unaffected in normal male rats following administration of Meth
(Marchlewska-Koj and Krulich, 1975) and, surprisingly, the levels of
PRL were even transiently enhanced by Meth in ovariectomized female
rats (Gallo et al., 1975) and by Meth and Cy in female monkeys (Gala
et al., 1978). For Meth, an antidopaminergic action directly on the
AP (Mc Leod and Login, 1977) has been invoked to explain the PRL-releasing effect.

A further approach for studying the effect of functional abla-
tion of central 5-HT system on circulating plasma PRL levels is that
of limiting in the diet the presence of the 5-HT precursor amino
acid, i.e., Trp. Plasma receives TP from two sources only, dietary Trp
and secretion from the free amino acid pools of various tissues
(Wurtman, 1974). Rats given free access to a Trp-deficient diet develop decreased plasma and, ultimately, brain Trp levels (Gál and
Drewes, 1962); this results in a decreased brain 5-HT turnover (Tagliamonte et al., 1971).

In our study, a group of rats was fed for up to 4 days, with a
diet made up of maize flour "polenta", the Trp content of which is
abnormally low (Cioffi and Fidanza, 1974). A second group of rats
received the same diet to which Trp (2.0 g/kg of diet) had been added and drank water supplemented with vitamins. As controls, rats
fed ad libitum with a standard laboratory diet or undernourished
rats, given half their usual food intake (about 10 g per rat per
day), were used. Plasma PRL levels in animals fed for 4 days with a
Trp-deficient diet were not different from PRL levels present in
normal controls. Addition of Trp to the Trp-deficient diet caused
a striking increase in PRL levels. In undernourished animals on
standard laboratory diet, PRL levels were slightly and not significantly lower than in ad libitum fed controls (Table 4). Brain 5-HT
and 5-HIAA levels were significantly decreased in rats on Trp-deficient diet, while no change in brain indole levels was present in
the undernourished rats. Addition of Trp to the Trp-deficient diet
increased brain 5-HT and 5-HIAA levels to 135 and 140%, respectively,
of values present in normal controls (Table 4). In our study, in
spite of a decrease in brain indole levels induced by ingestion of a
Trp deficient diet, there were no changes in baseline plasma PRL
levels. This might perhaps be due to the small and slowly occurring
depletion of brain indole levels induced by the diet regimen. However,
ingestion of the maize diet supplemented with TP, at a dose roughly
corresponding to the daily intake of Trp by rats fed with a standard

diet (unpusblished observations), resulted in a striking increase of baseline PRL levels over those present in both normal control and TP-deficient rats and in a concomitant rise in brain indole levels. These data may be interpreted to mean that addition of Trp to the maize diet elicited in the recipient rats an activation of 5-HT receptor sites, which was responsible for the stimulatory effect on PRL secretion.

In conclusion, the effects induced by a decrease in brain 5-HT activity on unstimulated PRL secretion are difficult to evaluate at present, even though it would appear that 5-HT neurons do not exert appreciable activity on tonic PRL secretion. Their action appears to be more relevant, instead, in conditions of stimulated PRL secretion. In male rats subjected to restraint stress, Mueller et al. (1976b) reported an increase in brain 5-HT turnover which was associated to increased plasma PRL and decreased TSH titers. An injection of Trp mimicked the effect of restraint stress on both brain indoles and plasma PRL and TSH concentrations. Pretreatment of normal male rats with Meth reportedly blocked the PRL-releasing effect of etherization and bleeding, whereas it did not inhibit the increase of serum PRL induced by either pimozide or α-MpT (Marchlewska-Koj and Krulich, 1975). These results would indicate that a 5-HT regulatory system activated by stress either stimulates a hypothalamic PRF or inhibits a PIF that could as well be the TIDA system. The observation that ether stress was equally effective in stimulating PRL release in untreated as in reserpine-treated animals in which a loss of tonic PIF control should have been present, led to postulate that the stress-induced release of PRL is not due to inhibition of PIF secretion but, most likely, to a PRL-releasing substance (Valverde et al., 1973). On the basis of the foregoing, an interaction can be envisaged by 5-HT system and PRF neurons which mediate the stress-induced PRL rise.

Table 4   Effect of Trp-deficient diet alone or supplemented with Trp on plasma prolactin and brain indoles levels[a,f]

Group	Plasma prolactin (ng/ml ± SE)	brain	
		5-HTP (μg/g ± SE)	5-HIAA (μg/g ± SE)
Standard pellet diet	14.9 ± 2.5	0.65 ± 0.01	0.46 ± 0.01
Trp-deficient diet	14.4 ± 1.9	0.47 ± 0.01[d]	0.27 ± 0.01[d]
Trp-deficient diet + Trp	35.0 ± 4.7[c]	0.88 ± 0.03[d,e]	0.64 ± 0.07[b,e,]
Underfed controls	10.4 ± 0.9	0.66 ± 0.02	0.46 ± 0.01

[a] 10 animals per group were used.

[b] p < 0.02   vs standard pellet diet.

[c] p < 0.01   vs all other groups.

[d] p < 0.001   vs standard pellet diet.

[e] p < 0.001   vs Trp-deficient diet.

[f] From Gil-Ad et al., Proc. Soc. Exp. Biol. Med. (1976a), by courtesy of Academic Press.

C. Other neurotransmitters

1. **γ-Aminobutyric Acid (GABA)**- Evidence, although inferential
for the participation of GABA-ergic neurons in the central mechani-
ism(s) controlling PRL secretion was derived from the observation that
the release of PRL induced by ovarian steroids in spayed female rats
was inhibited by picrotoxin,a GABA antagonist drug (Ito et al., 1968).
In addition 5-HT-induced PRL release was blocked by picrotoxin, whereas stimulation induced by TRH or α-MpT was unaffected (Caligaris
and Taleisnik, 1974). These data suggested that inhibition of PRL
secretion by the GABA antagonist took place at the CNS rather than
at the AP level, a view supported by the finding that huge doses of
GABA administered IVT induced PRL release in proestrus or ovariectomized female (Mioduszewski et al., 1976) or intact male rats (Ondo
and Pass, 1976), but were ineffective when injected systemically
into hypox-rats bearing an AP transplant, (Mioduszewski et al., 1976).
In contrast to these findings which indicate a PRL-releasing activity
of the amino acid, Schally et al. (1977) recently purified and isolated from pig hypothalami a neutral acidic fraction cromatographically
distinct from CA's which possessed PIF activity in many in vitro and
in vivo systems. The PIF active substance was identified as GABA.

In view of these discrepant results it was felt of interest to
investigate the effect of GABAergic neurotransmission on the secretion of PRL, using compounds with specific and stimulating actions
on the GABA-mediated synaptic function. For these experiments male
Sprague-Dawley rats (300-350 g BW) bearing chronic indwelling jugular
cannulae inserted into the right atrium for allowing repeated blood
sampling with minor stress were used. Table 5 shows the effect on
plasma PRL levels of muscimol (M), a psychotomimetic compound (Waser,
1967), which is a powerful GABA agonist with respect to bicuculline
sensitive postsynaptic receptors (Curtis et al., 1971). M administered at 11.30h induced a clear-cut reduction of plasma PRL starting
30 min after injection and still present at 210 min. Similar results
were obtained when M was administered at 15.30 h; its administration
was followed by a decrease in plasma PRL, which was evident at 30 min
postinjection. A comparison with plasma PRL values of control rats
showed the presence of significantly lower PRL levels in M-treated
rats at 30, 45, 60 and 90 min postinjection.

In a second series of experiments, the effect was tested of
guvacine, a potent substrate-competitive inhibitor of GABA uptake
(Johnston et al., 1975). Guvacine (40 mg/kg iv) caused a short-lived inhibition of plasma PRL levels at times (30-45 min) when a
slight reduction of spontaneous activity attributable to hyperfunct-
ioning of the GABAergic system was present (data not shown).

These data are in keeping with the findings of Schally et al.
(1977) and contrast the findings of those studies reporting a sti-

Table 5 Effect of iv administration of Muscimol on plasma PRL in intact male rats[a]

Treatment	PLASMA PRL (ng/ml)								
	$t_{-60\ min}$ (No. of rats)	t_0	$t_{15\ min}$	$t_{30\ min}$	$t_{45\ min}$	$t_{60\ min}$	$t_{90\ min}$	$t_{105\ min}$	$t_{210\ min}$
I Experiment Muscimol (2 mg/kg iv)	51.0±10.2 (7)	20.9± 2.1	20.0±2.5	12.5±0.8***	-	9.5±1.5**	7.5±1.8**	-	13.0±1.6●
II Experiment Saline	45.6±16.4 (3)	35.6± 14.1	25.3±9.3	27.3±8.4	21.6±5.3	20.3±6.1	16.0±3.1	18.0±2.0	-
Muscimol (2 mg/kg iv)	28.8± 9.3 (4)	19.8± 4.3	19.1±6.0	6.8±1.2●●	9.7±0.7●	10.4±1.9●	6.9±2.1●	11.5±7.4	-

●p < 0.05   vs t_0
●●p < 0.005 vs t_0
●●●p < 0.001 vs t_0
●p < 0.05 vs saline-injected rats.

[a]From Locatelli et al., Brain Res. (1978), by courtesy of Elsevier-North-Holland Biomedical Press.

mulatory effect of GABA on PRL release (see above). The reason for the disparity between the latter results and those of Schally et al. and ourselves is not presently clear; nevertheless, the possibility should be considered that the huge amounts of IVT-injected GABA which have been used may induce non-specific stimulation of other neurotransmitter system(s). In this context it is noteworthy that a small dose of IVT-injected GABA lowered plasma PRL levels (Libertun and Mc Cann, 1976). It must be noted, however, that, occasionally, both M and guvacine induced in our experiments early and transient rises in plasma PRL levels (data not shown). Thus, the possibility of a biphasic effect of these compounds on PRL secretion, necessitates clarification. In this context it is noteworthy that GABA antagonist drugs, i.e., picrotoxin and bicuculline (Ito et al., 1968; Curtis et al., 1971) injected systemically into normal male or spayed estrogen treated female rats, are unable to counteract the PRL-lowering effect of systemically-administered M and induce per se a clear-cut inhibition of plasma PRL (Müller et al., 1978).

Granted that GABAergic neurotransmission plays a physiologic role in the control of PRL secretion in rodents it would appear that the primary action on plasma PRL of GABA-mimetic drugs is exerted at AP level. Recent observations obtained in our laboratory have shown in fact that: 1) M given iv is capable of reducing the elevated baseline PRL levels of hypox-rats bearing an ectopic AP; 2) M impairs PRL release from AP incubated in vitro and 3) blocks the rise in plasma PRL evoked by α-MpT, a drug which blocks DA neurotransmission at CNS level (for details see Müller et al., 1978). Supporting the view of a direct AP site of action for the PRL-lowering effect of systemically administered M is the observation that less than .1% of ^{3}H-M is detectable in the hypothalamus after its iv administration (Enna et al., 1978). The recent demonstration of an intrahypothalamic GABAergic pathway which projects to the ME (Tappaz and Brownstein, 1977) might provide the neuroanatomical substrate for a direct inhibitory control over PRL secretion exerted by GABA at the level of the AP.

2. Enkephalins and Endorphins - Recent isolation and identification from porcine brain of endogenous peptides with opioid activity
i.e. enkephalins, H-Try-Gly-Gly-Phe-$\frac{Met}{Leu}$ OH (Hughes et al., 1975), and
elucidation of large peptides derived from pituitary β lipotropin,
i.e., β , α , and γ endorphins (Li and Chung, 1976; Lazarus et al.,
1976), also endowed with morphinomimetic properties, have provided
a new series of putative brain neurotransmitters and stimulated the
search for their potential neuroendocrine effects. It was known in
fact that morphine is a strong GH (Kokka et al., 1973) and PRL releaser (Ojeda et al., 1974). It is not at surprise therefore that
both enkephalins and endorphins, i.e., β endorphin, have proven to
be potent GH and PRL releasers when given either systemically or
intracerebroventricularly (Lien et al., 1976; Dupont et al., 1977;
Rivier et al., 1977; Cocchi et al., 1977). The GH and PRL responses
to both morphine, enkephalins and β endorphin are absent in vitro,
can be blocked by the specific morphine antagonist naloxone and
are present (GH) also after concomitant administration of antiserum
to somatostatin, thus suggesting an increased release of hypothalamic GRF by these substances (Labrie et al., 1977).

Despite the claim of Lien et al. (1976) that Leu^5-enkephalin
increases PRL release also from rat pituitary cell culture, an effect which is antagonized by naloxone, it must be noted that morphine
(Rivier et al., 1976), β -endorphin and Met-^{5}enkephalin (Rivier et
al., 1977) and /D-Ala/^{2}-Met-enkephalinamide (Shaar et al., 1977) were
found ineffective to stimulate PRL secretion in vitro, suggesting a
CNS site of action for these substances.

A CNS mediated mechanism for the PRL-releasing effect of opiate
peptides should encompass either a direct action on hypothalamic
regulatory hormones for PRL control i.e., PIF or PRF or the involvement of brain monoamines, notably 5-HT and/or DA, which provide the
functional link between CNS and neurosecretory neurons for PRL control (see section II A and B).

In the here-below reported study, we investigated whether or not
brain monoamine neurotransmission was involved in the stimulatory
effect of enkephalins on PRL release. In addition to Met^{2}enkephalin
(Met-enk), the most important brain enkephalin (Yang et al., 1977),
/D-Met^2-Pro^5/-enkephalinamide (EKNH$_2$), an analog resistant to proteo-
lytic degradation, possessing a long-lasting analgesic activity
(Székely et al., 1977) was used. The experimental model used, i.e.,
freely moving rats prepared with chronic indwelling jugular cannulae
has been already described in this paper. Rats received IVT injection
of Met-enk or iv injection of its analog. Fig. 6 reports data obtained in a preliminary experiment. Met-enk injected at 2 dose levels
(200 and 400 µg/rat) induced, 5 and 10 min after injection, a marked
and dose related increase in plasma PRL concentrations. EKNH$_2$ (.2
mg/kg) induced a rise in plasma PRL rather similar to that evoked
by the higher Met-enk dose (400 µg/rat).

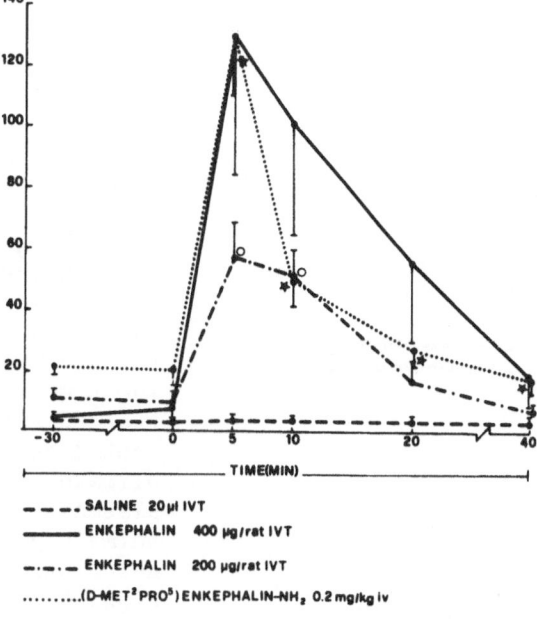

Fig. 6 Effect of Met-enkephalin and $/\overline{D}$-Met^{2}Pro$\overline{^{5}/}$-enkephalinamide on
       plasma PRL levels in the rat. Data are presented as mean +
       SEM of duplicate measurements of plasma samples from 5 to 6
       animals per group. Symbols indicate a significant difference
       vs the corresponding time interval in saline-treated controls.
       In this and the following figure for the sake of cla-
       rity, Met-enkephalin is indicated as enkephalin.

        Two drugs, Meth and MCE (see pag.9) were given for impairing
5-HT neurotransmission.

        Administration of MCE (1 mg/kg iv), 1h before, significantly
reduced the PRL-releasing effect of 200 μg of Met-enk (Fig. 7, upper
part). These results were duplicated by pretreating rats with Meth
(2.5 mg/kg iv, 1h before); in this instance the PRL-releasing effect
of Met-enk was completely abolished (data not shown). MCE (Fig. 7,
lower part) and Meth (data not shown) administered at the same doses
and schedule of the preceding experiment, were also capable of reducing significantly the increase in plasma PRL concentrations evoked
by EKNH$_{2}$ (.2 mg/kg iv).

        Although MCE and Meth share in common a major neuropharmacologic
action on the 5-HT system, as for other neuroactive drugs, their
neuronal specificity is not absolute. Thus, MCE, when administered
parenterally in the rat, is capable of exhibiting DA agonist properties at AP level on PRL secretion (Cocchi et al., 1978); for Meth
data have been obtained that suggest that the drug may act as a DA

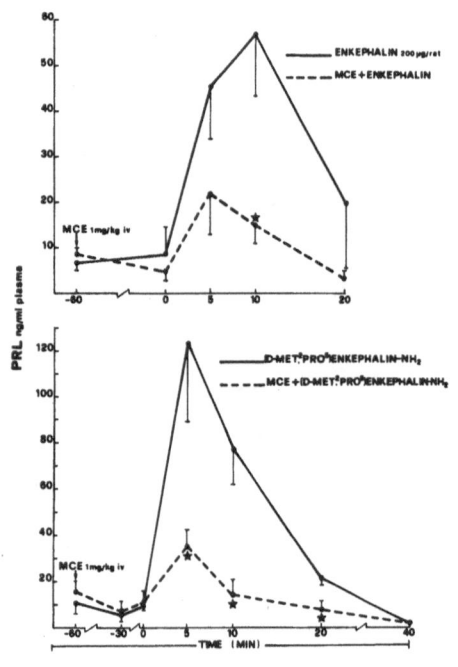

Fig. 7 Effect of pretreatment with metergoline (MCE) on the PRL-releasing effect of Met-enkephalin (upper part) or /D-Met-Pro-/-enkephalina-mide (lower part). Asterisk indicates a significant difference vs the corresponding time interval in diluent-treated rats. Data are presented as mean ± SEM of duplicate measurements of plasma samples from 5 to 6 animals per group. The same description applies also to Figs. 8, 9.

agonist following its injection into rats (Mac Leod and Login, 1977) (but see also pag.16).

However, on considering also the results obtained in our studies with the use of 5,6-DHT, a drug neurotoxic to the indoleaminergic system (Baumgarten et al., 1971), the likelihood is small that MCE and Meth counteracted enkephalin-induced PRL release via a dopaminergic mechanism (see also below). Fig. 8 shows the results of an experiment in which rats pretreated or not with 5,6-DHT (50 μg IVT, 6 days before) were given an ip injection of $EKNH_2$ (1.0 mg/kg). It is apparent that pretreatment with 5,6-DHT almost completely abolished the rise in plasma PRL evoked by the opiate.

In a following experiment the possibility was investigated that a DA-mediated mechanism might also be involved in the PRL-releasing effect of $EKNH_2$. Fig. 9 shows that administration of α-MpT (200 mg/kg ip) induced a clear-cut rise in plasma PRL; $EKNH_2$ (.2 mg/kg iv) given 30 min after α-MTp was capable to induce a rise in plasma PRL similar to that evoked in rats pretreated only with saline. These findings are not in accordance with the report of Ferland et al., (1976), who, on the basis of CA turnover studies in the ME of rats infused with Meth-enk, postulated that enkephalin-induced PRL rise is mediated at least in part by a reduction of CA, notably DA, activity in the CNS.

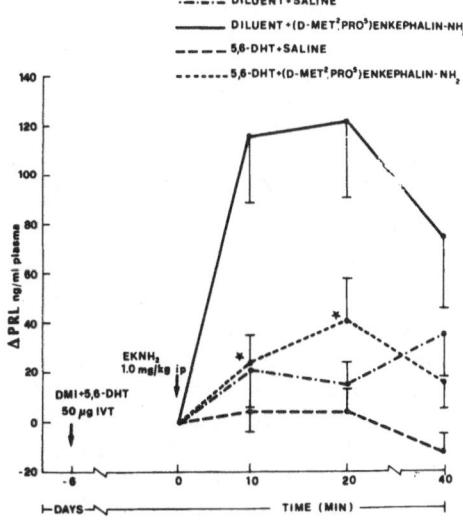

Fig. 8 Effect of pretreatment with 5,6-dihydroxytryptamine + desmethylimipramine (5,6 DHT + DMI) on the PRL releasing effect of $(D-Met^2-Pro^5)$-enkephalinamide. Data are expressed as Δ values from time 0, when the opiate peptide was administered. 7 - 8 animals per group were used. ✹$p < .05$ vs Diluent + $EKNH_2$.

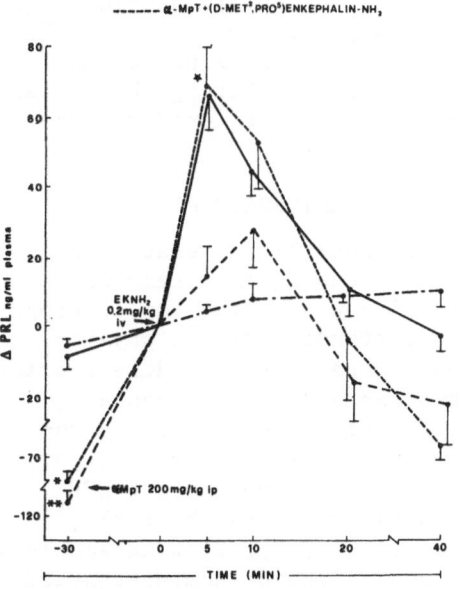

Fig. 9 Effect of pretreatment with α-metyl-p-thyrosine (α - MpT) on the PRL releasing effect of $(D-Met^2-Pro^5)$-enkephalinamide. Data are expressed as Δ values from time 0, when the opiate peptide was administered. 7, 8 animals per group were used. ✹$p < .05$ vs α-MpT + $EKNH_2$; *$p < .01$ vs Saline + $EKNH_2$; **$p < .001$ vs Saline + Saline.

Collectively, these data suggest that enkephalins exert their PRL-releasing effect likely via serotoninergic mediation, a view which is supported also by the finding that pretreatment with quipazine, a direct 5-HT agonist (Honing et al., 1969; Rodriguez et al., 1973), which per se induced a clear-cut plasma PRL rise, almost completely abolished the PRL-releasing effect of the opiate peptide (data not shown). It must be recalled, however, that unexpectedly both blockade of 5-HT biosynthesis by PCPA or impairment of 5-HT neurotransmission by a Trp-deficient diet failed to block the PRL-releasing effect of $EKNH_2$ (data not shown).

The demonstration offered by the above neuropharmacologic studies that enkephalins exert their PRL releasing effect likely via serotoninergic mediation underlines the complexity of the neural circuits involved in this neuroendocrine response. In fact a functional interaction between peptide and neurotransmitter molecules and between the latter and neurosecretory (PRF?) neurons has to be postulated. Further studies are needed to substantiate these findings and to establish the anatomical site(s) in the brain where these interactions may occur. In this context, a candidate area is likely to be the MBH as implied by the ability of the opiate antagonist naltrexone to inhibit PRL release in rats with hypothalamic deafferentation or with implants of β-endorphin in the MBH (Grandison and Guidotti, 1977).

On recalling the role played by 5-HT neurotransmission in conditions of stimulated PRL secretion e.g., stress (see Section II, B) and its mediation of the PRL releasing effect of opiate peptides (see above), one would expect that also enkephalins may play a role in the stress-induced increase in PRL secretion. The finding that naloxone is capable of completely or partially inhibiting the stress-induced rise in plasma PRL (Van Vugt et al., 1978) is evidence in favor of such view.

III. INTERACTIONS BETWEEN TRH AND DOPAMINERGIC DRUGS

The last part of this paper considers possible interactions between hypothalamic hormones, notably TRH, and dopaminergic drugs in phatologic conditions of the animal and man, and the implications of these findings with regard to the understanding of the physiophatology of GH and PRL control in human disease states. Here a brief parenthesis is needed with regard to two anomalous GH responses, originally described in acromegalic patients; 1) TRH, which is unable to elicit GH release in normal human subjects (Anderson et al., 1971; Schalch et al., 1972) is capable of evoking consistent hGH rises in more than 50% of acromegalic subjects (Schalch et al., 1972; Irie and Tsushima, 1972; Faglia et al., 1973); 2) a paradoxical lowering of hGH secretion has been induced following administration of DA-like drugs (see Müller et al., 1977a). L-DOPA, Apo or bromocriptine, consistently inhibited the elevated hGH levels in about 60% of the acromegalic studied (Fig. 10).

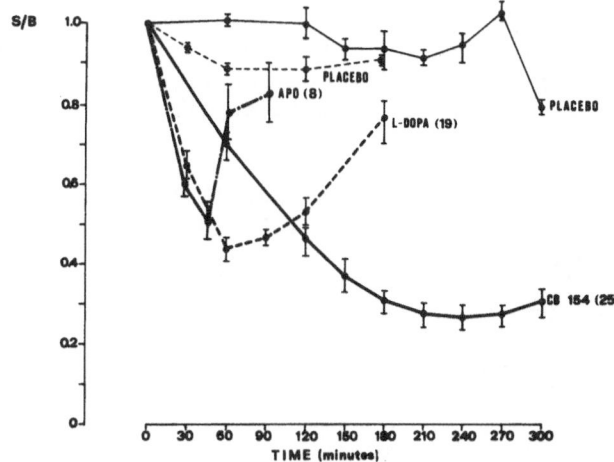

Fig. 10 Plasma hGH values in "responder" acromegalic patients after
administration of CB 154 (2.5 mg po), L-DOPA (500 mg po) or
apomorphine (APO,0.75 mg po). Values are expressed as ratio
of suppressed (S) to baseline (B) + SE. Number of patients
in parentheses. (From Liuzzi et al. "Growth Hormone and Related Peptides", 1976, by courtesy of Excerpta Medica, and
from Chiodini et al., J. Clin. Endocrinol. Metab. 1974, by
courtesy of J.B. Lippincott Co.,).

      The time course of drug action on hGH levels correlated well
with the time course of the activation of DA receptors. Thus, Apo
which induces a short-lived activation of DA receptors (Andén et al.,
1967) and prompt GH release in normal subjects (Lal et al., 1972),
induced a transient inhibition of hGH levels (Chiodini et al., 1974)
while the inhibition with CB 154 was of considerably longer duration
(Liuzzi et al., 1974a). Short-term administration of pimozide effectively counteracted in some patients the GH-lowering effect of either
L-DOPA or CB 154 (Müller et al., 1976b), thus stressing the participation of a dopaminergic mechanism in the aforesaid event.

      However, a rise in GH following TRH is not prerequisite of acrome-
galy, but is present also in other disease states, such as renal failure (Gonzales-Barcena et al., 1973), mental depression (Maeda et al.,
1975), anorexia nervosa (Maeda et al., 1976) and primary hypothyroidism (Hamada et al., 1976). Recently, a rise in GH following TRH has
been observed by us in another pathologic condition of man i.e. severe liver disease (Table 6).

      In animals, increased GH release following TRH has been report-
ed in sheep (Takahara et al., 1974b) or bovine (Machlin et al., 1974)
pituitaries in vitro; in the rat, positive GH responses to the tripeptide involve large doses of TRH infused directly into a hypohys-
eal portal vessel (Takahara et al., 1974c), perfused on hemipituita-

ries in vitro (Carlson et al., 1974), administered in vivo to hypox
rats bearing an ectopic pituitary (Udeschini et al., 1976) or into
ME-lesioned rats (Fig. 11). A clear-cut GH rise after TRH has been
observed also in the infant rat (Gil-Ad et al., 1976b).

Table 6 Individual plasma GH values (ng/ml) in 10 patients with
severe liver disease before (-30 and 0 min) and at intervals
after 400 µg TRH iv[a]

Cases	time (min)								Basal values (t_0)	Peak values	
	-30	0	15	30	45	60	90	120			
1	6.9	6.8	13.7	32.0	30.0	26.5	33.0	11.0	6.8	33.0	(90 min)*
2	2.0	5.2	2.1	5.1	11.4	17.8	14.1	6.5	5.2	17.8	(60 min)
3	0.4	2.8	8.7	2.0	1.5	2.3	1.5	0.1	2.8	8.7	(15 min)
4	8.3	7.2	8.5	10.5	7.6	8.6	1.6	2.3	7.2	10.5	(30 min)
5	8.0	7.7	14.8	28.7	10.9	5.7	2.7	4.6	7.7	28.7	(30 min)
6	0.0	0.3	0.6	0.0	3.1	0.4	5.1	10.3	0.3	10.3	(120 min)
7	1.7	6.7	16.8	11.3	10.3	7.4	9.6	6.6	6.7	16.8	(30 min)
8	0.0	1.1	1.0	0.9	0.6	1.0	13.0	21.7	1.1	21.7	(120 min)
9	1.1	0.4	0.3	0.5	1.0	0.9	4.0	3.8	0.4	4.0	(90 min)
10	1.4	0.1	0.2	1.3	1.2	4.0	10.2	11.4	0.1	11.4	(120 min)
									$\bar{x} \pm SE$ 3.8 $\pm$ 1.4	16.3 $\pm$ 2.9	
Normal subjects	-	-	-	-	-	-	-	-	$\bar{x} \pm SE$ 0.8 $\pm$ 0.4	1.9 $\pm$ 0.3	

*Time at which peaks occurred.

[a]From Panerai et al., J. Clin. Endocrinol. Metab. (1977), by courtesy of J.B. Lippincott Co.

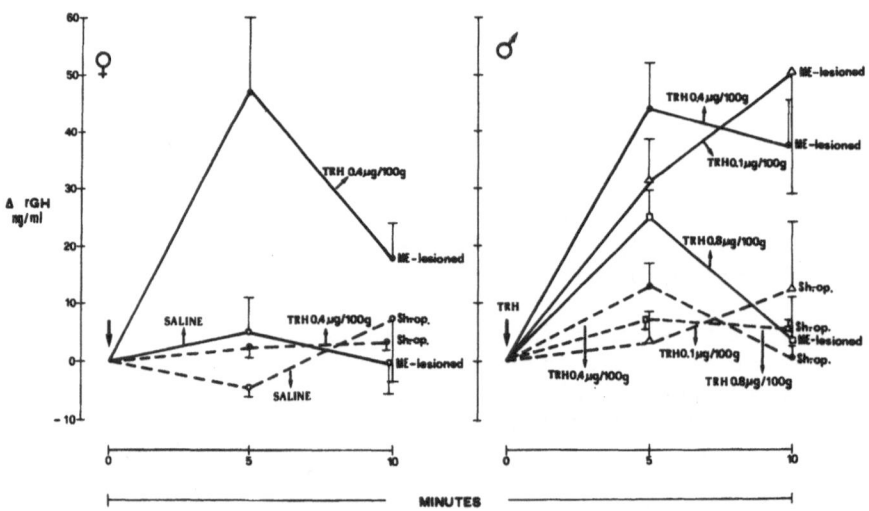

Fig. 11 Effect of TRH on plasma rGH in female (left) or male (right)
rats with ME lesions or in sham-operated controls 15 days
after surgery. Data are expressed as Δ values from baseline.
Mean + SEM (vertical lines) of 6 determinations for each
point. (From Müller et al., Endocrinology, 1977b, by courtesy of J.B. Lippincott Co.,).

Table 7 reports experimental conditions of animals and man in which TRH has shown capable of stimulating GH discharge.

Table 7   GH-releasing effect of TRH

Animal Species	Experimental Conditions	Effect	Authors
Rat (♂ S.D.)	Hypophysial portal vessels infusion	↑	Takahara et al., 1974c
(♂ S.D.)	"in vitro" perfusion	↑	Carlson et al., 1974
(♂ Wistar)	Urethane anesthetized	↑	Kato et al., 1975
(♂ Holtzman + E)	Urethane anesthetized	→	Szabo and Frohman, 1976
(♀ S.D.)	Hypox-bearing a pituitary transplant – Urethane anesthetized	↑	Udeschini et al., 1976
(♀ S.D.)	ME-lesioned-urethane anesthetized*	↑	Miller et al., 1977c
(♀ + ♂ S.D.)	Unanesthetized infant	↑	Gil-Ad et al., 1976b
Sheep	"in vivo"	↑	Takahara et al., 1974b
Cow	"in vitro"	↑	Machlin et al., 1974
	"in vitro" infusion	↑	Convey et al., 1973
Human	Acromegaly	↑	Irie and Tsushima, 1972 Schalch et al., 1972 Faglia et al., 1973
	Uremia	↑	Gonzales-Barcena et al., 1973
	Mental depression	↑	Maeda et al., 1975
	Anorexia nervosa	↑	Maeda et al., 1976
	Primary hypothyroidism	↑	Hamada et al., 1976 Collu et al., 1977
	Liver cirrhosis	↑	Panerai et al., 1977 Zanoboni and Zanoboni-Muciaccia, 1977

S.D. = Sprague Dawley   *Bilateral electrolytic lesioning

E = Estrogens   ↑ Means stimulation

→ Means no effect

Since most of the pathologic conditions of the human in which a nonspecific GH-release is evoked by TRH, have impaired GH responses to CNS-acting stimuli (Cryer and Daughaday, 1969; Sachar et al., 1976; Brauman and Gregoire, 1975), by analogy with experimental models of CNS-AP disconnection, a functional neuroendocrine disturbance has been postulated to occur in these cases (Liuzzi et al., 1976). It seemed therefore of interest to investigate whether the functional analogy between acromegaly on one side and experimental models of CNS-AP disconnection and the human diseases on the other was confined to the nonspecific GH releasing effect or TRH on may comprise also the "paradoxical" GH lowering effect of dopaminergic drugs. It is noteworthly in this context that combined application of CB 154 testing and administration of TRH has revealed most of the acromegalics to be either "responsive" to both stimuli or to none. Although rough, the distinction between "responders" and "nonresponders" was unfeasible on the basis of the more classical functional tests (Liuzzi et al., 1974b).

Therefore, the effect of dopaminergic drugs on plasma GH levels was evaluated both in patients with epathic cirrhosis and in rats with lesions of the ME. In addition, with the aim to elucidate the relationship between GH response to TRH and to dopaminergic drugs, the TRH-induced GH rise was evaluated in the cirrhotic patients, the rat with ME lesions, and in a group of acromegalis subjects, before and after administration of DA stimulant drugs. In these experimental models the interactions between TRH and dopaminergic drugs with respect to PRL secretion were also examined.

Fig. 12 shows the experiment in which the GH response to TRH or saline was evaluated in rats bearing electrolytic lesions of the ME, which did or did not receive infusion of DA. As expected, administration of TRH induced a striking rise in plasma GH at 5 and 10 min post-injection (p < .05 vs saline). Infusion of DA (100 µg /100 g BW in 25 min) did not modify the GH rise due to TRH (p < .05 vs saline at 5 and 10 min; p < .05 vs DA at 10 min). There was a slight and not significant rise in plasma GH in the later period of DA infusion. TRH induced a rise in plasma PRL at 5 and 10 min (p< .05 vs saline), which was almost completely abolished by infusion of DA (p < .05 vs TRH) (data not shown).

Administration of L-DOPA (500 mg po) in 5 male and 1 female cirrhotic patients, previously found to be responsive to TRH administration, induced a significant rise in plasma GH at 30, 40, 60 and 120 min (data not shown). Fig. 13 (upper panel) shows that in 5 cirrhotic patients, DA infusion (20 mg/60 min) did not affect either baseline GH values or the GH response to TRH. As expected, DA infusion significantly reduced plasma PRL levels and completely suppressed the PRL response to TRH (Fig. 13, lower panel).

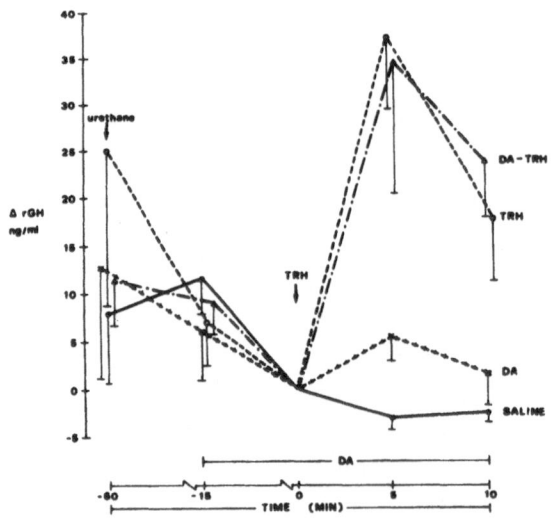

Fig. 12

Effect of TRH on plasma GH in rats bearing ME lesions infused or not with DA. Data are expressed as Δ values from T_0 (time of TRH administration). Mean + SEM (vertical lines) of 6 determinations for each point.

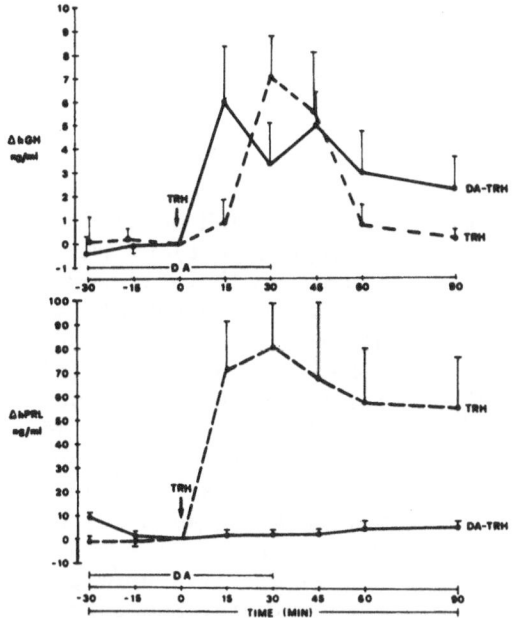

Fig. 13

Effect of TRH on plasma GH (upper part) or PRL (lower part) in 5 cirrhotic patients during saline _ _ _ _ _ or DA _____ infusion. See legend to Fig. 12 for further details.

In 12 acromegalic patients, selected out of a larger group on the basis of both GH responsiveness to TRH (increase of at least 100% over the baseline) and to acute administration of CB 154 (decrease of at least 50% below the baseline), pretreatment GH and PRL levels were 41.1 ± 14.6 ng/ml and 49.0 ± 18.3 ng/ml, respectively). Administration of TRH (200 μg iv) induced a clear-cut GH rise which was still evident 90 min later; DA infusion significantly reduced the TRH induced GH rise at 30, 60 and 90 min (p < .05) (paired t test) (Fig. 14, upper panel). The same pattern was present for the PRL response; also in this instance the TRH-induced plasma PRL rise was suppressed by DA infusion (p < .05 at 15-90 min). Differently from DA, chronic treatment with CB 154 (baseline GH and PRL levels 5.9 ± 2.3 and 5.8 ± 1.3 ng/ml, respectively) did not reduce significantly the TRH-induced GH rise, even though it suppressed the TRH-induced PRL rise (p < .05 at 15-90 min) (Fig. 14, lower panel).

In all these data do indicate that the functional analogy between experimental models of CNS-AP disconnection and human neuroendocrine disturbances on one side and acromegaly on the other side, ceases when another anomalous GH response of the latter, i.e., the "paradoxical" fall in GH after dopaminergic stimulation is considered. An iv loading of DA which is capable of suppressing the elevated plasma GH levels in acromegaly (Verde et al., 1976; Camanni et al., 1977) did not suppress GH secretion in rats with ME lesions. Similarly, subcutaneous administration of CB 154 or infusion of DA were unable to modify GH levels in hypox rats bearing one or two AP gland under the kidney capsule (unpublished results). In cirrhotic patients, the TRH-induced GH rise was concomitant to a normal

GH response to L-DOPA and no significant changes in GH levels after
DA.

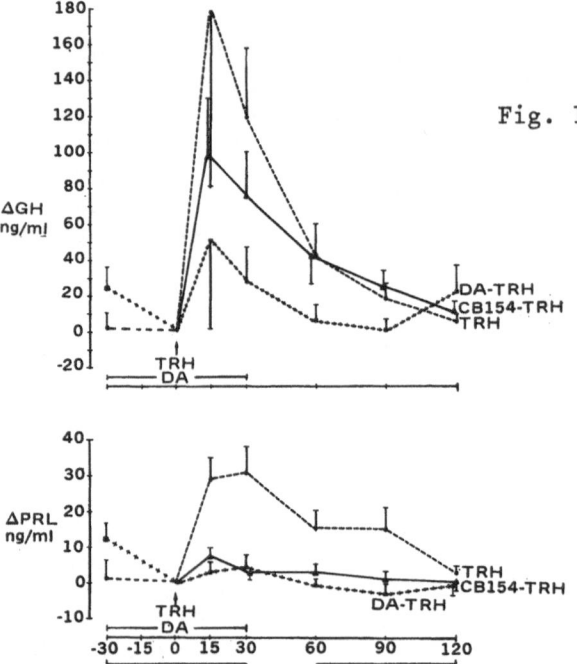

Fig. 14 Effect of TRH on plasma
GH (upper part) or PRL
(lower part) in 12 acro-
megalic subjects before or
after treatment with dopaminergic drugs. Data ref-
fering to GH and PRL va-
lues 2h after last CB 154
administration are reported in the text. GH and PRL
values at t_0 were in the
DA experiment 21.1 + 6.7
and 35.0 + 19.0 ng/ml,
respectively. See legend
to Fig. 12 for further
details.

A TRH-induced GH rise unaccompanied by an anomalous response to
dopaminergic drugs is also present in uremic (Ramirez and Bloomer,
1975), depressed (Sachar et al., 1976) and anorectic (Imura et al.,
1973) patients.

Thus, it would appear that whereas the positive GH response to
TRH is a label of pathologic conditions which share in common the
presence of an anatomical or functional interruption between the CNS
and an "intact" AP, the anomalous GH response to dopaminergic stimulation is instead a hallmark of acromegaly and likely reflects
the presence of receptors for DA located on the hyperplastic or tumoral somatotroph. In spite of the fact that evaluation of results
of these studies tends to dissociate between the anomalous GH response to TRH and that to dopaminergic stimuli, in acromegaly the
existence of a common physiophatologic trait is suggested by the
reported homogeneity (Liuzzi et al., 1974; Schwinn and Dirks, 1976;
Faglia et al., 1978). Incidentally, the homogeneity found between
GH response to TRH and to dopaminergic stimuli was taken to indicate
that in acromegaly the latter were acting directly at AP level (Liuzzi et al., 1976; Müller et al., 1977b), a view recently substantiated by the finding that bromocriptine added to human anterior pituitary tumors in culture inhibits the rate of GH release (Mashiter et
al., 1977).

There is no valid explanation for the mechanism(s) which under-lies the mutually antagonistic action of TRH and DA in acromegaly; it can be only said that likely pituitary somatotrophs comprising the tumor favor the development of or unmask receptors for both stimuli, resulting in aberrant GH responses. If the hypothesis pre-viously proposed, that DA is actually inhibitory to GH release also in the human is valid (see pag.6), the lowering effect of dopamin-ergic compounds in acromegaly should not be considered paradoxical; it would reflect only the ability of these compounds to stimulate postsynaptic pituitary DA receptors, functionally behaving as the postsynaptic DA receptors located in the CNS.

In contrast to acromegaly, the functional interaction between TRH and dopaminergic stimuli is lacking in the rat with CNS-AP disconnection and in the cirrhotic patients, being the GH rise in-duced by the tripeptide unopposed by DA pretreatment. This fact, coupled with the inability of DA to modify baseline GH levels , is consistent with the idea that there are no DA receptors for GH on an "intact" AP gland.

In our study, differently from infusion of DA, administration of bromocriptine did not induce in acromegalic subjects, a signifi-cant reduction of the TRH-induced GH elevation. This finding which is in agreement with recent reports of the literature (Sachdev et al., 1975; Ishibashi et al., 1977), is not attributable to an in-sufficient dosage of the drug being the TRH test performed in pa-tients treated chronically with and highly responsive to bromocript-ine. We do not know at present if this event implies that the mecha-nism whereby DA and CB 154 exert their GH-lowering effect in acro-megaly is different; both drugs, showed the same trend towards lower GH levels. Clarification of the different ability of DA and bromocriptine to interact with the GH-releasing action of TRH may require better biochemical and neuropharmacologic characterization of the DA and ergot drug receptors (Schmidt and Hill, 1977) on the tumourous somatotrophs. Differently from the striatum (Kebabian, 1972), at AP level the DA receptor and the-DA-sensitive AC do not appear to be one and the same. Both DA and ergot drugs are unable to stimulate pituitary AC (Clement-Cormier et al., 1977), an event which appears instead related to stimulation of PRL secretion (Na-gasawa and Yanai, 1972; Daniès et al., 1974).

ACKNOWLEDGMENTS
The skilful technical assistance of Miss Clara Frigerio, Mr.Pie-ro Crestani and Mr. Enrico Galbiati is gratefully acknowledged. The studies reported in this work have been supported in part by C.N.R. contract no. CT76.01552.04 and by Progetto Finalizzato "Biology of Reproduction", C.N.R., Rome, Italy.

REFERENCES

Andén N.-E., S.G. Butcher, H. Corrodi, K. Fuxe and U. Ungerstedt. Eur. J. Pharmacol. 11: 303-314, 1970.

Andén N.-E., A. Rubenson, K. Fuxe and T. Hökfelt. J. Pharm. Pharmacol. 19: 627-629, 1967 .

Anderson M.S., C.Y. Bowers, A.J. Kastin, D.S. Schalch, A.V. Schally, P.J. Snyder, R.D. Utiger, J.R. Wilber and A.J. Wise. New Eng. J. Med. 285: 1279, 1971.

Barrachlough C.A. and G.H. Sawyer. Endocrinology 65: 563-571, 1976.

Baumgarten H.G., A. Björklund, L. Lachenmayer, A. Nobin and U. Stenevi. Acta Physiol. Scand. Suppl. 373: 1-15, 1971.

Ben-Jonathan N., R.S. Mical and J.C. Porter. Endocrinology 96: 375-383, 1975.

Ben-Jonathan N.,C. Oliver, H.J. Weiner, R.S. Mical and J.C. Porter. Endocrinology 100: 452-458, 1977.

Benkert O., G. Laackman, Souvatzoglou and K. von Werder. J. Neural. Trans. 34: 291-299, 1973.

Bennet G.W., J.A. Edwardson, D.T. Holland, S.L. Jeffcoate and N. White. Nature 257: 323-324, 1975.

Beretta C., R. Ferrini and A.H. Glässer. Nature 207: 421-422, 1965.

Besser G.M. and C.H. Mortimer. in "Frontiers in Neuroendocrinology" (Martini L. and W.H. Ganong eds.) 227-254. Raven Press, New York.

Birge C.A., L.S. Jacobs, C.T. Hammer and W.H. Daughaday. Endocrinology 86: 120-130, 1970.

Bohnet H.G. and H.G. Friesen. in "Hypothalamus and Endocrine Functions" (Labrie F., J. Meites and G. Pelletier eds.) 257-282. Plenum Press, New York.

Brauman H. and F. Gregoire. Europ. J. Clin. Invest. 5: 289-296, 1975.

Brazeau P., W. Vale, R. Burgus, N. Ling, M. Butbher, J. Rivier and R. Guillemin. Science 179: 77-79, 1973.

Brown G.M., P. Seeman and T. Lee. Endocrinology 99: 1407-1410, 1976.

Brown W.A., M.H. Van Voert and L.M. Ambani. J. Clin. Endocrinol. Metab. 37: 463-465, 1973.

Bunney B.S. and G.K. Aghajanian. in "Pre-and Post-synaptic Receptors" (Usdin E. and W.E. Bunney eds.) 89. Marcel Dekker Inc. New York.

Caligaris L., and S. Taleisnik. J. Endocrinol. 62; 25-33, 1974.

Camanni F., F. Massara, L. Belforte and G.M. Molinatti. J. Clin. Endocrinol. Metab. 40: 363-366, 1975.

Camanni F., F. Massara, L. Belforte, A. Rosatello and G.M. Molinatti. J. Clin. Endocrinol. Metab. 44: 465-473, 1977.

Carlson H.E., I.K. Mariz and W.H. Daughaday. Endocrinology 94: 1709-1713, 1974.

Carlsson A. in "Pre-and Post-synaptic Receptors" (Usdin E. and W.E. Bunney eds.) 49. Marcel Dekker Inc., New York.

Chadwick A., S.J. Folley and C.A. Gemzell. Lancet II: 241-243, 1961.

Chambers J.W., and G.M. Brown. Endocrinology 98: 420-428, 1976.

Chen H.J., and J. Meites. Endocrinology 96: 10-14, 1975.

Chen H.J., G.P. Mueller and J. Meites. Endocrine Res. Commun.1: 283-291, 1974.

Chiodini P.G., A. Liuzzi, L. Botalla, G. Cremascoli and F. Silvestrini. J. Clin. Endocrinol. Metab. 38: 200-205, 1974.

Cioffi, L. and F. Fidanza. in "Lineamenti di Nutrizione Umana" (Fidanza F., G. Liguori and F. Mancini eds.) 50. Idelson, Naples.

Clemens J.A. in "Hypothalamus and Endocrine Functions" (Labrie, F., J. Meites and G. Pelletier eds.) 283-302. Plenum Press, New York.

Clemens J.A. and J. Meites. in "Hormonal Proteins and Peptides" (Li C.H. ed.) 139-174. Academic Press, New York.

Clement-Cormier Y.C. and G.A. Robison. Biochem. Pharmacol. 26: 1719-1722, 1977.

Clement-Cormier Y.C., J.J. Heindel and G.A. Robison. Life Sci. 21: 1357-1363, 1977.

Cocchi D., I. Gil-Ad, A.E. Panerai, V. Locatelli and E.E. Müller. Life Sci. 19: 825-836, 1976a.

Cocchi D., I. Gil-Ad, A.E. Panerai, V. Locatelli and E.E. Müller. Neuroendocrinology 24: 1-13, 1977a.

Cocchi D., A. Santagostino, I. Gil-Ad, S. Ferri and E.E. Müller. Life Sci. 20:2041-2046, 1977b.

Cocchi D., V. Locatelli, R. Carminati and E.E. Müller. Life Sci. in press, 1978.

Collu D., F. Fraschini, P. Visconti and L. Martini. Endocrinology 90: 1231-1237, 1972.

Collu R., G. Leboeuf, J. Letarte and J.R. Ducharme. J. Clin. Endocrinol. Metab. 44: 743-747, 1977.

Convey E.M., H.A. Tucher, V.G. Smith and J. Zolman. Endocrinology 92: 471-476, 1973.

Coppola J.A., R.G. Leonard , W. Lippman, J. Perrine and I. Ringler. Endocrinology 77: 485-490, 1965.

Corrodi H., K. Fuxe and T. Hökfelt. J. Pharm. Pharmacol. 19: 433-438, 1967.

Corrodi H., K. Fuxe, B. Hamberger and A. Ljungdall. Europ. J. Pharmacol. 12: 145-155, 1970.

Corrodi H., K. Fuxe, T. Hökfelt, P. Lidbrink and U. Ungerstedt. J. Pharm. Pharmacol. 25: 409-411, 1973.

Crosignani P.G., and C. Robyn. Prolactin and Human Reproduction. Academic Press, New York, 1977.

Cryer P.E. and W.H. Daughaday. J. Clin. Endocrinol. Metab. 29: 386-393, 1969.

Curtis, D.R., A.W. Duggan, D. Felix and C.A.R. Johnston. Brain Res. 32: 69-96, 1971.

Dannies P., K. Gautwik and A. Tashjian. 56th Ann. Meet. Endocr. Soc. Atlanta, Georgia Abs. 191: 1974.

Davis S.L. and M.L. Borger. Endocrinology 92: 303-309, 1973.

Di Chiara G., M.L. Porceddu, L. Vargiu, A. Argiolas and G.L. Gessa. Nature 264: 564-567, 1976.

Diefenbach W.M.P., P.W. Carmel, A.G. Frantz and M. Ferin. J. Clin. Endocrinol. Metab. 43: 638-642, 1976.

Donoso, A.O., W. Bishop, C.P. Fawcett, L. Krulich and S.M. Mc Cann. Endocrinology 89: 774-784, 1971.

Dupont A. and T.W. Redding. 57th Ann. Meet. Endocr. Soc. New York, Abs. 85: 1975.

Dupont A., L. Cusan, F. Labrie, D.H. Coy and C.H. Li. Biochem. Biophys. Res. Commun. 15: 76-82, 1977.

Durand D., J.B. Martin and P. Brazeau. Endocrinology 100: 722-728, 1977.

Enjalbert A., F. Moos, L. Carbonell, M. Priam and C. Kordon. Neuro-endocrinology 24: 174-161, 1977.

Enna S.J., J. Ferkany and P. Krogsgaard-Larsen. in "Gaba-Neurotrans-mitters" (Kofod H., P. Krogsgaard-Larsen and J. Scheel-Kruger eds.) Munksgaard, Copenhagen, in press.

Ettigi P., S. Lal, J.B. Martin and H.G. Friesen. J. Clin. Endocrinol. Metab. 40: 1094-1098, 1975.

Faglia G., P. Beck-Peccoz, C. Ferrari, P. Travaglini, B. Ambrosi and A. Spada. J. Clin. Endocrinol. Metab. 37: 338-340, 1973.

Faglia G., A. Paracchi, P. Beck-Peccoz, A. Spada and C. Ferrari. in "European Workshop on Treatment of Pituitary Adenomas" (Fahl-bush R. and K. von Werder eds.) G. Thomas, in press.

Fang S., S. Refetoff and R.L. Rosenfield. Endocrinology 95: 991-998, 1974.

Ferland L., K. Fuxe, P. Eneroth, J.-A.Gustafsson and P. Skett. Europ. J. Pharm. 43: 89-90, 1977.

Floss H.G., J.M. Cassidy and J.E. Robberts. J. Pharm. Sci. 62: 699, 1973.

Franks S., and H.S. Jacobs. in "Prolactin and Human Reproduction" (Crosignani P.G. and C. Robyn eds.) 245-258. Academic Press, New York, 1977.

Fuxe K. Z. Zellforsch 65: 573-596, 1965.

Fuxe K. and T. Hökfelt. in "Frontiers in Neuroendocrinology" (Martini L. and W.H. Ganong eds.) 47-96. Oxford University Press, New York, 1969.

Fuxe K., T. Hökfelt, G. Jonsson and A. Löfstrom. in "Neurosecretion, The final Neuroendocrine Pathway" (Knowles F. and L. Wollrath eds.) 269-275. Springer Verlag, Berlin, 1974.

Gal E.M. and P.A. Drewes. Proc. Soc. Exp. Biol. Med. 110: 369-371, 1962.

Gala R.R., J.A. Peters, D.R. Piper and M.D. Campbell. Life Sci. 22: 25-30, 1978.

Gallo R.V., J.Rabii and G.P. Möberg. Endocrinology 97: 1096-1105, 1975.

Ganong W.F. in "Hypothalamic Hormones" (Motta M., P.G. Crosignani and L. Martini eds.) 261-273. Academic Press, New York, 1975.

Gil-Ad I., F. Zambotti, M.O. Carruba, L. Vicentini and E.E. Müller. Proc. Soc. Exp. Biol. Med. 151: 512-518, 1976a.

Gil-Ad I., D. Cocchi, A.E. Panerai, V. Locatelli, P. Mantegazza, and E.E. Müller. Neuroendocrinology 21: 366-378, 1976b.

Gonzales-Barcena E., A.J. Kastin, D.S. Schalch, M. Torres-Zamora, E. Perez-Pasten, A. Kato and A.V. Schally. J. Clin. Endocrinol. Metab. 36: 117-120, 1973.

Gräf K.J., R. Horowski and M.F. El-Etreby. Acta Endocrinol (Kbh)85: 267-278, 1977.

Gräf K.J., F. Neumann and R. Horowski. Endocrinology 98: 598-605, 1976.

Grandison L. and A. Guidotti. Nature 270: 357-357, 1977.

Geibrokk T., J. Hansen, R. Knudsen, Y.K. Lam and F. Folkers. Biochem. Biophys. Res. Commun. 67: 338-344, 1975.

Gudelski G.A.and K.E. Moore. J. Neural. Trans. 38: 95-105, 1976.

Gudelski G.A., J. Simpkins, G.P. Mueller, J. Meites and K.E. Moore. Neuroendocrinology 22: 206-215, 1977.

Hamada N., K. Uoi, Y. Nishizawa, T. Okamoto, K. Hasegawa, H. Morü and M. Wada. Endocrinol. Jap. 23: 5-10, 1976.

Handwerger S., J.W. Plonk, H.E. Lebovitz, C.H. Bivens and J.M. Feldman. Horm. Res. 7: 214-217, 1975.

Hidaka H. in "Frontiers in Catecholamine Research" (Usdin E.and S. Snyder eds.) 87-90. Pergamon Press, Oxford, 1973.

Hidaka H., A. Nagasawa and A. Takeda. J. Clin. Endocrinol. Metab. 37: 145-147, 1973.

Honing E., F.L. Sancilio, R. Vargas and E.G. Pardo. Europ. J. Pharmacol. 6: 274-280, 1969.

Hughes J., T.W. Smith, H.W. Kosterlitz, L.A. Fothergill, B.A. Morgan and H.R/ Morris. Nature 258: 577-579, 1975.

Illner P., R.A. Steiner and C.C. Gale. Fed . Proc. Am. Soc. Exp. Biol. 35: Abs 3171, 1976.

Illner P., R.A. Steiner and C.C. Gale. A. J. Physiol. in press, 1977.

Imura H., Y. Nakai, S. Matsukara and H. Matsuyama. Horm. Metab. 5: 41-45, 1973.

Irie M., and T. Tsushima. J. Clin. Endocrinol. Metab. 35: 97-100, 1972.

Ishibashi M., T. Yamaji and K. Kosaka. J. Clin. Endocrinol. Metab. 45: 275-279, 1977.

Ito M., S.M. Highstein and T. Tsuchiya. Brain. Res. 17: 520-523, 1968.

Iversen L.L. The uptake and storage of noradrenaline in sympathetic nerves. Cambridge, Univ. Press, London, 1967.

Jacoby J.H?, M. Greenstein, J.F. Sassin and E.D. Weitzman. Neuroendocrinology 14: 95-102, 1974.

Johston G.A.R., P. Krogsgaard-Larsen and A. Stephanson. Nature 258: 627-628, 1975.

Kamberi I.A., R.S. Mical and J.C. Porter . Experientia 26: 1150-1151, 1970.

Kamberi I.A., R.S. Mical and J.C. Porter. Endocrinology 88: 1012-1020, 1971a.

Kamberi I.A., R.S. Mical and J.C. Porter. Endocrinology 88: 1288-1293, 1971b.

Kansal P.C., J. Buse, O.R. Talbert and M.G. Buse. J. Clin. Endocrinol. Metab. 34: 99-105, 1972.

Kato Y., K. Chihara, K. Maeda, S. Ohgo, Y. Okanishi and H. Imura. Endocrinology 96: 1114-1118, 1975.

Kato Y., J. Dupré and J.C. Beck. Endocrinology 93: 135-146, 1973.

Kato Y., Y. Nakai, H. Imura, K. Chihara and S. Ohgo. J. Clin. Endoc-
    rinol. Metab. 38: 695-697, 1974.
Kebabian J.W. The Mammalian Nervous Systems. Ph. D. Thesis, Yale
    University, 1972.
Keohg H.J., R.H. Johnson, R.N. Nanda and W.R. Sulaiman. J. Neurol.
    Neurosurg. Psychiatry 39: 244-248, 1976.
Kim J.S., L. Sherman, H.D. Kolodny, F. Benjamin and A. Singh. Clin.
    Res. 15: 718, 1971.
Kimura H. and R.M. Mc Leod. Abstr. Book Int. Congr. Pharmacol. 6th.
    88, 1975.
Kokka N., J.F. Garcia and H.W. Elliot. Prog. Brain Res. 39: 347-358,
    1973.
Kordon C., C.A. Blake, J. Terkel and C.H. Sawyer. Neuroendocrinology
    13: 213-223, 1973-1974.
Krulich L. Life Sci. 17: 1141-1144, 1975.
Labrie F., L. Cusan, L. Ferland, A. Dupont, C.H. Li, D.H. Coy, A.
    Arimura and A.V. Schally. Proc. Natl. Acad. Sci. U.S.A., in
    press, 1978.
Lal S., C.E. de la Vega, T.L. Sourkes and H.G. Friesen. Lancet II:
    661, 1972.
Lal S., C.E. de la Vega, T.L. Sourkes and H.G. Friesen. J. Clin.
    Endocrinol. Metab. 37: 719-724, 1973.
Lal S. , G. Tolis, J.B. Martin, G.M. Brown and H. Guyda. J. Clin.
    Endocrinol. Metab. 41: 827-832, 1975.
Lancranjan J. and P. Marbach. Metabolism 26: 1225-1230, 1977.
Lancranjan J., A. Wirz-Justice, W. Pühringer and E. Del Pozo. J.
    Clin. Endocrinol. Metab. 45: 588-593, 1977.
Lawson D.M. and R.R. Gala. Endocrinology 96: 313-318, 1975.
Lazarus L.H., A. Ling and R. Guillemin. Proc. Natl. Acad. Sci. U.S.A.,
    73: 2156-2159, 1976.
Li C.H. and D. Chung. Proc. Natl. Acad. Sci. U.S.A., 73: 1145-1148,
    1976.
Libertun C. and S.M. Mc Cann. IRCS 4: 734, 1976.
Lien E.L., R.L. Fenichel, V. Garsky, D. Sarantakis and N.H. Grant.
    Life Sci. 19: 837-840, 1976.
Liuzzi A., P.G. Chiodini, L. Botalla, G. Cremascoli, E.E. Müller and
    F. Silvestrini. J. Clin. Endocrinol. Metab. 38: 910-913, 1974a.
Liuzzi A., P.G. Chiodini, L. Botalla, F. Silvestrini and E.E. Müller.
    J. Clin. Endocrinol. Metab. 39: 871-876, 1974b.
Liuzzi A., A.E. Panerai, P.G. Chiodini,C. Secchi, D. Cocchi, L. Bo-
    talla, F. Silvestrini and E.E. Müller. in "Growth Hormone and
    Related Peptides" (Pecile A. and E.E. Müller eds.) 236-251.
    Excerpta Medica, Amsterdam, 1976.
Lyons W.R., C.H. Li and R.E. Johnson. Program 43rd Meet. Endocr. Soc.
    4. Thomas, Springfield Ill.
Locatelli V., D. Cocchi, G. Racagni, F. Cattabeni, A. Maggi, P.
    Krogsgaard-Larsen and E.E. Müller. Brain Res. 145: 173-179, 1978.
Lovinger R., J. Holland, S. Kaplan, M.M. Grumbach, A.T Boryczka, R.
    Shackelford, J. Salmon, I.A. Reid and W.F. Gannong. Neuroscience
    1: 443-450, 1976.

Lu H.K. and J. Meites. Proc. Soc. Exp. Biol. Med. 137: 480-483, 1971.

Lu H.K. and J. Meites. Endocrinology 91: 868-872, 1972.

Lu H.K. and J. Meites. Endocrinology 93: 152-155, 1973.

Lu H.K., H.T. Chen, H.H. Huang, L. Grandison, S. Marshall and J. Meites. J. Endocrinol. 68: 241-250, 1976.

Mc Indoe J.H. and R.V. Turkington. J. Clin. Invest. 52: 1972-1978, 1973.

Mac Leod R.M. and J.E. Lehemeyer.in "Lactogenic Hormones" Ciba Found. Symp. (Wolstenholeme G.E. and J. Knight eds.) 53-76. Churchill Livingstone, Edinburgh, 1972.

Mac Leod R.M. and J.E. Lehemeyer. Endocrinology 94: 1077-1085, 1974.

Mac Leod R.M. and I.S. Login. in "Nonstriatal Dopaminergic Neurons" (Costa E. and G.L. Gessa eds.) Adv. in Biochem. Psychoparm. 16: 147-157, Raven Press, New York, 1977.

Mac Leod R.M., H. Kimura and I. Login. in "Growth Hormone and Related Peptides" (Pecile A. and E.E. Müller eds.) 433-453. Excerpta Medica, Amsterdam, 1976.

Machlin L.S., L.S. Jacobs, U. Cirulis, R. Kimes and R. Miller. Endocrinology 95: 1350-1358, 1974.

Maeda K., Y. Kato, S. Ohgo, K. Chihara, Y. Yoshimoto, N. Yamaguchi, S. Kuromaru and H. Imura. J. Clin. Endocrinol. Metab. 40: 501-509, 1975.

Maeda K., Y. Kato, N. Yamaguchi, K. Chihara, S. Ohgo, Y. Okanishi, Y. Yoshimoto, K. Moridera, S. Kuromaru and H. Imura. Acta Endocrinol. (Kbh) 81: 1-8, 1976.

Marchlewska-Koy A. and L. Krulich. Fed. Proc. Fed. Am. Soc. Exp. Biol. 34: 252, 1975.

Martin J.B., G. Tannenbaum, J.O. Willoughby, L.P. Renaud and P. Brazeau. in "Hypothalamic Hormones " (Motta M., P.G. Crosignanai and L. Martini eds.) 217-235. Academic Press, New York, 1975.

Mashiter K. E. Adams, M. Beard and A. Holley. Lancet II: 197-198, 1977.

Meites J. in "Human Prolactin" (Pasteels J.L. and C. Robyn eds.) 105-118. Excerpta Medica, Amsterdam, 1973.

Meites J. in "Nonstriatal Dopaminergic Neurons" (Costa E. and G.L. Gessa eds.) Adv. in Biochem. Psychopharm. 16: 149-156. Raven Press, New York, 1977.

Meites J., H.K. Lu, W. Wuttke, G.W. Welsch, H. Nagasawa and S.K. Quadri. Recent Prog. Horm. Res. 28: 471-516, 1972.

Menon M.K., R.M. Fleming and W.G. Clark. Biochem. Pharmacol. 23: 879-885, 1974.

Mioduszewski R., L. Grandison and J. Meites. Proc. Soc. Exp. Biol. Med. 151: 44-46, 1976.

Molinatti G.M., F. Massara, E. Strumia, F. Pennisi, G.A. Scassellati and L. Vancheri. J. Nucl. Biol. Med. 13: 26-31, 1969.

Mueller G.P., J. Simpkins, J. Meites and K.E. Moore. Neuroendocrinology 20: 121-135, 1976a.

Mueller G.P., C.P. Twohy, H.T. Chen, J.P. Advis and J. Meites. Life Sci. 18: 715-724, 1976b.

Müller E.E., P. Dal Prà and A. Pecile. Endocrinology 83: 893-896, 1968.

Müller E.E., G. Nisticò and U. Scapagnini. in "Neurotransmitters and Anterior Pituitary Functio" Academic Press, New York, 1978.

Müller E.E., F. Brambilla, F. Cavagnini, M.Peracchi and A.E. Panerai. J. Clin. Endocrinol. Metab. 39: 1-5, 1974.

Müller E.E., D. Cocchi, H. Jalanbo and G. Udeschini. Endocrinology 92: A248, 1973.

Müller E.E., G. Udeschini, C. Secchi, F. Zambotti, L. Vicentini, A.E. Panerai, F. Cocola and P. Mantegazza. Acta Endocrinol. 82: 71-91, 1976a.

Müller E.E., I. Gil-Ad, A.E. Panerai, C. Secchi, A. Liuzzi, P.G. Chiodini and F. Silvestrini.in "The Basic and Clinical Uses of Hypothalamic Hormones" (Charro Salgado A.L., R. Fernadez Durango and G. Lopéz del Campo eds.) 211-231. Excerpta Medica, Amsterdam, 1976b.

Müller E.E., A. Liuzzi, D. Cocchi, A.E. Panerai, F. Oppizzi, V. Locatelli, F. Silvestrini,P. Mantegazza and P.G. Chiodini. in "Nonstriatal Dopaminergic Neurons" (Costa E. and G.L. Gessa eds.) Adv. in Biochem. Psychopharm. 16: 127-138. Raven Press, New York, 1977a.

Müller E.E., A.E. Panerai, D. Cocchi, I. Gil-Ad, G.L. Rossi and W.R. Olgiati. Endocrinology 100: 1663-1671, 1977b.

Müller E.E., D. Cocchi, V. Locatelli, P. Krogsgaard-Larsen, F. Cattabeni and G. Racagni. in "Gaba-Neurotransmitters" (Kofod H., P. Krogsgaard-Larsen and J. Scheel-Krüger eds.) Munksgaard, Copenhagen, in press, 1978.

Nagasawa H. and R. Yanai. J. Endocrinol. 55: 215-216, 1972.

Nencioni T., A. Miragoli, F. Dorato and F. Polvani. J. Endocrinol. Invest. 1: 65-67, 1978.

Niswender G.D., C.L. Chen, A.R. Midgley Jr., J. Meites and S. Ellis. Proc. Soc. Exp. Biol. Med. 130: 793-797, 1969.

Niswender G.D., A.R. Midgley Jr., S.E. Monroe and L.E. Reichert. Proc. Soc. Exp. Biol. Med. 128: 807-811, 1968.

Ng L.K.Y., T.N. Chase, R.W. Colburn and I.J. Kopin. Brain Res. 45: 499-505, 1972.

Ojeda S.R., P.G. Harms and S.M. Mc Cann. Endorinology 95: 1694-1703, 1974.

Ondo J.P. and K.A. Pass. Endocrinology 98: 1248-1252, 1976.

Panerai A.E., F. Salerno, M. Manneschi, D. Cocchi and E.E. Müller. J. Clin. Endocrinol. Metab. 45: 134-140, 1977.

Pasteels J.L. and C. Robyn (eds.) Human Prolactin Excerpta Medica, Amsterdam, 1973.

Pecile A. and E.E. Müller (eds.) Growth Hormone, Excerpta Medica, Amsterdam, 1972.

Pecile A. and E.E. Müller (eds.) Growth Hormone and Related Peptides, Excerpta Medica, Amsterdam 1976.

Ramirez G. and H.A. Bloomer. Abs. 8th Ann. Meet. Am. Soc. Nephrology, Washington, 21, 1975.

Rivera E.M., I.A. Forsyth and S.J. Folley. Lactogenic activity of mammalian growth hormones. Proc. Soc. Exp. Biol. Med. 124: 859-865, 1967.

Rivier C., M. Brown and W. Vale. 58th Ann. Meet. Endocr. Soc. S. Francisco, Calif. Abs. 126, 1976.

Rivier C., W. Vale, N. Ling, M. Brown and R. Guillemin. Endocrinology 100: 238-241, 1977.

Rodriguez R., J.A. Rojas-Ramirez and R.R. Drucker Colin. Europ. J. Pharmacol. 24: 164-171, 1973.

Rotsztejn W.G., J.L. Charli, E. Patton, J. Epelbaum and C. Kordon. Endocrinology 99: 1663-1666, 1976.

Ruch W., A.L. Jaton, B. Bucher, P. Marbach and W. Doepfner. Experientia 32: 529-531, 1976.

Ruch W., D.J. Koerker, M. Carino, D.S. Johnsen, B.R. Webster, J.W. Ensink, C.J. Goodner and C.C. Gale. in "Advances in Human Growth Hormone Research" (Raiti S. ed.) 271-289. DHEW Publication No. (NIH), 1973.

Ruch W., R.C. Mixter, R.M. Russel, J.F. Garcia and C.C. Gale. Am. J. Physiol. 233: E61, 1977.

Sachar E.J., P.H. Rolffwarg, P.N. Gruen, N. Altman and J. Sassin. Pharmakopsych. 9: 11-20, 1976.

Sachdev Y., W.M.G. Turnbridge, D.R. Weightman, A. Gomez-Pan, A. Duns and R. Hall. Lancet II: 1164-1168, 1975.

Sawyer C.H., J. Hilliard, S. Kanematsu, R. Scaramuzzi and C.A. Blake. Neuroendocrinology 15: 328-337, 1974.

Shaar C.J. and A.J. Clemens. Fed. Proc. Fed. Am. Soc. Exp. Biol. 35: Abs. 546, 1976.

Shaar C.J., R.C.A. Frederickson, N.B. Dininger, J.A. Clemens and R.H. Hull. Life Sci. 21: 853-860, 1977.

Schalch D.S. and S. Reichlin. Endocrinology 79: 275-280, 1966.

Schalch D.S. and S. Reichlin. in "Growth Hormone" (Pecile A. and E.E. Müller eds.) 211-225. Excerpta Medica, Amsterdam, 1968.

Schalch D.S., D. Gonzales-Barcena, A.J. Kastin; A.V. Schally and L.A. Lee. J. Clin. Endocrinol. Metab. 35: 609-615, 1972.

Schally A.V., T.W. Redding, A. Arimura, A. Dupont and G.L. Linthicum. Endocrinology 100: 681-691, 1977.

Sherman G.P., S. Kim, F. Benjamin and K.D. Kolodny. New Engl. J. Med. 384: 72-74, 1971.

Schmidt M.J. and L.E. Hill. Life Sci. 20: 789-797, 1977.

Shome B. and A.F. Parlow. J. Clin. Endocrinol. Metab. 45: 1112-1115, 1977.

Schwinn G. and H. Dirks. Abs. V Int. Congr. Endocrinol. Hamburg, Abs. 420, 1976.

Smalstig E.B., B.D. Sawyer and J.A. Clemens. Endocrinology 95: 123-129, 1974.

Smythe G.A., P.J. Compton and L. Lazarus. in "Growth Hormone and Related Peptides" (Pecile A. and E.E. Müller eds.) 222-235. Excerpta Medica, Amsterdam, 1976.

Steiner R.A., P. Illner, A.D. Rolfs, P.T.K. Toivola and C.C. Gale. Am. J. Physiol., in press, 1977.

Szabo M. and L.A. Frohman. Endocrinology 96: 955-961, 1975.

Székely S.I., A.Z. Ronai, Z. Dunai-Kovacs, E. Miglécz, I. Berzétri, S. Bajusz and L. Gräf. Europ. J. Pharmacol. 43: 293-294, 1977.

Tagliamonte A., P. Tagliamonte, J. Perez-Cruet, S. Stern and G.L. Gessa. J. Pharmacol. Exp. Ther. 177: 475-480, 1971.

Thorner M.O., J.A.H. Wass, A. Jones, S.R. Bloom and R.M. Mc Leod. Abs. Book Int. Cong. Endocrinol. 5th, Abs. 860, 1976.

Takahara J., A. Arimura and A.V. Schally. Endocrinology 95: 462-465, 1974a.

Takahara J., A. Arimura and A.V. Schally. Endocrinology 95: 1490-1494 1974b.

Takahara J., A. Arimura and A.V. Schally. Proc. Soc. Exp. Biol. Med. 146: 831-835, 1974c.

Takahara J., S. Yunoki, W. Yakushji, J. Yamaguchi, Y. Yamane and T. Ofuji. J. Clin. Endocrinol. Metab. 44: 1014-1017, 1977.

Takahashi K., T. Tsushima and M. Irie. Endocrinol. Japan Study, 323, 1973.

Tannembaum G.S. and J.B. Martin. Endocrinology 98: 562-570, 1976.

Tappaz M.L. and M.J. Brownstein. Brain Res. 132: 371-379, 1977.

Toivola P.T.K. and C.C. Gale. Neuroendocrinology 6: 210-219, 1970.

Toivola P.T.K. and C.C. Gale. Endocrinology 90: 895-902, 1972.

Toivola P.T.K., C.C. Gale, C.J. Goodner and J.H. Werrbach. Hormones 3: 193-213, 1972.

Udeschini G., D. Cocchi, A.E. Panerai, I. Gil-Ad, I. Rossi, P.G. Chiodini, A. Liuzzi and E.E. Müller. Endocrinology 98: 807-814, 1976.

Valverde R.C., C.V. Chieffo and S. Reichlin. Life Sci. 12: 327-335, 1973.

Van Maanen T.H. and P.G. Smelik. Neuroendocrinology 3: 177-186, 1968.

Van Vugt D.A., J.F. Bruni and J. Meites. Life Sci. 22: 85-90, 1978.

Verde G., G. Oppizzi, G.Colussi, G. Cremascoli, L. Botalla, E.E. Müller, F. Silvestrini, P.G. Chiodini and A. Liuzzi. Clin. Endocrinol. 5: 419-423, 1976.

Waser P.G. in "Ethnopharmacologic Search for Psychoactive Drugs" (Efron D.H., B. Holmsted and N.S. Kline eds.) U.S. 419-439. Public Health Service Publ. No. 1645, Washington, 1967.

Weiner R.J. in "Hypothalamic Hormones" (Motta M., P.G. Crosignani and L. Martini eds.) 249-253. Academic Press, New York, 1975.

Werrbach J.H., C.C. Gale, C.J. Goodner and M.J. Conway. Endocrinology 86: 77-82, 1970.

Wong D.T., F.P. Bymaster, J.S. Hong and B.B. Molloy. J. Pharmac. Exp. Ther. 193: 804-811, 1975.

Woolf P.D., L.S. Jacobs, R. Donofrio, S.Z. Burday and D.S. Schalch. J. Clin. Endocrinol. Metab. 38: 71-75, 1974.

Wurtman, R.J. in "Frontiers in Neurology and Neuroscience Research" (Seeman P. and G.M. Brown eds.) 16-25. University of Toronto, Press, Toronto, 1974.

Wuttke W., E. Casell and J. Meites. Endocrinology 88: 738-741, 1971.

Yang H.Y., J.S. Hong and E. Costa. Neuropharmacology 16: 303-307,1977.

Zanoboni A. and W. Zanoboni-Muciaccia. J. Clin. Endocrinol. Metab. 45: 576-578, 1977.

CLINICAL NEUROENDOCRINE RELATIONSHIPS IN NORMAL AND DISORDERED

PROLACTIN SECRETION

G.M. Besser, T. Yeo, G. Delitala, Ann Jones, W.A. Stubbs,
J.A.H. Wass, and M.O. Thorner

Department of Endocrinology, St. Bartholomew's Hospital,
London, EC1A 7BE, England.

Fortunately the establishment of the existence of prolactin
in man coincided with the clarification the neurohumoral control
of prolactin secretion in subhuman species in the decade around
1970 (see review by Thorner, 1977); this allowed the rapid extension of these studies into the clinical sphere. In general terms
the mechanisms controlling secretion of prolactin appear similar
in man and other mammals and this has allowed the successful exploitation of basic neuroendocrine advances to the development of drugs
of great value for the clinical control of the common hyperprolactinaemic states and these will be reviewed here.

NORMAL CONTROL OF PROLACTIN SECRETION

Unlike the other anterior pituitary hormones, in man as in
the rat prolactin secretion is under tonic inhibitory control.
If the connection between the hypothalamus and the pituitary is
interrupted then secretion of all the anterior pituitary hormones
is reduced except for prolactin which is then secreted in excess.
If after attempted hypophysectomy a few normal pituitary cells
remain within the fossa then prolactin secretion continues, and
may indeed be excessive, despite deficiency of all other hypophyseal
secretions. It is clear that the hypothalamus exerts an inhibitory effect on the release of prolactin under normal basal conditions,
and it is presumed that increased secretion usually occurs as
a result of diminution of the hypothalamic inhibitory influence.
However, known factors will increase prolactin release by direct
actions on the pituitary. Precise relationships between prolactin
releasing and prolactin-release inhibiting influences on the hypothalamus are not finally decided.

457

Prolactin in man is secreted in a pulsatile fashion without fixed periodicity. Variations in the overall secretory rate are associated with stress, suckling, coitus and falling asleep; levels fall during the periods of sleep associated with rapid eye movements, and also during the early morning period of the waking day. Although studies in rats and man have suggested that serotonin may be involved in at least some of these physiological stimuli to prolactin release (Kamberi, et al, 1971, Turkington and MacIndoe, 1972, Kordon et al, 1973, 1974, Gallo et al, 1975, Mendelson et al, 1975, Kato et al, 1974), considerable doubt must exist about the validity of some of these experiments since the compounds used to block receptors to serotonin such as methysergide can have variable actions on the other neurotransmitters depending upon the circumstances and concentrations used; thus the same agent may be a serotonin antagonist, dopamine agonist or dopamine receptor blocker (MacLeod, 1977).

Oestrogens seem to interact with the neurotransmitter and neurohumoral control of prolactin release to increase prolactin secretion. Thus in the foetus, at puberty in girls and particularly during pregnancy, prolactin levels rise. In addition the normal human pituitary almost doubles its size during pregnancy as a result of a dramatic increase in the number of prolactin secreting cells, and this too is an effect of oestradiol.

Factors Increasing Prolactin Secretion

Thyrotrophin releasing hormone (TRH): TRH releases prolactin as well as TSH in man and in other animals (Bowers et al, 1971, Jacobs et al, 1971, Noel et al, 1974, Mortimer, et al, 1974). Most authors believe this to be of doubtful physiological significance and that it is unlikely that TRH acts as a physiological prolactin releasing factor. Evidence cited in support of this view includes the dissociation which can be seen between TSH and prolactin secretion as during suckling or the circadian patterns of hormone release and prolactin releasing activity which can be found in hypothalamic extracts separate from thyrotrophin releasing activity (Labella et al, 1972, Valverde-R et al, 1972, Frohman and Szabo, 1975). However, the responses of prolactin as well as TSH are suppressed after TRH administration in thyrotoxicosis and are usually excessive in hypothyroidism. Indeed in about a third of patients with hypothyroidism prolactin levels are high and slowly fall to normal with thyroxine replacement albeit more slowly than the TSH. The question of the relevance of TRH to normal physiological control of prolactin release must remain open since it need not be assumed, necessarily, that if TRH does play a role in prolactin release then TSH and prolactin should be simultaneously secreted. Presumably release of TSH or prolactin in response to TRH must represent an interaction at the pituitary receptors on the thyrotroph and mammotroph cells of TRH with circulating neurotransmitter substances, and the feed-back effects of thyroid hormones and possibly also prolactin. These variables will interact independently with the effects of the incident TRH and the effects of any hypothalamic inhibiting factors such as the prolactin inhibiting factor and somatostatin which suppresses TSH with little effect on prolactin.

Enkephalin: Recently it has become clear that there may be another stimulatory factor of hypothalamic origin involved in the control of prolactin release, enkephalin. This pentapeptide identical with a sequence within β-endorphin and β-lipotrophin has profound endocrine and metabolic effects in man. Enkephalin itself is very short lived in the circulation and cannot be used effectively for this reason. We have therefore administered an analogue we refer to as DAME which contains D-alanyl and methylphenylalanyl substitutions in the amino acids 2 and 4, and a sulphoxyl substitution in the methionine of met-enkephalin (Sandoz, FK 33-824). 250μg administered as a 2.5 minute infusion into normal volunteers caused a rapid rise in prolactin which fell again to normal in about 3 hours. This effect was blunted by pre-treatment with the opiate receptor antagonist naloxone. In addition growth hormone secretion was stimulated and LH and FSH inhibited. The effect of the enkephalin analogue appears not to be directly on the pituitary cells since we have been unable to demonstrate any action when the enkephalin analogue is added to the medium perfusing columns of isolated rat pituitary cells. The physiological or pathophysiological significance of enkephalin in normal or disordered prolactin secretion is not yet clear but it is of interest that naloxone alone at current therapeutic doses does not lower elevated prolactin levels in patients with prolactin secreting tumours but that in a heroin addict who had received a dose of heroin one hour before, DAME had no effect on growth hormone or prolactin suggesting that the heroin had already occupied common receptors involved in enkephalin-mediated adenohypophyseal hormone response.

Prolactin Release Inhibiting Factor (PIF)

Data in man are consistent with the more precise evidence in rats. Dopamine and dopamine agonists can act directly at the pituitary level to inhibit prolactin secretion (animal data reveiwed by Thorner, 1977). Thus infusions of dopamine itself or L-dopa, or administration of dopamine agonists such as piribedil, or the ergot derivatives bromocriptine, lergotrile and lisuride all reduce circulating prolactin levels whether these are normal or elevated; dopamine receptor blocking agents such as metoclopramide administered with them inhibit their action. Direct evidence in the human for their action at the pituitary level is not secure but is suggested by animal data. However, dopamine itself does not cross the blood brain barrier and can only be available to act on mechanisms either in the pituitary or the median eminence.

Since this agent lowers prolactin in the human rapidly and very
effectively, this may be taken as pointing to similarlity of
the actions of the dopamine agonists with those in the rat.   In
this species it can clearly be shown that dopamine will rapidly
and reversibly reduce prolactin secretion by a direct action on
isolated pituitary cells (Figure 1; review Thorner, 1977).   The
effects of many drugs which raise prolactin levels can be understood only in terms of their action on dopamine or pituitary dopamine receptors. While evidence exists for additional non-dopamine
prolactin inhibiting factors and these may well be present (Dupont
and Redding, 1975; Enjalbert et al, 1977), there are no data which
suggest this is necessarily true for man.

DRUGS WHICH RAISE CIRCULATING PROLACTIN LEVELS IN MAN

    Drugs are among the most common causes of hyperprolactinaemia
in man.   They may be classified into three main groups: the dopamine
receptor blocking drugs, the dopamine depleting drugs and those
that do not clearly act through dopaminergic system (Table 1).

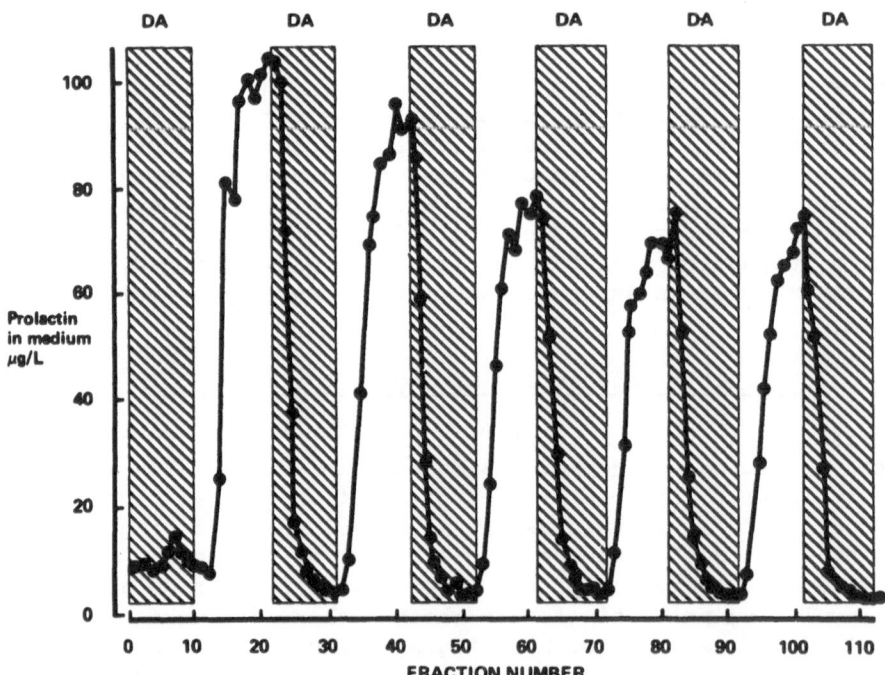

Figure 1   Effect of 10^{-5} molar dopamine (DA) on prolactin secretion
           by isolated rat pituitary cells supported in columns by
           Biogel beads and perfused continuously.

Table 1   Drugs Commonly Elevating Circulating
                              Prolactin Levels

1.   DA receptor blocking drugs:

     Phenothiazines   e.g. chlorpromazine, perphenazine
     Butyrophenones   e.g. haloperidol
     Pimozide
     Benzamides   e.g. metoclopramide, sulpiride

2.   DA depleting agents:

     Alpha-methyl-dopa
     Reserpine

3.   Non-catacholamine dependent causes:

     Oestrogens
     TRH

     Originally it was thought that the dopamine receptor blocking
drugs acted at the hypothalamic level to prevent dopamine induced
PIF secretion. Since the realization that the inhibitory control
of prolactin secretion from the pituitary is not simply due to
a hypothalamic peptide PIF secreted under dopaminergic control,
these views have had to be revised. While these dopamine receptor
antagonists may act at the mid brain level, it can be simply shown
that they will act directly at the pituitary level. Thus in our
system of perfused rat isolated pituitary cells, dopamine directly
inhibits prolactin secretion but this action is blocked by the
introduction of metoclopramide, haloperidol or chlorpromazine.
It is clearly unwarranted to assume an action on prolactin secretion
above the level of the pituitary for these agents. Dopamine depleting compounds presumably reduce the dopamine secreted from the
median eminence into the portal capillary system thereby lowering
the PIF effect on the pituitary cells and allowing prolactin secretion to rise. Oestrogens and TRH seem to act directly at the pituitary cells and there is no definite evidence that the dopamine receptor mechanisms are involved. The opiate agonists which raise
prolactin secretion appear to act via the hypothalamic mechanisms
rather than directly at the pituitary level.

DISORDERS OF PROLACTIN SECRETION

Hyperprolactinaemic States

     Since prolactin is under tonic inhibitory control from the
hypothalamus any disease of the hypothalamus or destruction of
the hypothalamic-hypophyseal portal circulation may lead to hyper-

prolactinaemia. Craniopharyngiomas and hypothalamic tumours or
granulomas are often associated with hyperprolactinaemia as is
either surgical or traumatic stalk section.   Similarly pituitary
tumours which may vary between large masses producing gross distort-
ion or ballooning of the pituitary fossa on skull X-ray, and lesions
a few millimetres in diameter producing no or minimal changes on
X-ray (the microadenomas) may themselves secrete prolactin and
be composed of mammotrophs. Alternatively truely non-functioning
tumours exist which seem to distort the pituitary stalk anatomy
sufficiently to deny the normal pituitary cells their supply of
PIF so that hyperprolactinaemia occurs, the hormone coming from
the normal pituitary cells which no longer are under the inhibitory
control of the hypothalamus. Primary hypothyroidism is associated
in at least one third of patients with hyperprolactinaemia. Chronic
renal failure is also associated with hyperprolactinaemia and appears to be directly related to renal function. Dialysis itself
does not alter prolactin levels but successful renal transplantation
is associated with a fall in prolactin. In addition a significant
proportion of patients present with clinical sequelae of hyperprolactinaemia in whom no clear cause can be found. These are labelled
'idiopathic' but undoubtedly some of these patients have microadenomas
of their pituitaries which are not yet big enough to show up on
X-ray of their skull.

     The clinical result of the elevated prolactin is usually im-
paired gonadal function irrespective of the cause of the hyperprolactinaemia. A small proportion of patients, about one third,
have inappropriate lactation although in many patients it may only
be found on expression of the breast. Men rarely present with
gynaecomastia the most common symptom being relative or absolute
impotence although some patients note a reduction in seminal volume
before impotence develops (Thorner et al, 1977). Women present
with amenorrhoea, dysfunctional uterine bleeding (irregular periods,
scanty periods or sometimes menorrhagia), or even regular menstruation
with infertility. In the latter group the patients either do
not have the normal ovulatory LH peak, or they do ovulate but have
a slow or inadequate rise in progesterone due to hyperprolactin-
aemia-induced dysfunction of the corpus luteum. In some women
hyperprolactinaemia is associated with mild virilism and increased
secretion of adrenal androgens such as dehydroepiandrosterone
and its sulfate and the clinical picture of the polycystic ovary
syndrome. It may be that prolactin alters the adrenal metabolic
pathways in some patients since these abnormalities in androgen
secretion appear to be both ACTH and prolactin dependent.

     The mechanism of the hypogonadism in patients with hyperpro-
lactinaemia is not certain. There is evidence that elevated prolactin levels partially block the actions of the gonadotrophins
on the gonad and this presumably accounts at least in part for
the impaired ovarian and testicular function. Thus exogenous

gonadotrophin administration is associated with impaired testosterone responses in males and an impaired oestradiol response in females with hyperprolactinaemia. In vitro culture of human granulosa
cells show an impaired progesterone secretion in response to gonadotrophin when increasing amounts of prolactin are added and hyperprolactinaemic patients often show impaired progesterone production
from the corpus luteum in vivo (McNatty et al, 1974; Del Pozo et
al, 1976; Seppala et al, 1976). In addition it has been proposed
that there is an alteration in the functioning of the hypothalamic
centres responsible for cyclical release of gonadotrophin in the
human since hyperprolactinaemic women do not show the normal LH
surge in response to exogenous oestrogen administration (Glass
et al, 1975). There is however no deficiency of readily releasable
LH and FSH within the pituitary in the majority of patients with
hyperprolactinaemia (Mortimer et al, 1973; Thorner et al, 1974
and 1977).

MANAGEMENT OF PATIENTS WITH HYPERPROLACTINAEMIA

     Hyperprolactinaemia can be treated by medical measures in
the majority of patients. The use of dopamine agonists with a
long length of action lowers prolactin levels and is effective in
correcting any associated hypogonadism and galactorrhoea during
the time that the agents are given, even when large pituitary tumours are present. Inevitably if a patient has a pituitary or hypothalamic tumour big enough to produce visual field defects then
that patient will require surgical decompression of the tumour usually followed by external pituitary irradiation to prevent recur-
rence. Some hypothalamic tumours cannot be operated on and irradiation is the only therapy possible; this is then followed by dopamine agonist therapy to correct the hyperprolactinaemia. The problem
of the microadenoma without any local complications is a more difficult one, particularly if the patient is a women who wishes to
become pregnant. The principal problem lies in the undoubted risk
of rapid enlargement of any pituitary tumour in a patient who does
succeed in becoming pregnant since such an enlargement may produce
optic chiasmal compression and visual field defects. Hyperprolactinaemic patients who have their prolactin levels reduced with dopamine
agonists become highly fertile and the risk is real. The principal
aim in the management of these patients is:
1. Reduction of the growth potential, and destruction or removal
of any adenoma or microadenoma to prevent its enlarging spontaneously
during pregnancy.
2. Lowering of prolactin levels to normal to allow gonadal function
to return to normal, to reverse any galactorrhoea or virilism,
3. Preservation of the remaining anterior pituitary function.

Neurotransmitter Therapy

     The ergot alkaloids were first shown to lower prolactin sec-

retion in the human in 1972 (Besser et al, 1972). Bromocriptine
(2-brom-α-ergocryptine) was the drug used and this has since come
into regular use throughout the world for this purpose. Although
it was known from the work of Flückiger and others that this agent,
like many other ergot alkaloids, was a specific prolactin inhibitor
working directly at the pituitary cells the precise mode of action
was not known when it was first used (Flückiger and Wagner, 1968).
Only later it was shown to be a dopamine receptor agonist (Corrodi
et al, 1973, Fuxe et al, 1974). Unlike L-dopa, bromocriptine acts
itself whereas L-dopa has to be converted into dopamine before it
is effective and has only a short duration of action. Bromocriptine
has a much longer duration of action which appears to be due to
a persisting effect at the receptor site. In vivo, bromocriptine
is well absorbed and has a duration of action in suppressing prolactin lasting 6 - 12 hours after a single oral dose (Figure 2).

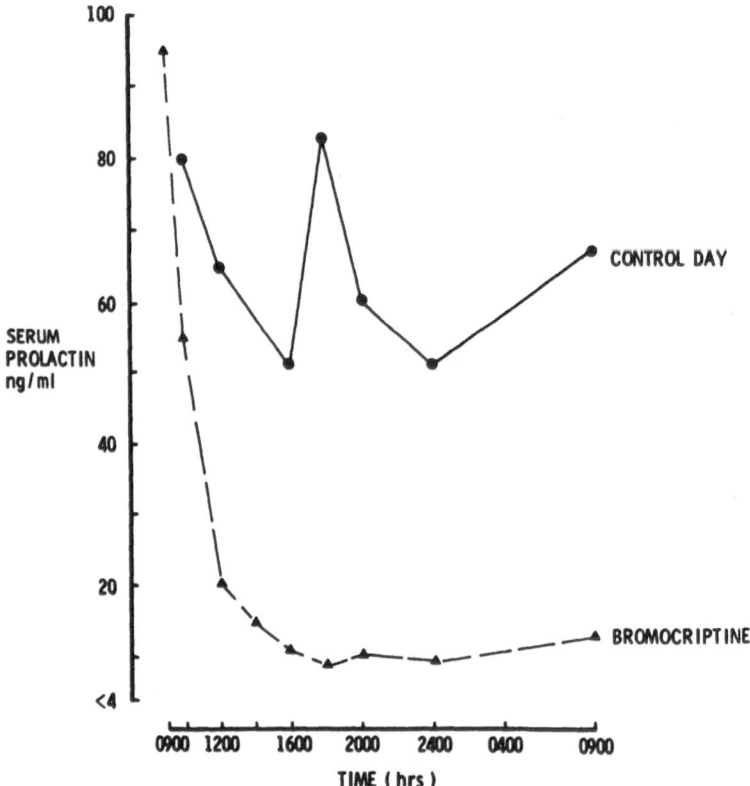

Figure 2   Changes in circulating prolactin concentration in a women
            with idiopathic hyperprolactinaemia and galactorrhoea
            during a control day (●─●) and after the first 5mg dose
            of bromocriptine (▲─▲) at 09.00 and 2.5mg at 18.00 hrs.

Bromocriptine has to be given at the start by slowly increasing doses and during food in order to avoid the initiating side effects of postural hypotension, nausea and vomiting. 1.25mg is given on retiring and then every three or four days the dose is increased until a dose of 2.5mg three times a day is reached. This is usually sufficient to lower prolactin levels to normal but occasionally it is necessary to increase the dose to between 15 and 40mg a day or more in divided doses in patients who have aggressive pituitary tumours. For the suppression of post partum lactation the drug is started on the day of delivery in the full dose of 2.5mg 3 times a day and only in this group of patients no side effects are noted.

In the patients with pathological hyperprolactinaemia, galactorrhoea if present decreases within a few days or weeks from the start of bromocriptine therapy. Normal menstruation is rapidly restored in premenopausal women usually within eight weeks although rarely it has taken longer, up to $7\frac{1}{2}$ months in our hands. In only 8 of the first 51 women did normal cyclical ovarian function fail to return in our patients. Three of these had premature ovarian failure, another four had pituitary tumours, three of whom were gonadotrophin deficient following partial hypophysectomy. Only rarely (3 out of 51 patients) have we found that doses of more than 40mg per day are required to lower prolactin levels to normal in hyperprolactinaemic patients. With the resumption of normal cyclical ovarian function libido usually improves, dyspareunia and depression disappear, and the patients frequently lose the excessive weight that is sometimes a feature in this condition. They become highly fertile.

Bromocriptine and problems in pregnancy. In the past the treatment of infertility in this group of patients has been a major problem since it has been largely dependent on the use of exogenous human gonadotrophins, although occasionally patients did respond to clomiphene. Bromocriptine therapy in patients with infertility and hyperprolactinaemia irrespective of its aetiology usually results in rapid return of normal gonadal function and fertility. It is now most unusual to have to use gonadotrophin therapy.

In our own practice we have had 53 pregnancies in 44 women to date and all 31 babies born so far have been normal. In a recently published review of 448 pregnancies associated with bromocriptine therapy the incidence of major or minor abnormalities has been precisely that expected from this number of patients and there is no evidence that bromocriptine is teratogenic (Griffiths et al, 1978). It is still normal practice, however, to advise women to stop bromocriptine therapy as soon as they know they are pregnant. The incidence of spontaneous abortion also appears to be that seen in the general population.

We have previously discussed the risks associated with induced
pregnancy in patients with pituitary tumours (Thorner et al.,
1975) and it is possible that patients with hyperprolactinaemia
who do not have even minor radiological abnormalities of the pituitary on skull X-ray still have a microadenoma and run the same
risks. In all these patients the risk of the development of visual
field defects during pregnancy is present since pituitary tumours
however small may rapidly swell during pregnancy under the influence
of increasing circulating oestrogen levels. We have adopted a
cautious and anticipatory policy. If we suspect the presence
of a pituitary tumour all our patients receive a course of external
pituitary irradiation (4500 rads from 15 MeV linear accelerator
in 25 fractions over 35 days) at least 3 months before allowing
the patient to become pregnant. By doing this we hope to prevent
swelling of the pituitary gland during the pregnancy and the development of visual problems. To date none of our 16 patients with
pituitary tumours have shown any evidence of pituitary enlargement
associated with pregnancy.

Other workers adopt different views. In a recent review,
Bergh et al, (1978) have followed 17 pregnancies in women with
pituitary tumours which were not treated before allowing the women
to become pregnant. In these pregnancies four patients showed
evidence of clear enlargement of the pituitary tumour with change
in radiological appearance, severe headache or serious visual
field defects. This study has confirmed that the risk is present
in bromocriptine induced pregnancy as with gonadotrophin induced
ovulation. Centres other than our own have used other forms of
irradiation e.g. cobalt irradiation, local implantation of radioactive yttrium, or proton beam therapy and an alternative approach
is the selective removal of a microadenoma by transphenoidal surgery
(Hardy, 1973). It is not possible to decide at this time which
form of therapy is better. However, it is clear that transphenoidal
micro-surgery carries a risk of approximately 20% of inducing
hypogonadotrophic hypogonadism as well as the persistence of
the hyperprolactinaemia.

Treatment of men with hyperprolactinaemia. Men, too, rapidly
respond to bromocriptine therapy. Galactorrhoea ceases and potency
is restored with an increase in seminal volume. The possibility
that the dopamine agonist may act directly on neural mechanisms
involved in producing erections cannot be excluded.

Bromocriptine and tumour growth. Little direct evidence
exists about the effects of dopamine agonists on the growth of
human pituitary tumours. In studies in rats on transplanted pituitary
tumours, ectopic pituitary autographs, and oestrogen induced
mitoses in pituitaries in situ, dopamine agonists have been shown
not only to reduce the rate of mitosis but also lead to shrinkage
of the tumour (Davies et al, 1974, Lloyd et al, 1975, Lu and Meites,

1971, MacLeod and Lehmeyer, 1973, Quadri and Meites, 1973). We
have now similar but indirect evidence in the human since we have
seen two patients in whom pituitary fossa X-rays have markedly
improved to near normality after four years' treatment with bromocriptine, and two other patients whose visual field defects have
disappeared. Ezrin and colleagues have also had similar experiences
(Cornblum et al, 1975). One may therefore presume that in man
as in rats bromocriptine and other dopamine agonists reduce pituitary
tumours size and growth.

Dopamine Agonists Other than Bromocriptine

Other ergot related dopamine agonists such as lergotrile
and lisuride have been investigated for their efficacy in the
management of hyperprolactinaemic states in man. Lisuride is
also a serotonin antagonist although there is no evidence at present
that this action is essential to its prolactin lowering effects
in man. The non ergot alkaloid, dopamine agonist piribedil has
also been studied for its prolactin lowering effects.

Lergotrile has had the most extensive use and has a similar
action to bromocriptine but is of shorter duration (Figure 3,
Thorner et al, 1978). It acts as a dopamine receptor agonist
in vivo and in vitro. In clinical practice it differs from bromocriptine only that it is somewhat sedating whereas bromocriptine
if anything is a central stimulant. The drug has run into major
problems of toxicity and it has been withdrawn from clinical practice. Lisuride too has actions similar to that of bromocriptine
although it has only had preliminary endocrine investigations.
It is, however, approximarly 100 times more potent than either
bromocriptine or lergotrile and this may be economically important
in the long run. Piribedil also lowers prolactin levels in normal
subjects and in patients with pathological hyperprolactinaemia.
Piribedil is, however too short acting for regular clinical use
but has the advantage over the ergot alkaloids in that it can
be given easily by intravenous infusion (Figure 5).

Prolactin deficiency

Prolactin deficiency is rare and is most often seen after
pituitary necrosis or infarction such as Sheehan's syndrome, in
which lactation does not occur. Following infarction of a pre-
existing pituitary tumour or after hypophysectomy for breast cancer
prolactin may be undetectable in the circulation. Patients with
acromegaly receiving high doses of bromocriptine for control of
growth hormone secretion may also have very low prolactin levels.
Apart from the inability to lactate, very low or undetectable
prolactin in the circulation is not clearly related to any endocrine dysfunction.

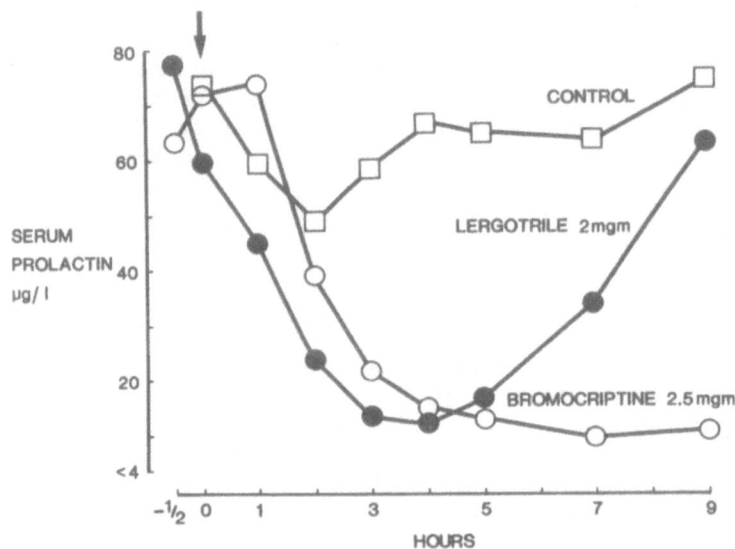

Figure 3 Effect of 2mg lergotrile compared with 2.5mg bromocriptine
         in an hyperprolactinaemic woman.

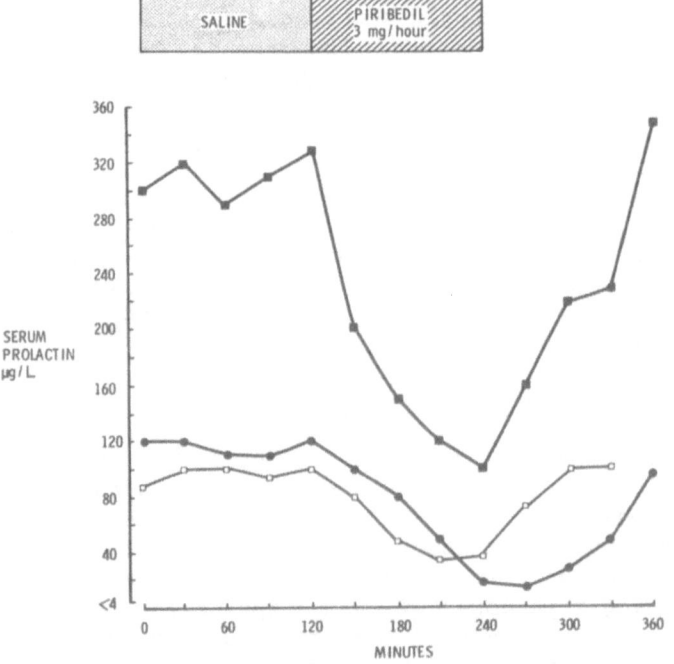

Figure 4 Effect of an intravenous piribedil on circulating prolactin
         infusion on 3 hyperprolactinaemic women.

CONCLUSION

Confirmation of the existence of prolactin as a normal pituitary hormone in the human was followed shortly by the realization that disorders of prolactin secretion were very common in the human and resulted in infertility, impotence and disturbed menstruation. Fortunately about the same time an agent which lowered prolactin secretion was introduced, the ergot alkaloid bromocriptine, and this reverted pituitary-gonadal function to normal. Although it was not initially realised, bromocriptine is a dopamine agonist and it is by this action that it lowers prolactin. Bromocriptine therapy is therefore an example of highly effective neurotransmitter therapy; it has revolutionised the management of pituitary-gonadal disorders and allowed simple treatment of conditions which before were difficult to manage. The problem of how to deal with pituitary tumours in patients who wish to become pregnant remains and hopefully we shall learn in the future how to select those patients who are most at risk of rapid enlargement of their tumour so that these may be selectively treated. The use of bromocriptine and other dopamine receptor agonists provides a clear example of the benefits which accrue in the clinical sphere when the basic observations made in animals are transferred to man. Its introduction together with the simultaneous introduction of the hypothalamic regulatory hormones for use by the clinician marked the emergence of the new discipline of clinical neuroendocrinology.

Acknowledgements

This work is supported by the Medical Research Council, the Joint Research Board of St. Bartholomew's Hospital, and the Peel Medical Trust. Dr. M.O. Thorner present address is: University of Virginia Medical School, Charlottesville, VA, U.S.A..

REFERENCES

Bowers, C.Y., Friesen, H., Hwang, P., Guyda, H., and Folkers, K. Biochem. Biophys. Res. Commun. 45, 1033-1041, (1971).

Berch, T., Nillius, S.J. and Wide, L. Brit. Med. J., 1, 875-880, (1978).

Besser, G.M., Parkes, L., Edwards, C.R.W., Forsyth, I.A. and McNeilly, A.S. Brit. Med. J. 3, 669-672, (1972).

Cornblum, B., Webster, B.R., Mortimer, C.B., and Ezrin, C. Clin. Res. 33, 614A, (1975).

Corrodi, H., Fuxe, K., Hokfelt, T., Lidbrink, P., and Ungerstedt, U. J. Pharmacol. 25, 409-412, (1973).

Davies, C., Jacobs, J., Lloyd, H.M., and Meares, J.D. J. Endocrinol, 61, 411-417, (1974).

Del Pozo, E., Wyss, H., Lancranyan, I., Obolensky, W., and Varga, L. In "Ovulation in the Human" (P.G. Crosignani and D.R. Mishell, eds), Serono Symp. No. 8, pp. 297-299. Academic Press, New York, (1976).

Dupont, A., and Redding, T.W. Program 57th Meet. Endocr. Soc., New York, Abstract No. 85, (1975).

Enjalbert, A., Moos, F., Carbonell, L., Priam, M. and Kordon, C. Neuroendocrinology, 24, 147-161, (1977).

Fluckiger, E., and Wagner, H. Experientia 24, 1130-1131, (1968).

Frohman, L.A., and Szabo, M. Program, 57th Annu. Meet. Endocr. Soc. , New York Abstract 86, (1975).

Fuxe, K., Corrodi, H., Hodfelt, T., Lidbrink, P., and Ungerstedt, U. Med. Biol. 52, 121-132, (1974).

Gallo, R.V., Rabii, J., and Moberg, G.P. Endocrinology 97, 1096-1105, (1975).

Glass, M.R., Shaw, R.W., Butt, W.R., Logan Edwards, R., and London, D.R. Brit. Med. J. 3, 274-275, (1975).

Griffiths, R.W., Turkalj, I., and Braun, P. Brit. J. Clin. Pharmac. 5, 227-231, (1978).

Hardy, J. In "Diagnosis and Treatment of Pituitary Tumours" (P.O. Kohler and G.T. Ross, eds.), Int. Congr. Ser. No. 303. pp. 179-194. Excerpta Med. Found., Amsterdam, (1973).

Jacobs, L.S., Snyder, P.J., Wilber, J.F., Utiger, R.D. and Daughaday, W.H. J. Clin. Endocr. Metab. 33, 996-998, (1971).

Kamberi, I.A., Mical, R.S., and Porter, J.C. Endocrinology 88, 1012-1020, (1971).

Kato, Y., Nakai, Y., Imura H., Chihara, K., and Ohgo, S. J. Clin. Endocr. Metab. 38, 695-697, (1974).

Kordon, C., Blake, C.A., Terkel, J., and Sawyer, C.H. Neuro-endocrinology 13, 213-223, (1973-1974).

Labella, F.S., Dular, R., and Vivian, S.R. Endocrinology International Congress, 4th Washington D.C. Excerpta Med. Int. Congr.
Ser. 256, 141 Abstr. 354, (1972).

Lloyd, H.M., Meares, J.D., and Jacobi, J. Nature (London) 255,
497-498, (1975).

Lu, K.H., and Meites, J. Proc. Soc. Exp. Biol. MED. 132, 480-
483, (1971).

MacLeod, R.M. Int. Semin. Reprod. Physiol. Sexual Endocrinol.,
6th Prog. Reprod. Biol. 2, 54-68, (1977).

MacLeod, R.M., and Lehmeyer, J.E. Cancer Res. 33, 849-855, (1973).

McNatty, K.P., Sawers, R.S., and McNeilly, A.S. Nature (London)
250, 653-655, (1974).

Mendelson, W.B., Jacobs, L.S., Reichman, J.D., Othmer, E., Cryer,
P.E., Trivedi, B., and Daughaday, W.H. J. Clin. Invest. 56, 690-
697, (1975).

Mortimer, C.H., Besser, G.M., McNeilly, A.S., Marshall, J.C.,
Harsoulis, P., Tunbridge, W.M.G., Gomez-Pan, A., and Hall, R.
Brit. Med. J. 4, 73-77, (1973).

Mortimer, C.H., Besser, G.M., Goldie, D.J., Hook, J.. and McNeilly,
A.S. Clin. Endocrinol. (Oxford) 3, 97-103, (1974).

Noel, G.L., Dimond, R.C. Wartofsky, L., Earll, J.M., and Frantz,
A.G. J. Clin. Endocr. Metab. 39, 6-17, (1974).

Quadri, S.K., and Meites, J. Proc. Soc. Exp. Biol. Med. 142,
837- 841, (1973).

Seppala, M., Hirvonen, E., and Ranta, T. Lancet 1, 229-230, (1976).

Thorner, M.O. In "Clinical Neuroendocrinology". Eds. Martini,
L. and Besser, G.M. Academic Press, New York, p.319-361, (1977).

Thorner, M.O. Lancet 1, 662-665, (1975).

Thorner, M.O., Besser, G.M., Hagen, C., and McNeilly, A.S. Brit.
Med. J. 2, 419-422, (1974).

Thorner, M.O., Besser, G.M., Jones, A., Dacie, J. and Jones, A.E.
Brit. Med. J. 4, 694-697, (1975).

Thorner, M.O., Edwards, C.R.W., Hanker, J.P., Abraham, G. and
Besser G.M.   In "The Testes in Normal and Infertile Men" (P.

Troen and H. Nankin, eds.) pp. 351-366, Raven Press, New York, (1977).

Thorner, M.O., Ryan, S.M., Wass, J.A.H., Jones, Ann, Bouloux, P., Shaw, J.E.H., Williams, Sally, Besser, G.M. The effect of the dopamine agonist, lergotrile mesylate on circulating anterior pituitary hormones in man. J. Clin. Endocr. Metab. In Press (1978).

Turkington, R.W. and MacIndoe, J.H. J. Clin. Invest. 51: 323 (abstr), (1972).

Valverdi-R, C., Chieffo, V., and Reichlin, S. Endocrinology 91, 982-993 (1972).

ROLE AND REGULATION OF NEUROPEPTIDE NEURONS

Claude Kordon

Unité 159 de Neuroendocrinologie de l'INSERM

2ter rue d'Alésia, 75014 PARIS - FRANCE -

Research on neuropeptides has resulted in an impressive amount of discoveries in the last few years. Attempts to detect new brain peptides led to the astonishing finding that a much larger number of such molecules are present in the Central Nervous System than was originally anticipated ; several of them have now been characterized (see review in 59), and the neurons which produce them have been mapped in many more areas of the brain than the hypothalamus and the structures directly involved in neuroendocrine regulation (see review in 25). Several peptides as well as potent analogs with agonist or antagonist activities have now been synthetized (59) with promising therapeutic applications. In parallel, the study of peptide biosynthesis and degradation also opened recently new areas of research.

Knowledge of the distribution of neuropeptides in the CNS provides a potent support for studies of their biological relevancy. Attempts to better understand the conditions under which they are produced and the regulation of their release will also help to elucidate their functional significance. In the present review, we will underline current concepts relating to these problems. The almost ubiquitous distribution of several peptides is problematic : should they be considered neurohormones, neurotransmitters or neuromodulators? We will try to answer this question in the first section of this chapter, and then review present knowledge of biosynthesis and degradation processes. Finally, experimental data concerning the regulation of neuropeptide release will be summarized.

I. <u>NEUROHORMONES, NEUROTRANSMITTERS OR NEUROMODULATORS</u> ?

One of the main recent discoveries in Neuroendocrinology is that the mediobasal hypothalamus contains far more neuropeptides than would be necessary to account for the neural regulation of the six pituitary hormones. In addition, almost all these peptides are also located in other areas of the brain, and even in many peripheral tissues. In fact, new data suggest that each of them not only influences the secretion of several pituitary hormones, but also serves a wide variety of non neuroendocrine brain functions.

In Table I, we have attempted to summarize current knowledge about the functions, or at least the sites of action, of those peptides or neurohormones which act directly on the pituitary. As one can see, these substances are quite variable in function. At one end of the spectrum is LHRH, a peptide found in a limited number of structures (25) and involved in a limited number of functions only. This molecule is mainly concerned with the regulation of sex hormones (46) and of sexual behaviour (47). At the other end of the scale, peptides of the opiate series are much more polyvalent. Found throughout the brain (27) but also in several peripheral organs, they have been shown to affect almost all pituitary functions (7, 10, 21, 37, 62), behavioural parameters and electrical activity in many structures of the CNS and under a large number of experimental conditions (6). But they also act directly on the gastrointestinal tract (26), the accessory sex organs, vasomotricity and the control of carbohydrate metabolism.

Other neuropeptides are involved in an intermediate number of functions. One can thus view these molecules as more or less universal keys, which, depending upon the way they have been selected in the course of phylogenetical development, can open a very variable number of doors. So far, several keys have been identified, and characterization of others such as CRF, GHRH or some of the prolactin inhibiting or stimulating factors, is well under way (59). Mapping of their pathways has also provided us recently with a fairly complete picture of their possible actions (25). Identification of their physiological role still requires extensive studies before their precise functions in the brain can be fully understood.

The relative ubiquity of brain peptides and the large variety of their effects does not really support the concept that some of them should be considered neurotransmitters or neuromediators, whereas others are neurohormones. In fact, the type of function exerted by a peptide seems to depend upon its localization, i.e. the level of its actions within the brain, rather than on its chemical structure. Depending upon where they are released and recognized, neuropeptides can either function as transmitters or as hormones. This is also the case of some transmitters, as dopamine, which has been clearly shown to act as a neurohormone affecting prolactin cells in the pituitary (41).

NEURO-PEPTIDE	MAIN DISTRIBUTION SITES		NEUROENDOCRINE EFFECTS	CENTRAL EFFECTS	PERIPHERAL EFFECTS
	IN BRAIN	IN PERIPHERY			
LHRH	Almost exclusively hypothalamic	None	FSH LH	Sex behaviour Some electrophysiological effects	None
ADH	Mainly hypothalamic	Some peripheral tumours	ACTH	Behaviour "Memory" Neuronal electrical activity	Diuresis Vasomotricity
TRH	Hypothalamus Cortex Brain stem Spinal cord	skin retina	TSH PRL	Behaviour Neurotransmitter metabolism Neuronal electrical activity	Not elucidated
SRIF	Hypothalamus Amygdala Spinal cord	Pancreas Gastrointestinal tract	GH TSH PRL	Neurotransmitter metabolism Neuronal electrical activity	Insulin Glucagon Gastrin
Opiates	Almost any structure in the CNS	Pituitary Kidney Gastrointestinal tract Accessory sex organs	LH GH TSH PRL ACTH	Pain Sleep Behaviour Neurotransmitter metabolism Neuronal electrical activity	Gastrointestinal tract Vasomotricity Kidney Accessory sex organs Carbohydrate metabolism

TABLE I : Distribution and actions of some neuropeptides involved in neuroendocrine regulation. Sources are quoted in the text or reviewed in references 14, 20, 25, 26, 27, 59 and 63.

The variable competence of neuropeptides illustrates well a thought expressed by François Jacob. Natural Selection, he wrote (28), " works like a tinkerer who does not know exactly what he is going to produce, but uses whatever he finds around him ". Depending upon the success or the usefulness of such tinkered tools, they can be applied to an increasing number of functions or perform only selected activities. Our present concepts of brain peptides are well accounted for by such view.

II. BIOSYNTHESIS AND DEGRADATION OF NEUROPEPTIDES

Real understanding of the physiological regulation of neuropeptide secretion, a topic dealt with in the next section of this chapter, will only be achieved when we know more about the way peptides are biosynthetized and enzymatically degraded in the brain. Only then will it be possible to distinguish factors affecting peptide production from those controlling release ; in addition, control of degradation processes is often a prerequisite for evaluation of the actual amount of peptide released in in vitro systems, for instance.

How peptides are synthetized in the brain is not yet clear. Several attempts to demonstrate incorporation of labelled precursor aminoacids in vitro have proven unconclusive (40). Though findings suggest that small peptides, like TRH, could be synthetized enzymatically in the brain (45) or in peripheral tissues (39, Schaeffer and Axelrod, unpublished results), arguments in favour of a non ribosomal synthesis of neuropeptides in the brain are rather unconvincing (40, 50). In contrast, there is increasing evidence that larger precursor molecules are present in brain extracts. This is particularly true in the case of posterior pituitary hormones, especially vasopressin and the neurophysins, since high molecular weight material has been recovered from neurosecretory granules (57) and shown to be convertible to the actual hormones under certain experimental conditions (19). In the case of LHRH (44) and somatostatin (20), hypothalamic fractions also seem to contain heavier molecules, which can compete in radioimmunoassay systems with the corresponding native peptides (44). These fractions can also yield peptide-like material when incubated in the presence of subcellular fractions of brain tissue. However, in those cases, the molecular weight of precursor forms is only slightly higher than that of the peptides themselves (44).

Several technical problems may explain why it is so difficult to detect neuropeptide precursors. The existence of very potent peptidasic activities in all brain structures tested so far (23, 31, 36) complicates extraction procedures. Unless peptidases are inactivated, conversion of possible precursors into peptides is likely to remain undetected. For this reason, most authors extract their material with acid solutions or organic solvents (40, 44). Under these conditions, however, large protein precursors are likely to precipitate and may thus no longer be present in chromatographed fractions. In addition, immunological cross-reactivity of possible precursors has usually been studied in radioimmunoassay systems (44), that is by means of a competition between the putative heavy precursor and the native peptide used as tracer. Under these conditions, larger molecules may not be recognized by antibodies against the peptide, in particular when the antigenic site involves both the C and N terminals of the peptide. Even if the precursor is recognized, one cannot assume that affinities of precursors and of native peptides for the antibody are identical. If the precursor displaces the tracer less than the peptide itself, only a large amount of precursors could be detected by this method. This situation is not likely to be met under normal conditions.

Aminoacid incorporation studies have also proven disappointing (39, 40). In the case of ribosomal synthesis within neurons, one might assume that biosynthesis takes place mainly in or near the cell body. Cleavage of precursors should then occur within neurosecretory granules during their transport to the nerve ending, or in the nerve ending itself. This may explain why in vitro incorporation of amino-acids is either absent or not proportional to the time of incubation (40) since axonal transport is certainly impaired in incubation systems. In the case of LHRH and the neurophysins, some incorporation could, however, be obtained when the precursor aminoacids were microinfused in discrete brain structures in vivo and the tissue dissected and incubated a certain time after the injection (19, 40, 57). Of course, the critical problem in incorporation experiments is the characterization of radioactive material (19, 39, 40, 50) ; incorporation of label in LHRH or neurophysins represents only 1 in 1000, or less, of the total incorporated radioactivity, so that the yield of the separation procedure needs to be particularly high for the neurohormone to be detectable. So far, immunoprecipitation procedures are not sufficient to demonstrate incorporation of labelled amino-acids into peptide molecules.

At the present stage, one may thus only state that arguments in favour of ribosomal synthesis of peptide precursors in the brain are slightly more conclusive than arguments against this hypothesis (40).

Studies on peptide degradation have demonstrated the existence of very potent peptidases in the brain (42). In some cases, cleavage sites seem to be specific (1, 41), though in brain homogenates or extracts, heterogenous populations of enzymes have been found. Incubation of synthetic molecules with such extracts results in several end products (42). Recently, an interesting hypothesis has been proposed : peptidasic activity in the brain could be affected by the endocrine condition of the animal and, consequently, peptidases would play a role in the physiological regulation of neuropeptide secretion (23, 29, 36). However, evidence in favour of this hypothesis remains controversial. In fact, the amount of peptidases present in the brain exceed by far the endogenous concentrations of neuropeptides (31). Enzymatic activity degrading a

given peptide is found throughout the brain, even in areas which
contain insignificant concentrations of the peptide itself (31).
In addition, subcellular distribution studies show that most peptides are almost entirely recovered from nerve endings (38). This
contrasts with the subcellular distribution of peptidases, which
are mainly recovered from high speed supernatant fractions (35)
(Fig. 1).

        In fact, in the Central Nervous System, catabolic enzymes seem
mainly concerned with the degradation of active molecules after
they have acted at their target site. There are only a few instances in which an enzyme can be clearly shown to be rate limiting
for the availability of an active substance. The conversion of
serotonin to melatonin in the pineal, or of dopamine to noradrenalin in noradrenergic neurons are such cases in which both the precursor and the end-product exhibit different biological activities;
enzymes which convert the one into the other regulate the available concentration of the end-product. Similar examples also exist
among the neuropeptides : when incubated with brain extracts, oxytocin gives raise to a tripeptide containing the C-terminal sequence of oxytocin ; this peptide has melanostimulating hormone inhibiting activity (MIF) and exhibits behavioral effects (59). A
natural derivative of TRH, histidyl proline-diketopiperazine, also
seems to belong to this category of end products. It is formed
under the influence of a pirrolidone carboxypeptidase, which is
present in hypothalamic extracts and possibly associated with a
membrane fraction and gives rise to a histidyl-prolinamide
dipeptide ; this intermediary product spontaneously undergoes cyclisation, with loss of a NH3, to yield the relatively stable histidyl-proline diketopiperazine (1) (Fig. 2). The effect of this
TRH catabolite on prolactin release is opposite to that of TRH itself. On both a cultivated clone of prolactin cells (2) and normal
half pituitaries (Enjalbert et al, unpublished results), it inhibits the in vitro release of prolactin at fairly low molecular
concentrations (Fig. 3). His-pro-diketopiperazine could thus be
another candidate to account for prolactin inhibiting activity
(PIF) of hypothalamic extracts, besides dopamine and GABA, whose
effects as PIF are now well documented (15, 41, 60, 62). In that
case, pirrolidone carboxypeptidasic activity can thus be assumed
to be rate limiting for the production of the active dipeptide
and to play a role in neuroendocrine control.

III. REGULATION OF NEUROPEPTIDE RELEASE

        Under a wide variety of physiological conditions, specific
stimuli are known to trigger or to inhibit neuropeptide release
from the median eminence and, hence, pituitary hormone secretion.
Such stimuli are involved, for example, in neuroendocrine reflexes,
such as the suckling reflex observed in lactating mammals (64) or
the cold-induced release of thyreotropin (65). Neurotransmitters,
such as dopamine, noradrenaline, serotonin, acetylcholine, and

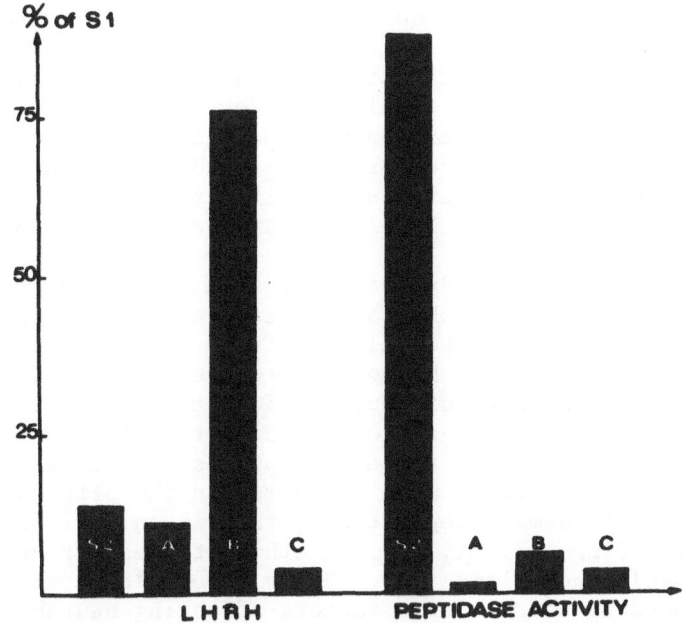

Fig. 1 : Comparative distribution of the neuropeptide LHRH and of the overall peptidasic activity degrading the neurohormone in different subcellular fractions.
S1, low speed supernatant ; S2, high speed (100 000 g x 30 mn) sucrose supernatant ; A, B, and C interfaces obtained from top to bottom of a discontinuous sucrose gradient and corresponding respectively to bands containing myelin, purified nerve endings (synaptosomes) and mitochondria (along with membrane debris). LHRH was measured by radioimmunoassay and peptidases by evaluation of the specific enzyme activity calculated from disappearance of immunoreactive material under saturating conditions of the substrate (from 38).

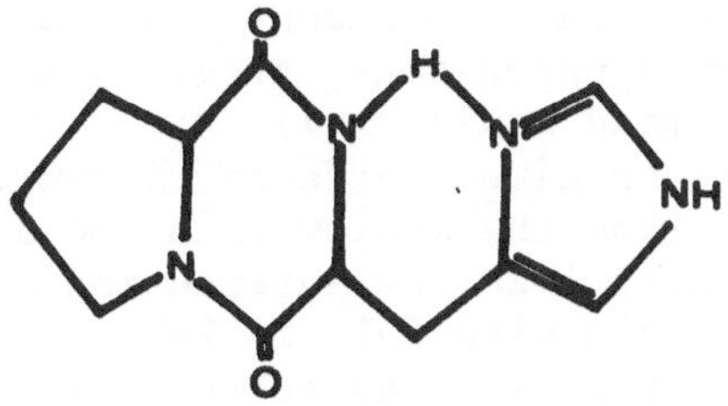

Fig. 2 : Histidyl-proline diketopiperazine. The molecule is formed from histidyl-prolinamide by spontaneous cyclisation (1).

others, play a role in this activation of neuropeptide release
(34). In most of these interactions, neurotransmitters act at the
level of complex neuronal circuits, which indirectly modulate
neuropeptide secretion (34). Only in a few cases have direct influences of neurotransmitters on neuropeptide secretion been clearly demonstrated, usually under in vitro conditions. In such cases,
the interaction probably involves specific transmitter receptors
located at the presynaptic level of neurosecretory nerve endings
(54).

Early attempts to study the regulation of neuropeptide release in vitro were based on the use of cascade systems, in which
fragments of mediobasal hypothalamic tissue were coincubated with
hemipituitaries. Pituitary hormones were then bioassayed or radio-
immunoassayed as an index of neuropeptide release resulting from
addition of neurotransmitters to the incubation medium (18, 61).
When radioimmunoassays for the neuropeptides themselves became
available, pioneer studies were performed on the effects of transmitters on synaptosomal preparations, that is on purified hypothalamic nerve endings (3, 24, 30). In other studies, fragments of
the mediobasal hypothalamus (32, 53) or hypothalamic slices (8)
have been used to characterize factors affecting neurohormone release from nerve endings.

Before valid conclusions can be drawn from such preparations,
their basic properties have to be carefully assessed. In particular, neurohormone release should satisfy the conditions defined by
Douglas and Poisner when they first proposed the stimulus secretion
coupling hypothesis (11) : basal neuropeptide release should be
enhanced upon incubation with high molarities of K^+, which depola-
rize nerve endings, and such effects should be Ca^{++}-dependent.
So far, incubation of fragments or slices of tissue containing
neurosecretory nerve endings were found to better satisfy these
requirements than purified nerve endings (35, 53). It has been
postulated that, in the course of the rather drastic fractionation
procedures used to prepare synaptosomes, neuropeptides stored in
granules may be displaced to a cytosoluble compartment and more
easily leak out during incubation, thus increasing non specific
background release rate. Membrane properties of nerve endings may
also be affected by the high sucrose molarities used in most
fractionation methods (35). However, recent data indicate that a
depolarization induced, Ca^{++}-dependent neuropeptide release can
also be obtained from synaptosomes, provided they are incubated in
an appropriately modified medium (67).

The principal known direct effects of neurotransmitters on
neuropeptide release from the medium eminence are summarized in
Table II. In some cases, like the interaction of acetylcholine with
LHRH or that of serotonin with LHRH or TRH release, the effect may
be artefactual, since the high concentrations of agonists used may
have non specific actions (5). Also, different preparations may
give different results (54). No clearcut direct effect of neurotransmitters on the release of posterior pituitary hormones from

NEUROHORMONE	NEUROTRANSMITTER	TYPE OF EFFECT ON RELEASE	OBSERVATIONS	REFERENCES
LHRH	Dopamine	Stimulatory	Only in intact or sex steroid treated animals	54,61
	Serotonin	Inhibitory		9
	Acetylcholine	Stimulatory or negative	Only at high concentrations	18
				9
TRH	Noradrenaline	Stimulatory	Only on release of a precursor-labelled immunoreactive form	23
	Histamine	Stimulatory		8,32
CRF	Acetylcholine	Stimulatory		14,24
	Serotonin	Stimulatory or negative	Possibly via ACh	30
				14
	GABA	Inhibitory	Only on ACh induced release	30
SRIF	Noradrenaline	Stimulatory	Not in median eminence, only in preoptic nuclei or cortex	*
				4

TABLE II : *In vitro effects of neurotransmitters on the release of neuropeptides having neuroendocrine actions from neurosecretory nerve endings. Many other neurotransmitter-neuropeptide interactions, which affect only indirectly neuroendocrine control, are not listed here (for review, see 34).*
* Epelbaum, unpublished results ; see Fig. 4.

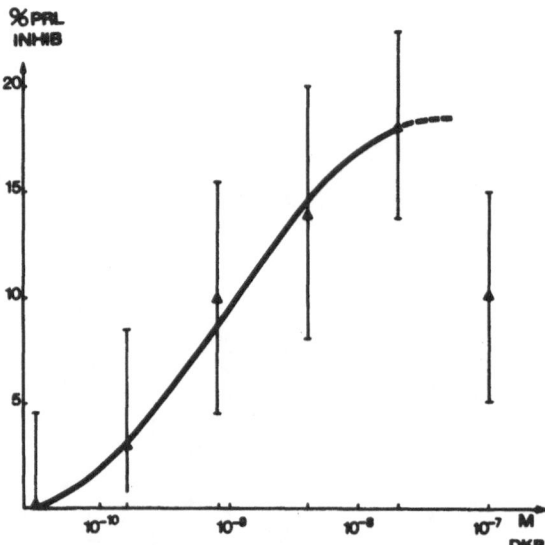

Fig. 3 : Effect of increasing concentrations of histidyl-proline diketopiperazine (DKP) on in vitro prolactin release from hemipituitaries of male rats (Enjalbert et al, unpublished data).

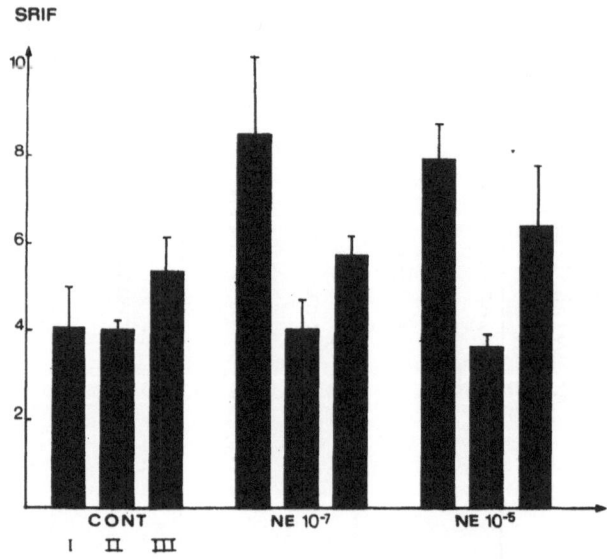

Fig. 4: Effect of noradrenalin (NE) (10^{-7} and 10^{-5}M) on the in vitro release of somatostatin from slices of amygdala (AMY), preoptic structures (APO) and mediobasal hypothalamus (MBH). Epelbaum, unpublished results. The results are expressed as pg SRIF released/pg SRIF content (%). I: POA; II: MBH; III: Amygdala.

neurohypophyseal nerve endings or of somatostatin from the median eminence have been detected so far. In the latter case, however, it seems that somatostatin release from other brain structures, in particular the preoptic area of the hypothalamus, the amygdala (Epelbaum, unpublished results) (Fig. 4) or the cortex (4) can be stimulated by noradrenalin. This suggests that neuropeptide release from different brain structures is not necessarily regulated in the same manner. This concept was also demonstrated in the case of LH-RH : dopamine triggers the release of this peptide from hypothalamic fragments containing the palisadic zone of the median eminence, but not from the tuberoinfundibular sulcus or the organum vasculosum of the lamina terminalis (54). In this case, a fair correlation between the endogenous dopamine concentration and the effectiveness of the transmitter to release the neuropeptide supports the presumption that the effect is physiologically relevant (54). Interestingly enough, this interaction has only been seen so far in intact animals or those treated with sex steroids, and not in castrated subjects. The interaction between sex steroids and dopamine receptors is presently being investigated in several laboratories (54, 61). Other neurotransmitters do not seem able to release given neuropeptides, but modulate their release stimulated by other agonists. The negative interaction between GABA (24) and serotonin (30) on the release of CRF induced by acetylcholine is an example of this. More recently, a similar situation has been described for morphine and morphinomimetic peptides. Morphine and Met-Enkephalin do not affect the basal release of LHRH from the palisadic zone of the median eminence by themselves ; however, when added to an incubation medium in presence of dopamine, they antagonize the stimulatory effect of the neurotransmitter (55, 56). This interaction does not seem to affect dopamine receptors themselves, since naloxone, an opiate antagonist, selectively blocks this inhibitory effect of morphine and Met-Enkephalin (55, 56) (Fig. 5).

Similarly, morphine and Met-Enkephalin, which have no effect on the in vitro release of prolactin by themselves (13, 15b) are able to antagonize the prolactin inhibitory action of dopamine when added together with the transmitter to pituitary incubation. Opiate antagonists reverse this effect, so that independent dopamine and opiate receptors can also be assumed to be present on pituitary prolactin cells (15b) (Fig. 6).

Such interactions of opiate peptides with specific agonists which trigger neurohormone release are interesting in view of the numerous neuroendocrine effects of these substances. Morphine, enkephalins or endorphins have been shown to inhibit the release of LH (10, 48) and TSH (7), and, on the contrary, to enhance secretion of growth hormone (13, 62) and of prolactin (21, 37, 62). Data from pituitary incubation experiments suggested so far that these effects take place at the level of the Central Nervous System rather than directly on the pituitary. In the case of prolactin release, this conclusion should now be reevaluated, since opiates have been

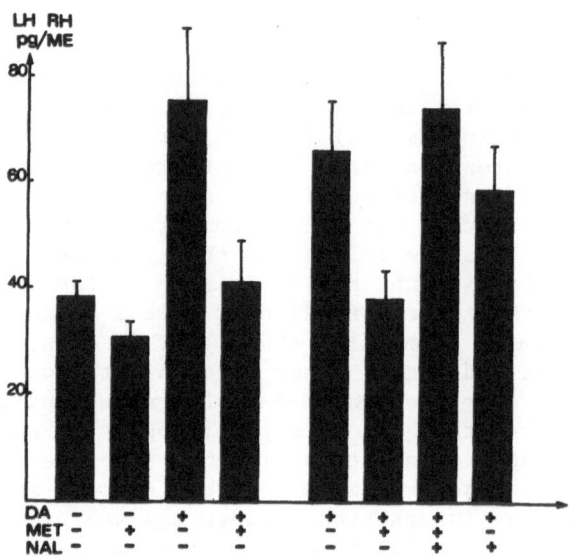

<u>Fig. 5</u> : Effects of dopamine (DA) (10^{-7}M), Met-Enkephalin (MET) (10^{-6}M) and naloxone (NAL) (10^{-6}M), alone or in combination, on <u>in vitro</u> release of LHRH from median eminence fragments (ME) of male rats (from 55).

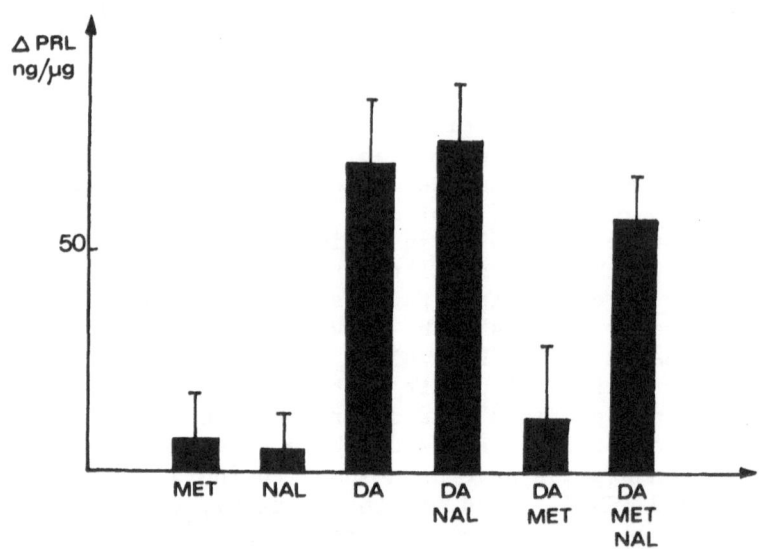

<u>Fig. 6</u> : Effect of dopamine (DA) (10^{-7}M), Met-Enkephalin (MET) (10^{-6}M) and naloxone (NAL) (5.10^{-6}M), alone or in combination, on <u>in vitro</u> prolactin release inhibition (expressed as the difference in hormone release between treated and corresponding control pituitaries) (15b).

found effective <u>in vitro</u>, but only in the presence of dopamine. At present, two hypotheses can account for the neuroendocrine effects of opiates. The first is based on the observation that the administration of morphine or morphinomimetic peptides inhibits the turnover of tuberoinfundibular dopamine (17). Similar effects have been documented in other dopamine-rich regions of the brain like the striatum (49). But we have seen that opiate actions may also take place at the " post-synaptic " level, that is directly on neuroendocrine targets. In that case, they seem to modulate responses to other agonists like dopamine, rather than to induce an endocrine response by themselves. In fact, the two theories are not incompatible ; opiates may well modulate the turnover rate as well as the receptor interactions of dopamine.

The fact that morphinomimetic peptides appear to be released during stress (52), suggests that several non specific endocrine consequences of stress, such as inhibition of LH and increase in prolactin secretion (16), could be accounted for by release of enkephalins or endorphins with ACTH. Reversal of stress-induced prolactin release by naloxone (66) is indeed in good agreement with this hypothesis. The neuroendocrine role of opiates might thus be to suspend normal regulatory mechanisms whenever the organism faces an agression or an emergency situation.

REFERENCES

1. Bauer, K. (1976). Nature 259, 591.
2. Bauer, K., Gräf, K. J., Faivre-Bauman, A., Beier, S., Tixier-Vidal, A., and Kleinkauf , M. (1978). Nature 274, 174.
3. Bennett, G. W., Edwardson, J. A., Holland, D. T., Jeffcoate, S. L., and White, N. (1975). Nature 257, 323.
4. Bennett, G. W., Edwardson, J. A., Marcano de Cotte, O., and Lloyd, H. (1978). Neurosci. Lett. 7, in press.
5. Bockaert, J., Premont, J., Glowinski, J., Tassin, J. P. and Thierry, A. M. (1977). In : Advances in Biochemical Pharmacology, ed. by E. Costa and G. I. Gessa, Raven Press, New York, Vol. 16, p. 29.
6. Bradley, P. B., and Dray, A. (1974). Br. J. Pharmacol. 50, 47.
7. Bruni, J. F., Van Vugt, D., Marshall, S., and Meites, J. (1977). Life Sci. 21, 461.
8. Charli, J. L., Joseph-Bravo, P., Palacios, J. M., and Kordon, C. (1978). Eur. J. Pharmacol., 49, in press.
9. Charli, J. L., Rotsztejn, W. H., Pattou, E., and Kordon, C. (1978). Neurosci. Lett. 10, 159.
10. Cicero, T. J., Meyer, E. R., Bell, R. D., and Koch, G. A. (1976). Endocrinology 98, 367.
11. Douglas, W. W., and Poisner, A. M. (1964). J. Physiol. (London) 172, 1.
12. Drouin, J., Lagacé, L., and Labrie, F. (1976). Endocrinology 99, 1477.
13. Dupont, A., Cusan, L., Caron, M., Labrie, F., and Li, C. H.

(1977). Proc. Nat. Acad. Sci., U. S. A. 74, 358.

14. Edwardson, J. A., and Bennett, G. W. (1977). In : Biologically
    Active Substances, Exploration and Exploitation, ed. by D. A.
    Mems , p. 281.

15a.Enjalbert, A., Moos, F. Carbonell, L., Priam, M., and Kordon,
    C. (1978). Neuroendocrinology 27, 24.

15b.Enjalbert, A., Ruberg, M., Fiore, L., Arancibia, S., Priam, M.,
    and Kordon, C. (1978). Pharmacol. Eur. J.

16. Euker, J. S., Meites, J., and Riegle, G. D. (1975). Endocrino-
    logy, 96, 85.

17. Ferland, L., Fuxe, K., Eneroth, P., Gustafsson, J. A., and
    Skett, P. (1977). Eur. J. Pharmacol. 43, 89.

18. Fiorindo, R. P. , and Martini, L. (1975). Neuroendocrinology
    18, 322.

19. Gainer, H., and Brownstein, M. J. (1978). In : Cell Biology of
    Hypothalamic Neurosecretion, ed. by J. D. Vincent and C.
    Kordon, CNRS, Paris, p.525

20. Gerich, J. E., and Lorenzi, M. (1978). In : Frontiers in Neuro-
    endocrinology, ed. by W. F. Ganong and L. Martini, Raven Press,
    New York, Vol. 5, p. 265.

21. Grandisson, L., and Guidotti, A. (1977). Nature 270, 357.

22. Griffiths, E. C., Hooper, K. C., Jeffcoate, S. L., and Holland,
    D. T. (1975). Brain Res. 88, 384.

23. Grimm-Jörgensen, Y., and Reichlin, S. (1973). Endocrinology
    93, 626.

24. Hillhouse, E. W., Burden, J., and Jones, M. T. (1975). Neuro-
    endocrinology 17, 1.

25. Hökfelt, T., Johansson, O., Ljungdahl, Å., Lundberg, J., Schultz-
    berg, M., Fuxe, K., Goldstein, M., Steinbusch, H., Verhofstad, A.,
    and Elde, R. (1978). In: Principles for the Central Neuronal
    Regulation of the Endocrine System, ed. by K. Fuxe, T. Hökfelt,
    and R. Luft, Plenum Press, New York, in press.

26. Hughes, J. (1975). Brain Res. 88, 295.

27. Hughes, J., Smith, T. W., Kosterlitz, H. W., Fothergill, L. A.,
    Morgan, B. A., and Morris, H. R. (1975). Nature 258, 577.

28. Jacob, F. (1977). Science 196, 1161.

29. Jeffcoate, S. L., White, N., Bennett, G. W., Edwardson, J. A.,
    Griffiths, E. C., Forbes, R., and Kelly, J. A. (1978). In :
    Principles for the Central Neuronal Regulation of the Endocrine
    System, ed. by K. Fuxe, T. Hökfelt, and R. Luft, Plenum Press,
    New York, in press.

30. Jones, M. T., Hillhouse, E. W., and Burden, J. (1976). J. En-
    docr. (London) 69, 1.

31. Joseph-Bravo, P., Loudes, C., Charli, J. L., and Kordon, C.
    (1978). Brain Res., in press.

32. Joseph-Bravo, P., Charli, J. L., Palacios, J. M., and Kordon,
    C. (1978). Endocrinology 103, in press.

33. Koch, Y., Boram, T., Chobaleng, P., and Fridken, M. (1974).
    Biol. Biochem. Res. Commun. 61, 1.

34. Kordon, C., Enjalbert, A., Héry, M., Joseph-Bravo, P., and Ruberg, M. (1979). In : Handbook of the Hypothalamus, ed. by P. Morgane, M. Dekker Inc., New York, in press.

35. Kordon, C., Epelbaum, J., and Gautron, J. P. (1978). Bull. Schweiz Akad. Med. Wiss., 34, 131.

36. Kuhl, H., Rosniatowski, C., and Taubert, H. D. (1977). Acta Endocr. (Kbh) 86, 80.

37. Lien, E. L., Fenichel, R. L., Gorsky, V., Sarantakis, D., and Grant, N. H. (1976). Life Sci. 19, 837.

38. Loudes, C., Joseph-Bravo, P., Leblanc, P., and Kordon, C. (1978).Biochem. Biophys. Res. Commun., in press.

39. McKelvy, J. F. (1974). Brain Res. 65, 489.

40. McKelvy, J. F. (1977). In : Hypothalamic Peptide Hormones and Pituitary Regulation, ed. by J. C. Porter, Plenum Press, New York, p. 77.

41. McLeod, R. M., and Lehmeyer, J. E. (1974). Endocrinology 94, 1077.

42. Marks, N. (1976). in : Subcellular Mechanisms in Reproductive Neuroendocrinology, ed. by F. Naftolin, K. J. Ryan, and I. J. Davies, Elesevier, Amsterdam, p. 129.

43. Marks, N., and Stein, F. (1974). Biochem. Biophys. Res. Commun. 61, 1458.

44. Millar,R. P., Aehnelt, C., and Rossier, G. (1977). Biochem. Biophys. Res. Commun. 74, 720.

45. Mitnick, M., and Reichlin, S. (1972). Endocrinology 91, 1145.

46. Matsuo, H., Baba, Y., Nair, R. M. G., Arimura, A., and Schally, A. V. (1971). Biochem. Biophys. Res. Commun. 43, 1334.

47. Moss, R. L., and McCann, S. M. (1973). Science 181, 177.

48. Pang, C. N., Zimmermann, E., and Sawyer, C. H. (1977). Endocrinology 101, 1726.

49. Pollard, H., Llorens-Cortes, C., and Schwartz, J. C. (1977). Nature 268, 745.

50. Reichlin, S. (1976). In : Subcellular Mechanisms in Reproductive Neuroendocrinology, ed. by F. Naftolin, K. Ryan and I. J. Davies, Elsevier, Amsterdam, p. 109.

51. Renaud, L. P., Martin, J. B., and Brazeau, P. (1975). Nature 255, 233.

52. Rossier, J., French, E. D., Rivier, C., Ling, N., Guillemin, R., and Bloom, F. E. (1977). Nature 270, 618.

53. Rotsztejn, W. H., Charli, J. L., Pattou, E., Epelbaum, J., and Kordon, C. (1976). Endocrinology 99, 1663.

54. Rotsztejn, W. H., Charli, J. L., Pattou, E., and Kordon, C. (1977). Endocrinology 101, 1475.

55. Rotsztejn, W. H., Drouva, S. V., Pattou, E., and Kordon, C. (1978). Nature, 274, 281.

56. Rotsztejn, W. H., Drouva, S. V., Pattou, E., and Kordon, C. (1978). Eur. J. Pharmacol. 50, 285.

57. Sachs, H., and Takabatake, Y. (1964). Endocrinology 75, 943.

58. Schaeffer, J. M., Axelrod, J., and Brownstein, M. J. (1977). Brain Res. 138, 571.

59. Schally, A. V., Arimura, A., Coy, D. H., Kastin, A. J., Meyers, C. A., Redding, T. W., Chihara, K., Huang, W. Y., Chang, R. C. C., Pedroza, E., and Vilchez, M. J. (1978). In : Principles for the Central Neuronal Regulation of the Endocrine System, ed. by K. Fuxe, T. Hökfelt, and R. Luft, Plenum Press, New York, in press.
60. Schally, A. V., Redding, T. W., Arimura, A., Dupont, A., and Linthicum, G. L. (1977). Endocrinology 100, 681.
61. Schneider, H. P. G., and McCann, S. M. (1969). Endocrinology 85, 121.
62. Shaar, C. J., Frederickson, C. A., Dininger, N. B., and Jackson, L. (1977). Life Sci. 21, 853.
63. Smith, T. W., Hughes, J., Kosterlitz, H. W., and Soza, R. P. (1976). In : Opiates and Endogenous opioid peptides, ed. by H. W. Kosterlitz, North Holland Publishing Company, Amsterdam, p. 57.
64. Terkel, J., Blake, C. A., and Sawyer, C. H. (1972). Endocrinology 91, 49.
65. Tuomisto, J., Ranta, A., Saarinen, P., Männistö, P., and Leppälnoto, J. (1973). Lancet 11, 510.
66. Van Vugt, D. A., Bruni, J. F., and Meites, J. (1977). Life Sci. 22, 85.
67. Warberg, J., Eskay, R. L., Barnea, A., Reynolds, R. C. and Porter, J. C. (1977). Endocrinology 100 : 814.

SOME PRINCIPLES OF NEURONAL REGULATION AT THE POSTSYNAPTIC LEVEL

Jean-Pierre Changeux

Laboratoire de Neurobiologie de l'Institut

Pasteur associé au CNRS et College De

France, Paris

To discuss in general terms five reports as different as those of Drs. Renaud, Labrie, Rodbell, Snyder and Terenius cannot be done without risks. The domains of research covered by these authors range from the physiology of the whole cell investigated by electrophysiological methods down to the characterization of receptor sites and receptor macromolecules. The danger is either to be too general and therefore useless or to be too precise but wrong. Our only hope is that some of the rules or principles uncovered with simple systems such as, at the cell level, the neuromuscular junction (1) or the fish electroplaque (2, 3) and, at the molecular level, the acetylcholine receptor protein may extend to the case of the CNS neuron and, more precisely, to the peptidergic neuron. This discussion is based on this analogy although it should be emphasized, at the very beginning, that important differences of course exist. What is true for the neuromuscular junction might be true for the chemical synapse in the CNS but distinctive features of the neuronal cell such as, for instance, its integrative properties, are certainly not shared by skeletal muscle or fish electroplaque.

I. Regulation of neuronal activity at the cell level

The electrophysiological properties of the hypothalamic peptidergic neurons studied by Renaud and collaborators are in

many instances similar to those of typical CNS neurons. Their
main characteristic feature is that they release peptides by
some of their axonal terminals at non-synaptic sites located in,
or close to, blood vessels. Apart from this, they establish chemical synapses by the rest of their axonal branches and, of course
receive a wide diversity of nerve inputs on their dendritic arbo-
risation.

The electrical "activity" of these neurohypophyseal or tuberoinfundibular neurons may be defined as the firing pattern
recorded at the single cell level via intra- or juxta- cellular
electrodes. As is generally the case in other systems, the
cell displays, under "resting" conditions, irregular low frequency discharges. This spontaneous activity does not contribute to
any significant release of circulating hormone. The activation
of the cell to a state where the hormone is liberated corresponds
to a replacement of the spontaneous spike discharge by bursts
lasting a few seconds. In many instances, these bursts are
evoked by peripheral stimulation such as, in the case of the
oxytocinergic neuron, the suckling activity of the pups. Afferent synapses on these neurons therefore enhance firing : they
are defined as activatory synapses. Others have the opposite
effect and are referred to as inhibitory. The firing pattern of
the neuron results from the integration (which can, in a few
cases, be calculated mathematically) of these facilitating and
depressing synaptic actions.

In Renaud's presentation, electrotonic coupling between
neurons has not been mentioned but, of course, might occur at
this level as it does in other systems (4, 5) to synchronize the
activity of an ensemble of nerve cells with similar, if not
identical, function. On the other hand, some evidence exists
that chemical synapses transmit the activatory and inhibitory
inputs received by the hypothalamic peptidergic neurons. First
of all, these synapses can be demonstrated anatomically, (see
Hokfelt, Fuxe, this volume) and also their effect can be
mimicked by microiontophoretic application of known neurotransmitters. Acetylcholine enhances firing, norepinephrine also but
may have the opposite effect. Dopamine and serotonine may either
depress or enhance firing. Without exception, glutamate has an
excitatory action and Gaba an inhibitory one. Peptides also
regulate the activity of the hypothalamic neurons. Oxytocine
enhances firing and vasopressin depresses the activity of the
neurohypophyseal neurons; LH, LHRH and somatostatin at physiological concentrations either enhance or depress firing of some
tuberoinfundibular neurons. These observations suggest (but

yet do not prove, that synaptic feedback loops can be established through retrograde collaterals of the neurosecretory neurons. Accordingly, the same peptide would both be secreted in the blood and act as a classical neurotransmitter at the synaptic level.

Steroid hormones also regulate the synthesis, storage and release of neurohypophyseal hormones but, most likely, at target sites entirely different from those of the neurotransmitters. Neurotransmitters are known to act at the membrane level. Would it be also the case for steroids ? One known effect of sex steroids on the electrical activity of some of the peptidergic neurons (oxytocinergic) is a modulation of spontaneous firing. Can this be considered as a membrane effect, or part of a more general regulation, at the gene level, mediated by the cytosolic receptor of these steroids (see Beaulieu for discussion) ?

In any case, a clearcut conclusion is that compounds with strikingly different structures may have the same effect on neuronal firing. No direct correlation exists between the sign of the effect and the chemical structure of the effector. The neurotransmitter (and/or peptide) does not take part in the chemical reaction it regulates nor is chemically transformed. It acts by the simple virtue of binding to a specific site. The transmission of neurotransmitter signals by the neuron must therefore proceed via some "indirect" mechanism.

II. Signal transduction by the neuronal membrane

The hypothesis that an "indirect" mechanism accounts for signal transduction means that at least two categories of topographically distinct membrane sites are involved in this regulation : 1) a site for the binding of the regulatory ligand e.g. the neurotransmitter, generally referred to as a receptor site. 2) a site responsible for the particular biological activity which, in the membrane, is under the control of neurotransmitter binding to its receptor site (Figure 1). Models for these indirect or "allosteric" (6, 7) interactions are offered by the classical bacterial enzymes : aspartate transcarbamylase (8), threonine deaminase (9) and many others. The two classes of sites, in these instances the catalytic and regulatory sites, might as well be carried by different protein units (e.g. aspartate transcarbamylase) or be part of the same polypeptide chain (e.g. threonine deaminase).

To account for the coupling between topographically distinct sites at the level of a regulatory protein it was postulated (6) that the protein complex (or "Regulator" (3)) carrying these

sites undergoes a conformational transition between at least two
discrete states : a resting and an active one. These states would
differ by their affinity for the considered ligand and by their
biological activity (Figure 1). This leads to the distinction of
two categories of pharmacological effectors. The agonists, like
the physiological neurotransmitter, trigger the transition from
the resting to the active state while the antagonists stabilize
the resting state. Both series of compounds bind to the same
site which therefore should exhibit a different structural
specificity in each of the two states.

Compounds acting on a given Regulator as antagonists are
either synthetic or natural substances (as for instance alcaloïds
from plants) but never physiological signals. On the other hand,
the same agonist may trigger activatory or inhibitory responses
depending on the target cell and therefore the membrane Regulator
to which it binds. Finally, the same antagonist (for instance
atropine) may block both the activatory and inhibitory responses
elicited by a given agonist (for instance acetylcholine). An
important question is then : are the same propositions valid in
the case of neuropeptides acting as transmitters or neuro-modu-
lators ? Most likely, the answer is yes. The known peptide antagonists are synthetic compounds (see for instance the antagonists
of LHRH in Schally's presentation). On the other hand, it would
be an interesting exception to this rule if it was discovered that
some natural peptides present in CNS were acting as antagonists
of a given (activatory or inhibitory) transmitter.

Two classical modes of signal transduction by biological
membranes are known. They differ by the nature of the "site of
biological activity" of the Regulator which can either be an
ion selective channel or ionophore or, the catalytic site of
a nucleotide cyclase (Figure 2).

In the case of a ionophoric-Regulator, and for both
activatory and inhibitory responses, agonist binding increases
the permeability of the associated ionophore. (A few exceptions
to this rule have, however, been reported where agonists close
channels which are opened at rest e.g. some photo-receptors
and neurons from vertebrate retina (10)). The sign of the effect
is therefore determined by the nature of the ion(s) transported
and the direction of the ionic flow accross the membrane. For
instance Gaba exerts inhibition by opening a gate selective for
chloride ions, thereby hyper-polarising the cell and raising its
firing threshold. On the other hand, acetylcholine has the
opposite effect at activatory synapses by increasing the permeability to Na^+ and K^+ (11).

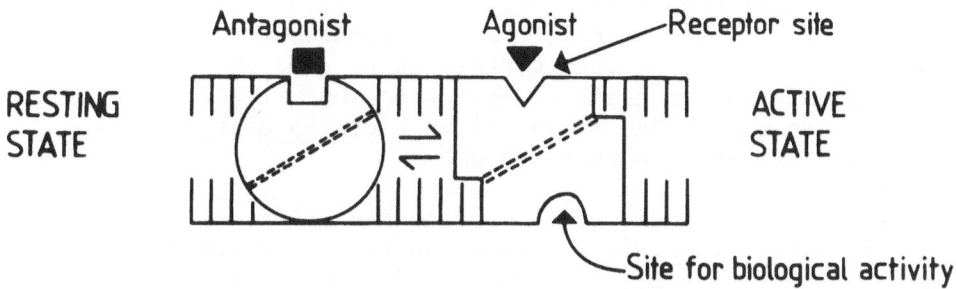

Fig. 1

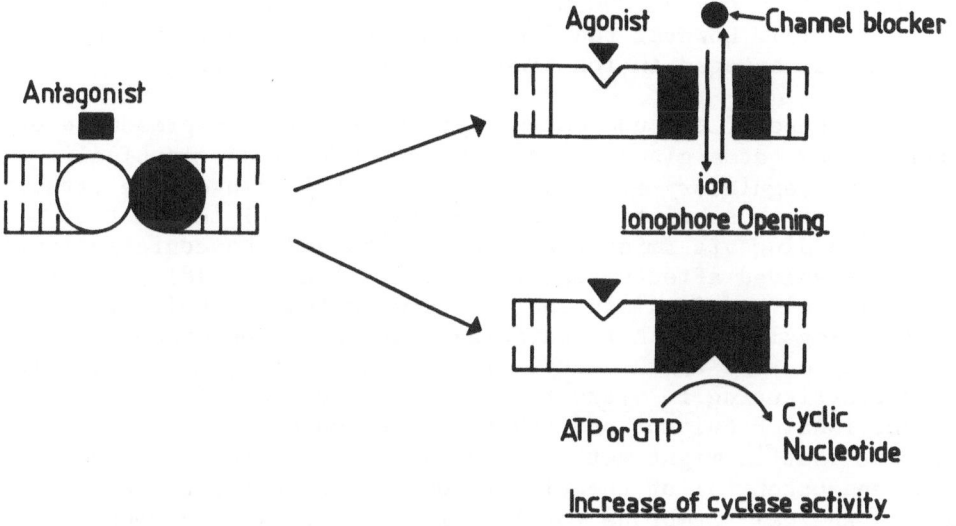

Fig. 2

The postulate of the ionophore as a distinct structural entity leads to the prediction of a third category of pharmacologically active agents. These compounds would alter the properties of the ion gate, for instance by blocking ion transport by steric hindrance. Examples are, in the case of the acetylcholine-Regulator, the local anesthetics (see 3) or the frog toxin histrinicotoxin (12) and, respectively, in the case of the glycine- and Gaba- Regulators, strychnine (13) and picrotoxin (14). Like the antagonists, none of these compounds has ever been found in vertebrate nervous system. Up to now, there is no existing information about a regulation where the active ligand, for instance a peptide, would act by blocking directly ion transport through an opened ionophore.

In the case of the acetylcholine-Regulator some of these blocking agents (histrionicotoxin and the fluorescent local anesthetic quinacrine (15), have been used in attempts to label and isolate the ionophore (16). Starting from a suspension of membrane fragments particularly rich in acetylcholine-receptor, a protein has, indeed, been purified which binds these two compounds and splits in a denaturing electrophoresis gel into a component of apparent molecular weight 43,000 significantly different from the 40,000 mol. w. band of the acetylcholine-receptor (17). However the final demonstration that this protein is, indeed, engaged in ion transport is still lacking.

The case of the nucleotide cyclase-Regulators presented by Rodbell may, at a glance, look similar to that of classical globular, regulatory enzymes. In this last instance, the catalytic and regulatory subunits (when they are distinct) interact by direct bonding via amino acid side chains and the conformational change involved affects subunits association (see 18). No direct experimental proof exists that the same situation holds for the cyclase-Regulators. Their striking temperature dependance, which has been related to the state of membrane lipids, the difficulty in reconstituting in vitro the isolated cyclase and receptor into a functional Regulator and other related observations, in fact, suggest that it might not necessarily be the case. Are the membrane potential or the electrochemical gradient established across the cell membrane involved in this coupling ? Does the recently discovered acetylation of lipids (see Axelrod, this Symposium), or other intermediary signal, mediate this interaction ? In any case, it seems clear that protein components binding GTP are also involved in this coupling making the cyclase-Regulator, if it exists as a separate entity, a rather complex macro-molecular edifice.

As for the ionophoric-Regulator, agonist binding to the cyclase-Regulator has, in a general manner, a positive effect. It causes an increase of the activity of the cyclase and as a consequence a rise of the cell content in cy AMP or cy GMP. The sign of the biological response is thus determined by the interaction of the cyclic nucleotide with its intracellular target rather than by the primary effect of the transmitter (or the hormone) on the Regulator itself.

At the cell level, a cascade of effects are, consequentially expected to occur. Coupling between different classes of Regulators may even take place : for example, Ca^{++} ion entering by a cation selective ionophore may regulate the activity of a topographically distant cyclase which itself regulates the transport properties of another ionophoric Regulator and so on. Networks of regulatory interactions are expected to take place within the cell via diffusible internal signals. From a theoretical point of view such cascade of effects might account for some of the so called "integrative" properties of the neuron. On the other hand, from a practical point of view, it complicates the identification of the primary target of a neurotransmitter or a peptide by simply looking at the response of a cell or a tissue.

III. Microheterogeneity of receptors

The binding curve of a radioactive ligand to a population of membrane fragments prepared, for instance, from brain tissue is expected to be complex. If a single class of independant sites was present and if the ligand was binding exclusively to these sites, then the binding curve should be fitted by the Langmuir isotherm i.e. by an hyperbola and a single dissociation constant would be determined. Reality is of course far from this ideal situation and deviation from a straightforward hyperbola generally is found. Causes of such deviations are multiple. A rather trivial one, is the so called non specific binding of the ligand. By "non specific" one most often means binding to a large number of sites with low affinity and a wide diversity of binding constants. One example is the incorporation of the ligand to the lipid phase of the membrane. In general, binding assays include a test to eliminate, or correct for, such non specific binding.

Another cause of deviation is the presence in the tissue of several populations of sites with which the ligand interacts "specifically". A classical example is that of the cholinergic synapse where two of them are present in almost equal amounts.

One is the enzyme acetylcholinesterase, the other the acetylcholine-receptor site. The distinction between these two classes
of sites solely on the basis of binding curve analysis would
have always been unconvincing despite the fact that these sites
differ by their dissociation constants for a wide variety of
ligands. The introduction of a third partner, an α-toxin from
snake venom (19), in fact lead to an unambiguous distinction
between these two classes of sites. This small protein (approx.
7,000 mol. w.) blocks exclusively the acetylcholine receptor site
but not the catalytic site of acetylcholinesterase. The fraction
of bound cholinergic ligand displaced by the α-toxin was therefore defined as specifically associated with the acetylcholine
receptor site (20), a proposition which was rapidly confirmed
with the purified protein (21).

In most instances, tools like the α-toxins are not available
and the authors often find sufficient to compare the thermodynamic binding constants and affinity order for a wide spectrum
of ligands with the apparent dissociation constants measured by
following biological activity. A close correlation between
biological activity and binding is thus taken as the unique proof
for the identification of a pharmacological receptor site (see
Snyder, this Symposium). In addition to the above mentioned
difficulty, another one should be kept in mind : under equilibrium conditions the ligand might bind to a state of the
receptor which is no longer biologically active (see next
chapter). The correlation between binding and biological activity can never be straightforward.

Nevertheless, these ligand binding studies have lead to the
identification of sites which possess the pharmacological properties expected for the physiological receptor sites of neurotransmitters and neuropeptides and, most likely, belong to these
macro-molecules (see Terenius, Snyder, this Symposium). The
striking heterogeneity found for the binding of a given neurotransmitter (and related compounds) to these receptor sites
supports the conclusion that different species of receptor
proteins for a given transmitter (or peptide) may exist in brain.
The difference of binding specificity of these iso-receptors
would be such that they should possess different primary structures (which, yet, is not shown). These are, for instance, the
μ and δ receptors for opiates (Terenius, this Symposium) or the
α_1 and α_2 adrenergic receptors (Snyder, this Symposium). The
regional distribution of these iso-receptors within the brain
often differ, which strongly supports the view that they are
indeed distinct molecular species. Reasons for this molecular

heterogeneity are not evident. For instance, are these different iso-receptors coupled with different biological units : ionophores or nucleotide cylases ... ?

Another cause of receptor microheterogeneity which should unambiguously be distinguished from the previous one is the possibility of interconversion of the same receptor molecule between different conformations. As already mentioned, these states might exhibit different "specificities" towards agonists and antagonists even though their primary structure is the same. The acetylcholine receptor protein from fish electric organ offers a striking illustration of this situation. The absolute value of the binding constants of cholinergic agonists for the α -toxin sensitive receptor site varied considerably, first from one research group to another, but also, and more consistently, between membrane-bound and detergent-extracted (or purified) receptor protein (for a review see 22). For acetylcholine it ranged between 10^{-8} M (high affinity sites) and 10^{-5} M or higher (low affinity sites), the range of variation for the antagonists being much smaller. Two observations lead to an unambiguous inter-pretation of this heterogeneity. 1) Dissolution of the receptor protein by cholate yieds sites with low (or intermediate) affi-nity but removal of cholate results in the conversion of these low affinity sites into high affinity ones without change in the total number of sites (23). 2) rapid mixing of membrane fragments rich in acetylcholine Regulator with a cholinergic agonist results in a time dependant change of affinity of the membrane bound receptor for agonists which can be monitored, indirectly by following α-toxin binding (24), or directly by spectroscopic methods (15, 25, 26). In the membrane at rest, the receptor protein, is present in a state of low affinity for agonists and, at equilibrium, the cholinergic agonists stabilize a state of high affinity for these ligands. The receptor protein therefore exists under several interconvertible states of affinity (at least two, most likely three). The fact that these conformational transitions primarily affect the binding for agonists rather than that for antagonists suggest that it is related to the physiolo-gical action of these compounds. Then, the question is : which one of these states is the physiologically "active" one ? In particular, what is the physiological significance of the high affinity state found under equilibrium conditions ?

IV. The three states model for "activation" and "desensitization"

To answer these questions, one may first compare the absolute value of the equilibrium dissociation constant for the

considered state with the concentration range of physiological
ligand eliciting the response in vivo. For example, during
transmission of the nerve inpulse at the neuromuscular junction,
the concentration of acetylcholine in the cleft rises transiently
up to 10^{-4}M (11). The thermodynamic dissociation constant of the
high affinity state for acetylcholine is therefore about four
orders of magnitude lower suggesting that this state is not
directly involved in the physiological mechanism of synaptic
transmission. This suggestion is further supported by the
following remark. Consider for a moment that the dissociation
constant of acetylcholine for the physiologically active state
is indeed 10^{-8}M. If one assumes that the on rate constant of
association of acetylcholine to the high affinity state is diffusion limited, the off rate constant should be several 10's
of milliseconds. Thus, repetitive firing of the synapse in the
millisecond range should be excluded. Of course, this is not the
case in vivo.

Another approach is to compare the time courses found in
vivo for the physiological response and in vitro for ligand
binding and affinity changes. The growth phase of the endplate
potential at the neuromuscular junction is shorter than 1 msec.
On the other hand, the transition towards the high affinity state
observed in vitro upon mixing membrane fragments rich in acetylcholine Regulator with cholinergic agonists takes place in the
second-minute range. Conclusively, the high affinity state of the
acetylcholine receptor does not correspond to the "active" state
of the acetylcholine-Regulator.

It is well known however that prolonged exposure of a
cholinergic synapse to an agonist causes a slow decrease (in the
second-minute range) of the amplitude of the response to the
same agonist. This phenomenon is referred to as pharmacological
desensitization (27). The most likely interpretation of the
above data is therefore that, in agreement with Katz & Thesleff
(27) model, the high affinity state of the acetylcholine receptor
coïncides with a "desensitized" state of the acetylcholine-
Regulator under which the ionophore is shut.

Accordingly, and by analogy with the known behavior of
regulatory enzymes, it has been postulated (see 3) that the
acetylcholine-Regulator exists under at least three discrete and
interconvertible conformational states. These states referred to
as resting R, active A and desensitized D would differ by a)
the affinity of the acetylcholine receptor site for cholinergic

ligands : the affinity for agonists increase from R (low affinity) to A (medium affinity) to D (high affinity); on the other hand, the affinity for antagonists is lower in the A than in the R or D states; "non-exclusive" binding of a given ligand to more than one state is expected to occur; b) the affinity of the ionophore for local anesthetics : in the D and A states the affinity is higher than in the R state; c) the ion gate is open only in the A state.

In the membrane at rest, the R state is favored; the fraction of A state is negligible that of the D state small but not negligible, at least in vitro. In the presence of a given concentration of ligand, the equilibrium is shifted in favor of the state to which it binds preferentially.

Agonists will shift the equilibrium towards the D state, antagonist to either the R or D state and local anesthetics (and multivalent cations) to either the D or A state. Upon rapid change of ligand concentration and depending on the nature of the ligand, some states can be transiently populated due to preferential pathways for the interconversions. The "activation" reaction is viewed as a transient population of the A state by acetylcholine and the "desensitization" as the final equilibration in the D state.

This model fits quantitatively the in vitro fast kinetic of interaction of a fluorescent cholinergic agonist with receptor-rich membrane fragments (26, 28). It accounts also, at least qualitatively, for the in vivo responses of "activation" and "desensitization". Finally, direct experimental evidence indicates that several of these states, at least the high-affinity one, preexist to agonist binding. In other words, the agonist would trigger the physiological responses not as a consequence of an "induced-fit" mechanism but by stabilising selectively preexisting conformations of the Regulator (26, 28).

The extension of this three-states model to systems different from the cholinergic synapse is tempting but nevertheless hazardous. First it would be of primary interest to test its validity in the case of ionophoric-Regulators responding to trans-mitters different from acetylcholine such as, for instance, the neuropeptides.

According to Rodbell (this Symposium) several features of the cyclase-Regulators might be accounted for by a similar model.

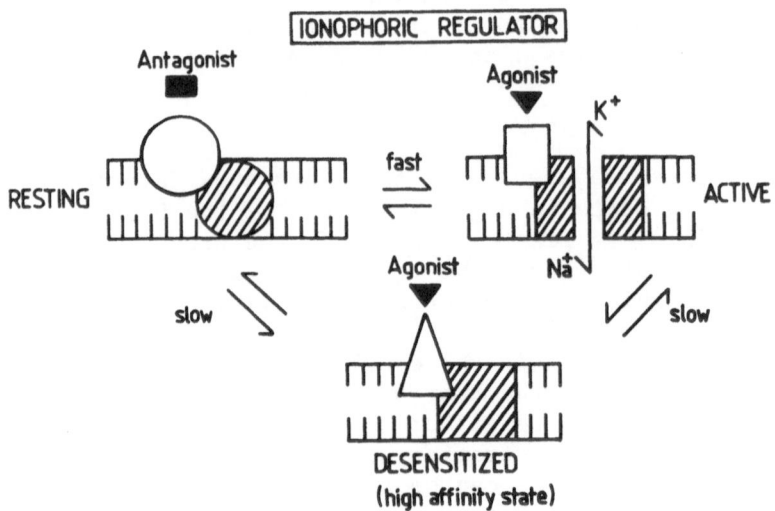

Fig. 3

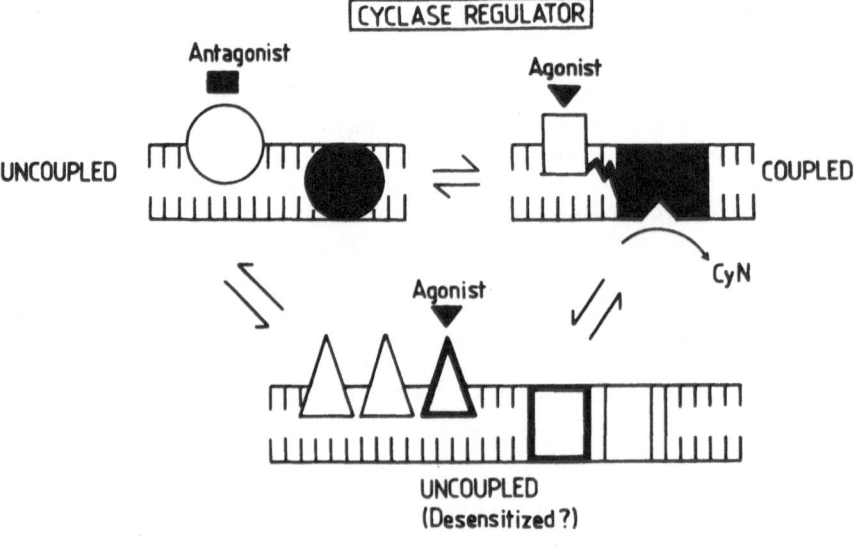

Fig. 4

At rest, receptor and cyclase would be uncoupled. Agonist would favor the coupling between cyclase and receptor, thereby causing the activation of the enzyme. Desensitization, if it has the same significance as for the cholinergic-Regulator, would result from the uncoupling of the cyclase-receptor complex accompanied by the stabilization of the receptor protein in a binding state with an affinity for the agonist higher than that of the active one. Under these conditions (and in the absence of GTP), target size analysis by high energy irradiation indicates that receptor and cyclase tend to segregate into high molecular weight agregates (Rodbell) (see Figure 4).

An interesting consequence of the three-states model is that it accounts for the effect of "allosteric" ligands known to bind to sites different from the receptor site for agonists.

For instance, in vitro, local anesthetics or the frog toxin histrionicotoxin stabilize the acetylcholine-Regulator under a high affinity state for agonists (even in the absence of agonists (3)). This effect might account for the acceleration of pharmacological densitization observed in vivo. Similarly GTP binding to protein entities different from the cyclase and receptor stabilizes a coupled state of these two proteins (see Rodbell and also Snyder).

Finally, it has been reported (see Snyder) that, in the case of the opiates receptor, Na^+ ions stabilize a state of high affinity for antagonists and low affinity for agonists. In vivo, opiate agonists block glutamate-induced depolarization by impairing the influx of Na^+ triggered at the post-synaptic membrane (29). Unambiguously some kind of competition exists between opiate agonists and Na^+ ions. Would opiate agonists binding, for instance, to the glutamate receptor site stabilize the glutamate-Regulator under a state where the Na^+ ionophore is shut i.e. under a resting state ? If this was the case, the so-called opiate "agonists" would, according to our definition, behave as antagonists. Alternative interpretations are of course plausible. For instance, the opiates could be ionophore-specific like the local anesthetics, strychnine or picrotoxin and plug directly the Na^+ ionophore in its open conformation. Finally, and most likely, the effect of Na^+ ions might not be directly relevant to the pharmacological action of opiates which would act through their own Regulator changing indirectly the permeability to Na^+ via some internal signal, such as Ca^{++} ions or cyclic nucleotides.

V. Regulation of receptor number

The above mentioned regulation of receptor conformation and coupling (with the relevant biologically active unit) takes place without change of the total number of receptor molecules or biologically active units. A well-understood case where the regulation affects the number of receptor molecules is that which follows the denervation of vertebrate skeletal muscle. After section of the motor nerve an hypersensitivity to acetylcholine develops in the extrajunctional area which corresponds to a 10-100-fold increase of the total number of acetylcholine receptor molecules (30) (as shown for instance by incorporation of labelled aminoacids into the neosynthetized receptor protein (31)). Reinnervation is followed by the disappearance of the extrajunctional sensitivity and a correlative decrease of the number of receptor molecules. Electrical stimulation of the denervated muscle has an effect which parallels that of reinnervation. Measurements of the life time of the receptor protein in extrajunctional areas during denervation and re-innervation does not reveal any change of degradation rate of the protein, which remains close to 18-20 hours. The regulation therefore affects the synthesis of the receptor molecule : electrical stimulation of the muscle "shuts off" the biosynthetic machinery which, on the other hand, is released by denervation. The internal signal involved in this regulation have not been identified.

Similar regulation at the level of receptor biosynthesis might take place in nerve or glial cells for other receptors (see Labrie, this Symposium). Alternatively, or in addition, a regulation of receptor degradation may happen. This might for instance be the case of the so-called "down regulation" of various peptide receptors where ligand binding is assumed to enhance the degradation of its specific receptor.

CONCLUSION

The regulation of neuronal activity via the postsynaptic membrane begins to be understood at the molecular level and more particularly at the level of the receptors for neurotransmitters (neuropeptides included). During the past ten years several of these receptors have been characterized, isolated and a few of them purified in milligram quantities. Change in the sensitivity of the cell to a neurotransmitter or a hormone may take place via the regulation of receptor number. Alternatively the amplitude of the response may vary without change in receptor number. This last mechanism begins to be understood with the membrane bound acetylcholine-Regulator from electric tissue or muscle and with cyclase regulators from various origins. It most likely involves the regulation of the coupling between the receptor <u>sensu stricto</u> and the associated biological unit mediated by a conformational transition which, in several instances, may resemble that found with typical regulatory proteins.

References

1. Katz, B., Mac-Graw-Hill Book Co., New-York, 1966.

   Katz, B., and Miledi, R., In "Motor innervation of Muscle" (S. Thesleff, ed.) Academic press, London, New-York, San-Francisco, pp. 31-50, 1976.

2. Nachmansohn, D., and Neumann, E., Chemical and Molecular basis of nerve activity, Academic Press, New-York, San-Francisco, London, 1975.

3. Heidmann, T., and Changeux, J.P., Ann. Rev. Biochem., 47: 371-441, 1978.

4. Bennett, M.V.L., Synaptic transmission and neuronal interaction, Raven Press, New-York, pp. 153-178, 1974.

5. Llinas, R., Baker, R., and Sotelo, C., J. Neurophysiol., 37: 560-571, 1974.

6. Monod, J., Changeux, J.P., and Jacob, F., J. Mol. Biol., 6: 306-328, 1963.

7. Changeux, J.P., In "Symmetry and Function of Biological Systems at the macromolecular Level" (A. Engström, and B. Strandberg, ed.), Nobel Symposium N° 11, Wiley Inter-Science Division, New York, pp. 235-256, 1969.

8. Gerhart, J.C., Curr. Top. Cell. Reg. 2: 275-325, 1970.

9. Changeux, J.P., Cold Spring Harb. Symp. Quant. Biol., 28: 497-504, 1963.

10. Gerschenfeld, H., In "Synapses" (G.A. Cottrell, P.N.R. Usherwood, and Blackie, ed.) Glasgow, London, 1977.

11. Kuffler, S.W., and Nicholls, J.G., Sinaner Associates, Inc. Publishers, Sunderland, Mass., 1977.

12. Daly, J.W., Karle, J., Myers, C.W., Tokuyama, T., Walters, J.A., and Witkop, B., Proc. Nat. Acad. Sci. USA, 68: 1870-1875, 1971.

13. Young, A.B., and Snyder, S.H., Proc. Nat. Acad. Sci. USA, 71, 4002-4005, 1974.

14. Olsen, R.W., In "Gaba in Nervous System Function" (E. Roberts, T.N., Chase, and D.B. Tower, ed.) Raven Press, New-York, 1976.

15. Grünhagen, H., and Changeux, J.P., J. Mol. Biol., 106: 497-535, 1976.

16. Sobel, A., Heidmann, T., and Changeux, J.P., C.R. Acad. Sci. Paris, 285 D : 1255-1258, 1977.

17. Sobel, A., Heidmann, T., Hofler, J., and Changeux, J.P., Proc. Nat. Acad. Sci. USA, 75: 510-514, 1978.

18. Monod, J., Wyman, J., and Changeux, J.P., J. Mol. Biol. 12, 88-118, 1965.

19. Lee, C.Y., and Chang, C.C., Mem. Inst. Butantan Symp. Intern., 33: 555-572, 1966.

20. Changeux, J.P., Kasai, M., and Lee, C.Y., Proc. Nat. Acad. Sci. USA, 67: 1241-1247, 1970.

21. Meunier, J.C., and Changeux, J.P., FEBS Letters, 32: 143-148, 1973.

22. Changeux, J.P., Benedetti, L., Bourgeois, J.P., Brisson, A., Cartaud, J., Devaux, P., Grünhagen, H., Moreau, M., Popot, J.L., Sobel, A., and Weber, M., In "The Synapse" Cold Spring Harbor Symp. on quantitative Biology, Vol. XL, pp. 211-230, 1975.

23. Sugiyama, H., and Changeux, J.P., Eur. J. Biochem., 55: 505-515, 1975.

24. Weber, M., David-Pfeuty, M.T., and Changeux, J.P., PNAS 72: 3443-3447, 1975.

25. Grünhagen, H., Iwatsubo, M., and Changeux, J.P., Eur. J. Biochem, 80: 225-242, 1977.

26. Heidmann, T., Iwatsubo, M., and Changeux, J.P., C.R. Acad. Sci. Paris, 284 D: 771-774, 1977.

27. Katz, B., and Thesleff, S., J. Physiol. London, 138: 63-80, 1957.

28. Heidmann, T., and Changeux, J.P., Eur. J. Biochem. (submitted), 1978.

29. Zieglgänsberger, W., and Bayerl, H., Brain Research, 115: 111-128, 1976.

30. Miledi, R., and Potter, L.T., Nature, 233: 599-603, 1971.

31. Hall, Z., and Reiness, C.G., Nature, 268: 655-657, 1977.

REPORTER'S REMARKS

Göran Sedvall

Laboratory of Experimental Psychiatry,
Dept. of Psychiatry,
Karolinska Institute, Stockholm, Sweeden

The organizing committee certainly took a great risk
in asking me to report on this session. Being an experimental psychiatrist I am far from an expert in this
highly developed field. But being a scholar I looked
upon this task as an opportunity to learn more about
a field that ultimately has to be of the greatest
importance to psychiatry.

The session I will report on was entitled "Transmitter and neuropeptide synaptic mechanisms". The organizers had selected such a broad title to cover the
presentations of six distinguished scientists, who have
made significant contributions to our understanding of
the basic mechanisms of brain function during the past
two or three decades. The session well illustrated the
interdisciplinary approach required to unravel the
complexity of neuronal systems. Methods from the fields
of biophysics, biochemistry, physiology, pharmacology,
neuroanatomy and behavioral sciences were presented
by the different speakers to transmit to the audience
knowledge of how information can be processed within
the brain and how nerve cells communicate with each
other.

In the first presentation Dr. Axelrod described some
of the mechanisms of information processing following
stimulation of β-adrenergic receptors in the brain. As
a model system for this receptor, which is stimulated
by activity in central noradrenergic neurons, he had

selected the pineal gland. Dr. Axelrod described how
small diurnal changes in the impulse activity to this
organ cause marked changes in the ability of the organ
to synthesize its effector hormone, melatonin. In fact
an only twofold increase in impulse activity during
the night results in a 50-fold increase in melatonin
synthesis. This tremendous amplification was shown to
be mediated by a cyclic AMP dependant mechanism stimulating the synthesis of the critical enzyme in melatonin
synthesis, serotonin-N-acetyl-transferase.

The β-receptor in the pineal was also shown to have
a memory function for the prior application of the
transmitter. Thus reduced release of noradrenaline
causes supersensitivity of the receptor whereas increased
release results in a rapid subsensitivity. The changes
in sensitivity of melatonin synthesis to receptor
activation was shown to be regulated at several steps
in the biochemical sequence. Thus receptor binding,
adenylate cyclase, and protein kinase activities were
all shown to be markedly influenced by the previous
exposure of the β-receptor to agonists. This multi-
locational regulation of the system was claimed to
explain the rapid acceleration of melatonin synthesis
during the night.

How the β-receptor may initiate such a series of
events was analysed by Dr. Axelrod using red cell ghosts.
In a fascinating series of slides he presented evidence
for the new hypothesis that the β-receptor stimulates
the methylation of phospholipids in the membrane. By
the stimulation of two methylating enzymes phospholipid
synthesis is accelerated, causing the translocation
of the phospholipids with a subsequent increase of
membrane fluidity or instability. This increases the
probability for interaction between the receptor and
adenylate cyclase in the membrane. In the agonist free
state the β-receptor may have an inhibitory effect on
phospholipid methylation. It seems possible, that the
discovery of this new mechanism in the erythrocyte
model may turn out to be of more general importance
not only for central β-receptors but also for other
transmitter induced membrane events in the nervous
system.

Dr. Greengard was next on the list of speakers. We
were presented with an excellent review of the biochemical mechanisms underlying pre- and postsynaptic
events following aminergic transmission in the brain,

based on results obtained in his laboratory, that were
subsequently to a large extent confirmed in other
laboratories.

For no less than 8, more or less established, central
neurotransmitter amines, acting at specific receptors,
postsynaptic effects were shown to be mediated by cyclic
AMP or cyclic GMP. For most of the major groups of
psychoactive compounds Dr. Greengard presented evidence
that their effect is mediated via transmitter sensitive
cyclic AMP generating mechanisms. Antischizophrenic
drugs were accordingly shown to interfere with dopamine
sensitive adenylate cyclases in the brain. Hallucinogenic
drugs like LSD blocks a 5-HT sensitive adenylate cyclase
in brain. Recently the whole series of antidepressant
drugs with different biochemical action profiles were
shown to block a histamine sensitive adenylate cyclase
with H_2 receptor properties in the brain. Most of the
drugs were effective at the very low concentrations
obtained in tissues of drug treated patients.

Dr. Greengard proceeded to show evidence for the
critical role of protein phosphorylation for a number
of regulatory agents in brain and peripheral tissues.
A cyclic AMP dependant phosphoprotein called protein I
has been shown to be uniquely present in the brain
where it is localized to pre- and postsynaptic structures.
The degree of phosphorylation of this protein was
markedly affected by in vivo treatment of animals with
sedative and central stimulating drugs. These results
indicate a close relationship between the control of
phosphorylation of protein I and brain excitability.
The findings of Dr. Greengard open up the possibility
that the phosphorylation of specific substrate proteins
by protein kinase represents a more general common
mechanism for the actions of numerous regulatory agents
besides the neurotransmitters acting through cyclic
AMP and cyclic GMP. Calcium ions and steroid hormones
are suggested to be examples of such agents.

Next on the scene was Dr. Floyd Bloom. On the basis
of the apparent multitude of established and putative
neurotransmitters in the brain he discussed the
contrasting principles of synaptic physiology between
peptidergic and non-peptidergic neurons. In spite of
his presentation of a list of about 15 putative
transmitters several members of the audience called for
an extension of the list. Dr. Bloom suggested the
classification of neuron populations in terms of spatial,

temporal and so called energy domains over which they
operate. By this multidimensional system anatomical,
electrophysiological and biochemical information can
be integrated in the attempt to understand the variance in
functional characteristics of different types of neurons.
He discussed as an example the noradrenergic locus
coeruleus neurons with their vast distribution to
different areas of the brain and long latencies and
duration of effects transmitted. In contrast the
peptidergic neurons, like those containing enkephalin
or β-endorphin, were shown to have a more localized
spatial domain. Whereas norepinephrine neurons were
suggested to have a biasing or enabling influence on
other transmitter functions the electrophysiological
characteristics and functional role of the peptidergic
neurons is so far poorly understood.

Dr. Iversen in his presentation showed us a list of
17 small peptides that by March -78 were suggested
to have a neuronal localization. Dr. Iversen made a
strong case for the transmitter function of several
of these, by demonstrating their release from brain
slices by depolarizing agents and also a calcium ion
dependence of the release. Radioimmunoassay was used
for detection of the peptides released. The occurrence
of differences between synaptic mechanisms in one type
of peptidergic neuron but in different locations of the
central nervous system, was illustrated by studying
release of substance P (SP) from substantia nigra (SN)
and the trigeminal nerve nucleus. Whereas the evoked
release of SP from sensory nerve terminals in slices
of the trigeminal nerve nucleus was inhibited by morphin
and other opiate analgesics, SP release from SN was
unaffected. GABA on the other hand had an inhibitory
effect in both regions. This finding directly illustrates
the occurrence of different types of interaction for
even a single class of peptidergic neurons in different
brain regions. In similar slice preparations from
other brain parts the release of enkephalins, somatostatin,
neurotensin and vasointestinal peptide, VIP, was demonstrated. Dr. Iversen also showed the possibility of
release of serotonin and dopamine from SN slices by SP.
The technique used may turn out to be useful not only
to study interaction between different neuron types in
central nervous system but also to find new principles
for manipulation of synaptic mechanisms by drugs.

Dr. Goldstein in his presentation summarized the
evidence for the presence of epinephrine (E) neurons in

the brain. By developing an immunofluorescence technique
in collaboration with Dr. Fuxe, the epinephrine forming
enzyme PNMT could be visualized and its distribution
in nerve terminals originating from the C_1 and C_2 cell
groups of medulla oblongata be verified. The functional
role of brain epinephrine was studied by following E
levels in brain after PNMT inhibition. Foot shock
stress, genetic factors and blood pressure were found
to influence epinephrine mechanisms in the brain. E
terminals were demonstrated in locus coeruleus (LC)
where the transmitter was suggested to have an inhibitory effect.

Dr. Carlsson's paper dealt with the effects of some
30 peptides on monoamine transmitter turnover and
behavior in rats. The peptides were administered
intraventricularly. The opiates increased the synthesis
of the catecholamines and serotonin and induced a
catatonia-like state. Naloxone blocked these effects
but also had effects when administered alone. SP and
its analogues also increased catecholamine turnover.
Naloxone counteracted this effect but potentiated
the behavioral action. TRH and specially its DOPA-
substituted analogue increased the synthesis of dopamine
and noradrenaline. After the DOPA-analogue the presence
of free DOPA could also be demonstrated in the brain.
Since the DOPA-substituted analogue but not DOPA resulted
in a type of "wet dog" behavioral syndrome it was
suggested that the specific localization of peptidases
in brain may cause the selective DOPA distribution to
specific brain sites when DOPA-containing peptides are
administered. The results obtained indicate the
possibility to use this principle to selectively adminis-
ter pharmacological agents to specific areas of the brain.

Now a few general comments. The Greek philosopher
Aristotle considered that the main function of the brain
was to be a cooling system for the blood. This reporter
certainly would have appreciated a cooling system for
his brain during the flooding of his information
processing transmitter mechanisms during this excellent
symposium.

If I should select two contributions from the session
that I found especially exciting and that may be of
special importance for further exploration as to
biological generality, I would mention Dr. Axelrod's
demonstration of β-receptor controlled phospholipid
methylation and Dr. Greengard's uncovering of phosphory-

lation of specific brain proteins for the mediation of
regulatory interactions between nerve cells. These
findings in addition to those of the other contributions
of this session will undoubtedly be of great value not
only for our understanding of brain function in general
but also in the formulation of new hypotheses regarding
pathophysiological mechanisms in patients and will make
it possible to manipulate these mechanisms therapeuti-
cally.

To an experimental psychiatrist like myself it is
embarrasing how little of the information presented
in this symposium that can yet be used and verified
clinically. In psychiatry therapeutic work has so far
been directed exclusively by behavioral and psychological
information. The coupling between central transmission
mechanisms and peripheral endocrine events as discussed
here indicates the indirect approach required.

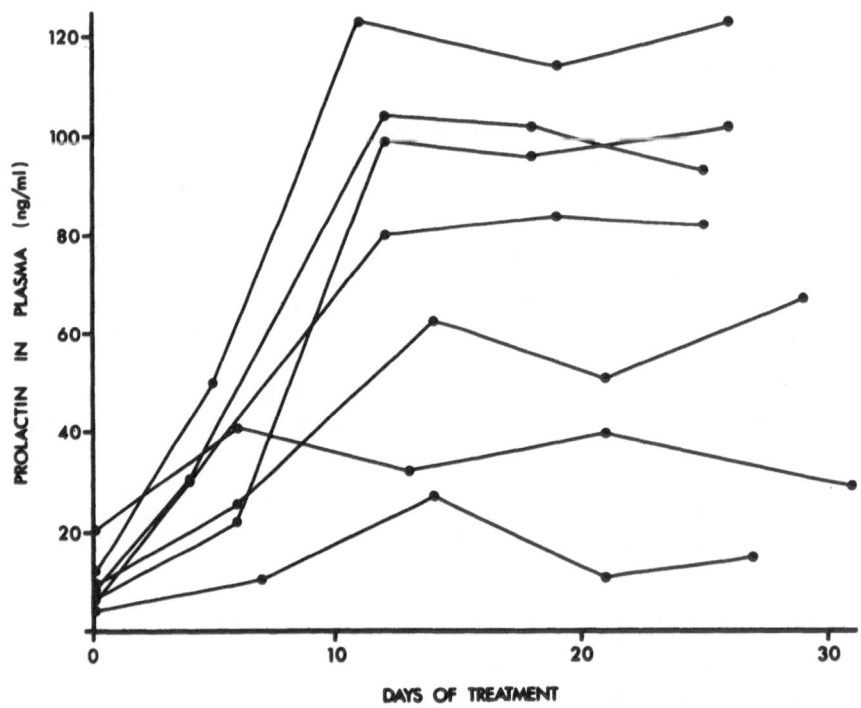

FIG. 1 Prolactin levels in plasma from psychotic
women treated with thiothixene (10 mg x 3).

The only example of this approach previously available
for psychiatry is the marked stimulation of prolactin
release in plasma of patients during treatment with
neuroleptic, dopamine receptor blocking, drugs. The
prolactin release induced by neuroleptics is mediated
by the interference with hypothalamic, and as demonstrated
by Dr. Besser today, by pituitary dopaminergic mechanisms.
As seen in Fig. 1 the effect is substantial and it can
be reliably measured. It is also fairly specifically
related to dopaminergic receptor interaction since
other psychoactive drugs interfering with other transmitter mechanisms do not have this effect (Table I).

TABLE I Effect of psychoactive drug treatment on levels
of prolactin-like material in cerebrospinal fluid of
female psychiatric patients.

| | n | Prolactin level in CSF (ng/ml) | |
		Before treatment	During treatment
Chlorpromazine	7	1.1 ± 0.2	5.6 ± 0.4[xxx]
Thiothixene	15	2.0 ± 0.3	14.4 ± 1.2[xxx]
Melperone	13	1.7 ± 0.2	5.6 ± 0.7[xxx]
Clomipramine	5	1.8 ± 0.3	2.7 ± 0.7
Lithium	5	1.0 ± 0.3	2.8 ± 0.8

Differs from before, xxx $p < 0.001$.

The patients were treated for 2 weeks with chlorpromazine (200-600 mg), thiothixene (30 mg), melperone
(300 mg), clomipramine (100 mg) or lithium (individual
dose, giving plasma concentration close to 1.0 meq/l).
(From Sedvall et al., Proc. Sixth Int. Congr. Pharmacol.
1975.)

It has also been suggested that this effect may be
useful for the prediction of the therapeutic effect of
antischizophrenic drugs. And as shown in Fig. 2 it is
possible to demonstrate direct correlations between
the reduction of psychotic morbidity and the prolactin
elevation in plasma of schizophrenic patients treated
with chlorpromazine. However, the individual variability
in prolactin secretion as influenced by sex, age,
estrogen secretion etc still makes this approach

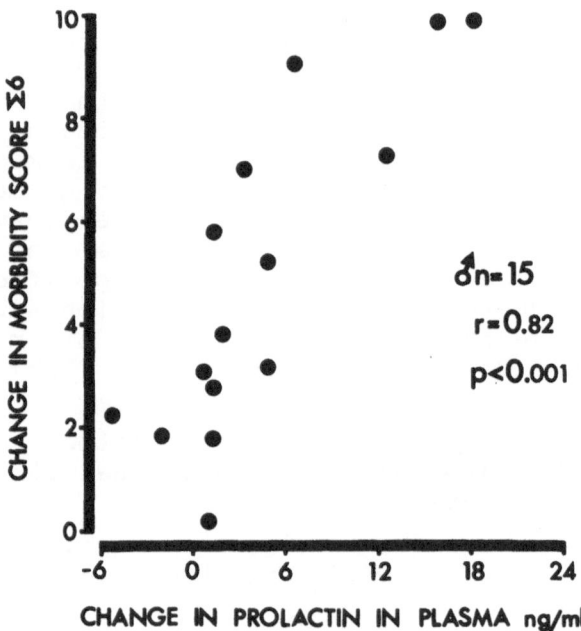

FIG. 2 Correlation in men between change
in prolactin in plasma and morbidity
score Σ6 (CPRS) after 4 weeks of chlorpromazine treatment. Spearman's rank correlation.
(From Wode-Helgodt, Thesis, Karolinska
Institutet, Stockholm, 1977.)

unsatisfactory for practical work with patients. It is
my conviction, however, that the elucidation of the
molecular events following transmitter and peptide
interaction in brain, that we have heard of here,
combined with the further refinement of quantitative
analytical techniques, will give us additional and
more useful tools to follow molecular events in the
central nervous system of patients in the near future.

OVERVIEW OF SESSION IV: INTERACTIONS BETWEEN HORMONES AND NEURO-

TRANSMITTERS IN THE CONTROL OF PEPTIDERGIC NEURONS

Reporter: Charles H. Sawyer

Department of Anatomy and Brain Research Institute
UCLA School of Medicine
Los Angeles, California 90024 U.S.A.

Drs. Fuxe and McCann have discussed many aspects of neuro-transmitter systems in the control of anterior pituitary function. It is apparent that acetylcholine, catecholamines, serotonin, histamine and GABA and their receptor stimulants or agonists, receptor blockers and synthesis inhibitors can stimulate or inhibit the secretion of many pituitary tropic hormones by influencing, in most cases, the secretion of peptide releasing hormones from hypothalamic neurons. The effects of catecholamines on gonadotropic function have been most thoroughly studied and perhaps for that reason have been the most controversial. Specifically, the most unresolved problem is the influence of dopamine on LHRH; everyone agrees that it inhibits prolactin secretion and many consider it PIF itself.

One of the difficulties is the definition of "stimulation". To Fuxe and his colleagues who have been looking at estrous cycles and the turn-over of median eminence catecholamines (15-17), and to those of us who are interested in ovulation (13, 14, 22, 23, 24, 28), stimulation implies the induction of an ovulatory surge of gonadotropin, and McCann and his associates (10, 11) employ the same meaning when considering effects of epinephrine or norepinephrine. However, when discussing effects of dopamine, McCann's laboratory (26, 30, 31) and Kordon's institute (21) look only for relatively small yet statistically significant changes such as may be induced in male rats. To our knowledge no one has ever induced an ovulatory surge of gonadotropin with dopamine at any dosage applied to any part of the animal body under any hormonal conditions. However, both epinephrine and norepinephrine infused intraventricularly have been highly effective in triggering an ovulatory surge of gonadotropin in the estrous rabbit and the

proestrous rat. Recent studies quoted by McCann confirm that under
certain conditions both in vivo and in vitro dopamine in weak dosages can activate the discharge of LHRH (21, 30). McCann's in
vitro preparation also responds to norepinephrine (26) but
Rotsztejn et al. (21) report that their dopamine-sensitive in vitro
preparation of male rat median eminence fails to release LHRH when
norepinephrine is added to the medium. In other studies on the
pituitary-median eminence unit in vitro Miyachi et al. (19) reported that dopamine inhibited release of gonadotropin into the medium, results consistent with Fuxe's hypothesis of inhibition of
LHRH discharge by dopamine.

Of critical importance to the triggering of a surge of gonadotropin by electrical or chemical stimulation is the estrous condition introduced by estrogen priming or natural proestrus. It is
practically impossible to ovulate an anestrous or pseudopregnant
rat or rabbit with electrical or chemical stimuli. Very dramatic
estrogen-induced reversals have been reported in plasma LH responses to arcuate nucleus stimulation in the ovariectomized rat--
depression changing to marked elevation following estrogen priming
to a state of artificial proestrus (8). Estrogen probably acts at
multiple sites to facilitate ovulation, e.g., brain, pituitary and
ovary. Even exogenous LHRH or gonadotropins are more effective in
an estrogen-primed animal. Because of the effects of specific
receptor blockers, receptor stimulators and synthesis inhibiting
agents, Gallo and his colleagues (1, 6) consider that the pulsatile
rhythm of LH in the gonadectomized rat is inhibited slightly by
endogenous dopamine and more effectively by serotonin: the depression resulting from dorsal raphe nucleus stimulation is blocked by
PCPA, an inhibitor of serotonin synthesis, and restored by 5HTP,
which bypasses the synthesis inhibition, and also blocked by metergoline, a serotonin receptor blocker (1). The inhibition of the
episodic rhythm by arcuate nucleus stimulation is reversed by PCPA
as effectively as by estrogen (7).

We have recently reported the noradrenergic stimulation of
electrical changes in the estrous rabbit rhinencephalon associated
with the induction of ovulation, changes which could not be induced
in the same animal by norepinephrine (NE) during pseudopregnancy
but which were restored when the doe again became estrous (24).
When a radioimmunoassay for rabbit LH was developed (25) we found
that intraventricular dopamine (DA) not only failed to induce ovulation in the estrous rabbit, it also failed in either high or low
dosage to elevate LH at all (23). When DA was given 2 hr before
the infusion of an ovulatory dose of NE none of the rabbits showed
a significant rise in serum LH or ovulation. Part of this blockade
of the NE effect may have been due to the stress of restraint and
intracardiac bleeding at the time of the DA infusion and not to the
effect of DA. In later experiments with induction of LH release
and ovulation by intraventricular $ACTH_{1-24}$ we have found that this

stimulus was blocked by the combination of restraint, bleeding and saline infusion 2 hr earlier (2). So our results do not affirm that DA is necessarily inhibitory, but that we have never found it stimulatory in vivo in either the rabbit or the rat.

McCann's laboratory (10, 11) has demonstrated that the induction of an ovulatory surge of LH by progesterone or electrochemical stimulation is blocked by diethyldithiocarbamate (DDC), an inhibitor of synthesis of NE; the latter is restored by dihydroxyphenyl-serine (DOPS), which bypasses the enzyme inhibition and permits direct synthesis of NE. This is among the best evidence of NE involvement in LH release. Another item is the differential elevation in NE in the suprachiasmatic nucleus on the afternoon of proestrus (27) consistent with Fuxe's NE turnover, and a third is the dose response curve for elevations of plasma LH on intraventricular injections of increasing dosages of NE to estrogen-progesterone primed ovariectomized rats (13). Comparable doses of epinephrine (E) overrode or bypassed pentobarbital blockade to trigger an ovulatory surge of LH at proestrus, but NE was no more effective than dopamine under pentobarbital (22). Vijayan and McCann (30) also note that E is the most effective catecholamine in stimulating LHRH release and Fuxe reemphasizes that E-containing neurons are present in the hypothalamus (9). Under urethane anesthesia both E and NE stimulated a biphasic elevation-depression sequence in median eminence multiunit activity (MUA) on the afternoon of proestrus. Dopamine failed to induce the electrical changes or trigger LH release (14, 33), even in dosages reported by Vijayan and McCann (30) to activate some LH output. E and NE both depressed MUA in the arcuate nucleus but intraventricular LHRH stimulated it (14).

According to Ungerstedt (29) noradrenergic neurons project from the hindbrain by dorsal and ventral bundles in the midbrain to the diencephalon and forebrain. Electrochemical stimulation in the dorsal midbrain tegmentum at the site of separation into dorsal and ventral bundles induced ovulation in constant estrous rats (4) and depressed MUA in the preoptic area (3). These effects were interpreted as representing depression of inhibitory interneurons inasmuch as the resultant activation of peptidergic LHRH neurons would appear to be a requisite of the ovulatory surge of LH. Interestingly, Fuxe also interposes an inhibitory interneuron between the ascending noradrenergic neuron and the peptidergic neuron to be discharged.

Recently in our laboratory Clifton (5) has severed the nor-adrenergic bundles at the midbrain level in cycling rats, but within 22 days all had resumed estrous cycles and ovulation in spite of an 83% drop in hypothalamic NE. Furthermore, treatment with DDC on the morning of proestrus blocked ovulation in 10/11 controls but in only 4/10 rats with transected adrenergic tracts.

It would appear that hypothalamic function including LHRH secretion
can adjust to the loss of most of its NE input. Similarly in rabbits, Rabii and Sawyer (unpublished) have made repeated injections
of 300 µg 6-hydroxydopamine (6-OHDA) into the third ventricle without blocking copulation-induced ovulation unless the injection produced a histologically-confirmed lesion. One interesting finding
was that the acute effect of 6-OHDA injection was stimulation of
LHRH neurons, i.e., LH release and ovulation. This observation
questions the interpretation of Martinovic and McCann (18) that
acute infusions of 6-OHDA into the region of the ventral noradrenergic bundle blocked progesterone-induced ovulation by virtue of
destroying the NE projection. Stimulation by 6-OHDA of the many
dopaminergic neurons in the vicinity of the infusion, some of which
project to the hypothalamus (12), would seem fully as logical an
explanation of the inhibition of LH release.

McCann's suggestion that histamine and GABA may be stimulatory
to LH release and Pass and Ondo's (20) report that GABA is effective only under pentobarbital anesthesia recall our 1955 report of
a synergistic effect of intraventricular histamine and systemic
pentobarbital resulting in rhinencephalic EEG changes and ovulation
in the rabbit (see 24). Whereas Vijayan and McCann report LH
release by GABA in the unanesthetized rat (31) and we have recently
stimulated LH release and ovulation with intraventricular aminoxy-
acetic acid in the unanesthetized rabbit (32), further investigations of interactions of pentobarbital with histamine and GABA are
indicated.

It is generally agreed that endogenous norepinephrine is instrumental in triggering an ovulatory surge of LHRH and LH in the
proestrous rat since the natural surge is prevented with α-adrenergic blocking agents and NE-synthesis inhibitors and stimulated
by exogenous NE or the restoration of endogenous NE. However,
since the ovulatory mechanism survives the elimination of most of
the hypothalamic content of NE, the latter must be relegated to
the role of modulator rather than essential stimulator of the surge.
The role of dopamine is still controversial as far as LHRH is concerned, but its inhibition of prolactin secretion is unquestionable.

REFERENCES

1. Arandash, G.W. and R.V. Gallo. _Endocrinology_ 102:1199-1206,
   1978.

2. Baldwin, D.M., C.K. Haun and C.H. Sawyer. _Brain Research_ 80:
   291-301, 1974.

3. Carrer, H.F. and C.H. Sawyer. _Exper. Neurol._ 52:525-534, 1976.

4. Carrer, H.F. and S. Taleisnik. J. Endocrinol. 48:527-539, 1970.

5. Clifton, D.K. and C.H. Sawyer. The Physiologist, in press 1978.

6. Drouva, S.V. and R.V. Gallo. Endocrinology 100:792-798, 1977.

7. Gallo, R.V. and G.P. Moberg. Endocrinology 100:945-954, 1977.

8. Gallo, R.V. and R.B. Osland. Endocrinology 99:659-668, 1976.

9. Hökfelt, T., K. Fuxe, M. Goldstein and O. Johansson. Brain Research 66:235-251, 1974.

10. Kalra, P.S., S.P. Kalra, L. Krulich, C.P. Fawcett and S.M. McCann. Endocrinology 90:1168-1176, 1972.

11. Kalra, S.P. and S.M. McCann. Endocrinology 93:356-362, 1973.

12. Kizer, J.S., M. Palkovits and M.J. Brownstein. Brain Research 108:363-370, 1976.

13. Krieg, R.J. and C.H. Sawyer. Endocrinology 99:411-419, 1976.

14. Krieg, R.J., O.P. Tandon, D.I. Whitmoyer and C.H. Sawyer. Neuroendocrinology 22:152-163, 1976.

15. Löfström, A., L.F. Agnati and K. Fuxe. Neuroendocrinology 24: 270-288, 1977.

16. Löfström, A., L.F. Agnati, K. Fuxe and T. Hökfelt. Neuroendocrinology 24:289-316, 1977.

17. Löfström, A., P. Eneroth, J.A. Gustafsson and P. Skett. Endocrinology 101:1559-1659, 1977.

18. Martinovic, J.V. and S.M. McCann. Endocrinology 100:1206-1213, 1977.

19. Miyachi, Y., R.S. Mecklenburg and M.B. Lipsett. Endocrinology 93:492-496, 1973.

20. Pass, K.A. and J.G. Ondo. Endocrinology 100:1437-1442, 1977.

21. Rotsztejn, W.H., J.L. Charli, E. Pattou and C. Kordon. Endocrinology 101:1475-1483, 1977.

22. Rubinstein, L. and C.H. Sawyer. Endocrinology 86:988-995, 1970.

23. Sawyer, C.H., J. Hilliard, S. Kanematsu, R. Scaramuzzi and
    C.A. Blake. Neuroendocrinology 15:328-337, 1974.

24. Sawyer, C.H. and H.M. Radford. Brain Research 146:83-93, 1978.

25. Scaramuzzi, R.J., C.A. Blake, H. Papkoff, J. Hilliard and
    C.H. Sawyer. Endocrinology 90:1285-1291, 1972.

26. Schneider, H.P.G. and S.M. McCann. Endocrinology 85:121-132,
    1969.

27. Selmanoff, M.K., M.H. Pramik-Holdaway and R.I. Weiner.
    Endocrinology 99:326-329, 1976.

28. Tima, L. and B. Flerkó. Neuroendocrinology 15:346-354, 1974.

29. Ungerstedt, W. Acta physiol. Scand. (Suppl.) 367:1-48, 1971.

30. Vijayan, E. and S.M. McCann. Neuroendocrinology 25:150-165,
    1978.

31. Vijayan, E. and S.M. McCann. Brain Research, in press 1978.

32. Wallis, C. and C.H. Sawyer. Fed. Proc. 37:225, 1978.

33. Weiner, R.I., C.A. Blake, L. Rubinstein and C.H. Sawyer.
    Science 171:411-412, 1971.

DISCUSSION OF CLINICAL NEUROENDOCRINOLOGY SECTION

Seymour Reichlin

Tufts University School of Medicine

Boston, Massachusetts

The three papers by Drs. Besser, Yen and Müller are elegant examples of current work in clinical neuroendocrinology. They illustrate the powerful insights and tools that basic research has given to the clinical investigator, outline the scope of the major questions in pituitary regulation that are still unresolved, and anticipate some of the clinical benefits that have followed elucidation of the mechanisms of hypothalamic-pituitary regulation. Dr. Besser ascribes the beginnings of clinical neuroendocrinology to the introduction of synthetic hypothalamic releasing hormones and of dopamine agonists into clinical medicine. To be sure, these events must be considered to be a watershed in development of clinical neuroendocrinology for they mark the introduction of specific pharmacological tools into diagnosis and treatment, but the origins of clinical neuroendocrinology reach back much further. Perhaps these origins can be found in the writings of Vesalius(1543) who described the drainage of cerebrospinal fluid into the nose, hence the term pituitary (mucous,pituita,Gr.) or even earlier, in the writings of Aristotle, (384-322,B.C.), a doctor's son who described pseudocyesis and the effects of castration on sex drive in animals and in man. But it would not be fair to date clinical neuroendocrinology to such early times because the general concept of hormones was not realized until the early part of this century, and the neuroendocrine doctrine that had emerged from the blending of the hypophysial-portal chemotransmitter hypothesis of Geoffrey Harris with the neurosecretion hypothesis of Ernst Scharrer is less than 40 years old. Some of this early history is recounted by Anderson and Haymaker (2), Greep (10), Saffran (26), Reichlin (25) and in two monographs entitled "Pioneers in Neuroendocrinol-

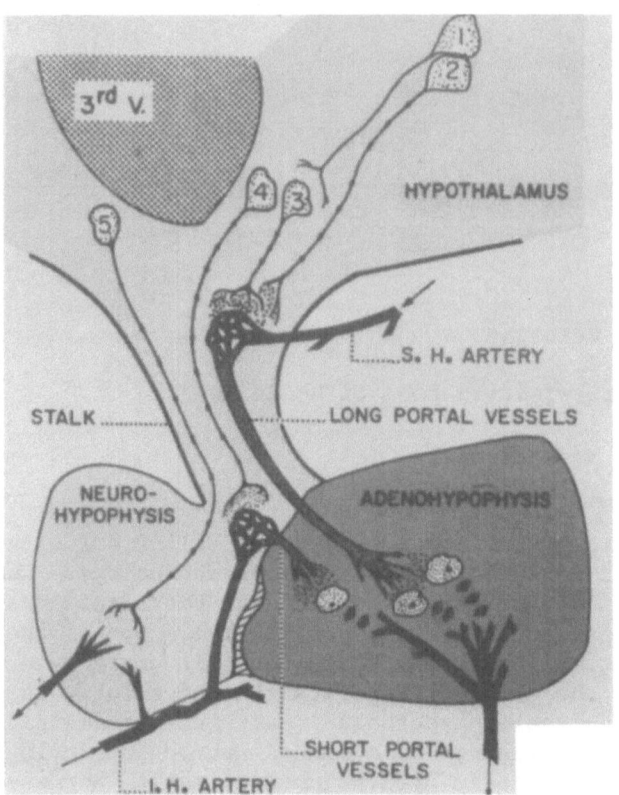

Fig. 1.   Schematic outline of hypothalamic-pituitary unit.
     This figure shows the types of neural inputs into pituitary reg-
ulation.   The neuron labelled (5) represents the supraoptico-hypo-
physial pathway of the neurohypophysis.   Neurons (3) and (4) represent the tuberohypophysial neurons that elaborate the hypophysiotropic factors.   As shown here, they may terminate in the median
eminence or low down in the stalk.   These hypophysiotropic neurons
in turn are regulated by other neurons, some of which may end as
axo-somatic synapses (1), or in the interstitial space of the stalk
median eminence region (2).   In this location there are few if any
axo-axonic synapses; instead, the secretions of tuberohypophysial
and bioaminergic neurons may come into contact with nerve endings.
From V.L.Gay, Fertility Sterility 23: 50-60, 1972.   Courtesy of
Williams & Wilkins, Baltimore, Md. (with permission).

ogy" (18,19.   Full treatises on clinical neuroendocrinology have
been published by Martin et al (16),and by Martini and Besser (17)
and a third is in press (Tolis et al 30).

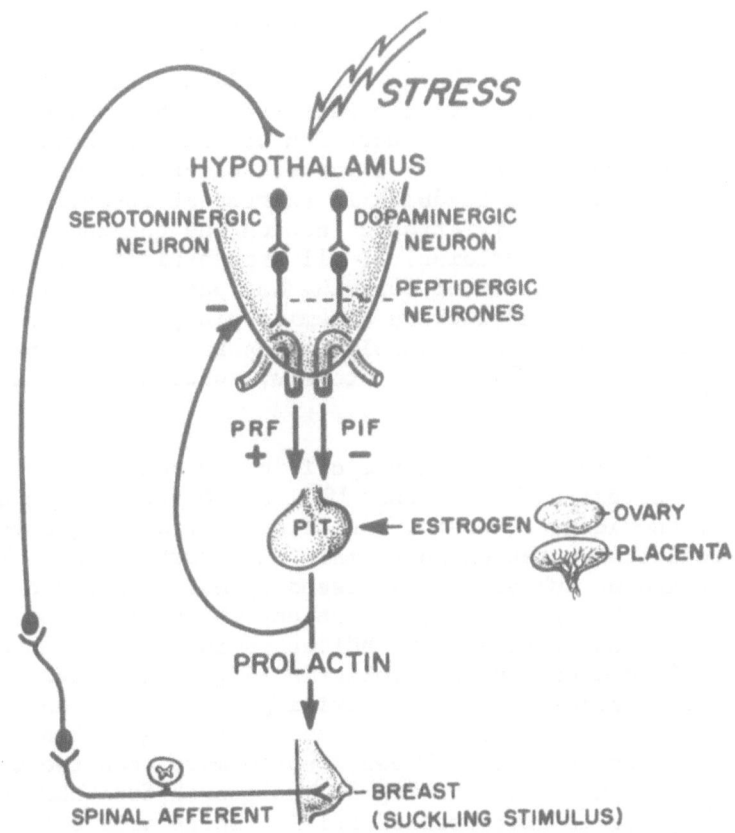

Fig. 2. Diagrammed here is the dual system for control of prolactin secretion.

Prolactin secretion is stimulated by one or more prolactin releasing factors (one of which is TRH), and is inhibited by one or
more PIF(s). Dopamine, which is a potent PIF, acts directly on the
pituitary, and may act indirectly through the release of one or more
PIF(s), the nature of which have not been identified. Direct dopa-
aminergic input comes through the tuberohypophysial dopaminergic
pathway that ends in the median eminence (not shown). Pituitary responses are modulated by hormones, the most important of which is
estrogen which sensitizes prolactinotropes, and also stimulates
growth and differentiation of these cells. From Clinical Neuroendocrinology, J.B.Martin,S.Reichlin and G.Brown, F.A.Davis, Philadelphia,
1977. (with permission) (16).

All the papers in this section take as dogma the schema out-
lined in Fig. 1. and presented in detail earlier in the symposium.

The crucial points to emphasize are 1) adenohypophysial secretion
is regulated by hypophysiotropic factors elaborated in the hypothalamus, and conducted by the portal vessels of the stalk to the
pituitary, 2) the hypophysiotropic factors are the product of
neurosecretory neurons, 3) the secretion of the hypophysiotropic
factors are in turn regulated by a variety of chemotransmitter-
coded neurons, 4) hormone feedback is exerted at several level:
the pituitary, the hypophysiotropic neuron and the chemotransmitter neurons of the hypothalamus, 5) all hypophysiotropic factors
with the possible exception of dopamine are peptides, 6) chemo-
transmitter-coded neurons include bioaminergic pathways: noradrenaline, serotonin, dopamine, possibly epinephrine, histamine; peptidergic pathways: endorphin, neurotensin, substance P, vasopressin;
GABAergic pathways.

The most ambitious of the three clinically related papers in
this section is that of Dr. Eugenio Müller who attempts to synthesize current knowledge of neurotransmitter control of growth hormone and prolactin secretion in animals with work in man. This is
a rapidly expanding subject as witnessed by Müller's recently published, detailed monograph on this subject which lists approximately 2800 references (22). Dr. Müller reminds us that the secretion of growth hormone, and of prolactin (Fig.2) are each regulated
by dual neural control systems, comprising stimulating and inhibitory hypophysiotropic factors. The chemistry of growth hormone
stimulating factor is still unknown despite more than two decades
of investigation but in the course of the search for GHRF, several
hypothalamic peptides have been identified which will stimulate GH
release in one or more assay systems. These include TRH (effective
in vitro in normal pituitaries, and in vivo in many patients with
growth hormone secreting tumors of the pituitary), and a molecule
that is identical with a portion of the β chain of porcine hemoglobin probably an artifact of the separation process. Other hypothalamic peptides that release GH, but only in vivo (and hence indirectly through hypothalamic mechanisms) are β-endorphin, substance
P, neurotensin, vasopressin and perhaps others. Several groups have
reported the presence of GH RF activity in crude hypothalamic extracts (14,15,23,27,33) which may or may not be one of the above.
In contrast to the uncertainties about the chemical nature of GH RF
is the certainty of the identity of somatostatin, a 14 amino acid
peptide first identified biologically by Krulich and McCann (14)
and chemically characterized by Brazeau and colleagues (6), working
in Guillemin's laboratory. Somatostatin, as every school boy knows
inhibits GH release regardless of the stimulus (31), is present in
a well defined somatostatinergic neural pathway ending in the median eminence in anatomical relationship to the interstitial space
and capillary loops of the primary hypophysial portal plexus (1,12)
is present in hypophysial portal blood (7) (thus fulfilling Harris'
requirements for proof of a bona fide hypophysiotrophic hormone).

That somatostatin is actually involved in regulation of GH secretion is shown by several studies in the rat using antisera against somatostatin. Neutralization of circulating somatostatin leads to a reversal of the usual stress induced inhibition of GH release (29), of starvation-induced GH inhibition (28), and increase in plasma TSH levels in a variety of experimental conditions (3). Whether somatostatin is involved in GH regulation in the human is unknown, but there are a number of clues that it may be. For one thing somatostatin is present in human hypothalamus and cerebrospinal fluid (24), although its concentration is unaltered in various kinds of pituitary disease (Urosa and Reichlin, unpublished).

The inhibition of pituitary TSH response to TRH in patients treated with growth hormone has been attributed by some to feedback activation of somatostatin release but this is a weak straw without documented proof. It is equally likely that GH acts directly on the pituitary (8). Because GH secretory failure is the final result of disease of the hypothalamus and stalk, it appears that the secretion of GH RF is the predominant factor in determining levels of GH secretion in man, although somatostatin secretion may modulate GH RF effects.

The state of knowledge of prolactin regulating factors is similar to that of growth hormone secretion. Prolactin inhibitory factor has been demonstrated. In fact there may be more than one such factor. Undoubtedly, dopamine, now recognized to be present in dopaminergic pathways terminating in the median eminence, is present in portal vessel blood in concentration sufficient to directly inhibit prolactin release and is therefore the most important PIF (9). But others have been described including GABA, and another putative peptide which is neither of the above. The chemical nature of prolactin releasing factor, PRF, is unknown. TRH is an authentic PRF, but that it is the physiological PRF is very unlikely. TRH, given to men, or women, releases prolactin as well as TSH, and the effects of this peptide are inhibited by thyroid hormone administration. In the rat, administration of antisera to TRH has been reported to modify plasma prolactin levels (13), in one study, but in others (11,21), no effects were observed. In the studies of Boyd et al (5), the PRF activity of hypothalamic extracts was found to be only partly due to TRH, but it is not known whether the rest of the effect can be attributed to β-endorphin, substance P, and neurotensin, all of which release prolactin when given in vivo to rats.

Lacking precise chemical markers of the respective GH and prolactin regulating hormones, neuroendocrinologists have been forced to utilize the output of the control system, namely blood levels of either prolactin or growth hormone to infer the various components of neurotransmitter control, both inhibitory and excit-

atory. When we come to man, the constraints set upon the clinician
to identify specific neuropharmacological regulators of the various
releasing and inhibitory factors becomes extraordinarily difficult.
It is this problem that Dr. Müller addresses.

Many experiments suggest that α-adrenergic receptors for norepinephrine are stimulators of GH release in both experimental
animals and man. On the other hand, there is some question about
the effects of dopamine (or other dopamine agonists). In man L-
Dopa almost always causes a release of GH, and this is true for
specific dopamine receptor agonists such as bromergocryptine. It
must be pointed out that Chihara and Arimura have recently reported that dopamine causes a release of somatostatin into the portal vessels of the rat (7), and this kind of response may be the
basis of the finding that DA receptors in monkeys inhibit GH secretion. Since dopamine also is converted to NE in the hypothalamus,
it is apparent that the overall effects of dopamine or L-Dopa will
depend upon the relative effects of NE (derived from dopamine) acting on receptors, and dopamine bringing about the release of somatostatin. Biphasic effects of dopamine are also evident in
Muller's studies in the rat: at a dose of 0.1 mg/kg, blood levels
of GH rose, but at a dose of 1 mg/kg, they were depressed. Other
workers report similar variability in dopamine responses. Müller
further introduces the complexity that dopaminergic agonists effects
may involve pre- as well as post-ganglionic DA receptors thereby
either turning off, or turning on dopaminergic tract activity.

Unfortunately the complexity of dopamine effects, i.e. pre-
vs post-synaptic, somatostatin regulatory vs GH RF regulatory has
made interpretation of L-Dopa responses in man extremely difficult.
How are we to explain the findings reported by Müller in patients
with Huntington's Chorea? It has been demonstrated that patients
with this disease have exaggeratd GH responses to insulin induced
hypoglycemia and higher GH responses to apomorphine stimulation.
Are these abnormalities of the DA receptor? Do these abnormalities
reflect an increase in GH RF release, or a deficiency of somatostatin? Of interest in this regard is our finding (E.A.Bird and S.
Reichlin, unpublished), that brain somatostatin concentration in
patients with Huntington's Chorea are the same as control brains,
but content is not an indicator of secretion rate.

Müller has analysed critically the evidence relating serotonin and GH release, and leaves us with the view that much of the
data derived from the use of tryptophan administration cannot be
taken as specific evidence that serotoninergic neurons release GH
although the work in man using blockers of serotonin action do
support a role of indolamines in GH action.

It is important to recognize that each of the three putative
neurotransmitters regulating GH release may be mobilized by dif-

ferent kinds of stimuli. For example, sleep related GH release is blocked by cyproheptadine, a serotonin blocker, and not by alpha blockers, whereas hypoglycemia induced GH release is markedly blunted by alpha blockers, and potentiated by beta adrenergic blockers. These finding suggest a variety of pharmacologically, perhaps anatomically, coded regulatory pathways.

Pharmacological studies of prolactin regulation, fortunately have given less ambiguous results, the main controversy in this area being the extent to which the PIF activities of the hypothalamus can be attributed to the tubero-infundibular dopaminergic pathway. All drugs that stimulate DA receptors inhibit prolactin release, and those that block endogenous dopamine stimulate prolactin release. Since the concentrations of dopamine present in portal vessel blood are sufficient to inhibit the pituitary when applied directly, we do not have to postulate other PIF's although others have been found, including GABA. Müller quite convincingly showed us that the DOPA agonist bromergocryptine can reduce prolactin secretion rate in rats with pituitary transplants, one more bit of evidence that dopamine receptors on the pituitary are prolactin inhibitory.

Data relating serotonergic activation to prolactin release are uniformly confirmatory, which has led to the view that the prolactin stimulatory component of neural activation is serotonergic, and the inhibitory is dopaminergic. Based on studies in reserpinized rats, Valverde and I have postulated that acute prolactin response to stimuli such as stress and suckling are due to release of a releasing hormone, and not to inhibition of secretion of an inhibitory hormone (32). Work with serotonin receptors summarized by Müller suggests that this stress induced release is mediated by serotonergic pathways.

Dr. Müller further summarized data relating other, less well established neurotransmitters to hypophysiotropic hormone secretion. These other candidate neurotransmitters, or neuromodulators including GABA and the endorphins, are present in relatively high concentration in the median eminence, and elsewhere in the hypothalamus, and conceivably are involved in either direct or indirect control. Again, as with the other pharmacological models, the fact that there are both inhibitory and excitatory hypophysiotropic hormones regulating GH and prolactin makes it difficult to be absolutely precise about the effects of GABA agonists like muscimol, or antagonists such as strychnine. Further, in the case of GABA, there is clear evidence for a direct influence on the pituitary. The bulk of data now available suggest that GABA, in modest dosage levels exerts an inhibitory effect on prolactin release at the hypothalamic level, and in large doses, the inhibitory action is exerted on the pituitary itself.

With respect to the endorphins, all the effects observed are clearly exerted within the central nervous system, and not on the pituitary. Like morphine, the endorphins administered systemically or intracerebroventricularly bring about the release of GH and PRL. Müller has provided data to suggest that the endorphin response is blocked by agents which block serotonin effects. I interpret these findings to mean that the endorphinergic system regulates serotonergic systems which in turn control secretion of GH RF.

Dr. Müller's introduction to the neuropharmacology of prolactin regulation provided an admirable basis for understanding of human prolactin secretory disorders, insightfully reviewed by Besser. He pointed out the fact that the predominant hypothalamic control of prolactin secretion is inhibitory, and gave detailed evidence for direct pituitary effects of dopamine agonists on prolactin release. Besser, in fact, believes that this is the major site of action, and takes a somewhat cautious view of the possible role of serotoninergic pathways in prolactin regulation. He reported on fascinating studies in man using a new analogue of met-enkephalin (Sandoz FK33-824) which contains "D-alanyl and methyl-phenylalanyl substitutions in the amino acids 2 and 4, and a sulphoxyl substitution in the methionine of met-enkephalin." This agent, dubbed DAME, causes a prompt rise in plasma prolactin, and growth hormone (as has previously been reported for morphine in man and animals). The bulk of Besser's talk dealt with the clinical problems of hyperprolactinemic states. It is an important and provocative fact that hyperprolactinemia is now the most common pituitary disorder seen by endocrinologists. In our own clinic, we follow approximately 200 such cases, seen from a referral area of an estimated 500,000 women or less. The minimal incidence would thus be 1:2,500, of whom at least one third have pituitary tumors. Hence the clinical delineation of this syndrome, (and the identification of occult tumors which underly the disease in many patients) is one of the most challenging problems to the clinician. (See Boyd and Reichlin, 1977) for further review.

Women present usually with galactorrhea, oligoamenorrhea or both, while men present with impotence, and almost always have much larger tumors than the women. There are intermediate grades of oligoamenorrhea including short luteal phase (due probably to direct effects of prolactin on the ovary), anovulation, and total gonadotropin failure. In addition to direct effect of prolactin on the ovary, there is undoubtedly an inhibitory effect on the secretion of LHRH. The latter view derives from the finding that almost all cases with galactorrhea-amenorrhea respond normally to injections of diagnostic test doses of LHRH, and from animal work by Meites and collaborators indicating that intrahypothalamic implantation of prolactin inhibits ovulation in the rat. Besser also deals with the overall problem of clinical management of the patient with prolactin hypersecretion. There is a role for surgery (usually transsphenoidal adenomectomy), especially if the desire of the patient is early pregnancy without risk of optic chiasmal compression from estrogen-induced tumor growth, but the use of dopamine agonists are of the greatest interest from the clinical investigative point of view. Bromocryptine in doses of 7.5mg per day, will, in most patients (up to 80% of cases) induce a fall in prolactin level, and a return to normal cyclical activity, the latter within 4 weeks in most patients. Prolonged treatment with this drug has proven to be relatively safe, although some recent controversial reports suggest that there may be an increased incidence of peptic ulcer, and/or gastrointestinal bleeding. In view of the enormous number of patients who may be treated with this agent, the possibility of side effects must be carefully considered. Besser also discusses a difficult problem, the use of bromergocryptine for the induction of fertility. Although current practice is to advise patients with adenomas not to become pregnant, many patients wish to take the drug primarily to restore normal ovulation so that they may become pregnant. Besser recommends that patients with known tumors have prophylactic radiation therapy, and that the drug be stopped as soon as the pregnancy is established. His experience with 52 pregnancies (31 babies) indicated no risk of teratogenesis, a conclusion supported by Griffiths' review of 448 pregnancies. There is a risk of tumor growth however, and clinicians are now debating the feasibility of continuing the drug throughout pregnancy in the hope that this agent will impair tumor (or normal pituitary) growth. That this is a possible alternative is suggested by animal work which indicates that dopamine agonists have striking effects on growth of pituitary tumors, and several patients have been reported to show a decrease in pituitary fossa size after bromocryptine therapy. One of the most important and provocative of Dr. Besser's findings (mentioned in his oral presentation), one which has important pathophysiological implications is that men with hyperprolactinemia, gonadotropin and testosterone deficiency and are impotent, may show a return of potency before there has been a significant increase in plasma testosterone. Dr. Ivor Jackson has studied such a patient at the New England Medical Center Hospital who was given testosterone without beneficial effect on his potency. However, when bromocryptine was added, there was a return of potency paripassu with the fall in plasma prolactin. Still obscure is the mechanism of this response. Is prolactin an anaphrodisiac, or is bromocryptine a stimulator of libido?

The third of the clinical papers was that of Dr. Yen who has attempted to delineate the neurotransmitter control mechanism of prolactin and gonadotropin secretion. Most attention has been given to dopamine as a pituitary regulator. The use of dopamine infusion (rather than L-Dopa or bromocryptine) localizes the action of the drug to regions outside the blood brain barrier, by which one infers the pituitary itself, and a portion of the stalk median eminence. In normal women, dopamine was found to suppress both prolactin and LH, without effects on FSH secretion. A direct effect on the pituitary was demonstrated using LHRH provocative tests. In hypogonadal women, as for example after menopause, the inhibitory effects of dopamine on LH are even more striking, and an effect on FSH can also be demonstrated. The potent DOPA agonist, bromocryptine, also causes FSH lowering. These findings suggest that there are receptors for dopamine on gonadotropin cells, and that the hypothalamic dopaminergic pathways may be inhibitory to gonadotropin release. This concept now expands the notion of a dual inhibitory and excitatory control (first demonstrated for GH and Prolactin secretion) to the regulation of gonadotropins. LHRH is stimulatory, and dopamine is inhibitory. Further evidence suggests that the secretion of LHRH is stimulated by noradrenergic neurotransmitter control, and inhibited by central dopaminergic fibers.

In earlier work, Dr. Yen showed that the effects of hypothalamic hormones on gonadotropin secretion are modified by estrogen. For example, he has demonstrated that estrogen sensitizes the pituitary to LHRH, and that the self-priming effects of LHRH on LH secretion are estrogen dependent. It was natural therefore for him to try to determine the interaction of estradiol with prolactin regulatory responses, particularly since estrogens sensitize the pituitary to the releasing effects of TRH, and stimulate prolactinotrope growth. The mid-cycle surge of LH and FSH coincides with the period of greatest sensitivity of the pituitary to the suppressive action of dopamine; similarly, hypogonadal women who also have hypersecretory states of LH and FSH also show marked dopamine sensitivity. Estrogen treatment, which stimulates prolactin release, and suppresses basal LH and FSH lowers pituitary sensitivity to dopamine. Yen proposes that dopamine lowers LH and FSH during the times of gonadotropin hypersecretion by reducing LHRH secretion. Although it is possible that this is an effect mediated at the level of the median eminence (accessible to dopamine), the demonstration of direct pituitary effects (by the LHRH test) in normal women makes it mandatory to carry out analogous studies in the hypersecretory states.

One of the most novel findings reported by Yen was the effect of progesterone in inducing release of prolactin. This effect requires a particular sequence of priming procedures. In ovariectomized women treated with estrogen, a single injection of progesterone induced an increase in prolactin after a latent period of about 4 hours which lasted for at least 5 hours. This rise accompanies the increase in plasma FSH and LH, which is well known from earlier studies of Odell and collaborators. If the sequence is reversed, the phenomenon does not occur. Although the mechanism of this effect is not known, Yen's speculation that progesterone inhibits hypothalamic dopamine secretion by reducing tyrosine hydroylase activity is supported by experimental data in the rat.

Yen, like the other speakers, also directed his attention to the important clinical problem of hyperprolactinemia. He had earlier postulated that, "hyperprolactinemia in humans, as in the rat, induces an elevation of endogenous DA in the tuberoinfundibular system through a short-loop feedback. The increased dopaminergic inhibition of LRF neuronal activity would then account for the low and acyclic gonadotropin secretion." He attempts to prove this point by showing that dopamine is less effective in lowering LH in hyperprolactinemic states than in normals. However, the fact that baseline levels of LH are lower in hyperprolactinemic women (not shown by the methods of plotting used) may make it difficult to compare magnitude of response, i.e., does one use absolute decrements, percentage decrements? Further, since dopamine may act at two sites (median eminence as well as pituitary) and the latter site is probably more important, it is difficult for me to draw hard and fast conclusions about the functional state of hypothalamic dopamine secretion from these studies.

Another fascinating observation made by Yen, and reported here for the first time is that prolonged LHRH infusion brings about a rise in prolactin levels in normal cycling women. He proposed that LHRH is stimulating prolactin release as a response to the ultra-short-loop feedback effects of LHRH on hypothalamic dopaminergic neurons. The idea of ultra-short loop control is attributed to Motta and Hyppa (20) who found that crude hypothalamic extracts administered to castrated, hypophysectomized male rats restored hypothalamic levels of bioassayable FSH RF. If one considers that this suggestion was based on work done before the isolation of LHRH, it reflects an extraordinary prescience, if Yen's suggestion should prove to be true. Yen's ideas also give us insight into the advances which have been made in clinical neuroendocrinology.

REFERENCES

1. Alpert, L.C., J.R.Brawer, Y.C.Patel, and S. Reichlin. Somato-statinergic neurons in anterior hypothalamus: Immunohistochemical localization. Endocrinology, 98: 255-258, 1976.
2. Anderson, E. and W. Haymaker. Breakthroughs in hypothalamic and pituitary research. Progress in Brain Research Vol. 31, 1-60,1974.
3. Arimura, A. and A. Schally. Increase in basal and thyrotropin-releasing hormone (TRH)-stimulated secretion of thyrotropin (TSH) by passive immunization with antiserum to somatostatin in rats. Endocrinology 98: 1069-1072, 1976.

4. Boyd,III,A.E., S. Reichlin, and R.N.Turksoy. Galactorrhea-amen-
    orrhea syndrome: Diagnosis and therapy. Ann.Int. Med. 87: 165-
    175, 1977.

5   Boyd,III,A.C.,E.Spencer,I.M.D.Jackson and S.Reichlin. Prolactin
    releasing factor (PRF) in porcine hypothalamic extract distinct
    from TRH. Endocrinology, 99: 861-871, 1976.

6. Brazeau, P., W. Vale, R. Burgus, N.Ling, M.Butcher, J. Rivier,
    and R.Guillemin. Hypothalamic peptide that inhibits the secre-
    tion of immunoreactive growth hormone. Science 179: 77-79, 1973.

7. Chihara, K. and A. Arimura. Effect of intraventricular injec-
    tion of dopamine, norepinephrine, 5-hydroxytryptamine and ace-
    tylchole on somatostatin levels of rat hypophyseal portal blood.
    Prog.60th Ann. Meeting, Endocrine Society, Americana Hotel,
    Miami, Florida, 1978, A241,pg. 195 (abst.)

8. Cobb, W.E., S. Reichlin, and I.M.D.Jackson. Importance of growth
    hormone (GH) in regulating the thyrotropin (TSH) response to thy-
    rotropin releasing hormone (TRH) in patients with hypothalamic-
    pituitary (HP) disease. Amer. Thyroid Assoc. 54th Meeting, Sept.
    13-16, 1978, Portland, Oregon T8 (abst)

9. Gibbs, D.M. and J. D. Neill. Dopamine levels in hypophysial
    stalk blood in the rat are sufficient to inhibit prolactin secre-
    tion. Endocrinology 102: 1895-1900, 1978.

10. Greep, R.O. History of research on anterior hypophysial hormones
    In: Handbook of Physiology, Sec.7, Endocrinology, Vol.iv, edited
    by R.O.Greep and E.B.Astwood, American Physiological Society,
    Washington,D.C., 1974, pg. 1-27.

11. Harris,A.R.C., D.Christianson, M.S.Smith,S-LFang, L.E.Braverman,
    and A.G.Vagenakis. The physiological role of thyrotropin-releas-
    ing hormone in the regulation of thyroid-stimulating hormone and
    prolactin secretion in the rat. J.Clin. Invest. 61: 441-448, 1978.

12. Knigge,K.M., S.A.Joseph, and G.E.Hoffman. Organization of LRF-
    and SRIF-neurons in the endocrine hypothalamus. ·In: The Hypo-
    thalamus, edited by S.Reichlin, R.J.Baldessarini and J.B. Martin,
    Raven Press, New York, 1978, pg. 49-67.

13. Koch, Y., Goldhaber, G., Fireman, I.,U.Zor,J.Shand, and E.Tal,
    Suppression of prolactin and thyrotropin secretion in the rat by
    antiserum to thyrotropin-releasing hormone. Endocrinology, 100:
    1476-1478, 1977.

14. Krulich, L., A.P.S. Dhariwal and S.M.McCann. Stimulatory and
    inhibitory effects of purified hypothalamic extracts on growth
    hormone release from rat pituitary in vitro. Endocrinology 83:
    783-790, 1968.

15. Malacara, J.M., R.C.Valverde, S. Reichlin, and J. Bollinger.
    Elevation of plasma radioimmunoassayable growth hormone in the
    rat induced by porcine hypothalamic extracts. Endocrinology 91:
    1189-1198, 1972.

16. Martin, J.B., S.Reichlin and G.M.Brown. Clinical Neuroendocrin-
    ology. F.A.Davis Publishers, Philadelphia, 1977.

17. Martini, L., and G.M.Besser, Eds. Clinical Neuroendocrinology.
    Academic Press, New York, 1977.

18. Meites, J. B.T.Donovan and S.M.McCann. Pioneers in Neuroendo-
    crinology. Plenum Press, New York, 1975.
19. Meites, J., B.T.Donovan and S.M.McCann, eds. Pioneers in Neu-
    roendocrinology, Vol. 2, Plenum Press, New York, 1978 (In press).
20. Motta, A., F. Fraschini and L. Martini. "Short" feedback mechan-
    isms in the control of anterior pituitary function. In: Fron-
    tiers in Neuroendocrinology, edited by W.F. Ganong and L.Martini
    Oxford Univ. Press, London, 1969, pg. 211-253.
21. Mueller, G.F. F.S.Franco, S.Reichlin, and I.M.D.Jackson. Ele-
    vated serum thyrotropin (TSH) of myxedema does not require con-
    tinuous thyrotropin-releasing hormone (TRH) secretion. Clin.
    Res. 25: 298A, 1977 (abst.)
22. Müller, E.E., G. Nistico, U.Scapagnini, eds. Neurotransmitters
    and anterior pituitary function. Academic Press,New York,1977.
23. Nair, R.M.G., C.deVillier, M.Barnes, J.Antalis, and D.L.Wilbur.
    A bovine hypothalamic peptide possessing immunoreactive growth
    hormone-releasing activity. Endocrinology 103: 112-120, 1978.
24. Patel, Y., K. Rao, and S. Reichlin. Somatostatin in human cer-
    ebrospinal fluid. New Engl. J. Med. 296: 529-533, 1977.
25. Reichlin, S. Anatomical and physiological basis of pituitary
    regulation. In: Clinical Neuroendocrinology, edited by G.Tolis
    F. Naftolin and J.B.Martin. Plenum Press, New York, 1978. (In
    press).
26. Saffran, M. Chemistry of hypothalamic hypophysiotropic factors.
    In: Handbook of Physiology, Sec.7, Endocrinology, Vol. IV, Part
    2, edited by R.O.Greep and E.B.Astwood, Washington, D.C., Amer.
    Physiology Soc., 1974, pg. 563-586.
27. Stachura, M.E., A.P.Dhariwal, and L.A.Frohman. Growth hormone
    synthesis and release in vitro: Effects of partially purified
    ovine hypothalamic extract. Endocrinology 91: 1071-1078, 1972.
28. Tannenbaum, G.S., J.Epelbaum, E.Colle, P.Brazeau, and J.B.
    Martin. Antiserum to somatostatin reverses starvation-induced
    inhibition of growth hormone but not insulin secretion. Endo-
    crinology (In press).
29. Terry, L.C., J.O.Willoughby, P. Brazeau, J.B.Martin and Y.Patel,
    Antiserum to somatostatin prevents stress-induced inhibition of
    growth hormone secretion in the rat. Science 192: 565-566,1976.
30. Tolis, G., F. Naftolin, and J.B.Martin. Clinical Endocrinology.
    Plenum Press, New York, in press.
31. Vale, W., C. Rivier, and M. Brown. Regulatory peptides of the
    hypothalamus, Ann. Rev. Physiol. 39: 473-527, 1977.
32. Valverde, R.C., V.Chieffo, and S. Reichlin. Failure of reser-
    pine to block ether-induced release of prolactin: Physiological
    evidence that stress induced prolactin release is not caused by
    acute inhibition of PIF secretion. Life Sciences 12: 327-335,
    1973.
33. Wilber, J.F., T. Nagel, and W.F.White. Hypothalamic growth
    hormone releasing activity: Characterization by the in vitro
    rat pituitary and radioimmunoassay. Endocrinology 89: 1419-
    1424, 1971.

PARTICIPANTS

AXELROD, Julius
National Institute of Mental
Health
9000 Rockville Pike
Bethesda
Maryland 20014, USA

BAULIEU, Etienne E.
Université de Paris-Sud
Départment de Chemie Biologique
Fac. de Medicine de Bicetre 78
Av. du General Le Clerc
94-Bicetre, France

BESSER, G. Michael
The Medical Professioral Unit
St Bartholomew's Hospital
London, England

BLOOM, Floyd E.
The Arthur V. Davis Center for
Behavioral Neurobiology
The Salk Institute
La Jolla
California 92037, USA

CARLSSON, Arvid
Department of Pharmacology
Medicinargathan 7
Göteborg University
Fack
S-400 33 Göteborg, Sweden

CHANGEUX, Jean-Pierre
Neurobiologie Moléculaire
Institut Pasteur
75024 Paris, Cedex 15, France

ENEROTH, Peter
Hormone Laboratory
Karolinska Hospital
S-104 01 Stockholm 60, Sweden

EULER, Ulf von
Department of Physiology
Karolinska Institute
S-104 01 Stockholm 60, Sweden

FUXE, Kjell
Department of Histology
Karolinska Institute
S-104 01 Stockholm 60, Sweden

GOLDSTEIN, Menek
Neurochemistry Laboratories
New York University Medical Center
New York
N.Y. 10016, USA

GREENGARD, Paul
Department of Pharmacology
Yale University School of Medicine
New Haven
Connecticut 06510, USA

GUSTAFSSON, Jan-Åke
Department of Chemistry
Karolinska Institute
S-104 01 Stockholm 60, Sweden

HÖKFELT, Tomas
Department of Histology
Karolinska Institute
S-104 01 Stockholm 60, Sweden

IVERSEN, Leslie L.
Department of Pharmacology
University of Cambridge
Cambridge, England

JEFFCOATE, Stephen L.
Department of Endocrinology
Chelsea Hospital for Women
London SW3, England

KORDON, Claude
Unité de Neurobiologie
de l'INSERM
2 ter rue d'Alésia
75014 Paris, France

LABRIE, Fernand
Group du Conseil de Recherches
Medicales en Endocrinologie
Moléculaire
Centre Hospitalier
de l'Université
Laval, Quebec, Canada

LUFT, Rolf
Department of Endocrinology
Karolinska Hospital
S-104 01 Stockholm 60, Sweden

MARTINI, Luciano
Università di Milano
Istituto di Endocrinologia
Via A. del Sarto 21
I-20129 Milano, Italy

MCCANN, Samuel M.
Department of Physiology
The University of Texas
Health Science Center
Dallas
Texas, USA

MCEWEN, Bruce S.
Rockefeller University
New York
N.Y. 10021, USA

MÜLLER, Eugenio E.
Università Degli Studi
Il Cattedra di Farmacologia
Facoltà di Medicina e Chirurgia
Via Vanvitelli 32
I-20129 Milano, Italy

RENAUD, Leo P.
Department of Medicine
Division of Neurology
Suite 753
Livingstone Hall
1650 Zedarave, Montreal
Quebec H3G1A4, Canada

REICHLIN, Seymour
Endocrinology Division
New England Medical Center
Hospital
171 Harrison Avenue
Boston
Massachusetts 02111, USA

RODBELL, Martin
Laboratory of Nutrition and
Endocrinology, NIAMDD
National Institutes of Health
Bethesda
Maryland 20014, USA

SAWYER, Charles H.
Department of Anatomy
University of California
School of Medicine
Los Angeles
90024 California, USA

SCHALLY, Andrew V.
Department of Medicine
Tulane University School of
Medicine
1430 Tulane Avenue
New Orleans
Louisiana 70146, USA

SEDVALL, Göran
Department of Psychiatry
Karolinska Hospital
S-104 01 Stockholm 60, Sweden

SNYDER, Solomon H.
Department of Pharmacology and
Experimental Therapeutics
Johns Hopkins School of Medicine
Baltimore
Maryland 21205, USA

TERENIUS, Lars
Department of Pharmacology
University of Uppsala
Box 579
S-751 23 Uppsala, Sweden

THORN, Niels A.
Institute of Medical Physiology C
University of Copenhagen
Copenhagen, Denmark

DE WIED, David
Rudolf Magnus Institute of
Pharmacology
University of Utrecht
Utrecht, The Netherlands

YEN, Samuel S.C.
Department of Reproductive
Medicine
University of California
San Diego, La Jolla
California 92093, USA

INDEX

Abortion, spontaneous, and
    bromocriptine, 465
Acetylcholine (Ach), 3, 4, 36,
    63, 492
  and ACTH release, 342
  and gonadotropin release, 333-
    334
  and growth hormone release,
    340
  and hypothalamic neuron firing,
    490
  and oxytocin release, 127
  and prolactin release, 337
  and vasopressin release, 127
Acetylcholine (muscarinic), and
    cyclic GMP, 160
Acetylcholine bromide, and TSH
    release, 341
Acetylcholine-regulator, 494
  effect of histrionicotoxin on,
    501
Acetylcholinesterase, 496
γ-Acetylenic GABA, influence on
    anterior pituitary func-
    tion, 371
Acromegaly
  and growth hormone, 440-447
  and prolactin, 440-447
ACTH analogues, amino acid
    sequences of, 300
Actinomycin D, and progesterone
    blockage, 267
Activation, three states model
    for, 497-499
Adenylate cyclase, 508
  and cyclic AMP, 152

Adenylate cyclase system, 71
  target size analysis of, 74-
    82
Adrenaline, 3, 4, 37-40
  effect on hypothalamic con-
    taining pathways, 349,
    350
Adrenaline systems
  and clonidine, 368-370
  neuroendocrine role of, 368-
    372
Adrenergic transmission, 3
α-Adrenergic receptors,
    pharmacological blockade
    of, 418
β-Adrenergic receptor(s)
  and information processing,
    507-508
  and ligand binding, 151-152
  and membrane lipid methyl-
    ation, 153-155
  in pineal gland, 150-153
  propranolol blockade of, 418
  in red cell membranes, 153-
    155
  regulation of sensitivity of,
    151-153
Adrenergics, isoreceptor sys-
    tems of, 139
Adrenocorticotropic hormone
    (ACTH)
  effect on learned behavior,
    297-298
  fragments and structure
    activity studies, 298-300
  and growth hormone, 365-366

synaptic transmitters involved
    in, 341–342
Adrenocorticotropin releasing
    factor (CRF), 5
Alatesin, and prolactin release,
    338–339
Alcaloids, 492
Alkaloid agonists, relative
    affinity of, 144
Alkaloid antagonists, relative
    affinity of, 144
Allosteric ligands, 501
Alpha–methyl–dopa, effect on
    prolactin levels, 461
Amine turnover, effects of
    endorphins on, 372
Amines
    coexistence with peptides, 40–
    44
    turnover changes in, 354–356
Androgen receptors (AR), 240
    a sexual dimorphism of, 324
Androgens
    and dopamine turnover, 361–364
    and FSH secretion, 88–89
Antagonist reversal, 140
Anterior pituitary function
    drugs influencing neuronal
        control of, 371
    neurotransmitters in control of,
    329–342
Anti–androgen, 264
Antidepressant drugs, and
        sensitive adenylate cyclase
        blockade, 509
Antiestrogen, 248–249, 263–264
Antihormone binding, 242
Antihormones, 248–250
Antischizophrenic drugs, and
        dopamine, 509
Anti–steroid hormone effect, 249
Apomorphine, 421
    as dopamine stimulant, 427
    GH–releasing effect of, 424
    influence on anterior pituitary
        function, 371
Arginine vasopressin (AVP), amino
        acid sequence of, 309
Arginine vasotocin (AVT)

amino acid sequence of, 309
    and gonadotropin release,
        334–335
Aromatization, 262, 267
Aspartate transcarbamylase,
    491
Atropine, 334, 492
    and ACTH release, 342
    and cyclic GMP, 160
    and neurosecretory neurons,
        127

Behavior, and pituitary neuro-
        peptides, 297–311
Behavior, learned
    effects of ACTH on, 297–298
    effects of peptides related
        to β–LPH on, 301–302
Behavior, rat, effect of
        peptides on, 223–237
Benserazide, inhibition of
        monoamine synthesis by,
        228
Benzamides, effect on prolactin
        levels, 461
Benzomorphan agonists, 138
Benzomorphans, 138
Bicuculline, and neurosecretory
        neurons, 127
Blood–brain barrier, and
        dopamine, 393
Bombesin
    and growth hormone release,
        340
    and prolactin release, 338–
        339
Brain, steroid hormone receptors
        in, 261–269
Brain stem, peptide secretion
        in, 67
Brain transmitter systems, and
        androgen exposure, 319
Bromocriptine, 431
    problems in pregnancy, 465–
        466
    and prolactin inhibition, 459
    and tumor growth, 466–467
    use in hyperprolactinaemia,
        464–465

Butyrophenones, effect on pro-
    lactin levels, 461

Calcium
  extracellular, 52
  messenger signals, 52-53
  protein phosphorylation
      regulation, 165-167
  and substance P, 191-193
Calcium-binding regulator pro-
      teins, and release mech-
      anisms, 53-55
Cascade effect, 495
Cascade systems, for neuro-
      peptide study, 480
Catecholamine-hormone inter-
      actions, 351, *illus.*
Catecholamine synthesis
  and growth hormone release, 340
  inhibition of, 332-333
Catecholamine turnover
  and clonidine, 369-370
  evaluation, 354
  importance, 354
  results, 354, 355 *illus.*
Catecholamines, 33, 427-431
  morphological studies, 350-354
  and prolactin release, 13-14
CB-154
  as dopamine stimulant, 427
  and luteinizing hormone release,
      390
Cell bodies, coexistence of
      amines and peptides in, 41-
      43
Central nervous system
  cyclic nucleotide mechanisms in,
      157
  extra-hypothalamic areas of,
      66-67
  protein phosphorylation mechan-
      isms in, 157-171
Cerebellum, peptide secretion in,
      67
Cerebral cortex, peptide secre-
      tion in, 66
Chloralhydrate, and protein I
      phosphorylation, 164-165
Chlorimipramine, 425

Chlorpromazine
  and adenylate cyclase acti-
      vation, 159
  effect on prolactin levels,
      461
Cholinergic transmission, 3-4
Cholinergic transmitter systems,
      sex-differentiated, 321
Cholinergics, isoreceptor
      systems of, 139
Circadian patterns, in pro-
      lactin release, 458
Clomiphene, 465
Clonidine, 217, 367
  and adrenaline systems, 368-
      370
  influence on anterior pitui-
      tary function, 371
  and TSH release, 341
Connectivity, principles of,
      180
Corticotropin releasing factor
      (CRF)
  isolation of, 10
  purification of, 11
Cyclase-regulator, model for,
      501
Cyclic AMP
  and N-acetyltransferase, 152
  and adenylate cyclase, 152-
      153
  and memory, 161
  neuronal function, 147-161
  and protein phosphorylation,
      168-171
  roles of, 160-161
Cyclic GMP
  neuronal function, 157-161
  and protein phosphorylation,
      167-168
  and synaptic transmission,
      160

DALA, effect on DOPA formation,
      227
Desensitization, three states
      model for, 497-499
Desglycinamide, arginine
      vasopressin (LVP), amino

acid sequence of, 309
Diethylstilbestrol (DES), 240
Diphenhydramine, and cyclic GMP,
    160
Down regulation, 243-244
DOPA, concentration in rat brain
    after peptide administra-
    tion, 233-234
L-dopa
  and luteinizing hormone
    release, 390
  and prolactin inhibition, 459
  variable response to, 392
Dopamine, 4, 37-40
  and ACTH release, 342
  action on isolated pituitary
    cells, 460
  action on LHRH nerve terminals,
    362, illus.
  and adenylate cyclase activa-
    tion, 159
  concentration in rat brain
    after peptide administra-
    tion, 233-234
  effect on neuropeptide release,
    478-480
  effect on somatostatin release,
    63
  and gonadotropin release, 330-
    332, 387-411
  and growth hormone, 340, 418-
    427
  and hypothalamic neuron firing,
    490
  inhibitory action of, 393-399
  and LHRH secretion, 354-355
  and prolactin control, 95-96,
    102-103
  and prolactin inhibition, 13-
    14, 350, 459-460
  and prolactin release, 86,
    335-336, 387-411
  and TSH release, 341, 382
Dopamine agonists
  apomorphine, 421
  effect of, 390-393
  for hyperprolactinemic therapy,
    467
  piribedil, 421

and prolactin inhibition,
    459-460
Dopamine-β-hydroxylase, organum
    vasculosum laminae
    terminalis incubated
    with antiserum to, illus.
    39
Dopamine cell bodies
  estrogen receptors in, 357
  in hypothalamus, 350
Dopamine depleting agents,
    effect on prolactin
    levels, 461
Dopamine infusion, effects of,
    388-390
Dopamine neurons, 349
Dopamine receptor(s)
  blockade of, 330
  influence of guanyl nucleo-
    tides on, 114-117
  "inhibitory," 421-422
  ionic influences on, 112-114
  location of, 393
  as mediators of growth
    hormone suppression, 419
  regulatory, 422
  specificity of, 99
  supersensitive, 422
Dopamine receptor agonists,
    331, 365
  and ovulation, 358-359
Dopamine receptor blocking
    drugs, effect on pro-
    lactin levels, 461
Dopamine stimulants, 427
Dopamine terminal systems,
    effects of hypophyseal
    hormones on, 364-368
Dopamine turnover
  and β-endorphin, 372-373
  and clonidine, 369-370
  effect of androgens on, 361-
    364
  and estrogen, 356-357
  and growth hormone, 365-366
  and met-enkephalin, 372-373
  and morphine, 372-373
  and prolactin, 364-365
  and steroid hormones, 356-364

Dopaminergic drugs
  effect on plasma GH levels, 444
  interaction with TRH, 440–447
Dopaminergic mechanism
  for luteinizing hormone release, 389
  for prolactin release, 389

Endogenous dopamine activity, assessments of, 405–408
β-Endorphin
  and dopamine turnover, 372–373
  influence on anterior pituitary function, 371
  relative affinity of, 144
β-Endorphin fragments
  amino acid sequence of, 302
  and avoidance behavior, 303–305
Endorphins, 141
  effect on DOPA formation, 226
  effect on substance P release, 194–196
  and growth hormone release, 340
  morphine-like activity of, 137–138, 301
  and prolactin secretion, 436
  and rat behavior patterns, 228–229
Enkephalin(s), 37, 141
  analogue of, 459
  effect on DOPA formation, 226–227
  and growth hormone release, 340
  and prolactin release, 436, 40, 459
  relative affinity of, 144
  in vitro release of, 198–202
Epinephrine
  distribution in medulla oblongata, 216
  effect of PNMT inhibition on, 212
  and growth hormone release, 340

  and prolactin release, 336
  and TSH release, 341
Epinephrine neuronal systems, 209–220
  effect of stress on, 210–213
  and hypertension, 214–217
Ergocornine, as dopamine stimulant, 427
Ergot drugs, influence on anterior pituitary function, 371
Estradiol, 239. See also Estrogen(s)
  antidopaminergic action of, 99–103
  effect on dopamine inhibition, 397
Estradiol binding protein, 240
Estradiol feedback, in estrous cycle, 357
Estradiol receptor (ER), 239
Estrogen(s)
  antidopaminergic activity of, 101–103
  cellular mechanism of, 265–266
  and dopamine turnover, 356–357
  effect on dopamine inhibition, 402
  effect on postsynaptic mechanisms, 359–361
  effects on progestin receptors, 266
  effect on prolactin levels, 461
  and prolactin release, 458
  and FSH secretion, 87–88
  and LHRH response, 287
  and LH secretion, 87–88
  localization action of, 264–265
  and noradrenaline turnover, 356–357
  and prolactin, 95–96, 356–357
  and thyroid hormone, 93–95
Estrogen receptors, 324
  and estrous cycle fluctuation,

263
  quantitation of, 262–264
  role in developing rodent
      brain, 268
  and sexual behavior, 263
Estrogen–sensitive neurons,
      activation of, 264–265
Estrogen treatment, effect on
      dopamine inhibition, 397–
      399
Estrous cycle, and positive
      estradiol feedback, 357
Ethinyl estradiol (EE), 286
  and gonadotropin secretion,
      401–402
Exocytosis, and vasopressin
      release, 52

Feedback effects
  of androgenic molecules, 282–
      283
  and neuroendocrine functions,
      273–291
Feminizing factor. *See*
      Feminotropin
Feminotropin, 318–319, 320
  control of secretion of, 321
Fluoxetine, 425, 431
Flutamide, 264
FSH secretion
  control of, 86–92
  and progesterone, 285–286

GABA (Gamma–aminobutyric acid),
      14
  and gonadotropin release, 334
  and hypothalamic neuron firing,
      490
  and neurosecretory neurons, 127
  and prolactin release, 338, 434
  and substance P, 191–193
  and TSH release, 341
GABA antagonist drugs
  bicuculline, 435
  picrotoxin, 435
GABA interneuron, 367
GABA synthesizing enzyme, 37
Gastrin
  and gonadotropin release, 334–

335
  and growth hormone release,
      340
  and prolactin release, 338–
      339
  and TSH release, 341
GEA 654, influence on anterior
      pituitary function, 371
Glucocorticosteroid receptor
      (GR), 240
Glutamate, and cyclic GMP, 160
L–Glutamate, and neurosecretory
      neurons, 127
Glycine
  and neurosecretory neurons,
      127
  and neurotransmitter recep-
      tors, 113–114
Glycine receptors, 114
Gonadotropin
  effects of progesterone on,
      89–90, 284–290
  feedback regulation of, 64
Gonadotropin release, synaptic
      transmitters controlling,
      329–335
Gonadotropin releasing factor
      (LRF), 5
Gonadotropin secretion, effects
      of testosterone on, 276–
      283
Grooming behavior, effect of
      peptides on, 305–306
Growth hormone
  α–adrenergic stimulation of,
      420
  control of, in human disease
      states, 440–447
  and L-dopa, 421
  and dopamine, 418–427
  monoaminergic control of, 418–
      427
  neurotransmitter control of,
      417–447
  and norepinephrine, 418–427
  synaptic transmitters in-
      volved in control of, 339–
      340
  TRH stimulation of, 443

Growth hormone release-inhibiting hormone (GH-RIH), 21-23
Growth hormone-releasing factor (GH-RF), 20
Growth hormone stimulating factor, chemistry, 524
GTP
  and adenylate cyclase, 71-82
  in transduction process, 72-74
Guanyl nucleotides, receptor regulation by, 114-117

Haloperidol, effect on prolactin levels, 461
Hepatic corticosteroid metabolism, sex related difference in, 315
Hepatic steroid metabolism
  and androgens, 317
  feminization of, 321
  hypothalamic control of, 318-319
  imprinting of, 325-326
Histamine, 36
  and adenylate cyclase activation, 159
  and cyclic GMP, 160
  and gonadotropin release, 334
  and prolactin release, 338
Histaminergics, isoreceptor systems of, 139
Histrinicotoxin, 494
Hormone metabolites, 247
Hormone secretion control, role of adrenaline pathways in, 368-372
Hormones
  anterior pituitary, 349-377
  binding properties, 246
  hypophyseal, 364-368
  hypothalamic, 85-103
  neurotransmitter mechanisms controlling, 349-377
  peripheral, 85-103
  physiological pharmacology of, 246-250
  and receptor plurality, 246
  synthetic analogs, 247-248
Huntington's chorea, and growth hormone secretion, 422

Hydroxyindole-O-methyltransferase, 149
5-Hydroxytryptamine
  hypothetical alternatives for storage compartments for, illus., 43
  raphe magnus nucleus incubated with antiserum to, illus., 42
  rat median eminence incubated with antiserum to, illus., 34, 35
Hyperprolactinemia
  anti-gonadotropic effect of, 429-431
  management of patients with, 463
  neurotransmitter therapy for, 463-465
  treatment of men with, 466
Hyperprolactinaemic states
  causes, 461-462
  clinical results of, 462-463
Hypertension, spontaneous, 214-218
Hypogonadism, mechanism in hyperprolactinaemia, 462-463
Hypophysectomy, and prolactin release, 457
Hypophysiotrophic area (HA), 6
Hypophysiotropic peptides, release of, 190
Hypothalamic catecholamines, 349
Hypothalamic hormone containing systems, morphological studies, 350-354
Hypothalamic hormones
  interaction with peripheral hormones, 85-103
  synthetic analogues of, 9-15
Hypothalamic peptidergic neurons, neurophysiology of, 119-136
Hypothalamic peptides, non-specificity of, 385
Hypothalamic-pituitary complex, sex hormones acting on,

273-274
Hypothalamic-pituitary unit,
    illus., 522
Hypothalamic-prolactin-
    gonadotropin relation-
    ship, 429-431
Hypothalamic releasing hormones,
    61-68
  behavioural profiles induced
    by, 235
Hypothalamo-pituitary-liver axis,
    sexual differentiating
    actions of steroids on,
    315-325
Hypothalamus, 62-66
  mediobasal, neuropeptides in,
    474
  and prolactin secretion, 86
Hypothalamus-neurohypophyseal
    system, activation of, 50
Hypothyroidism, 458
  primary, and prolactin levels,
    462

Immunocytochemistry, 31-45
Immunoreactivity, *illus.*, 352
  ACTH-like, *illus.*, 352 *illus.*,
    353
  prolactin-like, 352
Imprinting, specificity of, 319-
    320
Indoleaminergic system, 438
Indoleamines
  and GH secretion, 424-426
  and prolactinɡ secretion, 431-
    433
Infertility
  bromocriptine therapy for, 465
  gonadotropin therapy for, 465
Inhibin
  and FSH release, 90-92
  and gonadotropin secretion, 86-
    92
  and LH release, 90-92
Interneurons
  enkephalin, 350-352
  GABA, 350-352
Ionizing radiation, and adenylate
    cyclase system, 74-75

Ionophoric-regulator, 492-494,
    495
Islet of Langerhans, micro-
    graph after incubation
    with somatostatin anti-
    serum, 383
Isoreceptor systems, 139
Isoreceptors, 139-140

Lergotrile
  for hyperprolactinaemic
    therapy, 467
  and prolactin inhibition, 459
Ligand binding studies, 495-496
β-Lipotropin, 62-77
  neuroleptic-like activity of,
    303
Lisuride
  as dopamine stimulant, 427
  for hyperprolactinaemic
    therapy, 467
  and prolactin inhibition, 459
Liver, effects of androgens
    and estrogens on, 324
Lordosis behavior, 267
Lordosis response, 262
β-LPH related peptides and
    learned behavior, 301-
    302
LSD, and adenylate cyclase
    activation, 159
Luteinizing hormone, L-dopa
    and rebound release of,
    396
Luteinizing hormone-releasing
    hormone (LHRH), 37
  actions on pituitary, 15-16
  agonists, 18-20
  analogs, 17-20
  antifertility effects, 18-20
  and FSH-RH activity, 16
  immunological studies, 16-17
Luteinizing hormone-releasing
    hormone secretion
  and dopamine inhibition, 354,
    361
  and dopamine ratios, 355
  and noradrenaline, 354-355
Luteinizing hormone secretion

and androgens, 361–364
control of, 86–92
and dopamine receptor agonists, 357, 358
and prolactin, 364–365
regulation by estrogen, 360–361

Median eminence, coexistence of amines and peptides in, 40–44
Melanocyte stimulating hormone-release inhibiting factor (MIF), 14–15
Melatonin
and gonadotropin release, 333
synthesis of, 150–151
Melatonin synthesis, 508
Membrane fluidity, and phospholipid synthesis, 154, 155
Memory
effect of neurohypophyseal hormones on, 308–311
effect of oxytocin on, 307–308
effect of vasopressin on, 307–308
Met-enkephalin
and dopamine turnover, 372–373
opiate-like activity of, 303–304
Metergoline, specificity of, 437–438
Methysergide, 458
Metiamide, and adenylate cyclase activation, 159
Metoclopramide, and effects of prolactin release, 459, 461
Microiontophoresis, 127
Milk ejection reflex, 120–122
Mineralocorticosteroid receptor (MR), 240
Monoamine transmitter turnover, effect of peptides on, 511
Monoaminergic nuronal systems, 219
Monoamines, 4
catecholamines, 33
effects of neuropeptides on release of, 202–203

5-hydroxytryptamine, 33–36
synthesis, 236
Morphine
and dopamine turnover, 372–373
effect on DOPA formation, 224–225
influence on anterior pituitary function, 371
and rat behavior patterns, 228–229
receptor selectivity of, 142
Morphine-like activity, 137
Muscimol, 367–434
influence on anterior pituitary function, 371
Multiple opioid systems, 142

Naloxone, 138, 370, 459
antagonists in, 140
in chronic schizophrenia, 146
and DOPA formation, 226–227, 230
effect on rat motor activity, 232
and growth hormone release, 340
influence on anterior pituitary function, 371
and substance P release, 196
Naltrexone, antagonists in, 140
Nerve terminals
coexistence of amines and peptides in, 41–43
presynaptic peptidergic, 189
Neuroendocrine functions, control of, 273–291
Neurohormones, definition, 176–177
Neurohypophyseal fibres, axon collaterals in, 127
Neurohypophyseal hormones
effect on memory processes, 308–311
release mechanisms, 49–58
and steroid hormones, 491
Neurohypophyseal neurons

efferent connections of, 124–
126
electrical activity, 490
hormonal release by, 120–124
localization of, 120
pharmacology of, 126–128
Neurohypophyseal system, 119–128
Neurohypophysis
cAMP system in, 56–57
calcium regulating systems in,
52
cGMP system in, 56–57
membrane-bound calcium binding
proteins in, 55–56
soluble calcium-binding
protein in, 55
Neuromodulator, definition, 178–
179
Neuron populations, classification of, 509–510
Neuronal activity
and hormone release, 120
regulation at cell level, 489–
491
Neuronal function, and cyclic
nucleotides, 157–171
Neuronal regulation, at postsynaptic level, 489–503
Neurons, 173–184
Neuropeptide neurons, role of,
473–478
Neuropeptide precursors, detection of, 476–477
Neuropeptides, 37–40
action of neurotransmitters on,
480
biosynthesis of, 476–477
degradation of, 477–478
distribution patterns and, 31–
45
effects of neurotransmitters on,
481
main distribution sites of, 475
neuroendocrine effects of, 475
release regulation, 189–204,
478–485
varied functions of, 474–476
Neurosecretory granules, fusion
of, 53

Neurosecretory neurons, synaptic feedback loops in,
491
Neurotensin
and gonadotropin release,
334–335
and growth hormone release,
340
and prolactin release, 338–
339
in vitro release of, 202
and TSH release, 341
Neurotransmitter(s)
associated with cyclic AMP
system, 159
controlling anterior pituitary function, 329–342
definition, 177–178
distribution patterns and,
31–45
diversity of, 173–176
effects on neuropeptide
release, 481
peptidergic, 3
Neurotransmitter therapy, for
hyperprolactinaemia, 463–
465
Noradrenaline, 3, 4, 37–40
concentration in rat brain
after peptide administra-
tion, 233–234
effect on hypothalamic
containing pathways, 349,
350
effect on neuropeptide
release, 478–480
and growth hormone release,
340
and LHRH secretion, 354–355
presynaptic regulation of,
189
and prolactin release inhibition, 13–14
release inhibition, 189–190
and TSH release, 341, 382
Noradrenaline nerve terminals,
control of, 376–377
Noradrenaline terminal systems,
effects of hypophyseal

hormones on, 364–368
Noradrenaline turnover
  and estrogen, 356–357
  and steroid hormones, 356–364
Noradrenergic projection systems,
    182
β-Noradrenergic receptors, ionic
    influences on, 112–114
α-Noradrenergic receptors
  influence of guanyl nucleo-
      tides on, 114–117
  ionic influences on, 112–114
Norepinephrine (NE)
  and ACTH release, 342
  and adenylate cyclase activa-
      tion, 159
  and cyclic GMP, 160
  effect of PNMT inhibition on,
      213
  and gonadotropin release, 332–
      333
  and growth hormone, 418–427
  and hypothalamic neuron
      firing, 490
  and prolactin release, 336
  sodium regulation of, 112
Norepinephrine neuronal system
  effect of stress on, 210–213
  and PNMT inhibition, 210–211
Norepinephrine synthesis,
    blockage of, 333, 418–419
Nucleotide regulatory proteins,
    71

Octopamine, and adenylate
    cyclase activation, 159
Oestradiol. See Estradiol
Oestrogens. See Estrogens
Opiate drugs, effect on substance
    P release, 194–196
Opiate peptides
  interaction with neurohormone
      agonists, 483
  neurotransmitter role of, 375
  and prolactin release, 436
Opiate receptors
  cation regulation of, 111–112
  and enkephalin neurons, 109–
      110

influence of guanyl nucleo-
    tides on, 114–117
and noradrenaline systems,
    374–375
sodium effect on, 110–111
Opiates
  distribution sites of, 475
  neuroendocrine effects of,
      475
Opioiod system
  antagonists in, 140
  receptor heterogeneity in,
      141
Opioid receptors
  activity of, 137–138
  ligands, 137–147
Opioids, isoreceptor systems
    of, 139
Osmoreceptors, 50
Ovariectomy, and dopamine
    action, 397–398
Ovulation-inhibition
  by dopamine receptor agonists,
      357–359
  by noradrenaline receptor
      antagonists, 357
Oxytocin (OXT), 4
  amino acid sequence of, 309
  effect on memory processes,
      307–308
  and hypothalamic neuron
      firing, 490
  and labor, 122
  localization of, 120
  neurosecretory neurons re-
      sponsible for, 119–120
  as neurotransmitter, 127–128
Oxytocinergic neurons, 120–122
  activation mechanisms, 49–52

Peptidase activity, distribu-
    tion of, 236
Peptidases, 64–66
  age-dependent changes in,
      65–66
  effect of gonadal steroids,
      65
  and progesterone, 65
  sex differences in activity

of, 65
subcellular localization of,
  65
Peptide receptors, 62
  in brain, 109–117
Peptidergic neuronal systems, 219
Peptidergic neurons, 3–6
  in autonomic nervous system,
    119
  control in humans, 381–385
  endocrine signals received
    by, 273
  hypothalamic, electrophysio-
    logical properties of, 489–
    491
Peptidergic neurotransmitters,
  3
Peptidergic pathways, 349
Peptides
  coexistence with amines, 40–44
  effect on brain monoamines,
    223–237
  effect on monoamines and 5-
    hydroxyindoleacetic acid
    (5-H1AA) levels, 228
  and hypothalamic neuron
    firing, 490
  peptidase degradation of, 64
  transmitter function of, 510
Peripheral hormones, interaction
    with hypothalamic hormones,
    85–103
Perphenazine, effect on prolactin
    levels, 461
Pharmacological effectors,
    categories of, 492
Phenothiazines, effect on pro-
    lactin levels, 461
Phenoxybenzamine, influence on
    anterior pituitary
    function, 371
Phentolamine, 418
  and adenylate cyclase activa-
    tion, 159
  and cyclic GMP, 160
  and 5-HTP induced GH rise, 436
Phenylethanolamine-N-methyl
    transferase (PNMT)
  distribution in medulla

oblongata, 215
  and epinephrine distribu-
    tion, 219
  genetic variations in, 214–
    217
Phospholipids, role in re-
    ceptor action, 149
Picrotoxin, 434, 494
  and neurosecretory neurons,
    127
Pimozide, 441
  as dopamine receptor block-
    ing agent, 357
  effect on prolactin levels,
    461
  and 5-HTP induced GH rise,
    426
  influence on anterior
    pituitary function, 371
  and TSH release, 341
Pineal gland, 508
  regulation of β-adrenergic
    receptor in, 149–155
  response to noradrenergic
    stimulation, 150
Piperoxane, 370–371
  effect on epinephrine levels,
    217–218
  influence on anterior
    pituitary function, 371
Piribedil
  as dopamine stimulant, 427
  for hyperprolactinaemic
    therapy, 467
  and prolactin inhibition,
    459
Pituitary, steroid hormone
    receptors in, 261–269
Pituitary hormones, retro-
    grade transport of, 306–
    307
Pituitary neuropeptides, and
    behavior, 297–311
Pituitary transplants, 5
Plasma proteins, differential
    binding of, 247–248
PNMT inhibitor, influence on
    anterior pituitary
    function, 371

Postsynaptic receptors,
     bicuculline sensitive, 434
Potassium, and substance P, 191-
     193
Pregnancy, and prolactin secre-
     tion, 458
Pressinamide (PA), amino acid
     sequence of, 309
Progesterone
   biphasic effect on gonado-
       tropin secretion, 89-90
   effects on gonadotropin
       secretion, 284-290
   and FSH secretion, 285-286
   metabolism of, 274-275
   and Xenopus laevis oocyte,
       250-251
Progesterone effect, 243
Progesterone metabolites
   effects on gonadotropin
       secretion, 284-290
   and FSH secretion, 285-286
Progesterone receptor (PR), 240
Progestin receptors, estrogen
     effects on, 266-267
Prolactin
   CNS inhibition of, 427
   control of, 95-96
     in human disease states,
       440-447
   deficiency, 467
   and dopamine turnover, 364-365
   dopaminergic control of, 98-99
   drugs raising circulating
       levels of, 460-461
   and estrogen, 356-357
   and LH secretion, 364-365
   neurotransmitter control of,
       417-447
   release during LRF infusion,
       408-410
   release of, 95, 96-99
   specificity of, 96-99
   synthesis, 11
   and ultra-short loop control,
       531
   variations in secretion rate,
       458
Prolactin inhibitory factor, in

   medial palisade zone, 355
Prolactin receptors, types, 365
Prolactin release
   progesterone-induced, 402-
       405
   synaptic transmitters con-
       trolling, 335-339
Prolactin release-inhibiting
     factor (PIF), 13, 459-
     460
Prolactin releasing effect,
     and opiate peptides, 436
Prolactin releasing factor,
     12-13, 458
Prolactin secretion
   clinical neuroendocrine
       relationships in, 457-
       469
   disorders of, 461-468
   dual system for control of,
       illus., 523
   and estradiol, 93
   factors increasing, 458-459
   neurotransmitter control of,
       427-440
   normal control of, 457-458
   and piperoxane, 369
   and yohimbine, 369
Prolyl-argyl-glycinamide (PAG),
     amino acid sequence of,
     309
Propranolol, and adenylate
     cyclase activation, 159
Protein I
   biological properties of, 163
   chemical properties of, 164
Protein kinase activities, 508
Protein kinase hypothesis, 161
Protein phosphorylation
   and calcium, 165-167
   cyclic AMP-dependent, 161-165
   effect of cyclic AMP on,
       illus., 162
   regulation by cyclic GMP, 167-
       168
   regulation by steroid hor-
       mones, 168-171
Psychopathology, and endor-
     phins, 303-305

Radioisotopic methods, for
    estrogen quantitation, 262–
    263
Receptor(s)
  activation of, 241
  deactivation of, 241
  degradation of, 502
  development of concept of, 239–
    240
  microheterogeneity, 495–497
  recognition of ligand by, 150
Receptor-bound estradiol, ex-
    change assay for, 263
Receptor concentration, changes
    in, 248
Receptor exchange assays, for
    estrogen quantitation, 262
Receptor heterogeneity, 240–241
  methods for studying, 140–141
  in opioid system, 141–146
Receptor molecule, synthesis of,
    502
Receptor multiplicity, 137
Receptor number, regulation of,
    502
Receptor-second messenger
    systems, 157
Receptor selectivity, 141
Receptor site, identification
    by biochemical assay
    techniques, 110
Red cell membranes, regulation
    of β-adrenergic receptor
    in, 149–155
Releasing factors, 5
Renal failure, and prolactin
    levels, 462
Renal transplantation, and
    prolactin levels, 462
Reserpine
  and dopamine, 427
  effect on prolactin levels,
    461

Scopolamine, influence on
    anterior pituitary func-
    tion, 371
Secretory tissues, calcium-
    dependent regulator

    protein systems in, 56
Serotonin, 37–40
  and ACTH release, 342
  and adenylate cyclase activa-
    tion, 159
  and GH releasing mechanisms,
    424–425
  and gonadotropin release,
    333
  and growth hormone release,
    340
  and hypothalamic neuron
    firing, 490
  and prolactin release, 336–
    337
  and TSH release, 341
Serotonin agonists
  cyproheptadine, 424
  methysergide, 424
Serotonin N-acetyltransferase,
    149, 508
Serotoninergic system, and GH
    secretion, 424
Sertoli cell culture medium,
    and inhibin activity,
    90–91
Sex steroid binding plasma
    protein (SBP), 240
Sheehan's syndrome, 467
Signal transduction
  classical modes of, 492–495
  by neuronal membrane, 491–
    495
SKF 64139, 211
Sodium effect, and opiate
    receptor binding, 110–114
Somatostatin, 21, 37–40, 459
  analogs of, 22–23
  and central nervous system,
    22
  and DOPA formation, 234–235
  and GH secretion, 525
  and gut, 22
  and 5-HTP formation, 234–235
  and hypothalamic neuron
    firing, 490
  in vitro release of, 202
  median eminence incubated
    with antiserum to, illus.

38
  organum vasculosum laminae
    terminalis incubated with
    antiserum to, *illus.*, 39
  and pancreas, 21–22
  and pituitary, 21
  and stomach, 22
Somatostatin antiserum, immuno-
    fluorescence micrographs
    of tissues incubated in,
    383, 384
Steroid hormone binding pro-
    teins, 239–241
Steroid hormone-cell inter-
    actions, 239–260
Steroid hormone receptor com-
    plexes, in cell nucleus,
    241–243
Steroid hormones, and protein
    phosphorylation, 168–171
Steroid receptor concentrations
  double regulatory control of,
    243–244
  variation in, 245
Steroid receptors, structures of,
    241
Steroids, interaction at surface
    membrane level, 250–252
Stimulus-secretion coupling
    hypothesis, 480
Stress, effect on neuronal
    systems, 210–213
Strychnine, 494
  and neurosecretory neurons, 127
Substance P, 5, 37, 229. *See
    also* Peptidergic neuro-
    transmitters; SP
  analogues of, 229–230
  effect on rat motor activity,
    232
  and gonadotropin release,
    334–335
  and growth hormone release,
    340
  hypothetical alternatives for
    storage compartments for,
    illus., 43
  localization in small sensory

nerves, 196
  median eminence incubated
    with antiserum to, *illus.*,
    38
  opiate-induced suppression
    of, *illus.*,197, 196–198
  and potassium, 191–193
  and prolactin release, 338–
    339
  raphe magnus nucleus incu-
    bated with antiserum to,
    illus., 42
  release from rat brain, 190–
    198
  and TSH release, 341
Substantia nigra, and sub-
    stance P, 190–193
Sucrose, and enzyme system
    inhibition, 50
Sulpiride, effect on prolactin
    levels, 461
Synapses
  activatory, 490
  inhibitory, 490
Synaptic operations, domains
    of, 179–182
Synaptic physiology, contrast-
    ing principles of, 173–
    184
Synaptic transmission
  and cyclic AMP, 157–159
  modulation of, 157
Synaptosomes, 62–64
  effect of seasonal and
    cyclic changes, 63–64
  incubated *in vitro*, 62–64
  and neurotransmitter re-
    lease, 63
  peptide secretion by, 66, 67

Tamoxifen, 248–249
  and endometrial cancer, 249
Target cells, 239–241
  steroid hormone receptor
    concentrations in, 243–
    246
Testosterone
  action on developing rodent
    brain, 268

effects of chronic treatment
     with, 277
effects on prolactin secretion,
     283–284
and FSH release, 88–89, 282
implants of, 277–283
and LH release, 88
metabolism of, 274–275
metabolites of in general
     circulation, 275–276
negative feedback blocking,
     264
systemic injections and
     gonadotropin secretion,
     276–277
Testosterone metabolites
effects of chronic treatment
     with, 277
and FSH release, 282
implants of, 277–283
systemic injections and
     gonadotropin secretion,
     276–277
Thalamus, peptide secretion in,
     67
Threonine deaminase, 491
Thyroid hormone
and estrogen, 93–95
feedback effects of, 459
Thyroid stimulating hormone,
     synaptic transmitters
     controlling, 341
Thyrotoxicosis, 458
Thyrotropin (TSH), synthesis, 11
Thyrotropin releasing factor
     (TRF), 5
Thyrotropin releasing hormone
     (TRH), 37–40
analogs, 12
antidepressant activity, 11–12
effect on prolactin levels, 461
interaction with dopamine, 440–
     447
as neurotransmitter, 12
and prolactin, 11
and prolactin release, 458–459
and thyrotropin, 11
Tocinamide (TA), amino acid
     sequence of, 309

Tonic inhibitory control, of
     prolactin, 457
Transduction process, role of
     GTP in, 72–74
Trigeminal nucleus, spinal,
     and substance P, 194
Tryptophan
effect of various peptides
     on, 231
and LH-RH, 235
TSH secretion, 93–95
and dopamine, 382
and noradrenaline, 382
Tuberoinfundibular neurons.
     See also Neurohypophyseal
     neurons
definition, 128
pharmacology of, 131
Tuberoinfundibular system
and adenohypophyseal
     regulation, 128–131
efferent connections, 129–
     130
electrical properties of,
     128–129
localization of, 1280129
Tyrosine
effect of various peptides
     on, 231
and LH-RH, 235
Tyrosine hydroxylase, organum
     vasculosum laminae
     terminalis incubated with
     antiserum to, *illus.*, 39

Urethane, and protein I
     phosphorylation, 164–165

Vasoactive intestinal peptide
and gonadotropin release of,
     334–335
in vitro release of, 202
and prolactin release, 338–
     339
Vasopressin, 4
effect on memory processes,
     307–308
and extracellular Ca^{++}, 52
and hypothalamic neuron

firing, 490
localization of, 120
mechanism of, 52
neurosecretory neurons respon-
    sible for, 119-120
as neurotransmitter, 127
role of ATP in release of,
    57-58
Vasopressinergic neurons, 50-51,
    123-124
activation mechanisms, 49-52

Xenopus laevis oocytes, steroids
    reinitiating meiosis in,
    250-252

Yohimbine, 217, 370-371
influence on anterior pitui-
    tary function, 371

Zimelidine, influence on
    anterior pituitary
    function, 371